Model Predictive Control:
Theory and Design

ISBN 9780975937709

9 780975 937709

Model Predictive Control:
Theory and Design

James B. Rawlings
Department of Chemical and Biological Engineering
University of Wisconsin
Madison, Wisconsin, USA

David Q. Mayne
Department of Electrical and Electronic Engineering
Imperial College London
London, England

Nob Hill Publishing

Madison, Wisconsin

This book was set in Lucida using LaTeX, and printed and bound by Worzalla. It was printed on Forest Stewardship Council certified acid-free recycled paper.

Cover design by Cheryl M. and James B. Rawlings.

Nob Hill Publishing, LLC
Cheryl M. Rawlings, publisher
Madison, WI 53705
orders@nobhillpublishing.com
http://www.nobhillpublishing.com

Library of Congress Control Number: 2009904993

Rawlings, James B.
Model Predictive Control: Theory and Design /
by James B. Rawlings and David Q. Mayne.
p. cm.
Includes bibliographical references (p.) and index.
ISBN 978-0-975-93770-9 (cloth)
1. Predictive control. 2. Control theory.
3. Feedback control systems.
I. Mayne, David Q. II. Title.

Printed in the United States of America.

First Printing August 2009

FSC
Mixed Sources
Product group from well-managed forests, controlled sources and recycled wood or fiber

Cert no. BV-COC-080720
www.fsc.org
© 1996 Forest Stewardship Council

To Cheryl and Josephine,

for their love, encouragement, and patience.

Preface

Our goal in this text is to provide a comprehensive and foundational treatment of the theory and design of model predictive control (MPC). By now several excellent monographs emphasizing various aspects of MPC have appeared (a list appears at the beginning of Chapter 1), and the reader may naturally wonder what is offered here that is new and different. By providing a comprehensive treatment of the MPC foundation, we hope that this text enables researchers to learn and *teach* the fundamentals of MPC without continuously searching the diverse control research literature for omitted arguments and requisite background material. When teaching the subject, it is essential to have a collection of exercises that enables the students to assess their level of comprehension and mastery of the topics. To support the teaching and learning of MPC, we have included more than 200 end-of-chapter exercises. A complete solution manual (more than 300 pages) is available for course instructors.

Chapter 1 is introductory. It is intended for graduate students in engineering who have not yet had a systems course. But it serves a second purpose for those who have already taken the first graduate systems course. It derives all the results of the linear quadratic regulator and optimal Kalman filter using only those arguments that extend to the nonlinear and constrained cases to be covered in the later chapters. Instructors may find that this tailored treatment of the introductory systems material serves both as a review and a preview of arguments to come in the later chapters.

Chapters 2–4 are foundational and should probably be covered in any graduate level MPC course. Chapter 2 covers regulation to the origin for nonlinear and constrained systems. This material presents in a unified fashion many of the major research advances in MPC that took place during the last 20 years. It also includes more recent topics such as regulation to an unreachable setpoint that are only now appearing in the research literature. Chapter 3 addresses MPC design for robustness, with a focus on MPC using tubes or bundles of trajectories in place of the single nominal trajectory. This chapter again unifies a large body of research literature concerned with robust MPC. Chapter 4 covers state estimation with an emphasis on moving horizon estimation, but also covers extended and unscented Kalman filtering, and particle filtering.

Chapters 5–7 present more specialized topics. Chapter 5 addresses the special requirements of MPC based on output measurement instead of state measurement. Chapter 6 discusses how to design distributed MPC controllers for large-scale systems that are decomposed into many smaller, interacting subsystems. Chapter 7 covers the explicit optimal control of constrained linear systems. The choice of coverage of these three chapters may vary depending on the instructor's or student's own research interests.

Three appendices are included, again, so that the reader is not sent off to search a large research literature for the fundamental arguments used in the text. Appendix A covers the required mathematical background. Appendix B summarizes the results used for stability analysis including the various types of stability and Lyapunov function theory. Since MPC is an optimization-based controller, Appendix C covers the relevant results from optimization theory. In order to reduce the size and expense of the text, the three appendices are available on the web: `www.che.wisc.edu/~jbraw/mpc`. Note, however, that all material in the appendices is included in the book's printed table of contents, and subject and author indices. The website also includes sample exams, and homework assignments for a one-semester graduate course in MPC. All of the examples and exercises in the text were solved with Octave. Octave is freely available from `www.octave.org`.

JBR
Madison, Wisconsin, USA

DQM
London, England

Acknowledgments

Both authors would like to thank the Department of Chemical and Biological Engineering of the University of Wisconsin for hosting DQM's visits to Madison during the preparation of this monograph. Funding from the Paul A. Elfers Professorship provided generous financial support.

JBR would like to acknowledge the graduate students with whom he has had the privilege to work on model predictive control topics: Rishi Amrit, Dennis Bonné, John Campbell, John Eaton, Peter Findeisen, Rolf Findeisen, Eric Haseltine, John Jørgensen, Nabil Laachi, Scott Meadows, Scott Middlebrooks, Steve Miller, Ken Muske, Brian Odelson, Murali Rajamani, Chris Rao, Brett Stewart, Kaushik Subramanian, Aswin Venkat, and Jenny Wang. He would also like to thank many colleagues with whom he has collaborated on this subject: Frank Allgöwer, Tom Badgwell, Bhavik Bakshi, Don Bartusiak, Larry Biegler, Moritz Diehl, Jim Downs, Tom Edgar, Brian Froisy, Ravi Gudi, Sten Bay Jørgensen, Jay Lee, Fernando Lima, Wolfgang Marquardt, Gabriele Pannocchia, Joe Qin, Harmon Ray, Pierre Scokaert, Sigurd Skogestad, Tyler Soderstrom, Steve Wright, and Robert Young.

DQM would like to thank his colleagues at Imperial College, especially Richard Vinter and Martin Clark, for providing a stimulating and congenial research environment. He is very grateful to Lucien Polak and Graham Goodwin with whom he has collaborated extensively and fruitfully over many years; he would also like to thank many other colleagues, especially Karl Åström, Roger Brockett, Larry Ho, Petar Kokotovic and Art Krener, from whom he has learnt much. He is grateful to past students who have worked with him on model predictive control: Ioannis Chrysochoos, Wilbur Langson, Hannah Michalska, Sasa Raković and Warren Schroeder; Hannah Michalska and Sasa Raković, in particular, contributed very substantially. He owes much to these past students, now colleagues, as well as to Frank Allgöwer, Rolf Findeisen Eric Kerrigan, Konstantinos Kouramus, Chris Rao, Pierre Scokaert, and Maria Seron for their collaborative research in MPC.

Both authors would especially like to thank Tom Badgwell, Bob Bird, Eric Kerrigan, Ken Muske, Gabriele Pannocchia, and Maria Seron for their careful and helpful reading of parts of the manuscript. John Eaton again deserves special mention for his invaluable technical support during the entire preparation of the manuscript.

Contents

List of Figures

List of Examples and Statements

Notation

Mathematical notation

$\exists$	there exists
$\in$	is an element of
$\forall$	for all
$\implies \quad \impliedby$	implies; is implied by
$\not\Rightarrow \quad \not\Leftarrow$	does not imply; is not implied by
$:=$	equal by definition or is defined by
$\approx$	approximately equal
$V(\cdot)$	function V
$V : \mathbb{A} \to \mathbb{B}$	V is a function mapping set $\mathbb{A}$ into set $\mathbb{B}$
$x \mapsto V(x)$	function V maps variable x to value $V(x)$
x^+	value of x at next sample time (discrete time system)
$\dot{x}$	time derivative of x (continuous time system)
f_x	partial derivative of $f(x)$ with respect to x
∇	nabla or del operator
δ	unit impulse or delta function
$\lvert x \rvert$	absolute value of scalar; norm of vector (two-norm unless stated otherwise); induced norm of matrix
$\mathbf{x}$	sequence of vector-valued variable x, $\{x(0), x(1), \ldots, \}$
$\lVert \mathbf{x} \rVert$	sup norm over a sequence, $\sup_{i \geq 0} \lvert x(i) \rvert$
$\lVert \mathbf{x} \rVert_{a:b}$	$\max_{a \leq i \leq b} \lvert x(i) \rvert$
$\mathrm{tr}(A)$	trace of matrix A
$\det(A)$	determinant of matrix A
$\mathrm{eig}(A)$	set of eigenvalues of matrix A
$\rho(A)$	spectral radius of matrix A, $\max_i \lvert \lambda_i \rvert$ for $\lambda_i \in \mathrm{eig}(A)$
A^{-1}	inverse of matrix A
$A^\dagger$	pseudo-inverse of matrix A
A'	transpose of matrix A
inf	infimum or greatest lower bound
min	minimum
sup	supremum or least upper bound
max	maximum
arg	argument or solution of an optimization

$\mathbb{I}$	integers
$\mathbb{I}_{\geq 0}$	nonnegative integers
$\mathbb{I}_{n:m}$	integers in the interval $[n, m]$
$\mathbb{R}$	real numbers
$\mathbb{R}_{\geq 0}$	nonnegative real numbers
$\mathbb{R}^n$	real-valued n-vectors
$\mathbb{R}^{m \times n}$	real-valued $m \times n$ matrices
$\mathbb{C}$	complex numbers
$\mathcal{B}$	ball in $\mathbb{R}^n$ of unit radius
$x \sim p_x$	random variable x is distributed with probability density p_x
$\mathcal{E}(x)$	expectation of random variable x
$\mathrm{var}(x)$	variance of random variable x
$\mathrm{cov}(x, y)$	covariance of random variables x and y
$N(m, P)$	normal distribution with mean m and covariance P, $x \sim N(m, P)$
$n(x, m, P)$	normal probability density, $p_x(x) = n(x, m, P)$
$\emptyset$	the empty set
$\mathrm{aff}(\mathbb{A})$	affine hull of set $\mathbb{A}$
$\mathrm{int}(\mathbb{A})$	interior of set $\mathbb{A}$
$\mathrm{co}(\mathbb{A})$	convex hull of the set S
$\overline{\mathbb{A}}$	closure of set $\mathbb{A}$
$\mathrm{epi}(f)$	epigraph of function f
$\mathrm{lev}_a V$	sublevel set of function V, $\{x \mid V(x) \leq a\}$
$\mathbb{A} \oplus \mathbb{B}$	set addition of sets $\mathbb{A}$ and $\mathbb{B}$
$\mathbb{A} \ominus \mathbb{B}$	set subtraction of set $\mathbb{B}$ from set $\mathbb{A}$
$\mathbb{A} \setminus \mathbb{B}$	elements of set $\mathbb{A}$ not in set $\mathbb{B}$
$\mathbb{A} \cup \mathbb{B}$	union of sets $\mathbb{A}$ and $\mathbb{B}$
$\mathbb{A} \cap \mathbb{B}$	intersection of sets $\mathbb{A}$ and $\mathbb{B}$
$\mathbb{A} \subseteq \mathbb{B}$	set $\mathbb{A}$ is a subset of set $\mathbb{B}$
$\mathbb{A} \supseteq \mathbb{B}$	set $\mathbb{A}$ is a superset of set $\mathbb{B}$
$\mathbb{A} \subset \mathbb{B}$	set $\mathbb{A}$ is a proper (or strict) subset of set $\mathbb{B}$
$\mathbb{A} \supset \mathbb{B}$	set $\mathbb{A}$ is a proper (or strict) superset of set $\mathbb{B}$
$d(a, \mathbb{B})$	Distance between element a and set $\mathbb{B}$
$d_H(\mathbb{A}, \mathbb{B})$	Hausdorff distance between sets $\mathbb{A}$ and $\mathbb{B}$
$x \searrow y$	x converges to y from above
$x \nearrow y$	x converges to y from below
$\mathrm{sat}(x)$	saturation, $\mathrm{sat}(x) = x$ if $\lvert x \rvert \leq 1, -1$ if $x \leq -1, 1$ if $x \geq 1$

Symbols

A, B, C	system matrices, discrete time, $x^+ = Ax + Bu$, $y = Cx$
A_c, B_c	system matrices, continuous time, $\dot{x} = A_c x + B_c u$
A_{ij}	state transition matrix for player i to player j
A_i	state transition matrix for player i
A_{Li}	estimate error transition matrix $A_i - L_i C_i$
B_d	input disturbance matrix
B_{ij}	input matrix of player i for player j's inputs
B_i	input matrix of player i
C_{ij}	output matrix of player i for player j's interaction states
C_i	output matrix of player i
C_d	output disturbance matrix
$\mathcal{C}$	controllability matrix
C^*	polar cone of cone C
d	integrating disturbance
E, F	constraint matrices, $Fx + Eu \leq e$
f, h	system functions, discrete time, $x^+ = f(x, u)$, $y = h(x)$
$f_c(x, u)$	system function, continuous time, $\dot{x} = f_c(x, u)$
$F(x, u)$	difference inclusion, $x^+ \in F(x, u)$, F is set valued
G	input noise-shaping matrix
G_{ij}	steady-state gain of player i to player j
H	controlled variable matrix
$I(x, u)$	index set of constraints active at (x, u)
$I^0(x)$	index set of constraints active at $(x, u^0(x))$
k	sample time
K	optimal controller gain
$\ell(x, u)$	stage cost
$\ell_N(x, u)$	final stage cost
L	optimal estimator gain
m	input dimension
M	cross-term penalty matrix $x'Mu$
M	number of players, Chapter 6
$\mathcal{M}$	class of admissible input policies, $\boldsymbol{\mu} \in \mathcal{M}$
n	state dimension
N	horizon length
$\mathcal{O}$	observability matrix, Chapters 1 and 4
$\mathcal{O}$	compact robust control invariant set containing the origin, Chapter 3
p	output dimension

p	optimization iterate, Chapter 6
p_ξ	probability density of random variable ξ
$p_s(x)$	sampled probability density, $p_s(x) = \sum_i w_i \delta(x - x_i)$
P	covariance matrix in the estimator
P_f	terminal penalty matrix
$\mathcal{P}$	polytopic partition
$\mathbb{P}_N(x)$	MPC optimization problem; horizon N and initial state x
q	importance function in importance sampling
Q	state penalty matrix
r	controlled variable, $r = Hy$
R	input penalty matrix
s	number of samples in a sampled probability density
S	input rate of change penalty matrix
$S(x, u)$	index set of active polytopes at (x, u)
$S^0(x)$	index set of active polytopes at $(x, u^0(x))$
t	time
T	current time in estimation problem
u	input (manipulated variable) vector
$\widetilde{\mathbf{u}}^+$	warm start for input sequence
$\mathbf{u}^+$	improved input sequence
$\mathcal{U}_N(x)$	control constraint set
$\mathbb{U}$	input constraint set
v	output disturbance, Chapters 1 and 4
v	nominal control input, Chapters 3 and 5
$V_N(x, \mathbf{u})$	MPC objective function
$V_N^0(x)$	MPC optimal value function
$V_T(\chi, \boldsymbol{\omega})$	Full information state estimation objective function at time T with initial state χ and disturbance sequence $\boldsymbol{\omega}$
$\hat{V}_T(\chi, \boldsymbol{\omega})$	MHE objective function at time T with initial state χ and disturbance sequence $\boldsymbol{\omega}$
$V_f(x)$	terminal penalty
$\mathcal{V}_N(z)$	nominal control input constraint set
$\mathbb{V}$	output disturbance constraint set
w	disturbance to the state evolution
w_i	weights in a sampled probability density, Chapter 4
w_i	convex weight for player i, Chapter 6
$\overline{w}_i$	normalized weights in a sampled probability density
$\mathcal{W}$	class of admissible disturbance sequences, $\mathbf{w} \in \mathcal{W}$
$\mathbb{W}$	state disturbance constraint set

x	state vector
x_i	sample values in a sampled probability density
x_{ij}	state interaction vector from player i to player j
$\overline{x}(0)$	mean of initial state density
$X(k;x,\boldsymbol{\mu})$	state tube at time k with initial state x and control policy $\boldsymbol{\mu}$
$\mathcal{X}_j$	set of feasible states for optimal control problem at stage j
$\mathbb{X}$	state constraint set
$\mathbb{X}_f$	terminal region
y	output (measurement) vector
$\mathbb{Y}$	output constraint set
z	nominal state, Chapters 3 and 5
$Z_T(x)$	full information arrival cost
$\hat{Z}_T(x)$	MHE arrival cost
$\widetilde{Z}_T(x)$	MHE smoothing arrival cost
$\mathbb{Z}$	system constraint region, $(x,u) \in \mathbb{Z}$
$\mathbb{Z}_f$	terminal constraint region, $(x,u) \in \mathbb{Z}_f$
$\mathbb{Z}_N(x,\mathbf{u})$	constraint set for state and input sequence

Greek letters

$\Gamma_T(\chi)$	MHE prior weighting on state at time T
Δ	sample time
κ	control law
κ_j	control law at stage j
κ_f	control law applied in terminal region $\mathbb{X}_f$
$\mu_i(x)$	control law at stage i
$\boldsymbol{\mu}(x)$	control policy or sequence of control laws
ν	output disturbance decision variable in estimation problem
Π	cost to go matrix in regulator, Chapter 1
Π	covariance matrix in the estimator, Chapter 5
ρ_i	objective function weight for player i
Σ_i	Solution to Lyapunov equation for player i
$\phi(k;x,\mathbf{u})$	state at time k given initial state x and input sequence $\mathbf{u}$
$\phi(k;x,i,\mathbf{u})$	state at time k given state at time i is x and input sequence $\mathbf{u}$
$\phi(k;x,\mathbf{u},\mathbf{w})$	state at time k given initial state is x, input sequence is $\mathbf{u}$, and disturbance sequence is $\mathbf{w}$
χ	state decision variable in estimation problem
ω	state disturbance decision variable in estimation problem

Subscripts, superscripts and accents

$\hat{x}$	estimate
$\hat{x}^-$	estimate before measurement
$\tilde{x}$	estimate error
x_s	steady state
x_i	subsystem i in a decomposed large-scale system
x_{sp}	setpoint
V^0	optimal
V^{uc}	unconstrained
V^{sp}	unreachable setpoint

Acronyms

CLF	control-Lyapunov function
DARE	discrete algebraic Riccati equation
DP	dynamic programming
FLOP	floating point operation
FSO	final state observability
GAS	global asymptotic stability
GAS	global exponential stability
GPC	generalized predictive control
EKF	extended Kalman filter
i-IOSS	incrementally input/output-to-state stable
IOSS	input/output-to-state stable
ISS	input-to-state stable
KF	Kalman filter
KKT	Karush-Kuhn-Tucker
LAR	linear absolute regulator
LP	linear program
LQ	linear quadratic
LQG	linear quadratic Gaussian
LQR	linear quadratic regulator
MHE	moving horizon estimation
MPC	model predictive control
OSS	output-to-state stable
PF	particle filter
PID	proportional-integral-derivative
QP	quadratic program
RGA	relative gain array
RGAS	robust global asymptotic stability
RHC	receding horizon control
UKF	unscented Kalman filter

1

Getting Started with Model Predictive Control

1.1 Introduction

The main purpose of this chapter is to provide a compact and accessible overview of the essential elements of model predictive control (MPC). We introduce deterministic and stochastic models, regulation, state estimation, dynamic programming (DP), tracking, disturbances, and some important performance properties such as closed-loop stability and zero offset to disturbances. The reader with background in MPC and linear systems theory may wish to skim this chapter briefly and proceed to Chapter 2. Other introductory texts covering the basics of MPC include Maciejowski (2002); Camacho and Bordons (2004); Rossiter (2004); Goodwin, Seron, and De Doná (2005); Kwon (2005); Wang (2009).

1.2 Models and Modeling

Model predictive control has its roots in optimal control. The basic concept of MPC is to use a dynamic model to forecast system behavior, and optimize the forecast to produce the best decision — the control move at the current time. Models are therefore central to every form of MPC. Because the optimal control move depends on the initial state of the dynamic system, a second basic concept in MPC is to use the past record of measurements to determine the most likely initial state of the system. The state estimation problem is to examine the record of past data, and reconcile these measurements with the model to determine the most likely value of the state at the current time. Both the regulation problem, in which a model forecast is used to produce the optimal control action, and the estimation problem, in which the past record of measurements is used to produce an optimal state estimate, involve dynamic models and optimization.

We first discuss the dynamic models used in this text. We start with the familiar differential equation models

$$
\begin{aligned}
\frac{dx}{dt} &= f(x, u, t) \\
y &= h(x, u, t) \\
x(t_0) &= x_0
\end{aligned}
$$

in which $x \in \mathbb{R}^n$ is the state, $u \in \mathbb{R}^m$ is the input, $y \in \mathbb{R}^p$ is the output, and $t \in \mathbb{R}$ is time. We use $\mathbb{R}^n$ to denote the set of real-valued n-vectors. The initial condition specifies the value of the state x at time $t = t_0$, and we seek a solution to the differential equation for time greater than t_0, $t \in \mathbb{R}_{\geq t_0}$. Often we define the initial time to be zero, with a corresponding initial condition, in which case $t \in \mathbb{R}_{\geq 0}$.

1.2.1 Linear Dynamic Models

Time-varying model. The most general *linear* state space model is the time-varying model

$$
\begin{aligned}
\frac{dx}{dt} &= A(t)x + B(t)u \\
y &= C(t)x + D(t)u \\
x(0) &= x_0
\end{aligned}
$$

in which $A(t) \in \mathbb{R}^{n \times n}$ is the state transition matrix, $B(t) \in \mathbb{R}^{n \times m}$ is the input matrix, $C(t) \in \mathbb{R}^{p \times n}$ is the output matrix, and $D(t) \in \mathbb{R}^{p \times m}$ allows a direct coupling between u and y. In many applications $D = 0$.

Time-invariant model. If A, B, C, and D are time invariant, the linear model reduces to

$$
\begin{aligned}
\frac{dx}{dt} &= Ax + Bu \\
y &= Cx + Du \\
x(0) &= x_0
\end{aligned}
\tag{1.1}
$$

One of the main motivations for using linear models to approximate physical systems is the ease of solution and analysis of linear models.

Equation (1.1) can be solved to yield

$$x(t) = e^{At}x_0 + \int_0^t e^{A(t-\tau)}Bu(\tau)d\tau \tag{1.2}$$

in which $e^{At} \in \mathbb{R}^{n\times n}$ is the matrix exponential.[1] Notice the solution is a convolution integral of the entire $u(t)$ behavior weighted by the matrix exponential of At. We will see later that the eigenvalues of A determine whether the past $u(t)$ has more effect or less effect on the current $x(t)$ as time increases.

1.2.2 Input-Output Models

If we know little about the internal structure of a system, it may be convenient to take another approach in which we suppress the state variable, and focus attention only on the manipulatable inputs and measurable outputs. As shown in Figure 1.1, we consider the system to be the connection between u and y. In this viewpoint, we usually perform system identification experiments in which we manipulate u and measure y, and develop simple linear models for G. To take advantage of the usual block diagram manipulation of simple series and feedback connections, it is convenient to consider the Laplace transform of the signals rather than the time functions,

$$\overline{y}(s) := \int_0^\infty e^{-st}y(t)dt$$

in which $s \in \mathbb{C}$ is the complex-valued Laplace transform variable, in contrast to t, which is the real-valued time variable. The symbol := means "equal by definition" or "is defined by." The transfer function matrix is then identified from the data, and the block diagram represents the following mathematical relationship between input and output

$$\overline{y}(s) = G(s)\overline{u}(s)$$

$G(s) \in \mathbb{C}^{p\times m}$ is the transfer function matrix. Notice the state does not appear in this input-output description. If we are obtaining $G(s)$ instead from a state space model, then $G(s) = C(sI - A)^{-1}B$, and we assume $x(0) = 0$ as the system initial condition.

---

[1]We can define the exponential of matrix X in terms of its Taylor series,

$$e^X := \frac{1}{0!}I + \frac{1}{1!}X + \frac{1}{2!}X^2 + \frac{1}{3!}X^3 + \cdots$$

This series converges for all X.

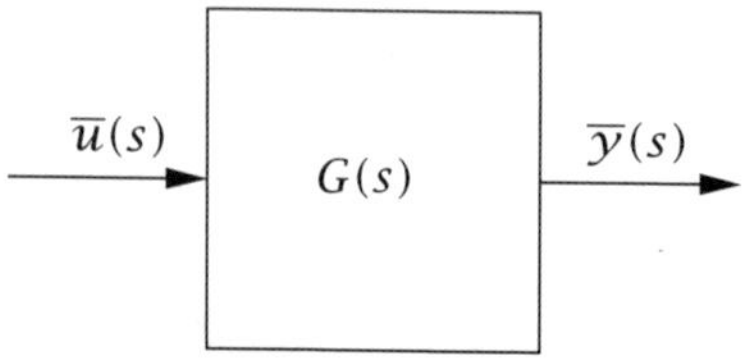

Figure 1.1: System with input $\overline{u}$, output $\overline{y}$ and transfer function matrix G connecting them; the model is $\overline{y} = G\overline{u}$.

1.2.3 Distributed Models

Distributed models arise whenever we consider systems that are not spatially uniform. Consider, for example, a multicomponent, chemical mixture undergoing convection and chemical reaction. The microscopic mass balance for species A is

$$\frac{\partial c_A}{\partial t} + \nabla \cdot (c_A v_A) - R_A = 0$$

in which c_A is the molar concentration of species A, v_A is the velocity of species A, and R_A is the production rate of species A due to chemical reaction, in which

$$\nabla := \delta_x \frac{\partial}{\partial x} + \delta_y \frac{\partial}{\partial y} + \delta_z \frac{\partial}{\partial z}$$

and the $\delta_{x,y,z}$ are the respective unit vectors in the (x, y, z) spatial coordinates.

We also should note that the distribution does not have to be "spatial." Consider a particle size distribution $f(r,t)$ in which $f(r,t)dr$ represents the number of particles of size r to $r + dr$ in a particle reactor at time t. The reactor volume is considered well mixed and spatially homogeneous. If the particles nucleate at zero size with nucleation rate $B(t)$ and grow with growth rate, $G(t)$, the evolution of the particle size distribution is given by

$$\begin{aligned} \frac{\partial f}{\partial t} &= -G \frac{\partial f}{\partial r} \\ f(r,t) &= B/G \qquad r = 0 \qquad t \geq 0 \\ f(r,t) &= f_0(r) \qquad r \geq 0 \qquad t = 0 \end{aligned}$$

Again we have partial differential equation descriptions even though the particle reactor is well mixed and spatially uniform.

1.2.4 Discrete Time Models

Discrete time models are often convenient if the system of interest is sampled at discrete times. If the sampling rate is chosen appropriately, the behavior between the samples can be safely ignored and the model describes exclusively the behavior at the sample times. The finite dimensional, linear, time-invariant, discrete time model is

$$\begin{aligned} x(k+1) &= Ax(k) + Bu(k) \\ y(k) &= Cx(k) + Du(k) \\ x(0) &= x_0 \end{aligned} \tag{1.3}$$

in which $k \in \mathbb{I}_{\geq 0}$ is a nonnegative integer denoting the sample number, which is connected to time by $t = k\Delta$ in which Δ is the sample time. We use $\mathbb{I}$ to denote the set of integers and $\mathbb{I}_{\geq 0}$ to denote the set of nonnegative integers. The linear discrete time model is a linear difference equation.

It is sometimes convenient to write the time index with a subscript

$$\begin{aligned} x_{k+1} &= Ax_k + Bu_k \\ y_k &= Cx_k + Du_k \\ x_0 &\ \ \text{given} \end{aligned}$$

but we avoid this notation in this text. To reduce the notational complexity we usually express (1.3) as

$$\begin{aligned} x^+ &= Ax + Bu \\ y &= Cx + Du \\ x(0) &= x_0 \end{aligned}$$

in which the superscript $^+$ means the state at the next sample time. The linear discrete time model is convenient for presenting the ideas and concepts of MPC in the simplest possible mathematical setting. Because the model is linear, analytical solutions are readily derived. The solution to (1.3) is

$$x(k) = A^k x_0 + \sum_{j=0}^{k-1} A^{k-j-1} Bu(j) \tag{1.4}$$

Notice that a convolution sum corresponds to the convolution integral of (1.2) and powers of A correspond to the matrix exponential. Because (1.4) involves only multiplication and addition, it is convenient to program for computation.

The discrete time analog of the continuous time input-output model is obtained by defining the Z-transform of the signals

$$\overline{y}(z) := \sum_{k=0}^{\infty} z^k y(k)$$

The discrete transfer function matrix $G(z)$ then represents the discrete input-output model

$$\overline{y}(z) = G(z)\overline{u}(z)$$

and $G(z) \in \mathbb{C}^{p\times m}$ is the transfer function matrix. Notice the state does not appear in this input-output description. We make only passing reference to transfer function models in this text.

1.2.5 Constraints

The manipulated inputs (valve positions, voltages, torques, etc.) to most physical systems are bounded. We include these constraints by linear inequalities

$$Eu(k) \leq e \qquad k \in \mathbb{I}_{\geq 0}$$

in which

$$E = \begin{bmatrix} I \\ -I \end{bmatrix} \qquad e = \begin{bmatrix} \overline{u} \\ -\underline{u} \end{bmatrix}$$

are chosen to describe simple bounds such as

$$\underline{u} \leq u(k) \leq \overline{u} \qquad k \in \mathbb{I}_{\geq 0}$$

We sometimes wish to impose constraints on states or outputs for reasons of safety, operability, product quality, etc. These can be stated as

$$Fx(k) \leq f \qquad k \in \mathbb{I}_{\geq 0}$$

Practitioners find it convenient in some applications to limit the rate of change of the input, $u(k) - u(k-1)$. To maintain the state space form of the model, we may augment the state as

$$\widetilde{x}(k) = \begin{bmatrix} x(k) \\ u(k-1) \end{bmatrix}$$

and the augmented system model becomes

$$\widetilde{x}^+ = \widetilde{A}\widetilde{x} + \widetilde{B}u$$
$$y = \widetilde{C}\widetilde{x}$$

in which

$$\widetilde{A} = \begin{bmatrix} A & 0 \\ 0 & 0 \end{bmatrix} \qquad \widetilde{B} = \begin{bmatrix} B \\ I \end{bmatrix} \qquad \widetilde{C} = \begin{bmatrix} C & 0 \end{bmatrix}$$

A rate of change constraint such as

$$\underline{\Delta} \leq u(k) - u(k-1) \leq \overline{\Delta} \qquad k \in \mathbb{I}_{\geq 0}$$

is then stated as

$$F\widetilde{x}(k) + Eu(k) \leq e \qquad F = \begin{bmatrix} 0 & -I \\ 0 & I \end{bmatrix} \qquad E = \begin{bmatrix} I \\ -I \end{bmatrix} \qquad e = \begin{bmatrix} \overline{\Delta} \\ -\underline{\Delta} \end{bmatrix}$$

To simplify analysis, it pays to maintain linear constraints when using linear dynamic models. So if we want to consider fairly general constraints for a linear system, we choose the form

$$Fx(k) + Eu(k) \leq e \qquad k \in \mathbb{I}_{\geq 0}$$

which subsumes all the forms listed previously.

When we consider nonlinear systems, analysis of the controller is not significantly simplified by maintaining linear inequalities, and we generalize the constraints to set membership

$$x(k) \in \mathbb{X} \qquad u(k) \in \mathbb{U} \qquad k \in \mathbb{I}_{\geq 0}$$

or, more generally,

$$(x(k), u(k)) \in \mathbb{Z} \qquad k \in \mathbb{I}_{\geq 0}$$

We should bear in mind one general distinction between input constraints, and output or state constraints. The input constraints often represent *physical limits*. In these cases, if the controller does not respect the input constraints, the physical system enforces them. In contrast, the output or state constraints are usually *desirables*. They may not be achievable depending on the disturbances affecting the system. It is often the function of an MPC controller to determine in real time that the output or state constraints are not achievable, and relax them in some satisfactory manner. As we discuss in Chapter 2, these considerations lead implementers of MPC often to set up the optimization problem using hard constraints for the input constraints and some form of soft constraints for the output or state constraints.

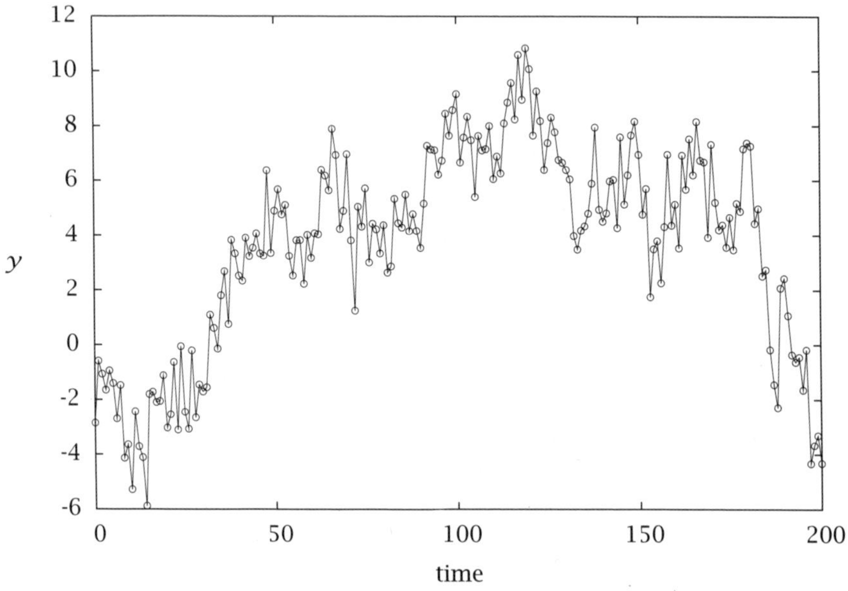

Figure 1.2: Output of a stochastic system versus time.

1.2.6 Deterministic and Stochastic

If one examines measurements coming from any complex, physical process, fluctuations in the data as depicted in Figure 1.2 are invariably present. For applications at small length scales, the fluctuations may be caused by the random behavior of small numbers of molecules. This type of application is becoming increasingly prevalent as scientists and engineers study applications in nanotechnology. This type of system also arises in life science applications when modeling the interactions of a few virus particles or protein molecules with living cells. In these applications there is no deterministic simulation model; the only system model available is stochastic.

Linear time-invariant models. In mainstream, classical process control problems, we are usually concerned with modeling, monitoring and controlling macroscopic systems, i.e., we are not considering systems composed of small numbers of molecules. So one may naturally ask (many do) what is the motivation for stochastic models in this arena? The motivation for stochastic models is to account for the unmodeled effects of the environment (disturbances) on the system under study. If we examine the measurement from any process control system of interest, no matter how "macroscopic," we are confronted with the physical reality that the measurement still looks a lot like Figure 1.2. If it is important to model the observed measurement fluctuations, we turn to stochastic models.

Some of the observed fluctuation in the data is assignable to the measurement device. This source of fluctuation is known as measurement "noise." Some of the observed fluctuation in the data is assignable to unmodeled disturbances from the environment affecting the state of the system. The simplest stochastic model for representing these two possible sources of disturbances is a linear model with added random variables

$$\begin{aligned} x^+ &= Ax + Bu + Gw \\ y &= Cx + Du + v \end{aligned}$$

with initial condition $x(0) = x_0$. The variable $w \in \mathbb{R}^g$ is the random variable acting on the state transition, $v \in \mathbb{R}^p$ is a random variable acting on the measured output, and x_0 is a random variable specifying the initial state. The random variable v is used to model the measurement noise and w models the process disturbance. The matrix $G \in \mathbb{R}^{n \times g}$ allows further refinement of the modeling between the source of the disturbance and its effect on the state. Often G is chosen to be the identity matrix with $g = n$.

1.3 Introductory MPC Regulator

1.3.1 Linear Quadratic Problem

We start by designing a controller to take the state of a deterministic, linear system to the origin. If the setpoint is not the origin, or we wish to track a time-varying setpoint trajectory, we will subsequently make modifications of the zero setpoint problem to account for that. The system model is

$$\begin{aligned} x^+ &= Ax + Bu \\ y &= Cx \end{aligned} \tag{1.5}$$

In this first problem, we assume that the state is measured, or $C = I$. We will handle the output measurement problem with state estimation in the next section. Using the model we can predict how the state evolves given any set of inputs we are considering. Consider N time steps into the future and collect the input sequence into $\mathbf{u}$,

$$\mathbf{u} = \{u(0), u(1), \ldots, u(N-1)\}$$

Constraints on the $\mathbf{u}$ sequence (i.e., valve saturations, etc.) are covered extensively in Chapter 2. The constraints are the main feature that distinguishes MPC from the standard linear quadratic (LQ) control.

We first define an objective function $V(\cdot)$ to measure the deviation of the trajectory of $x(k), u(k)$ from zero by summing the weighted squares

$$V(x(0), \mathbf{u}) = \frac{1}{2}\sum_{k=0}^{N-1} \left[x(k)'Qx(k) + u(k)'Ru(k)\right] + \frac{1}{2}x(N)'P_f x(N)$$

subject to

$$x^+ = Ax + Bu$$

The objective function depends on the input sequence and state sequence. The initial state is available from the measurement. The remainder of the state trajectory, $x(k), k = 1, \ldots, N$, is determined by the model and the input sequence $\mathbf{u}$. So we show the objective function's explicit dependence on the input sequence and initial state. The tuning parameters in the controller are the matrices Q and R. We allow the final state penalty to have a different weighting matrix, P_f, for generality. Large values of Q in comparison to R reflect the designer's intent to drive the state to the origin quickly at the expense of large control action. Penalizing the control action through large values of R relative to Q is the way to reduce the control action and slow down the rate at which the state approaches the origin. Choosing appropriate values of Q and R (i.e., tuning) is not always obvious, and this difficulty is one of the challenges faced by industrial practitioners of LQ control. Notice that MPC inherits this tuning challenge.

We then formulate the following optimal LQ control problem

$$\min_{\mathbf{u}} V(x(0), \mathbf{u}) \tag{1.6}$$

The Q, P_f and R matrices often are chosen to be diagonal, but we do not assume that here. We assume, however, that Q, P_f, and R are *real and symmetric*; Q and P_f are *positive semidefinite*; and R is *positive definite*. These assumptions guarantee that the solution to the optimal control problem exists and is unique.

1.3.2 Optimizing Multistage Functions

We next provide a brief introduction to methods for solving multistage optimization problems like (1.6). Consider the set of variables w, x, y, and z, and the following function to be optimized

$$f(w,x) + g(x,y) + h(y,z)$$

Notice that the objective function has a special structure in which each stage's cost function in the sum depends only on adjacent variable pairs. For the first version of this problem, we consider w to be a fixed parameter, and we would like to solve the problem

$$\min_{x,y,z} f(w,x) + g(x,y) + h(y,z) \qquad w \text{ fixed}$$

One option is to optimize simultaneously over all three decision variables. Because of the objective function's special structure, however, we can obtain the solution by optimizing a sequence of three single-variable problems defined as follows

$$\min_x \left[f(w,x) + \min_y \left[g(x,y) + \min_z h(y,z) \right] \right]$$

We solve the inner problem over z first, and denote the optimal value and solution as follows

$$\underline{h}^0(y) = \min_z h(y,z) \qquad \underline{z}^0(y) = \arg\min_z h(y,z)$$

Notice that the optimal z and value function for this problem are both expressed as a function of the y variable. We then move to the next optimization problem and solve for the y variable

$$\min_y g(x,y) + \underline{h}^0(y)$$

and denote the solution and value function as

$$\underline{g}^0(x) = \min_y g(x,y) + \underline{h}^0(y) \qquad \underline{y}^0(x) = \arg\min_y g(x,y) + \underline{h}^0(y)$$

The optimal solution for y is a function of x, the remaining variable to be optimized. The third and final optimization is

$$\min_x f(w,x) + \underline{g}^0(x)$$

with solution and value function

$$\underline{f}^0(w) = \min_x f(w,x) + \underline{g}^0(x) \qquad \underline{x}^0(w) = \arg\min_x f(w,x) + \underline{g}^0(x)$$

We summarize the recursion with the following annotated equation

$$\underbrace{\min_x \left[f(w,x) + \overbrace{\min_y \left[g(x,y) + \underbrace{\min_z h(y,z)}_{\underline{h}^0(y),\ \underline{z}^0(y)} \right]}^{\underline{g}^0(x),\ \underline{y}^0(x)} \right]}_{\underline{f}^0(w),\ \underline{x}^0(w)}$$

If we are mainly interested in the first variable x, then the function $\underline{x}^0(w)$ is of primary interest and we have obtained this function quite efficiently. This nested solution approach is an example of a class of techniques known as dynamic programming (DP). DP was developed by Bellman (Bellman, 1957; Bellman and Dreyfus, 1962) as an efficient means for solving these kinds of multistage optimization problems. Bertsekas (1987) provides an overview of DP.

The version of the method we just used is called *backward* DP because we find the variables in reverse order: first z, then y, and finally x. Notice we find the optimal solutions as *functions* of the variables to be optimized at the next stage. If we wish to find the other variables y and z as a function of the known parameter w, then we nest the optimal solutions found by the backward DP recursion

$$\underset{\sim}{y}^{\,0}(w) = \underline{y}^0(\underline{x}^0(w)) \qquad \underset{\sim}{z}^{\,0}(w) = \underline{z}^0(\,\underset{\sim}{y}^{\,0}(w)) = \underline{z}^0(\underline{y}^0(\underline{x}^0(w)))$$

As we see shortly, backward DP is the method of choice for the regulator problem.

In the state estimation problem to be considered later in this chapter, w becomes a variable to be optimized, and z plays the role of a parameter. We wish to solve the problem

$$\min_{w,x,y} f(w,x) + g(x,y) + h(y,z) \qquad z \text{ fixed}$$

We can still break the problem into three smaller nested problems, but the order is reversed

$$\underbrace{\min_y \left[h(y,z) + \overbrace{\min_x \left[g(x,y) + \underbrace{\min_w f(w,x)}_{\overline{f}^0(x),\, \overline{w}^0(x)} \right]}^{\overline{g}^0(y),\, \overline{x}^0(y)} \right]}_{\overline{h}^0(z),\, \overline{y}^0(z)} \tag{1.7}$$

This form is called *forward* DP because we find the variables in the order given: first w, then x, and finally y. The optimal value functions and optimal solutions at each of the three stages are shown in (1.7). This version is preferable if we are primarily interested in finding the final variable y as a function of the parameter z. As before, if we need the other optimized variables x and w as a function of the parameter z, we must insert the optimal functions found by the forward DP recursion

$$\widetilde{x}^0(z) = \overline{x}^0(\overline{y}^0(z)) \qquad \widetilde{w}^0(z) = \overline{w}^0(\widetilde{x}^0(z)) = \overline{w}^0(\overline{x}^0(\overline{y}^0(z)))$$

For the reader interested in trying some exercises to reinforce the concepts of DP, Exercise 1.15 considers finding the function $\widetilde{w}^0(z)$ with *backward* DP instead of forward DP as we just did here. Exercise C.1 discusses showing that the nested optimizations indeed give the same answer as simultaneous optimization over all decision variables.

Finally, if we optimize over all four variables, including the one considered as a fixed parameter in the two versions of DP we used, then we have two equivalent ways to express the value of the complete optimization

$$\min_{w,x,y,z} f(w,x) + g(x,y) + h(y,z) = \min_w \underline{f}^0(w) = \min_z \overline{h}^0(z)$$

The result in the next example proves useful in combining quadratic functions to solve the LQ problem.

Example 1.1: Sum of quadratic functions

Consider the two quadratic functions given by

$$V_1(x) = (1/2)(x-a)'A(x-a) \qquad V_2(x) = (1/2)(x-b)'B(x-b)$$

in which $A, B > 0$ are positive definite matrices and a and b are n-vectors locating the minimum of each function. Figure 1.3 displays the

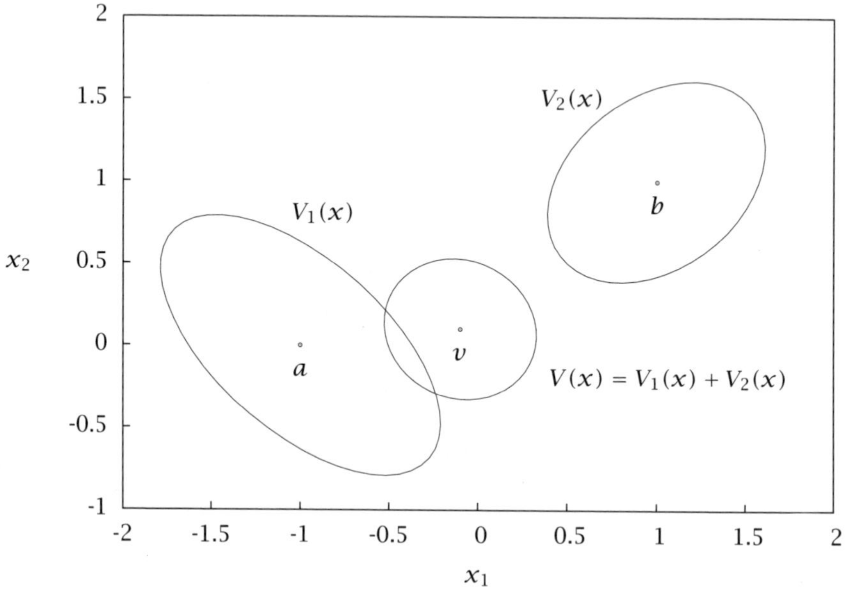

Figure 1.3: Two quadratic functions and their sum; $V(x) = V_1(x) + V_2(x)$.

ellipses defined by the level sets $V_1(x) = 1/4$ and $V_2(x) = 1/4$ for the following data

$$A = \begin{bmatrix} 1.25 & 0.75 \\ 0.75 & 1.25 \end{bmatrix} \qquad a = \begin{bmatrix} -1 \\ 0 \end{bmatrix}$$

$$B = \begin{bmatrix} 1.5 & -0.5 \\ -0.5 & 1.5 \end{bmatrix} \qquad b = \begin{bmatrix} 1 \\ 1 \end{bmatrix}$$

(a) Show that the sum $V(x) = V_1(x) + V_2(x)$ is also quadratic

$$V(x) = (1/2)(x - v)'H(x - v) + \text{constant}$$

in which

$$H = A + B \qquad v = H^{-1}(Aa + Bb)$$

and verify the three ellipses given in Figure 1.3.

(b) Consider a generalization useful in the discussion of the upcoming state estimation problem. Let

$$V_1(x) = (1/2)(x-a)'A(x-a) \qquad V_2(x) = (1/2)(Cx-b)'B(Cx-b)$$

Derive the formulas for H and v for this case.

(c) Use the matrix inversion lemma (see Exercise 1.12) and show that $V(x)$ can be expressed also in an inverse form, which is useful in state estimation problems

$$\begin{aligned} V(x) &= (1/2)(x-v)'\widetilde{H}^{-1}(x-v) + \text{constant} \\ \widetilde{H} &= A^{-1} - A^{-1}C'(CA^{-1}C' + B^{-1})^{-1}CA^{-1} \\ v &= a + A^{-1}C'(CA^{-1}C' + B^{-1})^{-1}(b - Ca) \end{aligned}$$

Solution

(a) The sum of two quadratics is also quadratic, so we parameterize the sum as

$$V(x) = (1/2)(x-v)'H(x-v) + d$$

and solve for v, H, and d. Comparing zeroth, first and second derivatives gives

$$\begin{aligned} V(v) &= d & &= V_1(v) + V_2(v) \\ V_x(x) &= H(x-v) & &= A(x-a) + B(x-b) \\ V_{xx} &= H & &= A + B \end{aligned}$$

Solving these gives

$$\begin{aligned} H &= A + B \\ v &= H^{-1}(Aa + Bb) \\ d &= V_1(v) + V_2(v) \end{aligned} \tag{1.8}$$

Notice that H is positive definite since A and B are positive definite. Substituting the values of a, A, b, and B, and setting $d = 0$ gives

$$V(x) = (1/2)(x-v)'H(x-v)$$

$$H = \begin{bmatrix} 2.75 & 0.25 \\ 0.25 & 2.75 \end{bmatrix} \qquad v = \begin{bmatrix} -0.1 \\ 0.1 \end{bmatrix}$$

$V(x) = 1/4$ is plotted for the choice of constant $d = 0$.

(b) Comparing zeroth, first and second derivatives gives

$$\begin{aligned} V(v) &= d & &= V_1(v) + V_2(v) \\ V_x(x) &= H(x-v) & &= A(x-a) + C'B(Cx-b) \\ V_{xx} &= H & &= A + C'BC \end{aligned}$$

Solving these gives

$$\begin{aligned} H &= A + C'BC \\ v &= H^{-1}(Aa + C'Bb) \\ d &= V_1(v) + V_2(v) \end{aligned}$$

Notice that H is positive definite since A is positive definite and $C'BC$ is positive semidefinite for any C.

(c) Define $\overline{x} = x - a$ and $\overline{b} = b - Ca$, and express the problem as

$$\begin{aligned} V(\overline{x}) &= \overline{x}'A\overline{x} + (C(\overline{x}+a) - b)'B(C(\overline{x}+a) - b) \\ &= \overline{x}'A\overline{x} + (C\overline{x} - \overline{b})'B(C\overline{x} - \overline{b}) \end{aligned}$$

Apply the solution of the previous part and set the constant to zero to obtain

$$\begin{aligned} V(\overline{x}) &= (\overline{x} - \overline{v})'H(\overline{x} - \overline{v}) \\ H &= A + C'BC \\ \overline{v} &= H^{-1}C'B\overline{b} \end{aligned}$$

Use the matrix inversion lemma's (1.55) on H and (1.56) on $\overline{v}$ to obtain

$$\begin{aligned} \widetilde{H} &= A^{-1} - A^{-1}C'(CA^{-1}C' + B^{-1})^{-1}CA^{-1} \\ \overline{v} &= A^{-1}C'(CA^{-1}C' + B^{-1})^{-1}\overline{b} \end{aligned}$$

The function V is then given by

$$\begin{aligned} V &= (1/2)(\overline{x} - \overline{v})'\widetilde{H}^{-1}(\overline{x} - \overline{v}) \\ V &= (1/2)(x - (a + \overline{v}))'\widetilde{H}^{-1}(x - (a + \overline{v})) \\ V &= (1/2)(x - v)'\widetilde{H}^{-1}(x - v) \end{aligned}$$

in which

$$v = a + A^{-1}C'(CA^{-1}C' + B^{-1})^{-1}\,(b - Ca)$$

□

1.3.3 Dynamic Programming Solution

After this brief introduction to DP, we apply it to solve the LQ control problem. We first rewrite (1.6) in the following form to see the structure clearly

$$V(x(0),\mathbf{u}) = \sum_{k=0}^{N-1} \ell(x(k),u(k)) + \ell_N(x(N)) \qquad \text{s.t. } x^+ = Ax + Bu$$

in which the *stage cost* $\ell(x,u) = (1/2)(x'Qx + u'Ru), k = 0,\ldots,N-1$ and the terminal stage cost $\ell_N(x) = (1/2)x'P_f x$. Since $x(0)$ is known, we choose *backward* DP as the convenient method to solve this problem. We first rearrange the overall objective function so we can optimize over input $u(N-1)$ and state $x(N)$

$$\min_{u(0),x(1),\ldots u(N-2),x(N-1)} \ell(x(0),u(0)) + \ell(x(1),u(1)) + \cdots + \min_{u(N-1),x(N)} \ell(x(N-1),u(N-1)) + \ell_N(x(N))$$

subject to

$$x(k+1) = Ax(k) + Bu(k) \qquad k = 0,\ldots N-1$$

The problem to be solved at the last stage is

$$\min_{u(N-1),x(N)} \ell(x(N-1),u(N-1)) + \ell_N(x(N)) \tag{1.9}$$

subject to

$$x(N) = Ax(N-1) + Bu(N-1)$$

in which $x(N-1)$ appears in this stage as a parameter. We denote the optimal cost by $V_{N-1}^0(x(N-1))$ and the optimal decision variables by $u_{N-1}^0(x(N-1))$ and $x_N^0(x(N-1))$. The optimal cost and decisions at the last stage are parameterized by the state at the previous stage as we expect in backward DP. We next solve this optimization. First we substitute the state equation for $x(N)$ and combine the two quadratic terms using (1.8)

$$\begin{aligned}
&\ell(x(N-1),u(N-1)) + \ell_N(x(N)) \\
&\quad = (1/2)\left(|x(N-1)|_Q^2 + |u(N-1)|_R^2 + |Ax(N-1) + Bu(N-1)|_{P_f}^2\right) \\
&\quad = (1/2)\left(|x(N-1)|_Q^2 + |(u(N-1) - v)|_H^2\right) + d
\end{aligned}$$

in which

$$\begin{aligned}
H &= R + B'P_f B \\
v &= K(N-1)x(N-1) \\
d &= (1/2)x(N-1)'\Big(K(N-1)'RK(N-1) + \\
&\quad (A + BK(N-1))'P_f(A + BK(N-1))\Big)x(N-1) \\
K(N-1) &= -(B'P_f B + R)^{-1}B'P_f A
\end{aligned} \tag{1.10}$$

Given this form of the cost function, we see by inspection that the optimal input for $u(N-1)$ is v defining the optimal control law at stage $N-1$ to be a linear function of the state $x(N-1)$. Then using the model equation, the optimal final state is also a linear function of state $x(N-1)$. The optimal cost is d, which makes the optimal cost a quadratic function of $x(N-1)$. Summarizing, for all x

$$\begin{aligned}
u^0_{N-1}(x) &= K(N-1)\, x \\
x^0_{N-1}(x) &= (A + BK(N-1))\, x \\
V^0_{N-1}(x) &= (1/2)x'\, \Pi(N-1)\, x \\
\Pi(N-1) &= Q + A'P_f A + \\
&\qquad K(N-1)'(B'P_f B + R)K(N-1) + 2K(N-1)'B'P_f A
\end{aligned}$$

We can rewrite $\Pi(N-1)$ using the result

$$\begin{aligned}
K(N-1)'(B'P_f B + R)K(N-1) + 2K(N-1)'B'P_f A = \\
- A'P_f B(B'P_f B + R)^{-1}B'P_f A
\end{aligned}$$

which is obtained by substituting (1.10) for K into the equation for $\Pi(N-1)$ and simplifying. Substituting this result into the previous equation gives

$$\Pi(N-1) = Q + A'P_f A - A'P_f B(B'P_f B + R)^{-1}B'P_f A$$

The function $V^0_{N-1}(x)$ defines the optimal *cost to go* from state x for the last stage under the optimal control law $u^0_{N-1}(x)$. Having this function allows us to move to the next stage of the DP recursion. For the next stage we solve the optimization

$$\min_{u(N-2),x(N-1)} \quad \ell(x(N-2), u(N-2)) + V^0_{N-1}(x(N-1))$$

subject to

$$x(N-1) = Ax(N-2) + Bu(N-2)$$

Notice that this problem is identical in structure to the stage we just solved, (1.9), and we can write out the solution by simply renaming variables

$$\begin{aligned}
u^0_{N-2}(x) &= K(N-2)\,x \\
x^0_{N-2}(x) &= (A + BK(N-2))\,x \\
V^0_{N-2}(x) &= (1/2)x'\,\Pi(N-2)\,x \\
K(N-2) &= -(B'\Pi(N-1)B + R)^{-1}B'\Pi(N-1)A \\
\Pi(N-2) &= Q + A'\Pi(N-1)A - \\
&\qquad A'\Pi(N-1)B(B'\Pi(N-1)B + R)^{-1}B'\Pi(N-1)A
\end{aligned}$$

The recursion from $\Pi(N-1)$ to $\Pi(N-2)$ is known as a backward Riccati iteration. To summarize, the backward Riccati iteration is defined as follows

$$\Pi(k-1) = Q + A'\Pi(k)A - A'\Pi(k)B\left(B'\Pi(k)B + R\right)^{-1}B'\Pi(k)A \qquad k = N, N-1, \ldots, 1 \tag{1.11}$$

with terminal condition

$$\Pi(N) = P_f \tag{1.12}$$

The terminal condition replaces the typical initial condition because the iteration is running backward. The optimal control policy at each stage is

$$u^0_k(x) = K(k)x \qquad k = N-1, N-2, \ldots, 0 \tag{1.13}$$

The optimal gain at time k is computed from the Riccati matrix at time $k+1$

$$K(k) = -\left(B'\Pi(k+1)B + R\right)^{-1}B'\Pi(k+1)A \qquad k = N-1, N-2, \ldots, 0 \tag{1.14}$$

and the optimal cost to go from time k to time N is

$$V^0_k(x) = (1/2)x'\Pi(k)x \qquad k = N, N-1, \ldots, 0 \tag{1.15}$$

1.3.4 The Infinite Horizon LQ Problem

Let us motivate the infinite horizon problem by showing a weakness of the finite horizon problem. Kalman (1960b, p.113) pointed out in his classic 1960 paper that optimality does not ensure stability.

> In the engineering literature it is often assumed (tacitly and incorrectly) that a system with optimal control law (6.8) is necessarily stable.

Assume that we use as our control law the first feedback gain of the finite horizon problem, $K(0)$,

$$u(k) = K(0)x(k)$$

Then the stability of the closed-loop system is determined by the eigenvalues of $A+BK(0)$. We now construct an example that shows choosing $Q > 0$, $R > 0$, and $N \geq 1$ does not ensure stability. In fact, we can find reasonable values of these parameters such that the controller destabilizes a stable system.[2] Let

$$A = \begin{bmatrix} 4/3 & -2/3 \\ 1 & 0 \end{bmatrix} \qquad B = \begin{bmatrix} 1 \\ 0 \end{bmatrix} \qquad C = [-2/3 \;\; 1]$$

This system is chosen so that $G(z)$ has a *zero* at $z = 3/2$, i.e., an unstable zero. We now construct an LQ controller that inverts this zero and hence produces an unstable system. We would like to choose $Q = C'C$ so that y itself is penalized, but that Q is only semidefinite. We add a small positive definite piece to $C'C$ so that Q is positive definite, and choose a *small* positive R penalty (to encourage the controller to misbehave), and $N = 5$,

$$Q = C'C + 0.001I = \begin{bmatrix} 4/9 + .001 & -2/3 \\ -2/3 & 1.001 \end{bmatrix} \qquad R = 0.001$$

We now iterate the Riccati equation four times starting from $\Pi = P_f = Q$ and compute $K(0)$ for $N = 5$; then we compute the eigenvalues of $A + BK(0)$ and achieve[3]

$$\text{eig}(A + BK_5(0)) = \{1.307, 0.001\}$$

[2]In Chapter 2, we present several controller design methods that prevent this kind of instability.

[3]Please check this answer with Octave or MATLAB.

Using this controller the closed-loop system evolution is $x(k) = (A + BK_5(0))^k x_0$. Since an eigenvalue of $A + BK_5(0)$ is greater than unity, $x(k) \to \infty$ as $k \to \infty$. In other words the closed-loop system is unstable.

If we continue to iterate the Riccati equation, which corresponds to increasing the horizon in the controller, we obtain for $N = 7$

$$\text{eig}(A + BK_7(0)) = \{0.989, 0.001\}$$

and the controller is stabilizing. If we continue iterating the Riccati equation, we converge to the following steady-state closed-loop eigenvalues

$$\text{eig}(A + BK_\infty(0)) = \{0.664, 0.001\}$$

This controller corresponds to an infinite horizon control law. Notice that it is stabilizing and has a reasonable stability margin. Nominal stability is a guaranteed property of infinite horizon controllers as we prove in the next section.

With this motivation, we are led to consider directly the infinite horizon case

$$V(x(0), \mathbf{u}) = \frac{1}{2} \sum_{k=0}^{\infty} x(k)' Q x(k) + u(k)' R u(k) \tag{1.16}$$

in which $x(k)$ is the solution at time k of $x^+ = Ax + Bu$ if the initial state is $x(0)$ and the input sequence is $\mathbf{u}$. If we are interested in a continuous process (i.e., no final time), then the natural cost function is an infinite horizon cost. If we were truly interested in a batch process (i.e., the process does stop at $k = N$), then stability is not a relevant property, and we naturally would use the finite horizon LQ controller and the *time-varying* controller, $u(k) = K(k)x(k), k = 0, 1, \ldots, N$.

In considering the infinite horizon problem, we first restrict attention to systems for which there exist input sequences that give bounded cost. Consider the case $A = I$ and $B = 0$, for example. Regardless of the choice of input sequence, (1.16) is unbounded for $x(0) \neq 0$. It seems clear that we are not going to stabilize an unstable system ($A = I$) without any input ($B = 0$). This is an example of an *uncontrollable* system. In order to state the sharpest results on stabilization, we require the concepts of controllability, stabilizability, observability, and detectability. We shall define these concepts subsequently.

1.3.5 Controllability

A system is *controllable* if, for any pair of states x, z in the state space, z can be reached in finite time from x (or x controlled to z) (Sontag, 1998, p.83). A *linear discrete time* system $x^+ = Ax + Bu$ is therefore controllable if there exists a finite time N and a sequence of inputs

$$\{u(0), u(1), \ldots u(N-1)\}$$

that can transfer the system from any x to any z in which

$$z = A^N x + \begin{bmatrix} B & AB & \cdots & A^{N-1}B \end{bmatrix} \begin{bmatrix} u(N-1) \\ u(n-2) \\ \vdots \\ u(0) \end{bmatrix}$$

We can simplify this condition by noting that the matrix powers A^k for $k \geq n$ are expressible as linear combinations of the powers 0 to $n-1$. This result is a consequence of the Cayley-Hamilton theorem (Horn and Johnson, 1985, pp. 86–87). Therefore the range of the matrix $\begin{bmatrix} B & AB & \cdots & A^{N-1}B \end{bmatrix}$ for $N \geq n$ is the same as $\begin{bmatrix} B & AB & \cdots & A^{n-1}B \end{bmatrix}$. In other words, for an unconstrained linear system, if we cannot reach z in n moves, we cannot reach z in any number of moves. The question of *controllability* of a linear time invariant system is therefore a question of *existence* of solutions to linear equations for an arbitrary right-hand side

$$\begin{bmatrix} B & AB & \cdots & A^{n-1}B \end{bmatrix} \begin{bmatrix} u(n-1) \\ u(n-2) \\ \vdots \\ u(0) \end{bmatrix} = z - A^n x$$

The matrix appearing in this equation is known as the *controllability matrix* C

$$C = \begin{bmatrix} B & AB & \cdots & A^{n-1}B \end{bmatrix} \tag{1.17}$$

From the fundamental theorem of linear algebra, we know a solution exists for all right-hand sides if and only if the *rows* of the $n \times nm$ controllability matrix are linearly independent.[4] Therefore, the system (A, B) is controllable if and only if

$$\operatorname{rank}(C) = n$$

[4]See Section A.4 of Appendix A or (Strang, 1980, pp.87–88) for a review of this result.

The following result for checking controllability also proves useful (Hautus, 1972).

Lemma 1.2 (Hautus Lemma for controllability). *A system is controllable if and only if*

$$\operatorname{rank}\begin{bmatrix} \lambda I - A & B \end{bmatrix} = n \qquad \text{for all } \lambda \in \mathbb{C} \tag{1.18}$$

in which $\mathbb{C}$ is the set of complex numbers.

Notice that the first n columns of the matrix in (1.18) are linearly independent if λ is not an eigenvalue of A, so (1.18) is equivalent to checking the rank at just the eigenvalues of A

$$\operatorname{rank}\begin{bmatrix} \lambda I - A & B \end{bmatrix} = n \qquad \text{for all } \lambda \in \operatorname{eig}(A)$$

1.3.6 Convergence of the Linear Quadratic Regulator

We now show that the infinite horizon regulator asymptotically stabilizes the origin for the closed-loop system. Define the infinite horizon objective function

$$V(x, \mathbf{u}) = \frac{1}{2}\sum_{k=0}^{\infty} x(k)'Qx(k) + u(k)'Ru(k)$$

subject to

$$\begin{aligned} x^+ &= Ax + Bu \\ x(0) &= x \end{aligned}$$

with $Q, R > 0$. If (A, B) is controllable, the solution to the optimization problem

$$\min_{\mathbf{u}} V(x, \mathbf{u})$$

exists and is unique for all x. We denote the optimal solution by $\mathbf{u}^0(x)$, and the first input in the optimal sequence by $u^0(x)$. The feedback control law $\kappa_\infty(\cdot)$ for this infinite horizon case is then defined as $u = \kappa_\infty(x)$ in which $\kappa_\infty(x) = u^0(x) = \mathbf{u}^0(0; x)$. As stated in the following lemma, this infinite horizon linear quadratic regulator (LQR) is stabilizing.

Lemma 1.3 (LQR convergence). *For (A, B) controllable, the infinite horizon LQR with $Q, R > 0$ gives a convergent closed-loop system*

$$x^+ = Ax + B\kappa_\infty(x)$$

Proof. The cost of the infinite horizon objective is bounded above for all $x(0)$ because (A, B) is controllable. Controllability implies that there exists a sequence of n inputs $\{u(0), u(1), \ldots, u(n-1)\}$ that transfers the state from any $x(0)$ to $x(n) = 0$. A zero control sequence after $k = n$ for $\{u(n+1), u(n+2), \ldots\}$ generates zero cost for all terms in V after $k = n$, and the objective function for this infinite control sequence is therefore finite. The cost function is strictly convex in $\mathbf{u}$ because $R > 0$ so the solution to the optimization is unique.

If we consider the sequence of costs to go along the closed-loop trajectory, we have

$$V_{k+1} = V_k - (1/2)\left(x(k)'Qx(k) + u(k)'Ru(k)\right)$$

in which $V_k = V^0(x(k))$ is the cost at time k for state value $x(k)$ and $u(k) = u^0(x(k))$ is the optimal control for state $x(k)$. The cost along the closed-loop trajectory is nonincreasing and bounded below (by zero). Therefore, the sequence $\{V_k\}$ converges and

$$x(k)'Qx(k) \to 0 \qquad u(k)'Ru(k) \to 0 \qquad \text{as } k \to \infty$$

Since $Q, R > 0$, we have

$$x(k) \to 0 \qquad u(k) \to 0 \qquad \text{as } k \to \infty$$

and closed-loop convergence is established. ■

In fact we know more. From the previous sections, we know the optimal solution is found by iterating the Riccati equation, and the optimal infinite horizon control law and optimal cost are given by

$$u^0(x) = Kx \qquad V^0(x) = (1/2)x'\Pi x$$

in which

$$\begin{aligned} K &= -(B'\Pi B + R)^{-1}B'\Pi A \\ \Pi &= Q + A'\Pi A - A'\Pi B(B'\Pi B + R)^{-1}B'\Pi A \end{aligned} \tag{1.19}$$

Proving Lemma 1.3 has shown also that for (A, B) controllable and $Q, R > 0$, a positive definite solution to the discrete algebraic Riccati equation (DARE), (1.19), exists and the eigenvalues of $(A + BK)$ are asymptotically stable for the K corresponding to this solution (Bertsekas, 1987, pp.58–64).

This basic approach to establishing regulator stability will be generalized in Chapter 2 to handle constrained and nonlinear systems, so it is helpful for the new student to first become familiar with these ideas in the unconstrained, linear setting. For linear systems, asymptotic convergence is equivalent to asymptotic stability, and we delay the discussion of stability until Chapter 2. In Chapter 2 the optimal cost is shown to be a Lyapunov function for the closed-loop system. We also can strengthen the stability for linear systems from asymptotic stability to exponential stability based on the form of the Lyapunov function.

The LQR convergence result in Lemma 1.3 is the simplest to establish, but we can enlarge the class of systems and penalties for which closed-loop stability is guaranteed. The system restriction can be weakened from controllability to *stabilizability*, which is discussed in Exercises 1.19 and 1.20. The restriction on the allowable state penalty Q can be weakened from $Q > 0$ to $Q \geq 0$ and (A, Q) *detectable*, which is also discussed in Exercise 1.20. The restriction $R > 0$ is retained to ensure uniqueness of the control law. In applications, if one cares little about the cost of the control, then R is chosen to be small, but positive definite.

1.4 Introductory State Estimation

The next topic is state estimation. In most applications, the variables that are conveniently or economically measurable (y) are a small subset of the variables required to model the system (x). Moreover, the measurement is corrupted with sensor noise and the state evolution is corrupted with process noise. Determining a good state estimate for use in the regulator in the face of a noisy and incomplete output measurement is a challenging task. That is the challenge of state estimation.

To fully appreciate the fundamentals of state estimation, we must address the fluctuations in the data. Probability theory has proven itself as the most successful and versatile approach to modeling these fluctuations. In this section we introduce the probability fundamentals necessary to develop an optimal state estimator in the simplest possible setting: a linear discrete time model subject to normally distributed process and measurement noise. This optimal state estimator is known as the Kalman filter (Kalman, 1960a). In Chapter 4 we revisit the state estimation problem in a much wider setting, and consider nonlinear models and constraints on the system that preclude an analytical solution such as the Kalman filter. The probability theory presented here is also preparation for understanding that chapter.

1.4.1 Linear Systems and Normal Distributions

This section summarizes the probability and random variable results required for deriving a linear optimal estimator such as the Kalman filter. We assume that the reader is familiar with the concepts of a random variable, probability density and distribution, the multivariate normal distribution, mean and variance, statistical independence, and conditional probability. Readers unfamiliar with these terms should study the material in Appendix A before reading this and the next sections.

In the following discussion let x, y, and z be vectors of random variables. We use the notation

$$x \sim N(m, P)$$
$$p_x(x) = n(x, m, P)$$

to denote random variable x is normally distributed with mean m and covariance (or simply variance) P, in which

$$n(x, m, P) = \frac{1}{(2\pi)^{n/2}(\det P)^{1/2}} \exp\left[-\frac{1}{2}(x-m)'P^{-1}(x-m)\right] \tag{1.20}$$

and $\det(P)$ denotes the determinant of matrix P. Note that if $x \in \mathbb{R}^n$, then $m \in \mathbb{R}^n$ and $P \in \mathbb{R}^{n \times n}$ is a positive definite matrix. We require three main results. The simplest version can be stated as follows.

Joint independent normals. If x and y are normally distributed and (statistically) independent[5]

$$x \sim N(m_x, P_x) \qquad y \sim N(m_y, P_y)$$

then their joint density is given by

$$p_{x,y}(x, y) = n(x, m_x, P_x)\, n(y, m_y, P_y)$$

$$\begin{bmatrix} x \\ y \end{bmatrix} \sim N\left(\begin{bmatrix} m_x \\ m_y \end{bmatrix}, \begin{bmatrix} P_x & 0 \\ 0 & P_y \end{bmatrix}\right) \tag{1.21}$$

Note that, depending on convenience, we use both (x, y) and the vector $\left[\begin{smallmatrix} x \\ y \end{smallmatrix}\right]$ to denote the pair of random variables.

Linear transformation of a normal. If x is normally distributed with mean m and variance P, and y is a linear transformation of x, $y = Ax$, then y is distributed with mean Am and variance APA'

$$x \sim N(m, P) \qquad y = Ax \qquad y \sim N(Am, APA') \tag{1.22}$$

[5] We may emphasize that two vectors of random variables are independent using *statistically independent* to distinguish this concept from linear independence of vectors.

Conditional of a joint normal. If x and y are jointly normally distributed as

$$\begin{bmatrix} x \\ y \end{bmatrix} \sim N\left(\begin{bmatrix} m_x \\ m_y \end{bmatrix} \begin{bmatrix} P_x & P_{xy} \\ P_{yx} & P_y \end{bmatrix}\right)$$

then the conditional density of x given y is also normal

$$p_{x|y}(x|y) = n(x, m, P) \tag{1.23}$$

in which the mean is

$$m = m_x + P_{xy}P_y^{-1}(y - m_y)$$

and the covariance is

$$P = P_x - P_{xy}P_y^{-1}P_{yx}$$

Note that the conditional mean m is itself a random variable because it depends on the random variable y.

To derive the optimal estimator, we actually require these three main results conditioned on additional random variables. The analogous results are the following.

Joint independent normals. If $p_{x|z}(x|z)$ is normal, and y is statistically independent of x and z and normally distributed

$$p_{x|z}(x|z) = n(x, m_x, P_x)$$
$$y \sim N(m_y, P_y) \qquad y \text{ independent of } x \text{ and } z$$

then the conditional joint density of (x, y) given z is

$$p_{x,y|z}(x, y|z) = n(x, m_x, P_x)\, n(y, m_y, P_y)$$

$$p_{x,y|z}\left(\begin{bmatrix} x \\ y \end{bmatrix} \middle| z\right) = n\left(\begin{bmatrix} x \\ y \end{bmatrix}, \begin{bmatrix} m_x \\ m_y \end{bmatrix}, \begin{bmatrix} P_x & 0 \\ 0 & P_y \end{bmatrix}\right) \tag{1.24}$$

Linear transformation of a normal.

$$p_{x|z}(x|z) = n(x, m, P) \qquad y = Ax$$
$$p_{y|z}(y|z) = n(y, Am, APA') \tag{1.25}$$

Conditional of a joint normal. If x and y are jointly normally distributed as

$$p_{x,y|z}\left(\begin{bmatrix} x \\ y \end{bmatrix} \middle| z\right) = n\left(\begin{bmatrix} x \\ y \end{bmatrix}, \begin{bmatrix} m_x \\ m_y \end{bmatrix}, \begin{bmatrix} P_x & P_{xy} \\ P_{yx} & P_y \end{bmatrix}\right)$$

then the conditional density of x given y, z is also normal

$$p_{x|y,z}(x|y,z) = n(x, m, P) \tag{1.26}$$

in which

$$m = m_x + P_{xy}P_y^{-1}(y - m_y)$$
$$P = P_x - P_{xy}P_y^{-1}P_{yx}$$

1.4.2 Linear Optimal State Estimation

We start by assuming the initial state $x(0)$ is normally distributed with some mean and covariance

$$x(0) \sim N(\overline{x}(0), Q(0))$$

In applications, we often do not know $\overline{x}(0)$ or $Q(0)$. In such cases we often set $\overline{x}(0) = 0$ and choose a large value for $Q(0)$ to indicate our lack of prior knowledge. This choice is referred to in the statistics literature as a *noninformative prior.* The choice of noninformative prior forces the upcoming $y(k)$ measurements to determine the state estimate $\hat{x}(k)$.

Combining the measurement. We obtain noisy measurement $y(0)$ satisfying

$$y(0) = Cx(0) + v(0)$$

in which $v(0) \sim N(0, R)$ is the measurement noise. If the measurement process is quite noisy, then R is large. If the measurements are highly accurate, then R is small. We choose a zero mean for v because all of the deterministic effects with nonzero mean are considered part of the model, and the measurement noise reflects what is left after all these other effects have been considered. Given the measurement $y(0)$, we want to obtain the conditional density $p_{x(0)|y(0)}(x(0)|y(0))$. This conditional density describes the change in our knowledge about $x(0)$ after we obtain measurement $y(0)$. This step is the essence of state estimation. To derive this conditional density, first consider the pair of variables $(x(0), y(0))$ given as

$$\begin{bmatrix} x(0) \\ y(0) \end{bmatrix} = \begin{bmatrix} I & 0 \\ C & I \end{bmatrix} \begin{bmatrix} x(0) \\ v(0) \end{bmatrix}$$

We assume that the noise $v(0)$ is statistically independent of $x(0)$, and use the independent joint normal result (1.21) to express the joint density of $(x(0), v(0))$

$$\begin{bmatrix} x(0) \\ v(0) \end{bmatrix} \sim N\left(\begin{bmatrix} \overline{x}(0) \\ 0 \end{bmatrix}, \begin{bmatrix} Q(0) & 0 \\ 0 & R \end{bmatrix}\right)$$

From the previous equation, the pair $(x(0), y(0))$ is a linear transformation of the pair $(x(0), v(0))$. Therefore, using the linear transformation of normal result (1.22), and the density of $(x(0), v(0))$ gives the density of $(x(0), y(0))$

$$\begin{bmatrix} x(0) \\ y(0) \end{bmatrix} \sim N\left(\begin{bmatrix} \overline{x}(0) \\ C\overline{x}(0) \end{bmatrix}, \begin{bmatrix} Q(0) & Q(0)C' \\ CQ(0) & CQ(0)C' + R \end{bmatrix}\right)$$

Given this joint density, we then use the conditional of a joint normal result (1.23) to obtain

$$p_{x(0)|y(0)}(x(0)|y(0)) = n(x(0), m, P)$$

in which

$$\begin{aligned} m &= \overline{x}(0) + L(0)\left(y(0) - C\overline{x}(0)\right) \\ L(0) &= Q(0)C'(CQ(0)C' + R)^{-1} \\ P &= Q(0) - Q(0)C'(CQ(0)C' + R)^{-1}CQ(0) \end{aligned}$$

We see that the conditional density $p_{x(0)|y(0)}$ is normal. The *optimal* state estimate is the value of $x(0)$ that maximizes this conditional density. For a normal, that is the mean, and we choose $\hat{x}(0) = m$. We also denote the variance in this conditional after measurement $y(0)$ by $P(0) = P$ with P given in the previous equation. The change in variance after measurement ($Q(0)$ to $P(0)$) quantifies the information increase by obtaining measurement $y(0)$. The variance after measurement, $P(0)$, is always less than or equal to $Q(0)$, which implies that we can only gain information by measurement; but the information gain may be small if the measurement device is poor and the measurement noise variance R is large.

Forecasting the state evolution. Next we consider the state evolution from $k = 0$ to $k = 1$, which satisfies

$$x(1) = \begin{bmatrix} A & I \end{bmatrix} \begin{bmatrix} x(0) \\ w(0) \end{bmatrix}$$

in which $w(0) \sim N(0, Q)$ is the process noise. If the state is subjected to large disturbances, then Q is large, and if the disturbances are small, Q

is small. Again we choose zero mean for w because the nonzero mean disturbances should have been accounted for in the system model. We next calculate the conditional density $p_{x(1)|y(0)}$. Now we require the conditional version of the joint density $(x(0), w(0))$. We assume that the process noise $w(0)$ is statistically independent of both $x(0)$ and $v(0)$, hence it is also independent of $y(0)$, which is a linear combination of $x(0)$ and $v(0)$. Therefore we use (1.24) to obtain

$$\begin{bmatrix} x(0) \\ w(0) \end{bmatrix} \sim N\left(\begin{bmatrix} \hat{x}(0) \\ 0 \end{bmatrix}, \begin{bmatrix} P(0) & 0 \\ 0 & Q \end{bmatrix}\right)$$

We then use the conditional version of the linear transformation of a normal (1.25) to obtain

$$p_{x(1)|y(0)}(x(1)|y(0)) = n(x(1), \hat{x}^-(1), P^-(1))$$

in which the mean and variance are

$$\hat{x}^-(1) = A\hat{x}(0) \qquad P^-(1) = AP(0)A' + Q$$

We see that forecasting forward one time step may increase or decrease the conditional variance of the state. If the eigenvalues of A are less than unity, for example, the term $AP(0)A'$ *may* be smaller than $P(0)$, but the process noise Q adds a positive contribution. If the system is unstable, $AP(0)A'$ *may* be larger than $P(0)$, and then the conditional variance definitely increases upon forecasting. See also Exercise 1.27 for further discussion of this point.

Given that $p_{x(1)|y(0)}$ is also a normal, we are situated to add measurement $y(1)$ and continue the process of adding measurements followed by forecasting forward one time step until we have processed all the available data. Because this process is recursive, the storage requirements are small. We need to store only the current state estimate and variance, and can discard the measurements as they are processed. The required online calculation is minor. These features make the optimal linear estimator an ideal candidate for rapid online application. We next summarize the state estimation recursion.

Summary. Denote the measurement trajectory by

$$\mathbf{y}(k) := \{y(0), y(1), \ldots y(k)\}$$

At time k the conditional density with data $\mathbf{y}(k-1)$ is normal

$$p_{x(k)|\mathbf{y}(k-1)}(x(k)|\mathbf{y}(k-1)) = n(x(k), \hat{x}^-(k), P^-(k))$$

and we denote the mean and variance with a superscript minus to indicate these are the statistics *before* measurement $y(k)$. At $k = 0$, the recursion starts with $\hat{x}^-(0) = \overline{x}(0)$ and $P^-(0) = Q(0)$ as discussed previously. We obtain measurement $y(k)$ which satisfies

$$\begin{bmatrix} x(k) \\ y(k) \end{bmatrix} = \begin{bmatrix} I & 0 \\ C & I \end{bmatrix} \begin{bmatrix} x(k) \\ v(k) \end{bmatrix}$$

The density of $(x(k), v(k))$ follows from (1.24) since measurement noise $v(k)$ is independent of $x(k)$ and $\mathbf{y}(k-1)$

$$\begin{bmatrix} x(k) \\ v(k) \end{bmatrix} \sim N\left(\begin{bmatrix} \hat{x}^-(k) \\ 0 \end{bmatrix}, \begin{bmatrix} P^-(k) & 0 \\ 0 & R \end{bmatrix} \right)$$

Equation (1.25) then gives the joint density

$$\begin{bmatrix} x(k) \\ y(k) \end{bmatrix} \sim N\left(\begin{bmatrix} \hat{x}^-(k) \\ C\hat{x}^-(k) \end{bmatrix}, \begin{bmatrix} P^-(k) & P^-(k)C' \\ CP^-(k) & CP^-(k)C' + R \end{bmatrix} \right)$$

We note $\{\mathbf{y}(k-1), y(k)\} = \mathbf{y}(k)$, and using the conditional density result (1.26) gives

$$p_{x(k)|\mathbf{y}(k)}\left(x(k)|\mathbf{y}(k)\right) = n\left(x(k), \hat{x}(k), P(k)\right)$$

in which

$$\begin{aligned} \hat{x}(k) &= \hat{x}^-(k) + L(k)\left(y(k) - C\hat{x}^-(k)\right) \\ L(k) &= P^-(k)C'(CP^-(k)C' + R)^{-1} \\ P(k) &= P^-(k) - P^-(k)C'(CP^-(k)C' + R)^{-1}CP^-(k) \end{aligned}$$

We forecast from k to $k+1$ using the model

$$x(k+1) = \begin{bmatrix} A & I \end{bmatrix} \begin{bmatrix} x(k) \\ w(k) \end{bmatrix}$$

Because $w(k)$ is independent of $x(k)$ and $\mathbf{y}(k)$, the joint density of $(x(k), w(k))$ follows from a second use of (1.24)

$$\begin{bmatrix} x(k) \\ w(k) \end{bmatrix} \sim N\left(\begin{bmatrix} \hat{x}(k) \\ 0 \end{bmatrix}, \begin{bmatrix} P(k) & 0 \\ 0 & Q \end{bmatrix} \right)$$

and a second use of the linear transformation result (1.25) gives

$$p_{x(k+1)|\mathbf{y}(k)}(x(k+1)|\mathbf{y}(k)) = n(x(k+1), \hat{x}^-(k+1), P^-(k+1))$$

in which

$$\hat{x}^-(k+1) = A\hat{x}(k)$$
$$P^-(k+1) = AP(k)A' + Q$$

and the recursion is complete.

1.4.3 Least Squares Estimation

We next consider the state estimation problem as a deterministic optimization problem rather than an exercise in maximizing conditional density. This viewpoint proves valuable in Chapter 4 when we wish to add constraints to the state estimator. Consider a time horizon with measurements $y(k), k = 0, 1, \ldots, T$. We consider the prior information to be our best initial guess of the initial state $x(0)$, denoted $\overline{x}(0)$, and weighting matrices $P^-(0)$, Q, and R for the initial state, process disturbance, and measurement disturbance. A reasonably flexible choice for objective function is

$$V_T(\mathbf{x}(T)) = \frac{1}{2}\Bigg(|x(0) - \overline{x}(0)|^2_{(P^-(0))^{-1}} + \sum_{k=0}^{T-1} |x(k+1) - Ax(k)|^2_{Q^{-1}} + \sum_{k=0}^{T} |y(k) - Cx(k)|^2_{R^{-1}} \Bigg) \tag{1.27}$$

in which $\mathbf{x}(T) := \{x(0), x(1), \ldots, x(T)\}$. We claim and then show that the following (deterministic) least squares optimization problem produces the same result as the conditional density function maximization of the Kalman filter

$$\min_{\mathbf{x}(T)} V_T(\mathbf{x}(T)) \tag{1.28}$$

Game plan. Using forward DP, we can decompose and solve recursively the least squares state estimation problem. To see clearly how the procedure works, first we write out the terms in the state estimation least squares problem (1.28)

$$\min_{x(0),\ldots,x(T)} \frac{1}{2}\Bigg(|x(0) - \overline{x}(0)|^2_{(P^-(0))^{-1}} + |y(0) - Cx(0)|^2_{R^{-1}} + |x(1) - Ax(0)|^2_{Q^{-1}} + |y(1) - Cx(1)|^2_{R^{-1}} + |x(2) - Ax(1)|^2_{Q^{-1}} + \cdots + |x(T) - Ax(T-1)|^2_{Q^{-1}} + |y(T) - Cx(T)|^2_{R^{-1}} \Bigg) \tag{1.29}$$

We decompose this T-stage optimization problem with forward DP. First we combine the prior and the measurement $y(0)$ into the quadratic function $V_0(x(0))$ as shown in the following equation

$$\min_{x(T),\ldots,x(1)} \overbrace{\min_{x(0)} \frac{1}{2}\Big(\underbrace{|x(0)-\overline{x}(0)|^2_{(P^-(0))^{-1}} + |y(0)-Cx(0)|^2_{R^{-1}}}_{\text{combine } V_0(x(0))} + |x(1)-Ax(0)|^2_{Q^{-1}}}^{\text{arrival cost } V_1^-(x(1))} + \\ |y(1)-Cx(1)|^2_{R^{-1}} + |x(2)-Ax(1)|^2_{Q^{-1}} + \cdots + \\ |x(T)-Ax(T-1)|^2_{Q^{-1}} + |y(T)-Cx(T)|^2_{R^{-1}} \Big)$$

Then we optimize over the first state, $x(0)$. This produces the arrival cost for the first stage, $V_1^-(x(1))$, which we will show is also quadratic

$$V_1^-(x(1)) = \frac{1}{2}\,|x(1)-\hat{x}^-(1)|^2_{(P^-(1))^{-1}}$$

Next we combine the arrival cost of the first stage with the next measurement $y(1)$ to obtain $V_1(x(1))$

$$\min_{x(T),\ldots,x(2)} \overbrace{\min_{x(1)} \frac{1}{2}\Big(\underbrace{|x(1)-\hat{x}^-(1)|^2_{(P^-(1))^{-1}} + |y(1)-Cx(1)|^2_{R^{-1}}}_{\text{combine } V_1(x(1))} + |x(2)-Ax(1)|^2_{Q^{-1}}}^{\text{arrival cost } V_2^-(x(2))} + \\ |y(2)-Cx(2)|^2_{R^{-1}} + |x(3)-Ax(2)|^2_{Q^{-1}} + \cdots + \\ |x(T)-Ax(T-1)|^2_{Q^{-1}} + |y(T)-Cx(T)|^2_{R^{-1}} \Big) \quad (1.30)$$

We optimize over the second state, $x(1)$, which defines arrival cost for the first two stages, $V_2^-(x(2))$. We continue in this fashion until we have optimized finally over $x(T)$ and have solved (1.29). Now that we have in mind an overall game plan for solving the problem, we look at each step in detail and develop the recursion formulas of forward DP.

Combine prior and measurement. Combining the prior and measurement defines V_0

$$V_0(x(0)) = \frac{1}{2}\Big(\underbrace{|x(0)-\overline{x}(0)|^2_{(P^-(0))^{-1}}}_{\text{prior}} + \underbrace{|y(0)-Cx(0)|^2_{R^{-1}}}_{\text{measurement}} \Big) \quad (1.31)$$

which can be expressed also as

$$V_0(x(0)) = \frac{1}{2}\Big(|x(0) - \overline{x}(0)|^2_{(P^-(0))^{-1}} + \\ |(y(0) - C\overline{x}(0)) - C(x(0) - \overline{x}(0))|^2_{R^{-1}} \Big)$$

Using the third form in Example 1.1 we can combine these two terms into a single quadratic function

$$V_0(x(0)) = (1/2)\ (x(0) - \overline{x}(0) - v)'\widetilde{H}^{-1}(x(0) - \overline{x}(0) - v) + \text{constant}$$

in which

$$v = P^-(0)C'(CP^-(0)C' + R)^{-1}C'R^{-1}\left(y(0) - C\overline{x}(0)\right)$$
$$\widetilde{H} = P^-(0) - P^-(0)C'(CP^-(0)C' + R)^{-1}CP^-(0)$$

and we set the constant term to zero because it does not depend on $x(1)$. If we define

$$P(0) = P^-(0) - P^-(0)C'(CP^-(0)C' + R)^{-1}CP^-(0)$$
$$L(0) = P^-(0)C'(CP^-(0)C' + R)^{-1}C'R^{-1}$$

and define the state estimate $\hat{x}(0)$ as follows

$$\hat{x}(0) = \overline{x}(0) + v$$
$$\hat{x}(0) = \overline{x}(0) + L(0)\left(y(0) - C\overline{x}(0)\right)$$

and we have derived the following compact expression for the function V_0

$$V_0(x(0)) = (1/2)\ |x(0) - \hat{x}(0)|^2_{P(0)^{-1}}$$

State evolution and arrival cost. Now we add the next term in (1.29) to the function $V_0(\cdot)$ and denote the sum as $V(\cdot)$

$$V(x(0), x(1)) = V_0(x(0)) + (1/2)\ |x(1) - Ax(0)|^2_{Q^{-1}}$$
$$V(x(0), x(1)) = \frac{1}{2}\left(|x(0) - \hat{x}(0)|^2_{P(0)^{-1}} + |x(1) - Ax(0)|^2_{Q^{-1}} \right)$$

Again using the third form in Example 1.1, we can add the two quadratics to obtain

$$V(x(0), x(1)) = (1/2)\ |x(0) - v|^2_{\widetilde{H}^{-1}} + d$$

in which

$$v = \hat{x}(0) + P(0)A\left(AP(0)A' + Q\right)^{-1}(x(1) - A\hat{x}(0))$$
$$d = (1/2)\left(|v - \hat{x}(0)|^2_{P(0)^{-1}} + |x(1) - Av|^2_{Q^{-1}}\right)$$

This form is convenient for optimization over the first decision variable $x(0)$; by inspection the solution is $x(0) = v$ and the cost is d. We define the arrival cost to be the result of this optimization

$$V_1^-(x(1)) = \min_{x(0)} V(x(0), x(1))$$

Substituting v into the expression for d and simplifying gives

$$V_1^-(x(1)) = (1/2)\,|x(1) - A\hat{x}(0)|^2_{(P^-(1))^{-1}}$$

in which

$$P^-(1) = AP(0)A' + Q$$

We define $\hat{x}^-(1) = A\hat{x}(0)$ and express the arrival cost compactly as

$$V_1^-(x(1)) = (1/2)\,\left|x(1) - \hat{x}^-(1)\right|^2_{(P^-(1))^{-1}}$$

Combine arrival cost and measurement. We now combine the arrival cost and measurement for the next stage of the optimization to obtain

$$V_1(x(1)) = \underbrace{V_1^-(x(1))}_{\text{prior}} + \underbrace{(1/2)\,\left|(y(1) - Cx(1))\right|^2_{R^{-1}}}_{\text{measurement}}$$

$$V_1(x(1)) = \frac{1}{2}\left(\,\left|x(1) - \hat{x}^-(1)\right|^2_{(P^-(1))^{-1}} + \left|y(1) - Cx(1)\right|^2_{R^{-1}}\right)$$

We can see that this equation is exactly the form as (1.31) of the previous step, and, by simply changing the variable names, we have that

$$P(1) = P^-(1) - P^-(1)C'(CP^-(1)C' + R)^{-1}CP^-(1)$$
$$L(1) = P^-(1)C'(CP^-(1)C' + R)^{-1}$$
$$\hat{x}(1) = \hat{x}^-(1) + L(1)(y(1) - C\hat{x}^-(1))$$

and the cost function V_1 is defined as

$$V_1(x(1)) = (1/2)(x(1) - \hat{x}(1))'P(1)^{-1}(x(1) - \hat{x}(1))$$

in which

$$\hat{x}^-(1) = A\hat{x}(0)$$
$$P^-(1) = AP(0)A' + Q$$

Recursion and termination. The recursion can be summarized by two steps. Adding the measurement at time k produces

$$\begin{aligned} P(k) &= P^-(k) - P^-(k)C'(CP^-(k)C' + R)^{-1}CP^-(k) \\ L(k) &= P^-(k)C'(CP^-(k)C' + R)^{-1} \\ \hat{x}(k) &= \hat{x}^-(k) + L(k)(y(k) - C\hat{x}^-(k)) \end{aligned}$$

Propagating the model to time $k+1$ produces

$$\begin{aligned} \hat{x}^-(k+1) &= A\hat{x}(k) \\ P^-(k+1) &= AP(k)A' + Q \end{aligned}$$

and the recursion starts with the prior information $\hat{x}^-(0) = \overline{x}(0)$ and $P^-(0)$. The arrival cost, V_k^-, and arrival cost plus measurement, V_k, for each stage are given by

$$\begin{aligned} V_k^-(x(k)) &= (1/2)\left|x(k) - \hat{x}^-(k)\right|_{(P^-(k))^{-1}} \\ V_k(x(k)) &= (1/2)\left|x(k) - \hat{x}(k)\right|_{(P(k))^{-1}} \end{aligned}$$

The process terminates with the final measurement $y(T)$, at which point we have recursively solved the original problem (1.29).

We see by inspection that the recursion formulas given by forward DP of (1.29) are the same as those found by calculating the conditional density function in Section 1.4.2. Moreover, the conditional densities before and after measurement are closely related to the least squares value functions as shown below

$$\begin{aligned} p(x(k)|\mathbf{y}(k-1)) &= \frac{1}{(2\pi)^{n/2}(\det P^-(k))^{1/2}} \exp(-V_k^-(x(k))) \qquad (1.32) \\ p(x(k)|\mathbf{y}(k)) &= \frac{1}{(2\pi)^{n/2}(\det P(k))^{1/2}} \exp(-V_k(x(k))) \end{aligned}$$

The discovery (and rediscovery) of the close connection between recursive least squares and optimal statistical estimation has not always been greeted happily by researchers:

> The recursive least squares approach was actually inspired by probabilistic results that automatically produce an equation of evolution for the estimate (the conditional mean). In fact, much of the recent least squares work did nothing more than rederive the probabilistic results (perhaps in an attempt to understand them). As a result, much of the least squares work contributes very little to estimation theory.
> — Jazwinski (1970, pp.152–153)

In contrast with this view, we find both approaches valuable in the subsequent development. The probabilistic approach, which views the state estimator as maximizing conditional density of the state given measurement, offers the most insight. It provides a rigorous basis for comparing different estimators based on the variance of their estimate error. It also specifies what information is required to define an optimal estimator, with variances Q and R of primary importance. In the probabilistic framework, these parameters should be found from modeling and data. The main deficiency in the least squares viewpoint is that the objective function, although reasonable, is ad hoc and not justified. The choice of weighting matrices Q and R is arbitrary. Practitioners generally choose these parameters based on a tradeoff between the competing goals of speed of estimator response and insensitivity to measurement noise. But a careful statement of this tradeoff often just leads back to the probabilistic viewpoint in which the process disturbance and measurement disturbance are modeled as normal distributions. If we restrict attention to unconstrained linear systems, the probabilistic viewpoint is clearly superior.

Approaching state estimation with the perspective of least squares pays off, however, when the models are significantly more complex. It is generally intractable to find and maximize the conditional density of the state given measurements for complex, nonlinear and constrained models. Although the state estimation problem can be stated in the language of probability, it cannot be solved with current methods. But reasonable objective functions can be chosen for even complex, nonlinear and constrained models. Moreover, knowing which least squares problems correspond to which statistically optimal estimation problems for the simple linear case, provides the engineer with valuable insight in choosing useful objective functions for nonlinear estimation. We explore these more complex and realistic estimation problems in Chapter 4. The perspective of least squares also leads to succinct arguments for establishing estimator stability, which we take up shortly. First we consider situations in which it is advantageous to use moving horizon estimation.

1.4.4 Moving Horizon Estimation

When using nonlinear models or considering constraints on the estimates, we cannot calculate the conditional density recursively in closed form as we did in Kalman filtering. Similarly, we cannot solve recursively the least squares problem. If we use least squares we must opti-

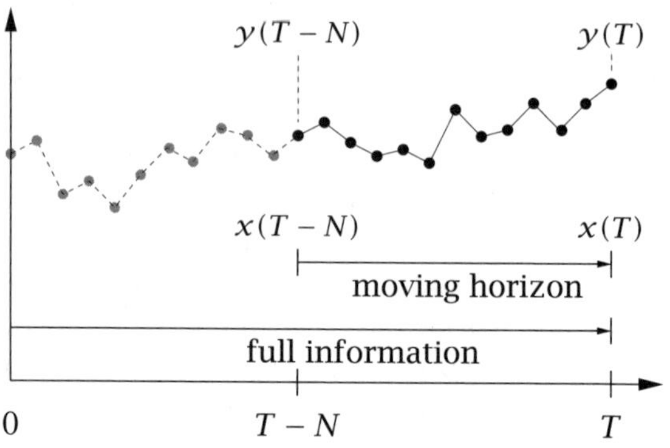

Figure 1.4: Schematic of the moving horizon estimation problem.

mize all the states in the trajectory $\mathbf{x}(T)$ simultaneously to obtain the state estimates. This optimization problem becomes computationally intractable as T increases. Moving horizon estimation (MHE) removes this difficulty by considering only the most recent N measurements and finds only the most recent N values of the state trajectory as sketched in Figure 1.4. The states to be estimated are $\mathbf{x}_N(T) = \{x(T-N), \ldots, x(T)\}$ given measurements $\mathbf{y}_N(T) = \{y(T-N), \ldots, y(T)\}$. The data have been broken into two sections with $\{\mathbf{y}(T-N-1), \mathbf{y}_N(T)\} = \mathbf{y}(T)$. We assume here that $T \geq N-1$ to ignore the initial period in which the estimation window fills with measurements and assume that the window is always full.

The simplest form of MHE is the following least squares problem

$$\min_{\mathbf{x}_N(T)} \hat{V}_T(\mathbf{x}_N(T)) \tag{1.33}$$

in which the objective function is

$$\hat{V}_T(\mathbf{x}_N(T)) = \frac{1}{2}\Bigg(\sum_{k=T-N}^{T-1} |x(k+1) - Ax(k)|^2_{Q^{-1}} + \sum_{k=T-N}^{T} |y(k) - Cx(k)|^2_{R^{-1}} \Bigg) \tag{1.34}$$

We use the circumflex (hat) to indicate this is the MHE cost function considering data sequence from $T-N$ to T rather than the full information or least squares cost considering the data from 0 to T.

MHE in terms of least squares. Notice that from our previous DP recursion in (1.30), we can write the full least squares problem as

$$V_T(\mathbf{x}_N(T)) = V_{T-N}^-(x(T-N)) + \frac{1}{2}\left(\sum_{k=T-N}^{T-1} |x(k+1) - Ax(k)|^2_{Q^{-1}} + \sum_{k=T-N}^{T} |y(k) - Cx(k)|^2_{R^{-1}}\right)$$

in which $V_{T-N}^-(\cdot)$ is the arrival cost at time $T-N$. Comparing these two objective functions, it is clear that the simplest form of MHE is equivalent to setting up a full least squares problem, but then setting the arrival cost function $V_{T-N}^-(\cdot)$ to zero.

MHE in terms of conditional density. Because we have established the close connection between least squares and conditional density in (1.32), we can write the full least squares problem also as an equivalent conditional density maximization

$$\max_{x(T)} p_{x(T)|\mathbf{y}_N(T)}(x(T)|\mathbf{y}_N(T))$$

with prior density

$$p_{x(T-N)|\mathbf{y}(T-N-1)}(x|\mathbf{y}(T-N-1)) = c\exp(-V_{T-N}^-(x)) \qquad (1.35)$$

in which the constant c can be found from (1.20) if desired, but its value does not change the solution to the optimization. We can see from (1.35) that setting $V_{T-N}^-(\cdot)$ to zero in the simplest form of MHE is equivalent to giving infinite variance to the conditional density of $x(T-N)|\mathbf{y}(T-N-1)$. This means we are using a noninformative prior for the state $x(T-N)$ and completely discounting the previous measurements $\mathbf{y}(T-N-1)$.

To provide a more flexible MHE problem, we therefore introduce a penalty on the first state to account for the neglected data $\mathbf{y}(T-N-1)$

$$\hat{V}_T(\mathbf{x}_N(T)) = \Gamma_{T-N}(x(T-N)) + \frac{1}{2}\left(\sum_{k=T-N}^{T-1} |x(k+1) - Ax(k)|^2_{Q^{-1}} + \sum_{k=T-N}^{T} |y(k) - Cx(k)|^2_{R^{-1}}\right)$$

For the linear Gaussian case, we can account for the neglected data exactly with no approximation by setting Γ equal to the arrival cost, or, equivalently, the negative logarithm of the conditional density of the state given the prior measurements. Indeed, there is no need to use

MHE for the linear Gaussian problem at all because we can solve the full problem recursively. When addressing nonlinear and constrained problems in Chapter 4, however, we must approximate the conditional density of the state given the prior measurements in MHE to obtain a computationally tractable and high-quality estimator.

1.4.5 Observability

We next explore the convergence properties of the state estimators. For this we require the concept of system observability. The basic idea of observability is that any two distinct states can be *distinguished* by applying some input and observing the two system outputs over some finite time interval (Sontag, 1998, p.262–263). We discuss this general definition in more detail when treating nonlinear systems in Chapter 4, but observability for linear systems is much simpler. First of all, the applied input is irrelevant and we can set it to zero. Therefore consider the linear time-invariant system (A, C) with zero input

$$\begin{aligned} x(k+1) &= Ax(k) \\ y(k) &= Cx(k) \end{aligned}$$

The system is observable if there exists a finite N, such that for every $x(0)$, N measurements $\{y(0), y(1), \ldots, y(N-1)\}$ distinguish uniquely the initial state $x(0)$. Similarly to the case of controllability, if we cannot determine the initial state using n measurements, we cannot determine it using $N > n$ measurements. Therefore we can develop a convenient test for observability as follows. For n measurements, the system model gives

$$\begin{bmatrix} y(0) \\ y(1) \\ \vdots \\ y(n-1) \end{bmatrix} = \begin{bmatrix} C \\ CA \\ \vdots \\ CA^{n-1} \end{bmatrix} x(0) \tag{1.36}$$

The question of *observability* is therefore a question of *uniqueness* of solutions to these linear equations. The matrix appearing in this equation is known as the *observability matrix* $\mathcal{O}$

$$\mathcal{O} = \begin{bmatrix} C \\ CA \\ \vdots \\ CA^{n-1} \end{bmatrix} \tag{1.37}$$

From the fundamental theorem of linear algebra, we know the solution to (1.36) is unique if and only if the *columns* of the $np \times n$ observability matrix are linearly independent.[6] Therefore, we have that the system (A, C) is observable if and only if

$$\text{rank}(\mathcal{O}) = n$$

The following result for checking observability also proves useful (Hautus, 1972).

Lemma 1.4 (Hautus Lemma for observability). *A system is observable if and only if*

$$\text{rank} \begin{bmatrix} \lambda I - A \\ C \end{bmatrix} = n \qquad \textit{for all } \lambda \in \mathbb{C} \tag{1.38}$$

in which $\mathbb{C}$ *is the set of complex numbers.*

Notice that the first n rows of the matrix in (1.38) are linearly independent if $\lambda \notin \text{eig}(A)$, so (1.38) is equivalent to checking the rank at just the eigenvalues of A

$$\text{rank} \begin{bmatrix} \lambda I - A \\ C \end{bmatrix} = n \qquad \text{for all } \lambda \in \text{eig}(A)$$

1.4.6 Convergence of the State Estimator

Next we consider the question of convergence of the estimates of several of the estimators we have considered. The simplest convergence question to ask is the following. Given an initial estimate error, and zero state and measurement noises, does the state estimate converge to the state as time increases and more measurements become available? If the answer to this question is yes, we say the estimates converge; sometimes we say the estimator converges. As with the regulator, optimality of an estimator does not ensure its stability. Consider the case $A = I, C = 0$. The optimal estimate is $\hat{x}(k) = \overline{x}(0)$, which does not converge to the true state unless we have luckily chosen $\overline{x}(0) = x(0)$.[7] Obviously the lack of stability is caused by our choosing an unobservable (undetectable) system.

We treat first the Kalman filtering or full least squares problem. Recall that this estimator optimizes over the entire state trajectory

[6] See Section A.4 of Appendix A or (Strang, 1980, pp.87–88) for a review of this result.
[7] If we could count on that kind of luck, we would have no need for state estimation.

$\mathbf{x}(T) := \{x(0), \ldots, x(T)\}$ based on all measurements $\mathbf{y}(T) := \{y(0), \ldots, y(T)\}$. In order to establish convergence, the following result on the optimal estimator cost function proves useful.

Lemma 1.5 (Convergence of estimator cost). *Given noise-free measurements* $\mathbf{y}(T) = \{Cx(0), CAx(0), \ldots, CA^T x(0)\}$, *the optimal estimator cost* $V_T^0(\mathbf{y}(T))$ *converges as* $T \to \infty$.

Proof. Denote the optimal state sequence at time T given measurement $\mathbf{y}(T)$ by

$$\{\hat{x}(0|T),\ \hat{x}(1|T),\ \ldots,\ \hat{x}(T|T)\}$$

We wish to compare the optimal costs at time T and $T-1$. Therefore, consider using the first $T-1$ elements of the solution at time T as decision variables in the state estimation problem at time $T-1$. The cost for those decision variables at time $T-1$ is given by

$$V_T^0 - \frac{1}{2}\left(|\hat{x}(T|T) - A\hat{x}(T-1|T)|^2_{Q^{-1}} - |y(T) - C\hat{x}(T|T)|^2_{R^{-1}} \right)$$

In other words, we have the full cost at time T and we deduct the cost of the last stage, which is not present at $T-1$. Now this choice of decision variables is not necessarily optimal at time $T-1$, so we have the inequality

$$V_{T-1}^0 \le V_T^0 - \frac{1}{2}\left(|\hat{x}(T|T) - A\hat{x}(T-1|T)|^2_{Q^{-1}} - |y(T) - C\hat{x}(T|T)|^2_{R^{-1}} \right)$$

Because the quadratic terms are nonnegative, the sequence of optimal estimator costs is nondecreasing with increasing T. We can establish that the optimal cost is bounded above as follows: at any time T we can choose the decision variables to be $\{x(0), Ax(0), \ldots, A^T x(0)\}$, which achieves cost $|x(0) - \overline{x}(0)|^2_{(P^-(0))^{-1}}$ independent of T. The optimal cost sequence is nondecreasing and bounded above and, therefore, converges. ∎

The optimal estimator cost converges regardless of system observability. But if we want the optimal estimate to converge to the state, we have to restrict the system further. The following lemma provides an example of what is required.

Lemma 1.6 (Estimator convergence). *For* (A, C) *observable,* $Q, R > 0$, *and noise-free measurements* $\mathbf{y}(T) = \{Cx(0), CAx(0), \ldots, CA^T x(0)\}$, *the optimal linear state estimate converges to the state*

$$\hat{x}(T) \to x(T) \quad \text{as } T \to \infty$$

Proof. To compress the notation somewhat, let $\hat{w}_T(j) = \hat{x}(T+j+1|T+n-1) - A\hat{x}(T+j|T+n-1)$. Using the optimal solution at time $T+n-1$ as decision variables at time $T-1$ allows us to write the following inequality

$$V_{T-1}^0 \leq V_{T+n-1}^0 - \frac{1}{2}\left(\sum_{j=-1}^{n-2} |\hat{w}_T(j)|^2_{Q^{-1}} + \sum_{j=0}^{n-1} |y(T+j) - C\hat{x}(T+j|T+n-1)|^2_{R^{-1}} \right)$$

Because the sequence of optimal costs converges with increasing T, and $Q^{-1}, R^{-1} > 0$, we have established that for increasing T

$$\begin{aligned} \hat{w}_T(j) &\to 0 \quad j = -1, \ldots, n-2 \\ y(T+j) - C\hat{x}(T+j|T+n-1) &\to 0 \quad j = 0, \ldots, n-1 \end{aligned} \tag{1.39}$$

From the system model we have the following relationship between the last n stages in the optimization problem at time $T+n-1$ with data $\mathbf{y}(T+n-1)$

$$\begin{bmatrix} \hat{x}(T|T+n-1) \\ \hat{x}(T+1|T+n-1) \\ \vdots \\ \hat{x}(T+n-1|T+n-1) \end{bmatrix} = \begin{bmatrix} I \\ A \\ \vdots \\ A^{n-1} \end{bmatrix} \hat{x}(T|T+n-1) + \begin{bmatrix} 0 & & & \\ I & 0 & & \\ \vdots & \vdots & \ddots & \\ A^{n-2} & A^{n-3} & \cdots & I \end{bmatrix} \begin{bmatrix} \hat{w}_T(0) \\ \hat{w}_T(1) \\ \vdots \\ \hat{w}_T(n-2) \end{bmatrix} \tag{1.40}$$

We note the measurements satisfy

$$\begin{bmatrix} y(T) \\ y(T+1) \\ \vdots \\ y(T+n-1) \end{bmatrix} = \mathcal{O}x(T)$$

Multiplying (1.40) by C and subtracting gives

$$\begin{bmatrix} y(T) - C\hat{x}(T|T+n-1) \\ y(T+1) - C\hat{x}(T+1|T+n-1) \\ \vdots \\ y(T+n-1) - C\hat{x}(T+n-1|T+n-1) \end{bmatrix} = \mathcal{O}(x(T) - \hat{x}(T|T+n-1)) -$$
$$\begin{bmatrix} 0 & & & \\ C & 0 & & \\ \vdots & \vdots & \ddots & \\ CA^{n-2} & CA^{n-3} & \cdots & C \end{bmatrix} \begin{bmatrix} \hat{w}_T(0) \\ \hat{w}_T(1) \\ \vdots \\ \hat{w}_T(n-2) \end{bmatrix}$$

Applying (1.39) to this equation, we conclude $\mathcal{O}(x(T) - \hat{x}(T|T+n-1)) \to 0$ with increasing T. Because the observability matrix has independent columns, we conclude $x(T) - \hat{x}(T|T+n-1) \to 0$ as $T \to \infty$. Thus we conclude that the *smoothed* estimate $\hat{x}(T|T+n-1)$ converges to the state $x(T)$. Because the $\hat{w}_T(j)$ terms go to zero with increasing T, the last line of (1.40) gives $\hat{x}(T+n-1|T+n-1) \to A^{n-1}\hat{x}(T|T+n-1)$ as $T \to \infty$. From the system model $A^{n-1}x(T) = x(T+n-1)$ and, therefore, after replacing $T+n-1$ by T, we have

$$\hat{x}(T|T) \to x(T) \quad \text{as } T \to \infty$$

and asymptotic convergence of the estimator is established. ■

This convergence result also covers MHE with prior weighting set to the exact arrival cost because that is equivalent to Kalman filtering and full least squares. The simplest form of MHE, which discounts prior data completely, is also a convergent estimator, however, as discussed in Exercise 1.28.

The estimator convergence result in Lemma 1.6 is the simplest to establish, but, as in the case of the LQ regulator, we can enlarge the class of systems and weighting matrices (variances) for which estimator convergence is guaranteed. The system restriction can be weakened from observability to *detectability*, which is discussed in Exercises 1.31 and 1.32. The restriction on the process disturbance weight (variance) Q can be weakened from $Q > 0$ to $Q \geq 0$ and (A, Q) *stabilizable*, which is discussed in Exercise 1.33. The restriction $R > 0$ remains to ensure uniqueness of the estimator.

1.5 Tracking, Disturbances, and Zero Offset

In the last section of this chapter we show briefly how to use the MPC regulator and MHE estimator to handle different kinds of control problems, including setpoint tracking and rejecting nonzero disturbances.

1.5.1 Tracking

It is a standard objective in applications to use a feedback controller to move the measured outputs of a system to a specified and constant setpoint. This problem is known as setpoint tracking. In Section 2.9 we consider the case in which the system is nonlinear and constrained, but for simplicity here we consider the linear unconstrained system in which y_{sp} is an arbitrary constant. In the regulation problem of Section 1.3 we assumed that the goal was to take the state of the system to the origin. Such a regulator can be used to treat the setpoint tracking problem with a coordinate transformation. Denote the desired output setpoint as y_{sp}. Denote a steady state of the system model as (x_s, u_s). From (1.5), the steady state satisfies

$$\begin{bmatrix} I - A & -B \end{bmatrix} \begin{bmatrix} x_s \\ u_s \end{bmatrix} = 0$$

For *unconstrained* systems, we also impose the requirement that the steady state satisfies $Cx_s = y_{\mathrm{sp}}$ for the tracking problem, giving the set of equations

$$\begin{bmatrix} I - A & -B \\ C & 0 \end{bmatrix} \begin{bmatrix} x_s \\ u_s \end{bmatrix} = \begin{bmatrix} 0 \\ y_{\mathrm{sp}} \end{bmatrix} \tag{1.41}$$

If this set of equations has a solution, we can then define deviation variables

$$\begin{aligned} \widetilde{x}(k) &= x(k) - x_s \\ \widetilde{u}(k) &= u(k) - u_s \end{aligned}$$

that satisfy the dynamic model

$$\begin{aligned} \widetilde{x}(k+1) &= x(k+1) - x_s \\ &= Ax(k) + Bu(k) - (Ax_s + Bu_s) \\ \widetilde{x}(k+1) &= A\widetilde{x}(k) + B\widetilde{u}(k) \end{aligned}$$

so that the deviation variables satisfy the same model equation as the original variables. The zero regulation problem applied to the system in deviation variables finds $\widetilde{u}(k)$ that takes $\widetilde{x}(k)$ to zero, or, equivalently, which takes $x(k)$ to x_s, so that at steady state, $Cx(k) = Cx_s = y_{\text{sp}}$, which is the goal of the setpoint tracking problem. After solving the regulation problem in deviation variables, the input applied to the system is $u(k) = \widetilde{u}(k) + u_s$.

We next discuss when we can solve (1.41). We also note that for *constrained* systems, we must impose the constraints on the steady state (x_s, u_s). The matrix in (1.41) is a $(n+p)\times(n+m)$ matrix. For (1.41) to have a solution for all y_{sp}, it is sufficient that the rows of the matrix are linearly independent. That requires $p \leq m$: we require at least as many inputs as outputs with setpoints. But it is not uncommon in applications to have many more measured outputs than manipulated inputs. To handle these more general situations, we choose a matrix H and denote a new variable $r = Hy$ as a selection of linear combinations of the measured outputs. The variable $r \in \mathbb{R}^{n_c}$ is known as the *controlled variable.* For cases in which $p > m$, we choose some set of outputs $n_c \leq m$, as controlled variables, and assign setpoints to r, denoted r_{sp}.

We also wish to treat systems with more inputs than outputs, $m > p$. For these cases, the solution to (1.41) may exist for some choice of H and r_{sp}, but cannot be unique. If we wish to obtain a unique steady state, then we also must provide desired values for the steady inputs, u_{sp}. To handle constrained systems, we simply impose the constraints on (x_s, u_s).

Steady-state target problem. Our candidate optimization problem is therefore

$$\min_{x_s,u_s} \frac{1}{2}\left(\left|u_s - u_{\text{sp}}\right|^2_{R_s} + \left|Cx_s - y_{\text{sp}}\right|^2_{Q_s}\right) \tag{1.42a}$$

subject to:

$$\begin{bmatrix} I-A & -B \\ HC & 0 \end{bmatrix}\begin{bmatrix} x_s \\ u_s \end{bmatrix} = \begin{bmatrix} 0 \\ r_{\text{sp}} \end{bmatrix} \tag{1.42b}$$

$$Eu_s \leq e \tag{1.42c}$$

$$FCx_s \leq f \tag{1.42d}$$

We make the following assumptions:

Assumption 1.7 (Target feasibility and uniqueness).

(a) The target problem is feasible for the controlled variable setpoints of interest r_{sp}.

(b) The steady-state input penalty R_s is positive definite.

Assumption 1.7 (a) ensures that the solution (x_s, u_s) exists, and Assumption 1.7 (b) ensures that the solution is unique. If one chooses $n_c = 0$, then no controlled variables are required to be at setpoint, and the problem is feasible for any $(u_{\text{sp}}, y_{\text{sp}})$ because $(x_s, u_s) = (0, 0)$ is a feasible point. Exercises 1.56 and 1.57 explore the connection between feasibility of the equality constraints and the number of controlled variables relative to the number of inputs and outputs. One restriction is that the number of controlled variables chosen to be offset free must be less than or equal to the number of manipulated variables and the number of measurements, $n_c \leq m$ and $n_c \leq p$.

Dynamic regulation problem. Given the steady-state solution, we define the following multistage objective function

$$V(\widetilde{x}(0), \widetilde{\mathbf{u}}) = \frac{1}{2} \sum_{k=0}^{N-1} |\widetilde{x}(k)|_Q^2 + |\widetilde{u}(k)|_R^2 \qquad \text{s.t. } \widetilde{x}^+ = A\widetilde{x} + B\widetilde{u}$$

in which $\widetilde{x}(0) = \hat{x}(k) - x_s$, i.e., the initial condition for the regulation problem comes from the state estimate shifted by the steady-state x_s. The regulator solves the following dynamic, zero-state regulation problem

$$\min_{\widetilde{\mathbf{u}}} V(\widetilde{x}(0), \widetilde{\mathbf{u}})$$

subject to

$$\begin{aligned} E\widetilde{u} &\leq e - Eu_s \\ FC\widetilde{x} &\leq f - FCx_s \end{aligned}$$

in which the constraints also are shifted by the steady state (x_s, u_s). The optimal cost and solution are $V^0(\widetilde{x}(0))$ and $\widetilde{\mathbf{u}}^0(\widetilde{x}(0))$. The moving horizon control law uses the first move of this optimal sequence, $\widetilde{u}^0(\widetilde{x}(0)) = \widetilde{\mathbf{u}}^0(0; \widetilde{x}(0))$, so the controller output is $u(k) = \widetilde{u}^0(\widetilde{x}(0)) + u_s$.

1.5.2 Disturbances and Zero Offset

Another common objective in applications is to use a feedback controller to compensate for an unmeasured disturbance to the system with the input so the disturbance's effect on the controlled variable is mitigated. This problem is known as disturbance rejection. We may wish to design a feedback controller that compensates for nonzero disturbances such that the selected controlled variables asymptotically approach their setpoints without offset. This property is known as zero offset. In this section we show a simple method for constructing an MPC controller to achieve zero offset.

In Chapter 5, we address the full problem. Here we must be content to limit our objective. We will ensure that *if the system is stabilized in the presence of the disturbance*, then there is zero offset. But we will not attempt to construct the controller that ensures stabilization over an interesting class of disturbances. That topic is treated in Chapter 5.

This more limited objective is similar to what one achieves when using the integral mode in proportional-integral-derivative (PID) control of an unconstrained system: either there is zero steady offset, or the system trajectory is unbounded. In a constrained system, the statement is amended to: either there is zero steady offset, or the system trajectory is unbounded, or the system constraints are active at steady state. In both constrained and unconstrained systems, the zero-offset property *precludes* one undesirable possibility: the system settles at an unconstrained steady state, and the steady state displays offset in the controlled variables.

A simple method to compensate for an unmeasured disturbance is to (i) model the disturbance, (ii) use the measurements and model to estimate the disturbance, and (iii) find the inputs that minimize the effect of the disturbance on the controlled variables. The choice of disturbance model is motivated by the zero-offset goal. To achieve offset-free performance we augment the system state with an *integrating* disturbance d driven by a white noise w_d

$$d^+ = d + w_d \tag{1.43}$$

This choice is motivated by the works of Davison and Smith (1971, 1974); Qiu and Davison (1993) and the Internal Model Principle of Francis and Wonham (1976). To remove offset, one designs a control system that can remove asymptotically constant, nonzero disturbances (Davison and Smith, 1971), (Kwakernaak and Sivan, 1972, p.278). To accomplish this end, the original system is augmented with a replicate of the constant, nonzero disturbance model, (1.43). Thus the states of the original system are moved onto the manifold that cancels the effect of the disturbance on the controlled variables. The augmented system model used for the state estimator is given by

$$\begin{bmatrix} x \\ d \end{bmatrix}^+ = \begin{bmatrix} A & B_d \\ 0 & I \end{bmatrix} \begin{bmatrix} x \\ d \end{bmatrix} + \begin{bmatrix} B \\ 0 \end{bmatrix} u + w \tag{1.44a}$$

$$y = \begin{bmatrix} C & C_d \end{bmatrix} \begin{bmatrix} x \\ d \end{bmatrix} + v \tag{1.44b}$$

and we are free to choose how the integrating disturbance affects the states and measured outputs through the choice of B_d and C_d. The only restriction is that the augmented system is detectable. That restriction can be easily checked using the following result.

Lemma 1.8 (Detectability of the augmented system). *The augmented system* (1.44) *is detectable if and only if the nonaugmented system* (A, C) *is detectable, and the following condition holds:*

$$\operatorname{rank} \begin{bmatrix} I - A & -B_d \\ C & C_d \end{bmatrix} = n + n_d \tag{1.45}$$

Corollary 1.9 (Dimension of the disturbance). *The maximal dimension of the disturbance d in* (1.44) *such that the augmented system is detectable is equal to the number of measurements, that is*

$$n_d \leq p$$

A pair of matrices (B_d, C_d) such that (1.45) is satisfied always exists. In fact, since (A, C) is detectable, the submatrix $\begin{bmatrix} I-A \\ C \end{bmatrix} \in \mathbb{R}^{(p+n)\times n}$ has rank n. Thus, we can choose any $n_d \leq p$ columns in $\mathbb{R}^{p+n}$ independent of $\begin{bmatrix} I-A \\ C \end{bmatrix}$ for $\begin{bmatrix} -B_d \\ C_d \end{bmatrix}$.

The state and the additional integrating disturbance are estimated from the plant measurement using a Kalman filter designed for the augmented system. The variances of the stochastic disturbances w and v may be treated as adjustable parameters or found from input-output measurements (Odelson, Rajamani, and Rawlings, 2006). The estimator provides $\hat{x}(k)$ and $\hat{d}(k)$ at each time k. The best forecast of the steady-state disturbance using (1.43) is simply

$$\hat{d}_s = \hat{d}(k)$$

The steady-state target problem is therefore modified to account for the nonzero disturbance $\hat{d}_s$

$$\min_{x_s, u_s} \frac{1}{2} \left(\left| u_s - u_{\text{sp}} \right|^2_{R_s} + \left| C x_s + C_d \hat{d}_s - y_{\text{sp}} \right|^2_{Q_s} \right) \tag{1.46a}$$

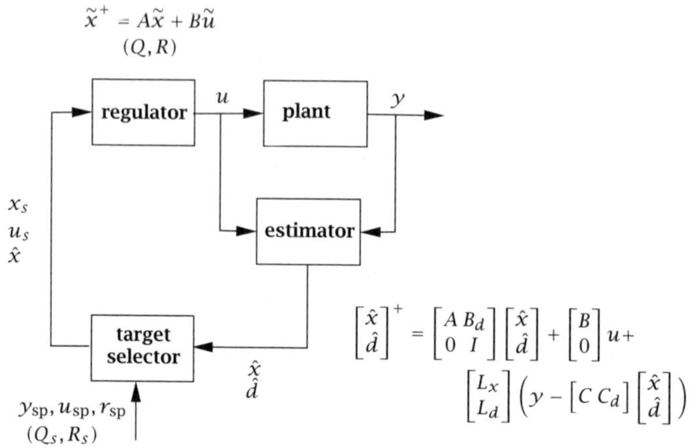

Figure 1.5: MPC controller consisting of: receding horizon regulator, state estimator, and target selector.

subject to:

$$\begin{bmatrix} I - A & -B \\ HC & 0 \end{bmatrix} \begin{bmatrix} x_s \\ u_s \end{bmatrix} = \begin{bmatrix} B_d \hat{d}_s \\ r_{\text{sp}} - HC_d \hat{d}_s \end{bmatrix} \tag{1.46b}$$

$$Eu_s \leq e \tag{1.46c}$$

$$FCx_s \leq f - FC_d \hat{d}_s \tag{1.46d}$$

Comparing (1.42) to (1.46), we see the disturbance model affects the steady-state target determination in four ways.

1. The output target is modified in (1.46a) to account for the effect of the disturbance on the measured output ($y_{\text{sp}} \rightarrow y_{\text{sp}} - C_d \hat{d}_s$).

2. The output constraint in (1.46d) is similarly modified ($f \rightarrow f - FC_d \hat{d}_s$).

3. The system steady-state relation in (1.46b) is modified to account for the effect of the disturbance on the state evolution ($0 \rightarrow B_d \hat{d}_s$).

4. The controlled variable target in (1.46b) is modified to account for the effect of the disturbance on the controlled variable ($r_{\text{sp}} \rightarrow r_{\text{sp}} - HC_d \hat{d}_s$).

Given the steady-state target, the same dynamic regulation problem as presented in the tracking section, Section 1.5, is used for the regulator. In other words, the regulator is based on the deterministic system

(A, B) in which the current state is $\hat{x}(k) - x_s$ and the goal is to take the system to the origin.

The following lemma summarizes the offset-free control property of the combined control system.

Lemma 1.10 (Offset-free control). *Consider a system controlled by the MPC algorithm as shown in Figure 1.5. The target problem* (1.46) *is assumed feasible. Augment the system model with a number of integrating disturbances equal to the number of measurements* ($n_d = p$)*; choose any* $B_d \in \mathbb{R}^{n \times p}$, $C_d \in \mathbb{R}^{p \times p}$ *such that*

$$\operatorname{rank} \begin{bmatrix} I - A & -B_d \\ C & C_d \end{bmatrix} = n + p$$

If the plant output $y(k)$ *goes to steady state* y_s*, the closed-loop system is stable, and constraints are not active at steady state, then there is zero offset in the controlled variables, that is*

$$Hy_s = r_{\text{sp}}$$

The proof of this lemma is given in Pannocchia and Rawlings (2003). It may seem surprising that the number of integrating disturbances must be equal to the number of *measurements* used for feedback rather than the number of *controlled variables* to guarantee offset-free control. To gain insight into the reason, consider the disturbance part (bottom half) of the Kalman filter equations shown in Figure 1.5

$$\hat{d}^+ = \hat{d} + L_d\left(y - \begin{bmatrix} C & C_d \end{bmatrix} \begin{bmatrix} \hat{x} \\ \hat{d} \end{bmatrix}\right)$$

Because of the integrator, the disturbance estimate cannot converge until

$$L_d\left(y - \begin{bmatrix} C & C_d \end{bmatrix} \begin{bmatrix} \hat{x} \\ \hat{d} \end{bmatrix}\right) = 0$$

But notice this condition merely restricts the output prediction error to lie in the nullspace of the matrix L_d, which is an $n_d \times p$ matrix. If we choose $n_d = n_c < p$, then the number of columns of L_d is greater than the number of rows and L_d has a nonzero nullspace.[8] In general, we require the output prediction error to be *zero* to achieve zero offset independently of the regulator tuning. For L_d to have only the zero vector in its nullspace, we require $n_d \geq p$. Since we also know $n_d \leq p$ from Corollary 1.9, we conclude $n_d = p$.

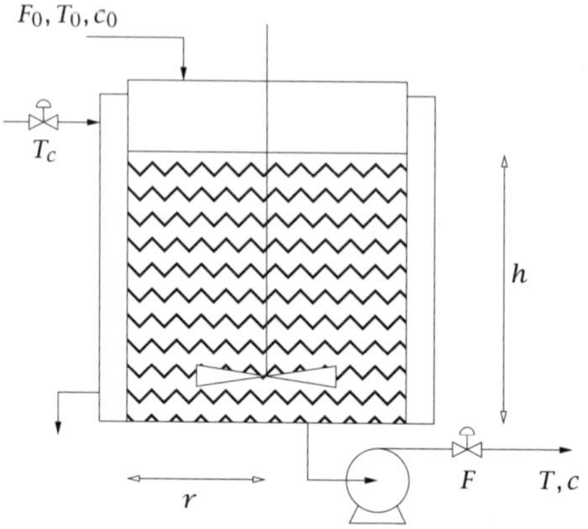

Figure 1.6: Schematic of the well-stirred reactor.

Parameter	Nominal value	Units
F_0	0.1	m^3/min
T_0	350	K
c_0	1	kmol/m^3
r	0.219	m
k_0	7.2×10^{10}	min^{-1}
E/R	8750	K
U	54.94	kJ/min·m^2·K
ρ	1000	kg/m^3
C_p	0.239	kJ/kg·K
ΔH	-5×10^4	kJ/kmol

Table 1.1: Parameters of the well-stirred reactor.

Notice also that Lemma 1.10 does not require that the plant output be generated by the model. The theorem applies regardless of what generates the plant output. *If the plant is identical to the system plus disturbance model assumed in the estimator*, then the conclusion can be strengthened. In the nominal case without measurement or process noise ($w = 0$, $v = 0$), *for a set of plant initial states*, the closed-loop system *converges to a steady state* and the feasible steady-state target is achieved leading to zero offset in the controlled variables. Characterizing the set of initial states in the region of convergence, and stabilizing

[8]This is another consequence of the fundamental theorem of linear algebra. The result is depicted in Figure A.1.

the system when the plant and the model differ, are treated in Chapters 3 and 5. We conclude the chapter with a nonlinear example that demonstrates the use of Lemma 1.10.

Example 1.11: More measured outputs than inputs and zero offset

We consider a well-stirred chemical reactor depicted in Figure 1.6, as in Pannocchia and Rawlings (2003). An irreversible, first-order reaction A⟶ B occurs in the liquid phase and the reactor temperature is regulated with external cooling. Mass and energy balances lead to the following nonlinear state space model:

$$\begin{aligned}
\frac{dc}{dt} &= \frac{F_0(c_0-c)}{\pi r^2 h} - k_0 c \exp\left(-\frac{E}{RT}\right) \\
\frac{dT}{dt} &= \frac{F_0(T_0-T)}{\pi r^2 h} + \frac{-\Delta H}{\rho C_p} k_0 c \exp\left(-\frac{E}{RT}\right) + \frac{2U}{r\rho C_p}(T_c - T) \\
\frac{dh}{dt} &= \frac{F_0 - F}{\pi r^2}
\end{aligned}$$

The controlled variables are h, the level of the tank, and c, the molar concentration of species A. The additional state variable is T, the reactor temperature; while the manipulated variables are T_c, the coolant liquid temperature, and F, the outlet flowrate. Moreover, it is assumed that the inlet flowrate acts as an unmeasured disturbance. The model parameters in nominal conditions are reported in Table 1.1. The open-loop stable steady-state operating conditions are the following:

$$c^s = 0.878\,\text{kmol/m}^3 \qquad T^s = 324.5\,\text{K} \qquad h^s = 0.659\,\text{m}$$
$$T_c^s = 300\,\text{K} \qquad F^s = 0.1\,\text{m}^3/\text{min}$$

Using a sampling time of 1 min, a linearized discrete state space model is obtained and, assuming that all the states are measured, the state space variables are:

$$x = \begin{bmatrix} c - c^s \\ T - T^s \\ h - h^s \end{bmatrix} \qquad u = \begin{bmatrix} T_c - T_c^s \\ F - F^s \end{bmatrix} \qquad y = \begin{bmatrix} c - c^s \\ T - T^s \\ h - h^s \end{bmatrix} \qquad p = F_0 - F_0^s$$

The corresponding linear model is:

$$\begin{aligned}
x(k+1) &= Ax(k) + Bu(k) + B_p p \\
y(k) &= Cx(k)
\end{aligned}$$

in which

$$A = \begin{bmatrix} 0.2681 & -0.00338 & -0.00728 \\ 9.703 & 0.3279 & -25.44 \\ 0 & 0 & 1 \end{bmatrix} \qquad C = \begin{bmatrix} 1 & 0 & 0 \\ 0 & 1 & 0 \\ 0 & 0 & 1 \end{bmatrix}$$

$$B = \begin{bmatrix} -0.00537 & 0.1655 \\ 1.297 & 97.91 \\ 0 & -6.637 \end{bmatrix} \qquad B_p = \begin{bmatrix} -0.1175 \\ 69.74 \\ 6.637 \end{bmatrix}$$

(a) Since we have two inputs, T_c and F, we try to remove offset in two controlled variables, c and h. Model the disturbance with *two* integrating output disturbances on the two controlled variables. Assume that the covariances of the state noises are zero except for the two integrating states. Assume that the covariances of the three measurements' noises are also zero.

Notice that although there are only two controlled variables, this choice of *two* integrating disturbances does not follow the prescription of Lemma 1.10 for zero offset.

Simulate the response of the controlled system after a 10% increase in the inlet flowrate F_0 at time $t = 10$ min. Use the nonlinear differential equations for the plant model. Do you have steady offset in any of the outputs? Which ones?

(b) Follow the prescription of Lemma 1.10 and choose a disturbance model with *three* integrating modes. Can you choose three integrating output disturbances for this plant? If so, prove it. If not, state why not.

(c) Again choose a disturbance model with three integrating modes; choose two integrating output disturbances on the two controlled variables. Choose one integrating input disturbance on the outlet flowrate F. Is the augmented system detectable?

Simulate again the response of the controlled system after a 10% increase in the inlet flowrate F_0 at time $t = 10$ min. Again use the nonlinear differential equations for the plant model. Do you have steady offset in any of the outputs? Which ones?

Compare and contrast the closed-loop performance for the design with two integrating disturbances and the design with three integrating disturbances. Which control system do you recommend and why?

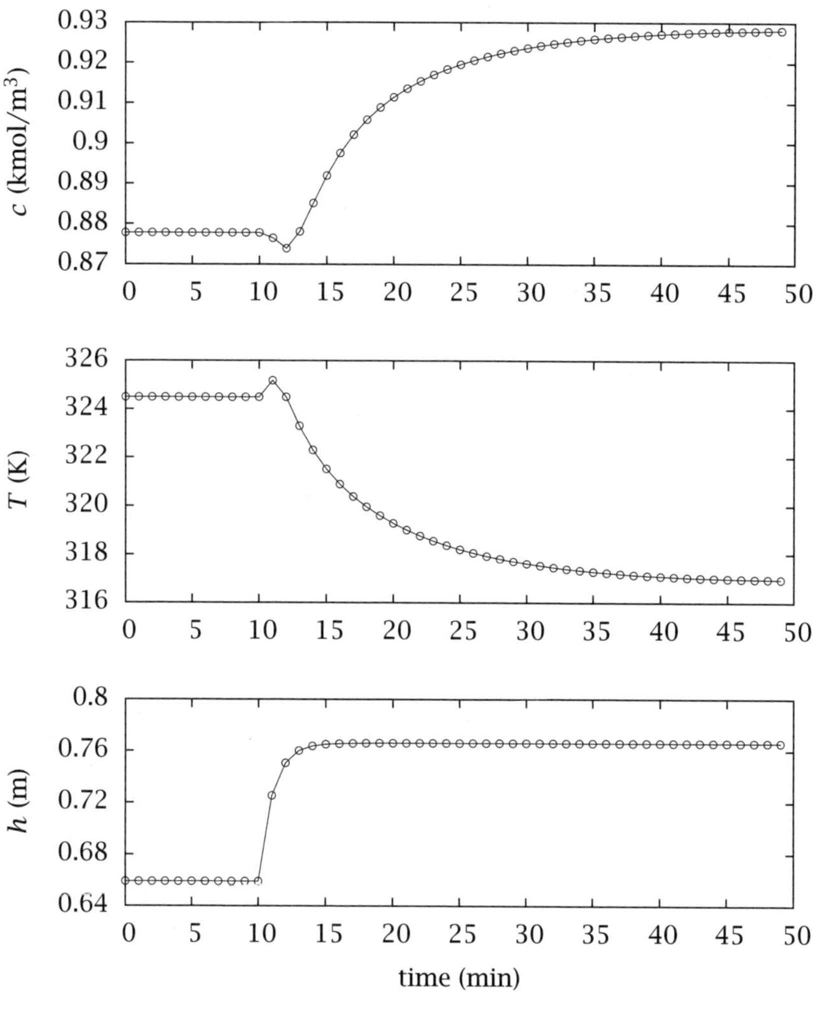

Figure 1.7: Three measured outputs versus time after a step change in inlet flowrate at 10 minutes; $n_d = 2$.

Solution

(a) Integrating disturbances are added to the two controlled variables (first and third outputs) by choosing

$$C_d = \begin{bmatrix} 1 & 0 \\ 0 & 0 \\ 0 & 1 \end{bmatrix} \qquad B_d = 0$$

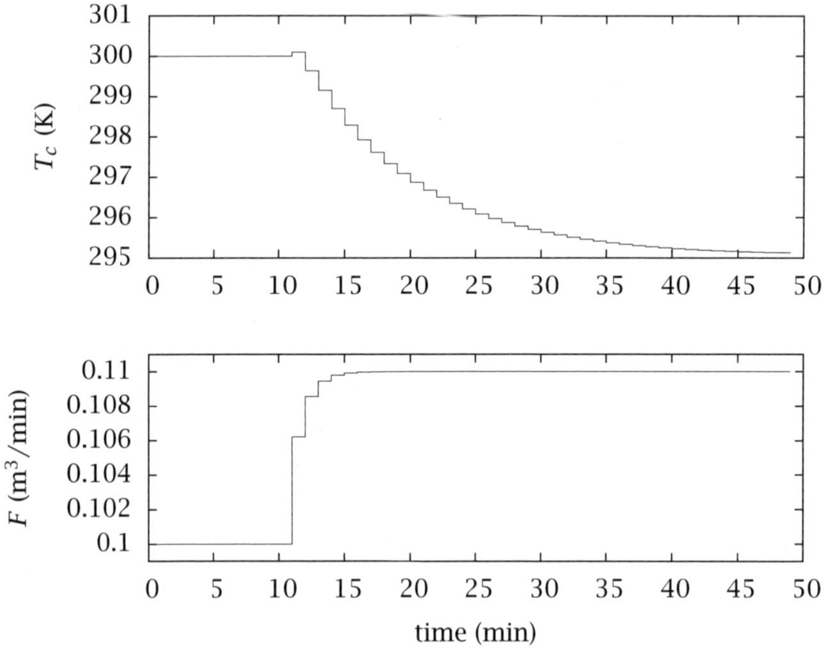

Figure 1.8: Two manipulated inputs versus time after a step change in inlet flowrate at 10 minutes; $n_d = 2$.

The results with two integrating disturbances are shown in Figures 1.7 and 1.8. Notice that despite adding integrating disturbances to the two controlled variables, c and h, both of these controlled variables as well as the third output, T, all display nonzero offset at steady state.

(b) A third integrating disturbance is added to the second output giving

$$C_d = \begin{bmatrix} 1 & 0 & 0 \\ 0 & 0 & 1 \\ 0 & 1 & 0 \end{bmatrix} \qquad B_d = 0$$

The augmented system is not detectable with this disturbance model. The rank of $\begin{bmatrix} I-A & -B_d \\ C & C_d \end{bmatrix}$ is only 5 instead of 6. The problem here is that the system level is itself an integrator, and we cannot distinguish h from the integrating disturbance added to h.

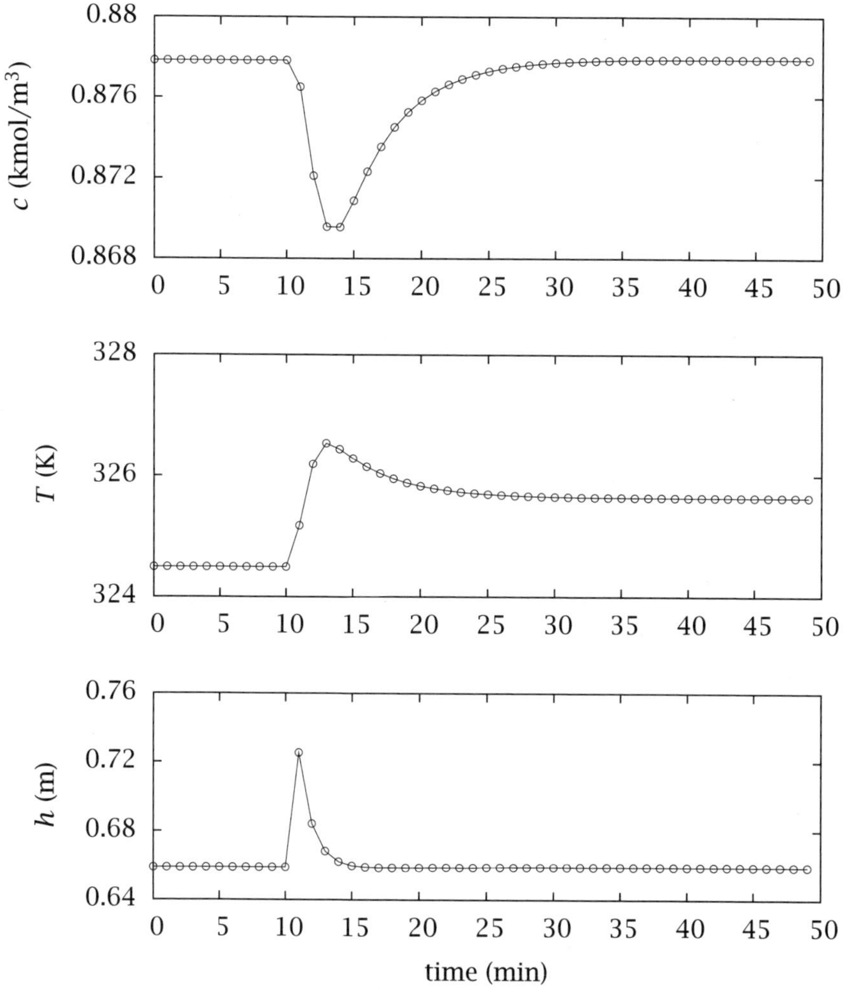

Figure 1.9: Three measured outputs versus time after a step change in inlet flowrate at 10 minutes; $n_d = 3$.

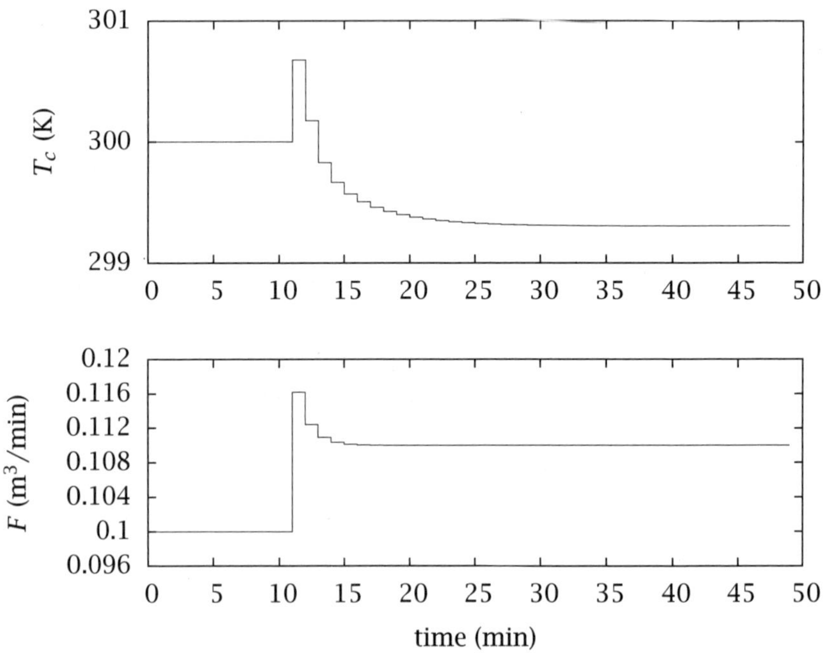

Figure 1.10: Two manipulated inputs versus time after a step change in inlet flowrate at 10 minutes; $n_d = 3$.

(c) Next we try three integrating disturbances: two added to the two controlled variables, and one added to the second manipulated variable

$$C_d = \begin{bmatrix} 1 & 0 & 0 \\ 0 & 0 & 0 \\ 0 & 1 & 0 \end{bmatrix} \qquad B_d = \begin{bmatrix} 0 & 0 & 0.1655 \\ 0 & 0 & 97.91 \\ 0 & 0 & -6.637 \end{bmatrix}$$

The augmented system is detectable for this disturbance model.

The results for this choice of three integrating disturbances are shown in Figures 1.9 and 1.10. Notice that we have zero offset in the two controlled variables, c and h, and have successfully forced the steady-state effect of the inlet flowrate disturbance entirely into the second output, T.

Notice also that the dynamic behavior of all three outputs is superior to that achieved with the model using two integrating disturbances. The true disturbance, which is a step at the inlet flowrate, is better represented by including the integrator in the outlet flowrate. With a more accurate disturbance model, better overall control is achieved. The controller uses smaller manipulated variable action and also achieves better output variable behavior. An added bonus is that steady offset is removed in the maximum possible number of outputs. □

Further notation

G	transfer function matrix
m	mean of normally distributed random variable
T	reactor temperature
$\tilde{u}$	input deviation variable
x, y, z	spatial coordinates for a distributed system
$\tilde{x}$	state deviation variable

1.6 Exercises

Exercise 1.1: State space form for chemical reaction model

Consider the following chemical reaction kinetics for a two-step series reaction

$$\mathrm{A} \xrightarrow{k_1} \mathrm{B} \qquad \mathrm{B} \xrightarrow{k_2} \mathrm{C}$$

We wish to follow the reaction in a constant volume, well-mixed, batch reactor. As taught in the undergraduate chemical engineering curriculum, we proceed by writing material balances for the three species giving

$$\frac{dc_A}{dt} = -r_1 \qquad \frac{dc_B}{dt} = r_1 - r_2 \qquad \frac{dc_C}{dt} = r_2$$

in which c_j is the concentration of species j, and r_1 and r_2 are the rates (mol/(time·vol)) at which the two reactions occur. We then assume some rate law for the reaction kinetics, such as

$$r_1 = k_1 c_A \qquad r_2 = k_2 c_B$$

We substitute the rate laws into the material balances and specify the starting concentrations to produce three differential equations for the three species concentrations.

(a) Write the linear state space model for the deterministic series chemical reaction model. Assume we can measure the component A concentration. What are x, y, A, B, C, and D for this model?

(b) Simulate this model with initial conditions and parameters given by

$$c_{A0} = 1 \quad c_{B0} = c_{C0} = 0 \qquad k_1 = 2 \quad k_2 = 1$$

Exercise 1.2: Distributed systems and time delay

We assume familiarity with the transfer function of a time delay from an undergraduate systems course

$$\overline{y}(s) = e^{-\theta s}\overline{u}(s)$$

Let's see the connection between the delay and the distributed systems, which give rise to it. A simple physical example of a time delay is the delay caused by transport in a flowing system. Consider plug flow in a tube depicted in Figure 1.11.

(a) Write down the equation of change for moles of component j for an arbitrary volume element and show that

$$\frac{\partial c_j}{\partial t} = -\nabla \cdot (c_j v_j) + R_j$$

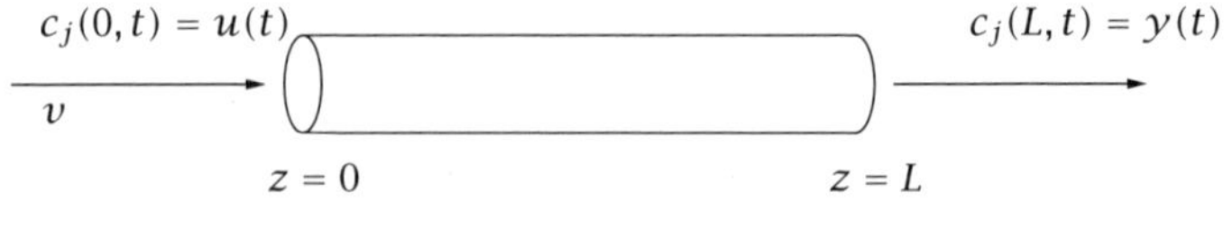

Figure 1.11: Plug-flow reactor.

in which c_j is the molar concentration of component j, v_j is the velocity of component j, and R_j is the production rate of component j due to chemical reaction.[9]

Plug flow means the fluid velocity of all components is purely in the z direction, and is independent of r and θ and, we assume here, z

$$v_j = v\delta_z$$

(b) Assuming plug flow and neglecting chemical reaction in the tube, show that the equation of change reduces to

$$\frac{\partial c_j}{\partial t} = -v\frac{\partial c_j}{\partial z} \tag{1.47}$$

This equation is known as a hyperbolic, first-order partial differential equation. Assume the boundary and initial conditions are

$$c_j(z,t) = u(t) \qquad 0 = z \qquad t \geq 0 \tag{1.48}$$

$$c_j(z,t) = c_{j0}(z) \qquad 0 \leq z \leq L \quad t = 0 \tag{1.49}$$

In other words, we are using the feed concentration as the manipulated variable, $u(t)$, and the tube starts out with some initial concentration profile of component j, $c_{j0}(z)$.

(c) Show that the solution to (1.47) with these boundary conditions is

$$c_j(z,t) = \begin{cases} u(t - z/v) & vt > z \\ c_{j0}(z - vt) & vt < z \end{cases} \tag{1.50}$$

(d) If the reactor starts out empty of component j, show that the transfer function between the outlet concentration, $y = c_j(L,t)$, and the inlet concentration, $c_j(0,t) = u(t)$, is a time delay. What is the value of θ?

Exercise 1.3: Pendulum in state space

Consider the pendulum suspended at the end of a rigid link depicted in Figure 1.12. Let r and θ denote the polar coordinates of the center of the pendulum, and let $p = r\delta_r$ be the position vector of the pendulum, in which δ_r and δ_θ are the unit vectors in polar coordinates. We wish to determine a state space description of the system. We are able to apply a torque T to the pendulum as our manipulated variable. The pendulum has mass m, the only other external force acting on the pendulum is gravity, and we neglect friction. The link provides force $-t\delta_r$ necessary to maintain the pendulum at distance $r = R$ from the axis of rotation, and we measure this force t.

(a) Provide expressions for the four partial derivatives for changes in the unit vectors with r and θ

$$\frac{\partial \delta_r}{\partial r} \qquad \frac{\partial \delta_r}{\partial \theta} \qquad \frac{\partial \delta_\theta}{\partial r} \qquad \frac{\partial \delta_\theta}{\partial \theta}$$

(b) Use the chain rule to find the velocity of the pendulum in terms of the time derivatives of r and θ. Do not simplify yet by assuming r is constant. We want the general result.

---

[9]You will need the Gauss divergence theorem and 3D Leibniz formula to go from a mass balance on a volume element to the equation of continuity.

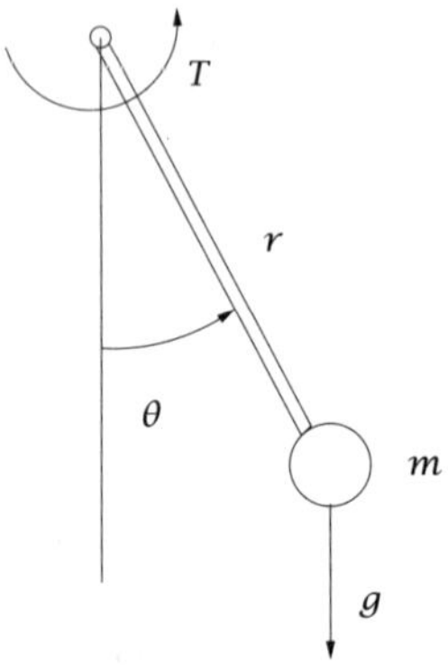

Figure 1.12: Pendulum with applied torque.

(c) Differentiate again to show that the acceleration of the pendulum is

$$\ddot{p} = (\ddot{r} - r\dot{\theta}^2)\delta_r + (r\ddot{\theta} + 2\dot{r}\dot{\theta})\delta_\theta$$

(d) Use a momentum balance on the pendulum mass (you may assume it is a point mass) to determine both the force exerted by the link

$$t = mR\dot{\theta}^2 + mg\cos\theta$$

and an equation for the acceleration of the pendulum due to gravity and the applied torque

$$mR\ddot{\theta} - T/R + mg\sin\theta = 0$$

(e) Define a state vector and give a state space description of your system. What is the physical significance of your state. Assume you measure the force exerted by the link.

One answer is

$$\begin{aligned} \frac{dx_1}{dt} &= x_2 \\ \frac{dx_2}{dt} &= -(g/R)\sin x_1 + u \\ y &= mRx_2^2 + mg\cos x_1 \end{aligned}$$

in which $u = T/(mR^2)$

Exercise 1.4: Time to Laplace domain

Take the Laplace transform of the following set of differential equations and find the transfer function, $G(s)$, connecting $\overline{u}(s)$ and $\overline{y}(s)$, $\overline{y} = G\overline{u}$

$$\begin{aligned} \frac{dx}{dt} &= Ax + Bu \\ y &= Cx + Du \end{aligned} \tag{1.51}$$

For $x \in \mathbb{R}^n$, $y \in \mathbb{R}^p$, and $u \in \mathbb{R}^m$, what is the dimension of the G matrix? What happens to the initial condition, $x(0) = x_0$?

Exercise 1.5: Converting between continuous and discrete time models

Given a prescribed $u(t)$, derive and check the solution to (1.51). Given a prescribed $u(k)$ sequence, what is the solution to the discrete time model

$$x(k+1) = \widetilde{A}x(k) + \widetilde{B}u(k)$$
$$y(k) = \widetilde{C}x(k) + \widetilde{D}u(k)$$

(a) Compute $\widetilde{A}, \widetilde{B}, \widetilde{C}$, and $\widetilde{D}$ so that the two solutions agree at the sample times for a zero-order hold input, i.e., $y(k) = y(t_k)$ for $u(t) = u(k),\ t \in (t_k, t_{k+1})$ in which $t_k = k\Delta$ for sample time Δ.

(b) Is your result valid for A singular? If not, how can you find $\widetilde{A}, \widetilde{B}, \widetilde{C}$, and $\widetilde{D}$ for this case?

Exercise 1.6: Continuous to discrete time conversion for nonlinear models

Consider the autonomous nonlinear differential equation model

$$\frac{dx}{dt} = f(x,u)$$
$$x(0) = x_0 \tag{1.52}$$

Given a zero-order hold on the input, let $s(t,u,x_0), 0 \le t \le \Delta$, be the solution to (1.52) given initial condition x_0 at time $t = 0$, and constant input u is applied for t in the interval $0 \le t \le \Delta$. Consider also the nonlinear discrete time model

$$x(k+1) = F(x(k), u(k))$$

(a) What is the relationship between F and s so that the solution of the discrete time model agrees at the sample times with the continuous time model with a zero-order hold?

(b) Assume f is linear and apply this result to check the result of Exercise 1.5.

Exercise 1.7: Commuting functions of a matrix

Although matrix multiplication does not commute in general

$$AB \neq BA$$

multiplication of functions of the same matrix do commute. You may have used the following fact in Exercise 1.5

$$A^{-1}\exp(At) = \exp(At)A^{-1} \tag{1.53}$$

(a) Prove that (1.53) is true assuming A has distinct eigenvalues and can therefore be represented as

$$A = Q\Lambda Q^{-1} \qquad \Lambda = \begin{bmatrix} \lambda_1 & 0 & \cdots & 0 \\ 0 & \lambda_2 & \cdots & 0 \\ \vdots & \vdots & \ddots & \vdots \\ 0 & 0 & \cdots & \lambda_n \end{bmatrix}$$

in which Λ is a diagonal matrix containing the eigenvalues of A, and Q is the matrix of eigenvectors such that

$$Aq_i = \lambda_i q_i, \qquad i = 1, \ldots, n$$

in which q_i is the ith column of matrix Q.

(b) Prove the more general relationship

$$f(A)g(A) = g(A)f(A) \tag{1.54}$$

in which f and g are any functions definable by Taylor series.

(c) Prove that (1.54) is true without assuming the eigenvalues are distinct.

Hint: use the Taylor series defining the functions and apply the Cayley-Hamilton theorem (Horn and Johnson, 1985, pp. 86–87).

Exercise 1.8: Finite difference formula and approximating the exponential

Instead of computing the exact conversion of a continuous time to a discrete time system as in Exercise 1.5, assume instead one simply approximates the time derivative with a first-order finite difference formula

$$\frac{dx}{dt} \approx \frac{x(t_{k+1}) - x(t_k)}{\Delta}$$

with step size equal to the sample time, Δ. For this approximation of the continuous time system, compute $\tilde{A}$ and $\tilde{B}$ so that the discrete time system agrees with the approximate continuous time system at the sample times. Comparing these answers to the exact solution, what approximation of $e^{A\Delta}$ results from the finite difference approximation? When is this a good approximation of $e^{A\Delta}$?

Exercise 1.9: Mapping eigenvalues of continuous time systems to discrete time systems

Consider the continuous time differential equation and discrete time difference equation

$$\frac{dx}{dt} = Ax$$

$$x^+ = \tilde{A}x$$

and the transformation

$$\tilde{A} = e^{A\Delta}$$

Consider the scalar A case.

(a) What A represents an integrator in continuous time? What is the corresponding $\tilde{A}$ value for the integrator in discrete time?

(b) What A give purely oscillatory solutions? What are the corresponding $\tilde{A}$?

(c) For what A is the solution of the ODE stable? Unstable? What are the corresponding $\tilde{A}$?

(d) Sketch and label these A and $\tilde{A}$ regions in two complex-plane diagrams.

Exercise 1.10: State space realization

Define a state vector and realize the following models as state space models **by hand**. One should do a few by hand to understand what the Octave or MATLAB calls are doing. Answer the following questions. What is the connection between the poles of G and the state space description? For what kinds of $G(s)$ does one obtain a nonzero D matrix? What is the order and gain of these systems? Is there a connection between order and the numbers of inputs and outputs?

(a) $G(s) = \dfrac{1}{2s+1}$

(b) $G(s) = \dfrac{1}{(2s+1)(3s+1)}$

(c) $G(s) = \dfrac{2s+1}{3s+1}$

(d) $y(k+1) = y(k) + 2u(k)$

(e) $y(k+1) = a_1 y(k) + a_2 y(k-1) + b_1 u(k) + b_2 u(k-1)$

Exercise 1.11: Minimal realization

Find minimal realizations of the state space models you found by hand in Exercise 1.10. Use Octave or MATLAB for computing minimal realizations. Were any of your hand realizations nonminimal?

Exercise 1.12: Partitioned matrix inversion lemma

Let matrix Z be partitioned into

$$Z = \begin{bmatrix} B & C \\ D & E \end{bmatrix}$$

and assume Z^{-1}, B^{-1} and E^{-1} exist.

(a) Perform row elimination and show that

$$Z^{-1} = \begin{bmatrix} B^{-1} + B^{-1}C(E - DB^{-1}C)^{-1}DB^{-1} & -B^{-1}C(E - DB^{-1}C)^{-1} \\ -(E - DB^{-1}C)^{-1}DB^{-1} & (E - DB^{-1}C)^{-1} \end{bmatrix}$$

Note that this result is still valid if E is singular.

(b) Perform column elimination and show that

$$Z^{-1} = \begin{bmatrix} (B - CE^{-1}D)^{-1} & -(B - CE^{-1}D)^{-1}CE^{-1} \\ -E^{-1}D(B - CE^{-1}D)^{-1} & E^{-1} + E^{-1}D(B - CE^{-1}D)^{-1}CE^{-1} \end{bmatrix}$$

Note that this result is still valid if B is singular.

(c) A host of other useful control-related inversion formulas follow from these results. Equate the (1,1) or (2,2) entries of Z^{-1} and derive the identity

$$(A + BCD)^{-1} = A^{-1} - A^{-1}B(DA^{-1}B + C^{-1})^{-1}DA^{-1} \tag{1.55}$$

A useful special case of this result is

$$(I + X^{-1})^{-1} = I - (I + X)^{-1}$$

(d) Equate the (1,2) or (2,1) entries of Z^{-1} and derive the identity

$$(A + BCD)^{-1}BC = A^{-1}B(DA^{-1}B + C^{-1})^{-1} \tag{1.56}$$

Equations (1.55) and (1.56) prove especially useful in rearranging formulas in least squares estimation.

Exercise 1.13: Perturbation to an asymptotically stable linear system

Given the system

$$x^+ = Ax + Bu$$

If A is an asymptotically stable matrix, prove that if $u(k) \to 0$, then $x(k) \to 0$.

Exercise 1.14: Exponential stability of a perturbed linear system

Given the system

$$x^+ = Ax + Bu$$

If A is an asymptotically stable matrix, prove that if $u(k)$ decreases exponentially to zero, then $x(k)$ decreases exponentially to zero.

Exercise 1.15: Are we going forward or backward today?

In the chapter we derived the solution to

$$\min_{w,x,y} f(w,x) + g(x,y) + h(y,z)$$

in which z is a fixed parameter using forward dynamic programming (DP)

$$\begin{aligned}
&\overline{y}^0(z) \\
&\tilde{x}^0(z) = \overline{x}^0(\overline{y}^0(z)) \\
&\tilde{w}^0(z) = \overline{w}^0(\overline{x}^0(\overline{y}^0(z)))
\end{aligned}$$

(a) Solve for optimal w as a function of z using backward DP.

(b) Is forward or backward DP more efficient if you want optimal w as a function of z?

Exercise 1.16: Method of Lagrange multipliers

Consider the objective function $V(x) = (1/2)x'Hx + h'x$ and optimization problem

$$\min_x V(x) \tag{1.57}$$

subject to

$$Dx = d$$

in which $H > 0$, $x \in \mathbb{R}^n$, $d \in \mathbb{R}^m$, $m < n$, i.e., fewer constraints than decisions. Rather than partially solving for x using the constraint and eliminating it, we make use of the method of Lagrange multipliers for treating the equality constraints (Fletcher, 1987; Nocedal and Wright, 1999).

In the method of Lagrange multipliers, we augment the objective function with the constraints to form the Lagrangian function, L

$$L(x,\lambda) = (1/2)x'Hx + h'x - \lambda'(Dx - d)$$

in which $\lambda \in \mathbb{R}^m$ is the vector of Lagrange multipliers. The necessary and sufficient conditions for a global minimizer are that the partial derivatives of L with respect to x and λ vanish (Nocedal and Wright, 1999, p. 444), (Fletcher, 1987, p.198,236)

(a) Show that the necessary and sufficient conditions are equivalent to the matrix equation

$$\begin{bmatrix} H & -D' \\ -D & 0 \end{bmatrix} \begin{bmatrix} x \\ \lambda \end{bmatrix} = - \begin{bmatrix} h \\ d \end{bmatrix} \tag{1.58}$$

The solution to (1.58) then provides the solution to the original problem (1.57).

(b) We note one other important feature of the Lagrange multipliers, their relationship to the optimal cost of the purely quadratic case. For $h = 0$, the cost is given by

$$V^0 = (1/2)(x^0)' H x^0$$

Show that this can also be expressed in terms of λ^0 by the following

$$V^0 = (1/2) d' \lambda^0$$

Exercise 1.17: Minimizing a constrained, quadratic function

Consider optimizing the positive definite quadratic function subject to a linear constraint

$$\min_x (1/2) x' H x \qquad \text{s.t. } Ax = b$$

Using the method of Lagrange multipliers presented in Exercise 1.16, show that the optimal solution, multiplier, and cost are given by

$$\begin{aligned} x^0 &= H^{-1} A' (A H^{-1} A')^{-1} b \\ \lambda^0 &= (A H^{-1} A')^{-1} b \\ V^0 &= (1/2) b' (A H^{-1} A')^{-1} b \end{aligned}$$

Exercise 1.18: Minimizing a partitioned quadratic function

Consider the partitioned constrained minimization

$$\min_{x_1, x_2} \begin{bmatrix} x_1 \\ x_2 \end{bmatrix}' \begin{bmatrix} H_1 & \\ & H_2 \end{bmatrix} \begin{bmatrix} x_1 \\ x_2 \end{bmatrix}$$

subject to

$$\begin{bmatrix} D & I \end{bmatrix} \begin{bmatrix} x_1 \\ x_2 \end{bmatrix} = d$$

The solution to this optimization is required in two different forms, depending on whether one is solving an estimation or regulation problem. Show that the solution can be expressed in the following two forms if both H_1 and H_2 are full rank.

- Regulator form

$$\begin{aligned} V^0(d) &= d' (H_2 - H_2 D (D' H_2 D + H_1)^{-1} D' H_2) d \\ x_1^0(d) &= \widetilde{K} d \qquad \widetilde{K} = (D' H_2 D + H_1)^{-1} D' H_2 \\ x_2^0(d) &= (I - D\widetilde{K}) d \end{aligned}$$

- Estimator form

$$\begin{aligned} V^0(d) &= d' (D H_1^{-1} D' + H_2^{-1})^{-1} d \\ x_1^0(d) &= \widetilde{L} d \qquad \widetilde{L} = H_1^{-1} D' (D H_1^{-1} D' + H_2^{-1})^{-1} \\ x_2^0(d) &= (I - D\widetilde{L}) d \end{aligned}$$

Exercise 1.19: Stabilizability and controllability canonical forms

Consider the partitioned system

$$\begin{bmatrix} x_1 \\ x_2 \end{bmatrix}^+ = \begin{bmatrix} A_{11} & A_{12} \\ 0 & A_{22} \end{bmatrix} \begin{bmatrix} x_1 \\ x_2 \end{bmatrix} + \begin{bmatrix} B_1 \\ 0 \end{bmatrix} u$$

with (A_{11}, B_1) controllable. This form is known as controllability canonical form.

(a) Show that the system is *not* controllable by checking the rank of the controllability matrix.

(b) Show that the modes x_1 can be controlled from any $x_1(0)$ to any $x_1(n)$ with a sequence of inputs $u(0), \ldots, u(n-1)$, but the modes x_2 *cannot* be controlled from any $x_2(0)$ to any $x_2(n)$. The states x_2 are termed the uncontrollable modes.

(c) If A_{22} is stable the system is termed *stabilizable.* Although not all modes can be controlled, the uncontrollable modes are stable and decay to steady state.

The following lemma gives an equivalent condition for stabilizability.

Lemma 1.12 (Hautus Lemma for stabilizability). *A system is stabilizable if and only if*

$$\operatorname{rank}\begin{bmatrix} \lambda I - A & B \end{bmatrix} = n \qquad \textit{for all } |\lambda| \geq 1$$

Prove this lemma using Lemma 1.2 as the condition for controllability.

Exercise 1.20: Regulator stability, stabilizable systems, and semidefinite state penalty

(a) Show that the infinite horizon LQR is stabilizing for (A, B) *stabilizable* with $R, Q > 0$.

(b) Show that the infinite horizon LQR is stabilizing for (A, B) stabilizable and $R > 0$, $Q \geq 0$, and (A, Q) detectable. Discuss what happens to the controller's stabilizing property if Q is not positive semidefinite or (A, Q) is not detectable.

Exercise 1.21: Time-varying linear quadratic problem

Consider the time-varying version of the LQ problem solved in the chapter. The system model is

$$x(k+1) = A(k)x(k) + B(k)u(k)$$

The objective function also contains time-varying penalties

$$\min_{\mathbf{u}} V(x(0), \mathbf{u}) = \frac{1}{2}\left(\sum_{k=0}^{N-1} \left(x(k)'Q(k)x(k) + u(k)'R(k)u(k)\right) + x(N)'Q(N)x(N)\right)$$

subject to the model. Notice the penalty on the final state is now simply $Q(N)$ instead of P_f.

Apply the DP argument to this problem and determine the optimal input sequence and cost. Can this problem also be solved in closed form like the time-invariant case?

Exercise 1.22: Steady-state Riccati equation

Generate a random A and B for a system model for whatever $n(\geq 3)$ and $m(\geq 3)$ you wish. Choose a positive semidefinite Q and positive definite R of the appropriate sizes.

(a) Iterate the DARE by hand with Octave or MATLAB until Π stops changing. Save this result. Now call the MATLAB or Octave function to solve the steady-state DARE. Do the solutions agree? Where in the complex plane are the eigenvalues of $A + BK$? Increase the size of Q relative to R. Where do the eigenvalues move?

(b) Repeat for a singular A matrix. What happens to the two solution techniques?

(c) Repeat for an unstable A matrix.

Exercise 1.23: Positive definite Riccati iteration

If $\Pi(k), Q, R > 0$ in (1.11), show that $\Pi(k-1) > 0$.
Hint: apply (1.55) to the term $(B'\Pi(k)B + R)^{-1}$.

Exercise 1.24: Existence and uniqueness of the solution to constrained least squares

Consider the least squares problem subject to linear constraint

$$\min_x (1/2)x'Qx \qquad \text{subject to} \quad Ax = b$$

in which $x \in \mathbb{R}^n$, $b \in \mathbb{R}^p$, $Q \in \mathbb{R}^{n\times n}$, $Q \geq 0$, $A \in \mathbb{R}^{p\times n}$. Show that this problem has a solution for every b and the solution is unique if and only if

$$\operatorname{rank}(A) = p \qquad \operatorname{rank}\begin{bmatrix} Q \\ A \end{bmatrix} = n$$

Exercise 1.25: Rate-of-change penalty

Consider the generalized LQR problem with the cross term between $x(k)$ and $u(k)$

$$V(x(0),\mathbf{u}) = \frac{1}{2}\sum_{k=0}^{N-1}\left(x(k)'Qx(k) + u(k)'Ru(k) + 2x(k)'Mu(k)\right) + (1/2)x(N)'P_f x(N)$$

(a) Solve this problem with backward DP and write out the Riccati iteration and feedback gain.

(b) Control engineers often wish to tune a regulator by penalizing the rate of change of the input rather than the absolute size of the input. Consider the additional positive definite penalty matrix S and the modified objective function

$$V(x(0),\mathbf{u}) = \frac{1}{2}\sum_{k=0}^{N-1}\left(x(k)'Qx(k) + u(k)'Ru(k) + \Delta u(k)'S\Delta u(k)\right)$$
$$+ (1/2)x(k)'P_f x(k)$$

in which $\Delta u(k) = u(k) - u(k-1)$. Show that you can augment the state to include $u(k-1)$ via

$$\widetilde{x}(k) = \begin{bmatrix} x(k) \\ u(k-1) \end{bmatrix}$$

and reduce this new problem to the standard LQR with the cross term. What are $\widetilde{A}$, $\widetilde{B}$, $\widetilde{Q}$, $\widetilde{R}$, and $\widetilde{M}$ for the augmented problem (Rao and Rawlings, 1999)?

Exercise 1.26: Existence, uniqueness and stability with the cross term

Consider the linear quadratic problem with system

$$x^+ = Ax + Bu \tag{1.59}$$

and infinite horizon cost function

$$V(x(0), \mathbf{u}) = (1/2) \sum_{k=0}^{\infty} x(k)'Qx(k) + u(k)'Ru(k)$$

The existence, uniqueness and stability conditions for this problem are: (A, B) stabilizable, $Q \geq 0$, (A, Q) detectable, and $R > 0$. Consider the modified objective function with the cross term

$$V = (1/2) \sum_{k=0}^{\infty} x(k)'Qx(k) + u(k)'Ru(k) + 2x(k)'Mu(k) \tag{1.60}$$

(a) Consider reparameterizing the input as

$$v(k) = u(k) + Tx(k) \tag{1.61}$$

Choose T such that the cost function in x and v does not have a cross term, and express the existence, uniqueness and stability conditions for the transformed system. Goodwin and Sin (1984, p.251) discuss this procedure in the state estimation problem with nonzero covariance between state and output measurement noises.

(b) Translate and simplify these to obtain the existence, uniqueness and stability conditions for the original system with cross term.

Exercise 1.27: Forecasting and variance increase or decrease

Given positive definite initial state variance $P(0)$ and process disturbance variance Q, the variance after forecasting one sample time was shown to be

$$P^-(1) = AP(0)A' + Q$$

(a) If A is stable, is it true that $AP(0)A' < P(0)$? If so, prove it. If not, provide a counterexample.

(b) If A is unstable, is it true that $AP(0)A' > P(0)$? If so, prove it. If not, provide a counterexample.

(c) If the magnitudes of *all* the eigenvalues of A are unstable, is it true that $AP(0)A' > P(0)$? If so, prove it. If not, provide a counterexample.

Exercise 1.28: Convergence of MHE with noninformative prior

Show that the simplest form of MHE defined in (1.33) and (1.34) is also a convergent estimator for an observable system. What restrictions on the horizon length N do you require for this result to hold?

Hint: you can solve the MHE optimization problem by inspection when there is no prior weighting of the data.

Exercise 1.29: Symmetry in regulation and estimation

In this exercise we display the symmetry of the backward DP recursion for regulation, and the forward DP recursion for estimation. In the regulation problem we solve at stage k

$$\min_{x,u} \ell(z,u) + V_k^0(x) \qquad \text{s.t. } x = Az + Bu$$

In backward DP, x is the state at the current stage and z is the state at the previous stage. The stage cost and cost to go are given by

$$\ell(z,u) = (1/2)(z'Qz + u'Ru) \qquad V_k^0(x) = (1/2)x'\Pi(k)x$$

and the optimal cost is $V_{k-1}^0(z)$ since z is the state at the previous stage.

In estimation we solve at stage k

$$\min_{x,w} \ell(z,w) + V_k^0(x) \qquad \text{s.t. } z = Ax + w$$

In forward DP, x is the state at the current stage, z is the state at the next stage. The stage cost and arrival cost are given by

$$\ell(z,w) = (1/2)(\,|y(k+1) - Cz|_{R^{-1}}^2 + w'Q^{-1}w) \qquad V_k^0(x) = (1/2)\,|x - \hat{x}(k)|_{P(k)^{-1}}^2$$

and we wish to find $V_{k+1}^0(z)$ in the estimation problem.

(a) In the estimation problem, take the z term outside the optimization and solve

$$\min_{x,w} \frac{1}{2}\left(w'Q^{-1}w + (x - \hat{x}(k))'P(k)^{-1}(x - \hat{x}(k))\right) \qquad \text{s.t. } z = Ax + w$$

using the inverse form in Exercise 1.18, and show that the optimal cost is given by

$$\begin{aligned} V^0(z) &= (1/2)(z - A\hat{x}(k))'(P^-(k+1))^{-1}(z - A\hat{x}(k)) \\ P^-(k+1) &= AP(k)A' + Q \end{aligned}$$

Add the z term to this cost using the third part of Example 1.1 and show that

$$\begin{aligned} V_{k+1}^0(z) &= (1/2)(z - \hat{x}(k+1))'P^{-1}(k+1)(z - \hat{x}(k+1)) \\ P(k+1) &= P^-(k+1) - P^-(k+1)C'(CP^-(k+1)C' + R)^{-1}CP^-(k+1) \\ \hat{x}(k+1) &= A\hat{x}(k) + L(k+1)(y(k+1) - CA\hat{x}(k)) \\ L(k+1) &= P^-(k+1)C'(CP^-(k+1)C' + R)^{-1} \end{aligned}$$

(b) In the regulator problem, take the z term outside the optimization and solve the remaining two-term problem using the regulator form of Exercise 1.18. Then add the z term and show that

$$\begin{aligned}
V_{k-1}^0(z) &= (1/2)z'\Pi(k-1)z \\
\Pi(k-1) &= Q + A'\Pi(k)A - A'\Pi(k)B(B'\Pi(k)B+R)^{-1}B'\Pi(k)A \\
u^0(z) &= K(k-1)z \\
x^0(z) &= (A+BK(k-1))z \\
K(k-1) &= -(B'\Pi(k)B+R)^{-1}B'\Pi(k)A
\end{aligned}$$

This symmetry can be developed further if we pose an output tracking problem rather than zero state regulation problem in the regulator.

Exercise 1.30: Symmetry in the Riccati iteration

Show that the covariance before measurement $P^-(k+1)$ in estimation satisfies an identical iteration to the cost to go $\Pi(k-1)$ in regulation under the change of variables $P^- \longrightarrow \Pi$, $A \longrightarrow A'$, $C \longrightarrow B'$.

Exercise 1.31: Detectability and observability canonical forms

Consider the partitioned system

$$\begin{bmatrix} x_1 \\ x_2 \end{bmatrix}^+ = \begin{bmatrix} A_{11} & 0 \\ A_{21} & A_{22} \end{bmatrix} \begin{bmatrix} x_1 \\ x_2 \end{bmatrix}$$

$$y = \begin{bmatrix} C_1 & 0 \end{bmatrix} \begin{bmatrix} x_1 \\ x_2 \end{bmatrix}$$

with (A_{11}, C_1) observable. This form is known as observability canonical form.

(a) Show that the system is *not* observable by checking the rank of the observability matrix.

(b) Show that the modes x_1 can be uniquely determined from a sequence of measurements, but the modes x_2 *cannot* be uniquely determined from the measurements. The states x_2 are termed the unobservable modes.

(c) If A_{22} is stable the system is termed *detectable.* Although not all modes can be observed, the unobservable modes are stable and decay to steady state.

The following lemma gives an equivalent condition for detectability.

Lemma 1.13 (Hautus Lemma for detectability). *A system is detectable if and only if*

$$\text{rank} \begin{bmatrix} \lambda I - A \\ C \end{bmatrix} = n \qquad \textit{for all } |\lambda| \geq 1$$

Prove this lemma using Lemma 1.4 as the condition for observability.

Exercise 1.32: Estimator stability and detectable systems

Show that the least squares estimator given in (1.28) is stable for (A, C) *detectable* with $Q > 0$.

Exercise 1.33: Estimator stability and semidefinite state noise penalty

We wish to show that the least squares estimator is stable for (A, C) detectable and $Q \geq 0$, (A, Q) stabilizable.

(a) Because Q^{-1} is not defined in this problem, the objective function defined in (1.27) requires modification. Show that the objective function with semidefinite $Q \geq 0$ can be converted into the following form

$$V(x(0), \mathbf{w}(T)) = \frac{1}{2}\Bigg(|x(0) - \overline{x}(0)|^2_{(P^-(0))^{-1}} + \sum_{k=0}^{T-1} |w(k)|^2_{\widetilde{Q}^{-1}} + \sum_{k=0}^{T} |y(k) - Cx(k)|^2_{R^{-1}} \Bigg)$$

in which

$$x^+ = Ax + Gw \qquad \widetilde{Q} > 0$$

Find expressions for $\widetilde{Q}$ and G in terms of the original semidefinite Q. How are the dimension of $\widetilde{Q}$ and G related to the rank of Q?

(b) What is the probabilistic interpretation of the state estimation problem with semidefinite Q?

(c) Show that (A, Q) stabilizable implies (A, G) stabilizable in the converted form.

(d) Show that this estimator is stable for (A, C) detectable and (A, G) stabilizable with $\widetilde{Q}, R > 0$.

(e) Discuss what happens to the estimator's stability if Q is not positive semidefinite or (A, Q) is not stabilizable.

Exercise 1.34: Calculating mean and variance from data

We are sampling a real-valued scalar random variable $x(k) \in \mathbb{R}$ at time k. Assume the random variable comes from a distribution with mean $\overline{x}$ and variance P, and the samples at different times are statistically independent.

A colleague has suggested the following formulas for estimating the mean and variance from N samples

$$\hat{x}_N = \frac{1}{N}\sum_{j=1}^{N} x(j) \qquad \hat{P}_N = \frac{1}{N}\sum_{j=1}^{N} (x(j) - \hat{x}_N)^2$$

(a) Prove that the estimate of the mean is unbiased for all N, i.e., show that for all N

$$\mathcal{E}(\hat{x}_N) = \overline{x}$$

(b) Prove that the estimate of the variance is not unbiased for any N, i.e., show that for all N

$$\mathcal{E}(\hat{P}_N) \neq P$$

(c) Using the result above, provide an alternative formula for the variance estimate that is unbiased for all N. How large does N have to be before these two estimates of P are within 1%?

Exercise 1.35: Expected sum of squares

Given that a random variable x has mean m and covariance P, show that the expected sum of squares is given by the formula (Selby, 1973, p.138)

$$\mathcal{E}(x'Qx) = m'Qm + \text{tr}(QP)$$

The trace of a square matrix A, written $\text{tr}(A)$, is defined to be the sum of the diagonal elements

$$\text{tr}(A) = \sum_i A_{ii}$$

Exercise 1.36: Normal distribution

Given a normal distribution with scalar parameters m and σ,

$$p_\xi(x) = \sqrt{\frac{1}{2\pi\sigma^2}} \exp\left[-\frac{1}{2}\left(\frac{x-m}{\sigma}\right)^2\right] \tag{1.62}$$

By direct calculation, show that

(a)

$$\mathcal{E}(\xi) = m$$
$$\text{var}(\xi) = \sigma^2$$

(b) Show that the mean and the maximum likelihood are equal for the normal distribution. Draw a sketch of this result. The maximum likelihood estimate, $\hat{x}$, is defined as

$$\hat{x} = \arg\max_x p_\xi(x)$$

in which arg returns the solution to the optimization problem.

Exercise 1.37: Conditional densities are positive definite

We show in Example A.44 that if ξ and η are jointly normally distributed as

$$\begin{aligned}\begin{bmatrix}\xi \\ \eta\end{bmatrix} &\sim N(m, P) \\ &\sim N\left(\begin{bmatrix} m_x \\ m_y \end{bmatrix}, \begin{bmatrix} P_x & P_{xy} \\ P_{yx} & P_y \end{bmatrix}\right)\end{aligned}$$

then the conditional density of ξ given η is also normal

$$(\xi|\eta) \sim N(m_{x|y}, P_{x|y})$$

in which the conditional mean is

$$m_{x|y} = m_x + P_{xy}P_y^{-1}(y - m_y)$$

and the conditional covariance is

$$P_{x|y} = P_x - P_{xy}P_y^{-1}P_{yx}$$

Given that the joint density is well defined, prove the marginal densities and the conditional densities also are well defined, i.e., given $P > 0$, prove $P_x > 0$, $P_y > 0$, $P_{x|y} > 0$, $P_{y|x} > 0$.

Exercise 1.38: Expectation and covariance under linear transformations

Consider the random variable $x \in \mathbb{R}^n$ with density p_x and mean and covariance

$$\mathcal{E}(x) = m_x \qquad \mathrm{cov}(x) = P_x$$

Consider the random variable $y \in \mathbb{R}^p$ defined by the linear transformation

$$y = Cx$$

(a) Show that the mean and covariance for y are given by

$$\mathcal{E}(y) = Cm_x \qquad \mathrm{cov}(y) = CP_xC'$$

Does this result hold for all C? If yes, prove it; if no, provide a counterexample.

(b) Apply this result to solve Exercise A.34.

Exercise 1.39: Normal distributions under linear transformations

Given the normally distributed random variable, $\xi \in \mathbb{R}^n$, consider the random variable, $\eta \in \mathbb{R}^n$, obtained by the linear transformation

$$\eta = A\xi$$

in which A is a nonsingular matrix. Using the result on transforming probability densities, show that if $\xi \sim N(m, P)$, then $\eta \sim N(Am, APA')$. This result basically says that linear transformations of normal random variables are normal.

Exercise 1.40: More on normals and linear transformations

Consider a normally distributed random variable $x \in \mathbb{R}^n$, $x \sim N(m_x, P_x)$. You showed in Exercise 1.39 for $C \in \mathbb{R}^{n \times n}$ invertible, that the random variable y defined by the linear transformation $y = Cx$ is also normal and is distributed as

$$y \sim N(Cm_x, CP_xC')$$

Does this result hold for all C? If yes, prove it; if no, provide a counterexample.

Exercise 1.41: Signal processing in the good old days — recursive least squares

Imagine we are sent back in time to 1960 and the only computers available have extremely small memories. Say we have a large amount of data coming from a process and we want to compute the least squares estimate of model parameters from these data. Our immediate challenge is that we cannot load all of these data into memory to make the standard least squares calculation.

Alternatively, go 150 years further back in time and consider the situation from Gauss's perspective,

> It occasionally happens that after we have completed all parts of an extended calculation on a sequence of observations, we learn of a new observation that we would like to include. In many cases we will not want to have to redo the entire elimination but instead to find the modifications due to the new observation in the most reliable values of the unknowns and in their weights.
> C.F. Gauss, 1823
> G.W. Stewart Translation, 1995, p. 191.

Given the linear model

$$y_i = X_i'\theta$$

in which scalar y_i is the measurement at sample i, X_i' is the independent model variable (row vector, $1 \times p$) at sample i, and θ is the parameter vector ($p \times 1$) to be estimated from these data. Given the weighted least squares objective and n measurements, we wish to compute the usual estimate

$$\hat{\theta} = (X'X)^{-1}X'y \tag{1.63}$$

in which

$$y = \begin{bmatrix} y_1 \\ \vdots \\ y_n \end{bmatrix} \qquad X = \begin{bmatrix} X_1' \\ \vdots \\ X_n' \end{bmatrix}$$

We do not wish to store the large matrices $X(n \times p)$ and $y(n \times 1)$ required for this calculation. Because we are planning to process the data one at a time, we first modify our usual least squares problem to deal with small n. For example, we wish to estimate the parameters when $n < p$ and the inverse in (1.63) does not exist. In such cases, we may choose to regularize the problem by modifying the objective function as follows

$$\Phi(\theta) = (\theta - \overline{\theta})'P_0^{-1}(\theta - \overline{\theta}) + \sum_{i=1}^{n}(y_i - X_i'\theta)^2$$

in which $\overline{\theta}$ and P_0 are chosen by the user. In Bayesian estimation, we call $\overline{\theta}$ and P_0 the prior information, and often assume that the prior density of θ (without measurements) is normal

$$\theta \sim N(\overline{\theta}, P_0)$$

The solution to this modified least squares estimation problem is

$$\hat{\theta} = \overline{\theta} + (X'X + P_0)^{-1}X'(y - X\overline{\theta}) \tag{1.64}$$

Devise a means to *recursively* estimate θ so that:

1. We never store more than one measurement at a time in memory.
2. After processing all the measurements, we obtain the same least squares estimate given in (1.64).

Exercise 1.42: Least squares parameter estimation and Bayesian estimation

Consider a model linear in the parameters

$$y = X\theta + e \tag{1.65}$$

in which $y \in \mathbb{R}^p$ is a vector of measurements, $\theta \in \mathbb{R}^m$ is a vector of parameters, $X \in \mathbb{R}^{p \times m}$ is a matrix of known constants, and $e \in \mathbb{R}^p$ is a random variable modeling the measurement error. The standard parameter estimation problem is to find the best estimate of θ given the measurements y corrupted with measurement error e, which we assume is distributed as

$$e \sim N(0, R)$$

(a) Consider the case in which the errors in the measurements are independently and identically distributed with variance σ^2, $R = \sigma^2 I$. For this case, the classic least squares problem and solution are

$$\min_{\theta} |y - X\theta|^2 \qquad \hat{\theta} = (X'X)^{-1}X'y$$

Consider the measurements to be sampled from (1.65) with true parameter value θ_0. Show that using the least squares formula, the parameter estimate is distributed as

$$\hat{\theta} \sim N(\theta_0, P_{\hat{\theta}}) \qquad P_{\hat{\theta}} = \sigma^2 \left(X'X\right)^{-1}$$

(b) Now consider again the model of (1.65) and a Bayesian estimation problem. Assume a prior distribution for the random variable θ

$$\theta \sim N(\overline{\theta}, \overline{P})$$

Compute the conditional density of θ given measurement y, show that this density is normal, and find its mean and covariance

$$p_{\theta|y}(\theta|y) = n(\theta, m, P)$$

Show that Bayesian estimation and least squares estimation give the same result in the limit of a noninformative prior. In other words, if the covariance of the prior is large compared to the covariance of the measurement error, show that

$$m \approx (X'X)^{-1}X'y \qquad P \approx P_{\hat{\theta}}$$

(c) What (weighted) least squares minimization problem is solved for the general measurement error covariance

$$e \sim (0, R)$$

Derive the least squares estimate formula for this case.

(d) Again consider the measurements to be sampled from (1.65) with true parameter value θ_0. Show that the weighted least squares formula gives parameter estimates that are distributed as

$$\hat{\theta} \sim N(\theta_0, P_{\hat{\theta}})$$

and find $P_{\hat{\theta}}$ for this case.

(e) Show again that Bayesian estimation and least squares estimation give the same result in the limit of a noninformative prior.

Exercise 1.43: Least squares and minimum variance estimation

Consider again the model linear in the parameters and the least squares estimator from Exercise 1.42

$$y = X\theta + e \qquad e \sim N(0, R)$$
$$\hat{\theta} = \left(X'R^{-1}X\right)^{-1} X'R^{-1}y$$

Show that the covariance of the least squares estimator is the smallest covariance of all linear unbiased estimators.

Exercise 1.44: Two stages are not better than one

We often can decompose an estimation problem into stages. Consider the following case in which we wish to estimate x from measurements of z, but we have the model between x and an intermediate variable, y, and the model between y and z

$$y = Ax + e_1 \qquad \text{cov}(e_1) = Q_1$$
$$z = By + e_2 \qquad \text{cov}(e_2) = Q_2$$

(a) Write down the optimal least squares problem to solve for $\hat{y}$ given the z measurements and the second model. Given $\hat{y}$, write down the optimal least squares problem for $\hat{x}$ in terms of $\hat{y}$. Combine these two results together and write the resulting estimate of $\hat{x}$ given measurements of z. Call this the two-stage estimate of x.

(b) Combine the two models together into a single model and show that the relationship between z and x is

$$z = BAx + e_3 \qquad \text{cov}(e_3) = Q_3$$

Express Q_3 in terms of Q_1, Q_2 and the models A, B. What is the optimal least squares estimate of $\hat{x}$ given measurements of z and the one-stage model? Call this the one-stage estimate of x.

(c) Are the one-stage and two-stage estimates of x the same? If yes, prove it. If no, provide a counterexample. Do you have to make any assumptions about the models A, B?

Exercise 1.45: Time-varying Kalman filter

Derive formulas for the conditional densities of $x(k)|\mathbf{y}(k-1)$ and $x(k)|\mathbf{y}(k)$ for the time-varying linear system

$$\begin{aligned} x(k+1) &= A(k)x(k) + G(k)w(k) \\ y(k) &= C(k)x(k) + v(k) \end{aligned}$$

in which the initial state, state noise and measurement noise are independently distributed as

$$x(0) \sim N(\overline{x}_0, Q_0) \qquad w(k) \sim N(0, Q) \qquad v(k) \sim N(0, R)$$

Exercise 1.46: More on conditional densities

In deriving the discrete time Kalman filter, we have $p_{x|\mathbf{y}}(x(k)|\mathbf{y}(k))$ and we wish to calculate recursively $p_{x|\mathbf{y}}(x(k+1)|\mathbf{y}(k+1))$ after we collect the output measurement at time $k+1$. It is straightforward to calculate $p_{x,y|\mathbf{y}}(x(k+1), y(k+1)|\mathbf{y}(k))$ from our established results on normal densities and knowledge of $p_{x|\mathbf{y}}(x(k)|\mathbf{y}(k))$, but we still need to establish a formula for pushing the $y(k+1)$ to the other side of the conditional density bar. Consider the following statement as a possible lemma to aid in this operation.

$$p_{a|b,c}(a|b,c) = \frac{p_{a,b|c}(a,b|c)}{p_{b|c}(b|c)}$$

If this statement is true, prove it. If it is false, give a counterexample.

Exercise 1.47: Other useful conditional densities

Using the definitions of marginal and conditional density, establish the following useful conditional density relations

1. $p_{A|B}(a|b) = \int p_{A|B,C}(a|b,c)p_{C|B}(c|b)dc$

2. $p_{A|B,C}(a|b,c) = p_{C|A,B}(c|a,b)\dfrac{p_{A|B}(a|b)}{p_{C|B}(c|b)}$

Exercise 1.48: Optimal filtering and deterministic least squares

Given the data sequence $\{y(0), \ldots, y(k)\}$ and the system model

$$\begin{aligned} x^+ &= Ax + w \\ y &= Cx + v \end{aligned}$$

(a) Write down a least squares problem whose solution would provide a good state estimate for $x(k)$ in this situation. What probabilistic interpretation can you assign to the estimate calculated from this least squares problem?

(b) Now consider the nonlinear model

$$\begin{aligned} x^+ &= f(x) + w \\ y &= g(x) + v \end{aligned}$$

What is the corresponding nonlinear least squares problem for estimating $x(k)$ in this situation? What probabilistic interpretation, if any, can you assign to this estimate in the nonlinear model context?

(c) What is the motivation for changing from these least squares estimators to the moving horizon estimators we discussed in the chapter?

Exercise 1.49: A nonlinear transformation and conditional density

Consider the following relationship between the random variable y, and x and v

$$y = f(x) + v$$

The author of a famous textbook wants us to believe that

$$p_{y|x}(y|x) = p_v(y - f(x))$$

Derive this result and state what additional assumptions on the random variables x and v are required for this result to be correct.

Exercise 1.50: Some smoothing

One of the problems with asking you to derive the Kalman filter is that the derivation is in so many textbooks that it is difficult to tell if you are thinking independently. So here's a variation on the theme that should help you evaluate your level of understanding of these ideas. Let's calculate a smoothed rather than filtered estimate and covariance. Here's the problem.

We have the usual setup with a prior on $x(0)$

$$x(0) \sim N(\overline{x}(0), Q_0)$$

and we receive data from the following system

$$\begin{aligned} x(k+1) &= Ax(k) + w(k) \\ y(k) &= Cx(k) + v(k) \end{aligned}$$

in which the random variables $w(k)$ and $v(k)$ are independent, identically distributed normals, $w(k) \sim N(0, Q)$, $v(k) \sim N(0, R)$.

(a) Calculate the standard density for the filtering problem, $p_{x(0),y(0)}(x(0)|y(0))$.

(b) Now calculate the density for the smoothing problem

$$p_{x(0)|y(0),y(1)}(x(0)|y(0),y(1))$$

that is, *not* the usual $p_{x(1)|y(0),y(1)}(x(1)|y(0),y(1))$.

Exercise 1.51: Alive on arrival

The following two optimization problems are helpful in understanding the arrival cost decomposition in state estimation.

(a) Let $V(x,y,z)$ be a positive, strictly convex function consisting of the sum of two functions, one of which depends on both x and y, and the other of which depends on y and z

$$V(x,y,z) = g(x,y) + h(y,z) \qquad V : \mathbb{R}^m \times \mathbb{R}^n \times \mathbb{R}^p \to \mathbb{R}_{\geq 0}$$

Consider the optimization problem

$$P1 : \min_{x,y,z} V(x,y,z)$$

The arrival cost decomposes this three-variable optimization problem into two, smaller dimensional optimization problems. Define the "arrival cost" $\widetilde{g}$ for this problem as the solution to the following single-variable optimization problem

$$\widetilde{g}(y) = \min_x g(x,y)$$

and define optimization problem $P2$ as follows

$$P2 : \min_{y,z} \widetilde{g}(y) + h(y,z)$$

Let (x',y',z') denote the solution to $P1$ and (x^0,y^0,z^0) denote the solution to $P2$, in which

$$x^0 = \arg\min_x g(x,y^0)$$

Prove that the two solutions are equal

$$(x',y',z') = (x^0,y^0,z^0)$$

(b) Repeat the previous part for the following optimization problems

$$V(x,y,z) = g(x) + h(y,z)$$

Here the y variables do not appear in g but restrict the x variables through a linear constraint. The two optimization problems are:

$$P1 : \min_{x,y,z} V(x,y,z) \qquad \text{subject to } Ex = y$$

$$P2 : \min_{y,z} \widetilde{g}(y) + h(y,z)$$

in which

$$\widetilde{g}(y) = \min_x g(x) \qquad \text{subject to } Ex = y$$

Exercise 1.52: On-time arrival

Consider the deterministic, full information state estimation optimization problem

$$\min_{x(0),\mathbf{w},\mathbf{v}} \frac{1}{2}\left(|x(0) - \overline{x}(0)|^2_{(P^-(0))^{-1}} + \sum_{i=0}^{T-1} |w(i)|^2_{Q^{-1}} + |v(i)|^2_{R^{-1}} \right) \tag{1.66}$$

subject to

$$\begin{aligned} x^+ &= Ax + w \\ y &= Cx + v \end{aligned} \tag{1.67}$$

in which the sequence of measurements $\mathbf{y}(T)$ are known values. Notice we assume the noise-shaping matrix, G, is an identity matrix here. Sec Exercise 1.53 for the general case. Using the result of the first part of Exercise 1.51, show that this problem is equivalent to the following problem

$$\min_{x(T-N),\mathbf{w},\mathbf{v}} V^-_{T-N}(x(T-N)) + \frac{1}{2} \sum_{i=T-N}^{T-1} |w(i)|^2_{Q^{-1}} + |v(i)|^2_{R^{-1}}$$

subject to (1.67). The arrival cost is defined as

$$V^-_N(a) = \min_{x(0),\mathbf{w},\mathbf{v}} \frac{1}{2}\left(|x(0) - \overline{x}(0)|^2_{(P^-(0))^{-1}} + \sum_{i=0}^{N-1} |w(i)|^2_{Q^{-1}} + |v(i)|^2_{R^{-1}} \right)$$

subject to (1.67) and $x(N) = a$. Notice that any value of N, $0 \leq N \leq T$, can be used to split the cost function using the arrival cost.

Exercise 1.53: Arrival cost with noise-shaping matrix G

Consider the deterministic, full information state estimation optimization problem

$$\min_{x(0),\mathbf{w},\mathbf{v}} \frac{1}{2}\left(|x(0) - \overline{x}(0)|^2_{(P^-(0))^{-1}} + \sum_{i=0}^{T-1} |w(i)|^2_{Q^{-1}} + |v(i)|^2_{R^{-1}} \right)$$

subject to

$$\begin{aligned} x^+ &= Ax + Gw \\ y &= Cx + v \end{aligned} \tag{1.68}$$

in which the sequence of measurements $\mathbf{y}$ are known values. Using the result of the second part of Exercise 1.51, show that this problem also is equivalent to the following problem

$$\min_{x(T-N),\mathbf{w},\mathbf{v}} V^-_{T-N}(x(T-N)) + \frac{1}{2}\left(\sum_{i=T-N}^{T-1} |w(i)|^2_{Q^{-1}} + |v(i)|^2_{R^{-1}} \right)$$

subject to (1.68). The arrival cost is defined as

$$V^-_N(a) = \min_{x(0),\mathbf{w},\mathbf{v}} \frac{1}{2}\left(|x(0) - \overline{x}(0)|^2_{(P^-(0))^{-1}} + \sum_{i=0}^{N-1} |w(i)|^2_{Q^{-1}} + |v(i)|^2_{R^{-1}} \right)$$

subject to (1.68) and $x(N) = a$. Notice that any value of N, $0 \leq N \leq T$, can be used to split the cost function using the arrival cost.

Exercise 1.54: Where is the steady state?

Consider the two-input, two-output system

$$A = \begin{bmatrix} 0.5 & 0 & 0 & 0 \\ 0 & 0.6 & 0 & 0 \\ 0 & 0 & 0.5 & 0 \\ 0 & 0 & 0 & 0.6 \end{bmatrix} \qquad B = \begin{bmatrix} 0.5 & 0 \\ 0 & 0.4 \\ 0.25 & 0 \\ 0 & 0.6 \end{bmatrix} \qquad C = \begin{bmatrix} 1 & 1 & 0 & 0 \\ 0 & 0 & 1 & 1 \end{bmatrix}$$

(a) The output setpoint is $y_{sp} = \begin{bmatrix} 1 & -1 \end{bmatrix}'$ and the input setpoint is $u_{sp} = \begin{bmatrix} 0 & 0 \end{bmatrix}'$. Calculate the target triple (x_s, u_s, y_s). Is the output setpoint feasible, i.e., does $y_s = y_{sp}$?

(b) Assume only the first input $u(1)$ is available for control. Is the output setpoint feasible? What is the target in this case using $Q_s = I$?

(c) Assume both inputs are available for control but only the first output has a setpoint, $y_{1t} = 1$. What is the solution to the target problem for $R_s = I$?

Exercise 1.55: Detectability of integrating disturbance models

(a) Prove Lemma 1.8; the augmented system is detectable if and only if the system (A, C) is detectable and

$$\text{rank} \begin{bmatrix} I - A & -B_d \\ C & C_d \end{bmatrix} = n + n_d$$

(b) Prove Corollary 1.9; the augmented system is detectable only if $n_d \leq p$.

Exercise 1.56: Unconstrained tracking problem

(a) For an *unconstrained* system, show that the following condition is *sufficient* for feasibility of the target problem for any z_{sp}.

$$\text{rank} \begin{bmatrix} I - A & -B \\ HC & 0 \end{bmatrix} = n + n_c \tag{1.69}$$

(b) Show that (1.69) implies that the number of controlled variables without offset is less than or equal to the number of manipulated variables and the number of measurements, $n_c \leq m$ and $n_c \leq p$.

(c) Show that (1.69) implies the rows of H are independent.

(d) Does (1.69) imply that the rows of C are independent? If so, prove it; if not, provide a counterexample.

(e) By choosing H, how can one satisfy (1.69) if one has installed redundant sensors so several rows of C are identical?

Exercise 1.57: Unconstrained tracking problem for stabilizable systems

If we restrict attention to stabilizable systems, the sufficient condition of Exercise 1.56 becomes a necessary and sufficient condition. Prove the following lemma.

Lemma 1.14 (Stabilizable systems and feasible targets). *Consider an unconstrained, stabilizable system (A, B). The target is feasible for any z_{sp} if and only if*

$$\text{rank}\begin{bmatrix} I - A & -B \\ HC & 0 \end{bmatrix} = n + n_c$$

Exercise 1.58: Existence and uniqueness of the unconstrained target

Assume a system having p controlled variables $z = Hx$, with setpoints z_{sp}, and m manipulated variables u, with setpoints u_{sp}. Consider the steady-state target problem

$$\min_{x,u}(1/2)(u - u_{sp})'R(u - u_{sp}) \qquad R > 0$$

subject to

$$\begin{bmatrix} I - A & -B \\ H & 0 \end{bmatrix}\begin{bmatrix} x \\ u \end{bmatrix} = \begin{bmatrix} 0 \\ z_{sp} \end{bmatrix}$$

Show that the steady-state solution (x, u) exists for any (z_{sp}, u_{sp}) and is unique if

$$\text{rank}\begin{bmatrix} I - A & -B \\ H & 0 \end{bmatrix} = n + p \qquad \text{rank}\begin{bmatrix} I - A \\ H \end{bmatrix} = n$$

Exercise 1.59: Choose a sample time

Consider the unstable continuous time system

$$\frac{dx}{dt} = Ax + Bu \qquad y = Cx$$

in which

$$A = \begin{bmatrix} -0.281 & 0.935 & 0.035 & 0.008 \\ 0.047 & -0.116 & 0.053 & 0.383 \\ 0.679 & 0.519 & 0.030 & 0.067 \\ 0.679 & 0.831 & 0.671 & -0.083 \end{bmatrix} \qquad B = \begin{bmatrix} 0.687 \\ 0.589 \\ 0.930 \\ 0.846 \end{bmatrix} \qquad C = I$$

Consider regulator tuning parameters and constraints

$$Q = \text{diag}(1, 2, 1, 2) \qquad R = 1 \qquad N = 10 \qquad |x| \leq \begin{bmatrix} 1 \\ 2 \\ 1 \\ 3 \end{bmatrix}$$

(a) Compute the eigenvalues of A. Choose a sample time of $\Delta = 0.04$ and simulate the MPC regulator response given $x(0) = \begin{bmatrix} -0.9 & -1.8 & 0.7 & 2 \end{bmatrix}'$ until $t = 20$. Use an ODE solver to simulate the continuous time plant response. Plot all states and the input versus time.

Now add an input disturbance to the regulator so the control applied to the plant is u_d instead of u in which

$$u_d(k) = (1 + 0.1w_1)u(k) + 0.1w_2$$

and w_1 and w_2 are zero mean, normally distributed random variables with unit variance. Simulate the regulator's performance given this disturbance. Plot all states and $u_d(k)$ versus time.

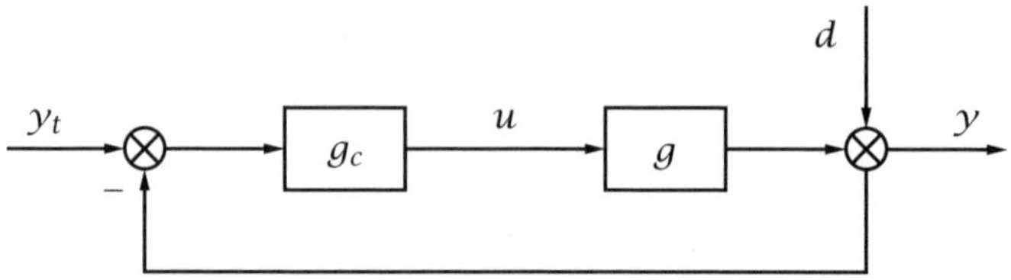

Figure 1.13: Feedback control system with output disturbance d, and setpoint y_{sp}.

(b) Repeat the simulations with and without disturbance for $\Delta = 0.4$ and $\Delta = 2$.

(c) Compare the simulations for the different sample times. What happens if the sample time is too large? Choose an appropriate sample time for this system and justify your choice.

Exercise 1.60: Disturbance models and offset

Consider the following two-input, three-output plant discussed in Example 1.11

$$x^+ = Ax + Bu + B_p p$$
$$y = Cx$$

in which

$$A = \begin{bmatrix} 0.2681 & -0.00338 & -0.00728 \\ 9.703 & 0.3279 & -25.44 \\ 0 & 0 & 1 \end{bmatrix} \qquad C = \begin{bmatrix} 1 & 0 & 0 \\ 0 & 1 & 0 \\ 0 & 0 & 1 \end{bmatrix}$$

$$B = \begin{bmatrix} -0.00537 & 0.1655 \\ 1.297 & 97.91 \\ 0 & -6.637 \end{bmatrix} \qquad B_p = \begin{bmatrix} -0.1175 \\ 69.74 \\ 6.637 \end{bmatrix}$$

The input disturbance p results from a reactor inlet flowrate disturbance.

(a) Since there are two inputs, choose two outputs in which to remove steady-state offset. Build an output disturbance model with two integrators. Is your augmented model detectable?

(b) Implement your controller using $p = 0.01$ as a step disturbance at $k = 0$. Do you remove offset in your chosen outputs? Do you remove offset in any outputs?

(c) Can you find any two-integrator disturbance model that removes offset in two outputs? If so, which disturbance model do you use? If not, why not?

Exercise 1.61: MPC, PID and time delay

Consider the following first-order system with time delay shown in Figure 1.13

$$g(s) = \frac{k}{\tau s + 1} e^{-\theta s}, \qquad k = 1, \tau = 1, \theta = 5$$

Consider a unit step change in setpoint y_{sp}, at $t = 0$.

(a) Choose a reasonable sample time, Δ, and disturbance model, and simulate an offset-free discrete time MPC controller for this setpoint change. List all of your chosen parameters.

(b) Choose PID tuning parameters to achieve "good performance" for this system. List your PID tuning parameters. Compare the performances of the two controllers.

Bibliography

R. E. Bellman. *Dynamic Programming.* Princeton University Press, Princeton, New Jersey, 1957.

R. E. Bellman and S. E. Dreyfus. *Applied Dynamic Programming.* Princeton University Press, Princeton, New Jersey, 1962.

D. P. Bertsekas. *Dynamic Programming.* Prentice-Hall, Inc., Englewood Cliffs, New Jersey, 1987.

E. F. Camacho and C. Bordons. *Model Predictive Control.* Springer-Verlag, London, second edition, 2004.

E. J. Davison and H. W. Smith. Pole assignment in linear time-invariant multivariable systems with constant disturbances. *Automatica*, 7:489–498, 1971.

E. J. Davison and H. W. Smith. A note on the design of industrial regulators: Integral feedback and feedforward controllers. *Automatica*, 10:329–332, 1974.

R. Fletcher. *Practical Methods of Optimization.* John Wiley & Sons, New York, 1987.

B. A. Francis and W. M. Wonham. The internal model principle of control theory. *Automatica*, 12:457–465, 1976.

G. C. Goodwin and K. S. Sin. *Adaptive Filtering Prediction and Control.* Prentice-Hall, Englewood Cliffs, New Jersey, 1984.

G. C. Goodwin, M. M. Seron, and J. A. De Doná. *Constrained control and estimation: an optimization approach.* Springer, New York, 2005.

M. L. J. Hautus. Controllability and stabilizability of sampled systems. *IEEE Trans. Auto. Cont.*, 17(4):528–531, August 1972.

R. A. Horn and C. R. Johnson. *Matrix Analysis.* Cambridge University Press, 1985.

A. H. Jazwinski. *Stochastic Processes and Filtering Theory.* Academic Press, New York, 1970.

R. E. Kalman. A new approach to linear filtering and prediction problems. *Trans. ASME, J. Basic Engineering*, pages 35–45, March 1960a.

R. E. Kalman. Contributions to the theory of optimal control. *Bull. Soc. Math. Mex.*, 5:102–119, 1960b.

H. Kwakernaak and R. Sivan. *Linear Optimal Control Systems.* John Wiley and Sons, New York, 1972.

W. H. Kwon. *Receding horizon control: model predictive control for state models.* Springer-Verlag, London, 2005.

J. M. Maciejowski. *Predictive Control with Contraints.* Prentice-Hall, Harlow, UK, 2002.

J. Nocedal and S. J. Wright. *Numerical Optimization.* Springer-Verlag, New York, 1999.

B. J. Odelson, M. R. Rajamani, and J. B. Rawlings. A new autocovariance least-squares method for estimating noise covariances. *Automatica*, 42(2):303–308, February 2006.

G. Pannocchia and J. B. Rawlings. Disturbance models for offset-free MPC control. *AIChE J.*, 49(2):426–437, 2003.

L. Qiu and E. J. Davison. Performance limitations of non-minimum phase systems in the servomechanism problem. *Automatica*, 29(2):337–349, 1993.

C. V. Rao and J. B. Rawlings. Steady states and constraints in model predictive control. *AIChE J.*, 45(6):1266–1278, 1999.

J. A. Rossiter. *Model-based predictive control: a practical approach.* CRC Press LLC, Boca Raton, FL, 2004.

S. M. Selby. *CRC Standard Mathematical Tables.* twenty-first edition, 1973.

E. D. Sontag. *Mathematical Control Theory.* Springer-Verlag, New York, second edition, 1998.

G. Strang. *Linear Algebra and its Applications.* Academic Press, New York, second edition, 1980.

L. Wang. *Model Predictive Control System Design and Implementation Using Matlab.* Springer, New York, 2009.

2

Model Predictive Control — Regulation

2.1 Introduction

In Chapter 1 we investigated a special, but useful, form of model predictive control (MPC); an important feature of this form of MPC is that there exists a set of initial states for which it is actually optimal for an *infinite horizon* optimal control problem and therefore inherits the associated advantages. Just as there are many methods other than infinite horizon linear quadratic control for stabilizing linear systems, however, there are alternative forms of MPC that can stabilize linear and even nonlinear systems. We explore these alternatives in the remainder of this chapter. But first we place MPC in a more general setting to facilitate comparison with other control methods.

MPC is, as we have seen earlier, a form of control in which the control action is obtained by solving *online*, at each sampling instant, a *finite horizon* optimal control problem in which the initial state is the current state of the plant. Optimization yields a finite control sequence, and the first control action in this sequence is applied to the plant. MPC differs, therefore, from conventional control in which the control law is precomputed offline. But this is not an essential difference; MPC implicitly implements a control law that can, in principle, be computed offline as we shall soon see. Specifically, if the current state of the system being controlled is x, MPC obtains, by solving an open-loop optimal control problem for this initial state, a specific control action u to apply to the plant.

Dynamic programming (DP) may be used to solve a feedback version of the same optimal control problem, however, yielding a receding horizon control *law* $\kappa(\cdot)$. The important fact is that if x is the current state, the optimal control u obtained by MPC (by solving an open-loop optimal control problem) satisfies $u = \kappa(x)$. For example, MPC computes the *value* $\kappa(x)$ of the optimal receding horizon control law for the current state x, while DP yields the control *law* $\kappa(\cdot)$ that can be used for *any* state. DP would appear to be preferable since it provides a control law that can be implemented simply (as a look-up table). Obtaining a DP solution is difficult, if not impossible, however, for most optimal control problems if the state dimension is reasonably high — unless the system is linear, the cost quadratic and there are no control or state constraints. The great advantage of MPC is that open-loop optimal control problems often can be solved rapidly enough, using standard mathematical programming algorithms, to permit the use of MPC even though the system being controlled is nonlinear, and hard constraints on states and controls must be satisfied. Thus MPC permits the application of a DP solution, even though explicit determination of the optimal control law is intractable. MPC is an effective *implementation* of the DP solution and not a new method of control.

In this chapter we study MPC for the case when the state is known. This case is particularly important, even though it rarely arises in practice, because important properties, such as stability and performance, may be relatively easily established. The relative simplicity of this case arises from the fact that if the state is known and if there are no disturbances or model error, the problem is *deterministic*, i.e., there is no uncertainty making feedback unnecessary in principle. As we pointed out previously, for deterministic systems the MPC action for a given state is identical to the receding horizon control law, determined using DP, and evaluated at the given state. When the state is *not* known, it has to be estimated and state estimation error, together with model error and disturbances, makes the system uncertain in that future trajectories cannot be precisely predicted. The simple connection between MPC and the DP solution is lost because there does not exist an open-loop optimal control problem whose solution yields a control action that is the same as that obtained by the DP solution. A practical consequence is that special techniques are required to ensure robustness against these various forms of uncertainty. So the results of this chapter hold when there is no uncertainty. We prove, in particular, that the optimal control problem that defines the model predictive control can always be solved if the initial optimal control problem can be solved, and that the optimal cost can always be reduced allowing us to prove asymptotic or exponential stability of the target state. We refer to stability in the absence of uncertainty as *nominal stability*.

When uncertainty is present, however, neither of these two assertions is necessarily true; uncertainty may cause the state to wander outside the region where the optimal control problem can be solved and may lead to instability. Procedures for overcoming the problems arising from uncertainty are presented in Chapters 3 and 5. In most of the control algorithms presented in this chapter, the decrease in the optimal cost, on which the proof of stability is founded, is based on the assumption that the next state is exactly as predicted and that the global solution to the optimal control problem can be computed. In the suboptimal control algorithm in Section 2.8, where global optimality is not required, the decrease in the optimal cost is still based on the assumption that the current state is exactly the state as predicted one step back in time.

2.2 Model Predictive Control

As discussed briefly in Chapter 1, most nonlinear system descriptions derived from physical arguments are continuous time descriptions in the form of nonlinear differential equations

$$\frac{dx}{dt} = f(x, u)$$

For this class of systems, the control law with arguably the best closed-loop properties is the solution to the following infinite horizon, constrained optimal control problem. The cost is defined to be

$$V_\infty(x, u(\cdot)) = \int_0^\infty \ell(x(t), u(t))dt$$

in which $x(t)$ and $u(t)$ satisfy $\dot{x} = f(x, u)$. The optimal control problem $\mathbb{P}(x)$ is defined by

$$\min_{u(\cdot)} V_\infty(x, u(\cdot))$$

subject to:

$$\dot{x} = f(x, u) \qquad x(0) = x_0$$
$$u(t) \in \mathbb{U} \qquad x(t) \in \mathbb{X} \qquad \text{for all } t \in (0, \infty)$$

If $\ell(\cdot)$ is positive definite, the goal of the regulator is to steer the state of the system to the origin.

We denote the solution to this problem (when it exists) and the optimal value function by

$$V_\infty^0(x) \qquad u_\infty^0(\cdot; x)$$

The closed-loop system under this optimal control law evolves as

$$\frac{dx(t)}{dt} = f(x(t), u_\infty^0(t; x))$$

We can demonstrate that the origin is an asymptotically stable solution for the closed-loop system as follows. If $f(\cdot)$ and $\ell(\cdot)$ satisfy certain differentiability and growth assumptions and there are no state constraints, then a solution to $\mathbb{P}(x)$ exists for all x; $V_\infty^0(\cdot)$ is differentiable and satisfies

$$\dot{V}_\infty^0(x) = -\ell(x, u_\infty^0(0; x))$$

Using this and upper and lower bounds on $V_\infty^0(\cdot)$ enables global asymptotic stability of the origin to be established.

Although the control law $u_\infty^0(0; \cdot)$ provides excellent closed-loop properties, there are several impediments to its use. A feedback, rather than an open-loop, control is usually necessary because of uncertainty. Solution of the optimal control problem $\mathbb{P}(x)$ yields the optimal control $u_\infty^0(0; x)$ for the state x but does not provide a control law. Dynamic programming may, in principle, be employed, but is generally impractical if the state dimension and the horizon are not small.

If we turn instead to an MPC approach in which we generate online only the value of $u_\infty^0(\cdot; x)$ for the currently measured value of x, rather than for all x, the problem remains formidable for the following reasons. First, we are optimizing over a time *function*, $u(\cdot)$, and functions are infinite dimensional. Secondly, the time interval of interest, $[0, \infty)$, is a semi-infinite interval, which poses other numerical challenges. Finally, the cost function $V(x, u(\cdot))$ is usually not a convex function of $u(\cdot)$, which presents significant optimization difficulties, especially in an online setting. Even proving existence of the optimal control in this general setting is a challenge.

Our task in this chapter may therefore be viewed as restricting the system and control parameterization to replace problem $\mathbb{P}(x)$ with a more easily computed approximate problem. We show how to pose various approximate problems for which we can establish existence of the optimal solution and asymptotic closed-loop stability of the resulting controller. For these approximate problems, we almost always replace the continuous time differential equation with a discrete time difference equation. We often replace the semi-infinite time interval with a finite time interval and append a terminal region such that we can approximate the cost to go for the semi-infinite interval once the system enters the terminal region. Although the solution of problem

$\mathbb{P}(x)$ in its full generality is out of reach with today's computational methods, its value lies in distinguishing what is *desirable* in the control problem formulation and what is *achievable* with available computing technology.

We develop here MPC for the control of constrained nonlinear time-invariant systems. The nonlinear system is described by the nonlinear difference equation

$$x^+ = f(x, u) \tag{2.1}$$

in which $x \in \mathbb{R}^n$ is the current state, u is the current control, and x^+ the successor state; $x^+ = f(x, u)$ is the discrete time analog of the continuous time differential equation $\dot{x} = f(x, u)$. The function $f(\cdot)$ is assumed to be continuous and to satisfy $f(0, 0) = 0$, i.e., 0 is an equilibrium point. Any solution $x(\cdot)$ of (2.1), if the initial state is $x(0) = x_0$ and the input (control) is $u(\cdot)$, satisfies

$$x(k+1) = f(x(k), u(k)) \qquad k = 0, 1, \ldots$$

and the initial condition $x(0) = x_0$.

We introduce here some notation that we employ in the sequel. The set $\mathbb{I}$ denotes the set of integers, $\mathbb{I}_{\geq 0} := \{0, 1, 2, \ldots\}$ and, for any two integers m and n satisfying $m \leq n$, $\mathbb{I}_{m:n} := \{m, m+1, \ldots, n\}$. We refer to the pair (x, i) as an event; an event (x, i) denotes that the state at time i is x. We use $\mathbf{u}$ to denote the possibly infinite control sequence $\{u(k) \mid k \in \mathbb{I}_{\geq 0}\} = \{u(0), u(1), u(2), \ldots\}$. In the context of MPC, $\mathbf{u}$ frequently denotes the finite sequence $\{u(0), u(1), \ldots, u(N-1)\}$ in which N is the control *horizon*. For any integer $j \in \mathbb{I}_{\geq 0}$, we employ $\mathbf{u}_j$ to denote the finite sequence $\{u(0), u(1), \ldots, u(j-1)\}$. Similarly $\mathbf{x}$ denotes the possibly infinite state sequence $\{x(0), x(1), x(2), \ldots\}$ and $\mathbf{x}_j$ the finite sequence $\{x(0), x(1), \ldots, x(j)\}$. When no confusion can arise we often employ, for simplicity in notation, $\mathbf{u}$ in place of $\mathbf{u}_N$ and $\mathbf{x}$ in place of $\mathbf{x}_N$. Also for simplicity in notation, $\mathbf{u}$, when used in algebraic expressions, denotes the column vector $(u(0)', u(1)', \ldots, u(N-1)')'$; similarly $\mathbf{x}$ in algebraic expressions denotes the column vector $(x(0)', x(1)', \ldots, x(N)')'$.

The solution of (2.1) at time k, if the initial state at time 0 is x and the control sequence is $\mathbf{u}$, is denoted by $\phi(k; x, \mathbf{u})$; the solution at time k depends only on $u(0), u(1), \ldots, u(k-1)$. Similarly, the solution of the system (2.1) at time k, if the initial state at time i is x and the control sequence is $\mathbf{u}$, is denoted by $\phi(k; (x, i), \mathbf{u})$. Because the system is time invariant, the solution does not depend on the initial time; if the initial state is x at time i, the solution at time $j \geq i$ is $\phi(j-i; x, \mathbf{u})$.

Thus the solution at time k if the initial event is (x, i) is identical to the solution at time $k - i$ if the initial event is $(x, 0)$. For each k, the function $(x, \mathbf{u}) \mapsto \phi(k; x, \mathbf{u})$ is continuous as we show next.

Proposition 2.1 (Continuous system solution). *Suppose the function $f(\cdot)$ is continuous. Then, for each integer $k \in \mathbb{I}$, the function $(x, \mathbf{u}) \mapsto \phi(k; x, \mathbf{u})$ is continuous.*

Proof.
Since $\phi(1; x, u(0)) = f(x, u(0))$, the function $(x, u(0)) \mapsto \phi(1; x, u(0))$ is continuous. Suppose the function $(x, \mathbf{u}_{j-1}) \mapsto \phi(j; x, \mathbf{u}_{j-1})$ is continuous and consider the function $(x, \mathbf{u}_j) \mapsto \phi(j+1; x, \mathbf{u}_j)$. Since

$$\phi(j+1; x, \mathbf{u}_j) = f(\phi(j; x, \mathbf{u}_{j-1}), u(j))$$

where $f(\cdot)$ and $\phi(j; \cdot)$ are continuous and since $\phi(j+1; \cdot)$ is the composition of two continuous functions $f(\cdot)$ and $\phi(j; \cdot)$, it follows that $\phi(j+1; \cdot)$ is continuous. By induction $\phi(k; \cdot)$ is continuous for any positive integer k. ■

The system (2.1) is subject to hard constraints which may take the form

$$u(k) \in \mathbb{U} \quad x(k) \in \mathbb{X} \qquad \text{for all } k \in \mathbb{I}_{\geq 0} \tag{2.2}$$

The constraint (2.2) does not couple $u(k)$ or $x(k)$ at *different* times; constraints that involve the control at several times are avoided by introducing extra states. Thus the common rate constraint $|u(k) - u(k-1)| \leq c$ may be expressed as $|u(k) - z(k)| \leq c$ where z is an extra state variable satisfying the difference equation $z^+ = u$ so that $z(k) = u(k-1)$. The constraint $|u - z| \leq c$ is an example of a *mixed* constraint, i.e., a constraint that involves both states and controls. Hence, a more general constraint formulation of the form

$$y(k) \in \mathbb{Y} \qquad \text{for all } k \in \mathbb{I}_{\geq 0} \tag{2.3}$$

in which the output y satisfies

$$y = h(x, u)$$

is sometimes required. A mixed constraint often is expressed in the form $Fx + Eu \leq e$, and may be regarded as a state dependent control constraint. Because the constraint (2.3) is more general, the constraint (2.2) may be expressed as $y(k) \in \mathbb{Y}$ by an appropriate choice of the output function $h(\cdot)$ and the output constraint set $\mathbb{Y}$ ($y = (x, u)$ and

$\mathbb{Y} = \mathbb{X} \times \mathbb{U}$). We assume in this chapter that the state x is known; if the state x is estimated, uncertainty (state estimation error) is introduced and *robust* MPC, discussed in Chapter 3, is required.

The next ingredient of the optimal control problem is the cost function. Practical considerations require that the cost be defined over a finite horizon N — to ensure the resultant optimal control problem can be solved sufficiently rapidly to permit effective control. We consider initially the regulation problem where the target state is the origin. If x is the current state and i the current time, then the optimal control problem may be posed as minimizing a cost defined over the interval from time i to time $N + i$. The optimal control problem $\mathbb{P}_N(x, i)$ at event (x, i) is the problem of minimizing the cost

$$\sum_{k=i}^{i+N-1} \ell(x(k), u(k)) + V_f(x(N + i))$$

with respect to the sequences $\mathbf{x} := \{x(i), x(i+1), \ldots, x(i+N)\}$ and $\mathbf{u} := \{u(i), u(i+1), \ldots, u(i+N-1)\}$ subject to the constraints that $\mathbf{x}$ and $\mathbf{u}$ satisfy the difference equation (2.1), the initial condition $x(i) = x$, and the state and control constraints (2.2). We assume that $\ell(\cdot)$ is continuous and that $\ell(0, 0) = 0$. The optimal control and state sequences, obtained by solving $\mathbb{P}_N(x, i)$, are functions of the initial event (x, i)

$$\mathbf{u}^0(x, i) = \{u^0(i; (x, i)), u^0(i + 1; (x, i)), \ldots, u^0(N - 1; (x, i))\}$$
$$\mathbf{x}^0(x, i) = \{x^0(i; (x, i)), x^0(i + 1; (x, i)), \ldots, x^0(N; (x, i))\}$$

where $x^0(i; (x, i)) = x$. In MPC, the first control action $u^0(i; (x, i))$ in the optimal control sequence $\mathbf{u}^0(x, i)$ is applied to the plant, i.e., $u(i) = u^0(i; (x, i))$. Because the system $x^+ = f(x, u)$, the stage cost $\ell(\cdot)$, and the terminal cost $V_f(\cdot)$ are all time invariant, however, the solution of $\mathbb{P}_N(x, i)$, for any time $i \in \mathbb{I}_{\geq 0}$, is identical to the solution of $\mathbb{P}_N(x, 0)$ so that

$$\mathbf{u}^0(x, i) = \mathbf{u}^0(x, 0)$$
$$\mathbf{x}^0(x, i) = \mathbf{x}^0(x, 0)$$

In particular, $u^0(i; (x, i)) = u^0(0; (x, 0))$, i.e., the control $u^0(i; (x, i))$ applied to the plant is equal to $u^0(0; (x, 0))$, the first element in the sequence $\mathbf{u}^0(x, 0)$. Hence we may as well merely consider problem $\mathbb{P}_N(x, 0)$ which, since the initial time is irrelevant, we call $\mathbb{P}_N(x)$. Similarly, for simplicity in notation, we replace $\mathbf{u}^0(x, 0)$ and $\mathbf{x}^0(x, 0)$ by, respectively, $\mathbf{u}^0(x)$ and $\mathbf{x}^0(x)$.

The optimal control problem $\mathbb{P}_N(x)$ may then be expressed as minimization of

$$\sum_{k=0}^{N-1} \ell(x(k), u(k)) + V_f(x(N))$$

with respect to the *decision variables* $(\mathbf{x}, \mathbf{u})$ subject to the constraints that the state and control sequences $\mathbf{x}$ and $\mathbf{u}$ satisfy the difference equation (2.1), the initial condition $x(0) = x$, and the state and control constraints (2.2). Here $\mathbf{u}$ denotes the control sequence $\{u(0), u(1), \ldots, u(N-1)\}$ and $\mathbf{x}$ the state sequence $\{x(0), x(1), \ldots, x(N)\}$. Retaining the state sequence in the set of decision variables is discussed in Chapter 6. For the purpose of analysis, however, it is preferable to constrain the state sequence $\mathbf{x}$ *a priori* to be a solution of $x^+ = f(x, u)$ enabling us to express the problem in the equivalent form of minimizing, with respect to the decision variable $\mathbf{u}$, a cost that is purely a function of the initial state x and the control sequence $\mathbf{u}$. This formulation is possible since the state sequence $\mathbf{x}$ may be expressed, via the difference equation $x^+ = f(x, u)$, as a function of $(x, \mathbf{u})$. The cost becomes $V_N(x, \mathbf{u})$ defined by

$$V_N(x, \mathbf{u}) := \sum_{k=0}^{N-1} \ell(x(k), u(k)) + V_f(x(N)) \tag{2.4}$$

where, now, $x(k) := \phi(k; x, \mathbf{u})$ for all $k \in \mathbb{I}_{0:N}$. Similarly the constraints (2.2), together with an additional terminal constraint

$$x(N) \in \mathbb{X}_f$$

where $\mathbb{X}_f \subseteq \mathbb{X}$, impose an implicit constraint on the control sequence of the form

$$\mathbf{u} \in \mathcal{U}_N(x) \tag{2.5}$$

in which the control constraint set $\mathcal{U}_N(x)$ is the set of control sequences $\mathbf{u} := \{u(0), u(1), \ldots, u(N-1)\}$ satisfying the state and control constraints. It is therefore defined by

$$\mathcal{U}_N(x) := \{\mathbf{u} \mid (x, \mathbf{u}) \in \mathbb{Z}_N\} \tag{2.6}$$

in which the set $\mathbb{Z}_N \subset \mathbb{R}^n \times \mathbb{R}^{Nm}$ is defined by

$$\mathbb{Z}_N := \big\{(x, \mathbf{u}) \mid u(k) \in \mathbb{U}, \quad \phi(k; x, \mathbf{u}) \in \mathbb{X}, \quad \forall k \in \mathbb{I}_{0:N-1}, \\ \text{and } \phi(N; x, \mathbf{u}) \in \mathbb{X}_f\big\} \tag{2.7}$$

The optimal control problem $\mathbb{P}_N(x)$, is, therefore

$$\mathbb{P}_N(x): \qquad V_N^0(x) := \min_{\mathbf{u}}\{V_N(x,\mathbf{u}) \mid \mathbf{u} \in \mathcal{U}_N(x)\} \tag{2.8}$$

Problem $\mathbb{P}_N(x)$ is a *parametric* optimization problem in which the decision variable is $\mathbf{u}$, and both the cost and the constraint set depend on the *parameter* x. The set $\mathbb{Z}_N$ is the set of admissible $(x,\mathbf{u})$, i.e., the set of $(x,\mathbf{u})$ for which $x \in \mathbb{X}$ and the constraints of $\mathbb{P}_N(x)$ are satisfied. Let $\mathcal{X}_N$ be the set of states in $\mathbb{X}$ for which $\mathbb{P}_N(x)$ has a solution

$$\mathcal{X}_N := \{x \in \mathbb{X} \mid \mathcal{U}_N(x) \neq \emptyset\} \tag{2.9}$$

It follows from (2.8) and (2.9) that

$$\mathcal{X}_N = \{x \in \mathbb{R}^n \mid \exists \mathbf{u} \in \mathbb{R}^{Nm} \text{ such that } (x,\mathbf{u}) \in \mathbb{Z}_N\}$$

which is the orthogonal projection of $\mathbb{Z}_N \subset \mathbb{R}^n \times \mathbb{R}^{Nm}$ onto $\mathbb{R}^n$. The domain of $V_N^0(\cdot)$, i.e., the set of states in $\mathbb{X}$ for which $\mathbb{P}_N(x)$ has a solution, is $\mathcal{X}_N$.

Not every optimization problem has a solution. For example, the problem $\min\{x \mid x \in (0,1)\}$ does not have a solution; $\inf\{x \mid x \in (0,1)\} = 0$ but $x = 0$ does not lie in the constraint set $(0,1)$. By Weierstrass's theorem, however, an optimization problem does have a solution if the cost is continuous (in the decision variable) and the constraint set compact (see Proposition A.7). This is the case for our problem as shown subsequently in Proposition 2.4. We assume, without further comment, that the following standing conditions are satisfied in the sequel.

Assumption 2.2 (Continuity of system and cost)**.** The functions $f(\cdot)$, $\ell(\cdot)$ and $V_f(\cdot)$ are continuous, $f(0,0) = 0$, $\ell(0,0) = 0$ and $V_f(0) = 0$.

Assumption 2.3 (Properties of constraint sets)**.** The sets $\mathbb{X}$ and $\mathbb{X}_f$ are closed, $\mathbb{X}_f \subseteq \mathbb{X}$ and $\mathbb{U}$ is compact; each set contains the origin.

The sets $\mathbb{U}$, $\mathbb{X}$ and $\mathbb{X}_f$ are assumed to contain the origin because the first problem we tackle is regulation to the origin. This assumption is modified when we consider the tracking problem.

Proposition 2.4 (Existence of solution to optimal control problem)**.** *Suppose Assumptions 2.2 and 2.3 hold. Then*

(a) The function $V_N(\cdot)$ *is continuous in* $\mathbb{Z}_N$*.*

(b) For each $x \in \mathcal{X}_N$*, the control constraint set* $\mathcal{U}_N(x)$ *is compact.*

(c) For each $x \in \mathcal{X}_N$, a solution to $\mathbb{P}_N(x)$ exists.

Proof.

(a) That $(x, \mathbf{u}) \mapsto V_N(x, \mathbf{u})$ is continuous follows from continuity of $\ell(\cdot)$ in Assumption 2.2, and the continuity of $(x, \mathbf{u}) \mapsto \phi(j; x, \mathbf{u})$ for each $j \in \mathbb{I}_{0:N-1}$, established in Proposition 2.1.

(b) We have to show that for each $x \in \mathcal{X}_N$, the set $\mathcal{U}_N(x)$ is closed and bounded. It is clearly bounded since $\mathcal{U}_N(x) \subseteq \mathbb{U}^N$, which is compact (bounded and closed) by Assumption 2.3. By Proposition 2.1, the function $\phi(j; \cdot)$ is continuous for any $j \in \mathbb{I}_{0:N}$. Since $\mathbb{U}$, $\mathbb{X}$ and $\mathbb{X}_f$ are all closed, any sequence $\{(x_i, \mathbf{u}_i)\}$ in $\mathbb{Z}_N$, defined in (2.7), that converges to, say, $(\bar{x}, \bar{\mathbf{u}})$ satisfies $\phi(j; \bar{x}, \bar{\mathbf{u}}) \in \mathbb{X}$ for all $j \in \mathbb{I}_{0:N-1}$, $\phi(N; \bar{x}, \bar{\mathbf{u}}) \in \mathbb{X}_f$ and $\bar{\mathbf{u}} \in \mathbb{U}^N$. Hence $(\bar{x}, \bar{\mathbf{u}}) \in \mathbb{Z}_N$ so that $\mathbb{Z}_N$ is closed. It follows that $\mathcal{U}_N(x) = \{\mathbf{u} \mid (x, \mathbf{u}) \in \mathbb{Z}_N\}$ is closed and, therefore, compact for all $x \in \mathcal{X}_N$.

(c) Since $V_N(x, \cdot)$ is continuous and $\mathcal{U}_N(x)$ is compact, by Weierstrass's theorem (Proposition A.7) a solution to $\mathbb{P}_N(x)$ exists for each $x \in \mathcal{X}_N$. ■

Although the function $(x, \mathbf{u}) \mapsto V_N(x, \mathbf{u})$ is continuous, the function $x \mapsto V_N^0(x)$ is not necessarily continuous; we discuss this possibility and its implications later. For each $x \in \mathcal{X}_N$, the solution of $\mathbb{P}_N(x)$ is

$$\mathbf{u}^0(x) = \arg\min_{\mathbf{u}} \{V_N(x, \mathbf{u}) \mid \mathbf{u} \in \mathcal{U}_N(x)\}$$

If $\mathbf{u}^0(x) = \{u^0(0; x), u^0(1; x), \ldots, u^0(N-1; x)\}$ is unique for each $x \in \mathcal{X}_N$, then $\mathbf{u}^0 : \mathbb{R}^n \to \mathbb{R}^{Nm}$ is a function; otherwise it is a set-valued function.[1] In MPC, the control applied to the plant is the first element $u^0(0; x)$ of the optimal control sequence. At the next sampling instant, the procedure is repeated for the successor state. Although MPC computes $\mathbf{u}^0(x)$ only for specific values of the state x, it could, in principle, be used to compute $\mathbf{u}^0(x)$ and, hence, $u^0(0; x)$ for every x for which $\mathbb{P}_N(x)$ is feasible, yielding the implicit MPC control law $\kappa_N(\cdot)$ defined by

$$\kappa_N(x) := u^0(0; x), \qquad x \in \mathcal{X}_N$$

MPC does *not* require determination of the control law $\kappa_N(\cdot)$, a task that is usually intractable when constraints or nonlinearities are present; it is this fact that makes MPC so useful.

[1] A set-valued function $\phi(\cdot)$ is a function whose value $\phi(x)$ for each x in its domain is a set.

If, at a given state x, the solution of $\mathbb{P}_N(x)$ is not unique, then $\kappa_N(\cdot) = u^0(0;\cdot)$ is set valued and the model predictive controller selects one element from the set $\kappa_N(x)$.

Example 2.5: Linear quadratic MPC

Suppose the system is described by

$$x^+ = f(x,u) := x + u$$

with initial state x. The stage cost and terminal cost are

$$\ell(x,u) := (1/2)(x^2 + u^2) \qquad V_f(x) := (1/2)x^2$$

The control constraint is

$$u \in [-1,1]$$

and there are no state or terminal constraints. Suppose the horizon is $N = 2$. Under the first approach, the decision variables are $\mathbf{u}$ and $\mathbf{x}$, and the optimal control problem is minimization of

$$\begin{aligned} &V_N(x(0),x(1),x(2),u(0),u(1)) = \\ &\qquad\qquad (1/2)\left(x(0)^2 + x(1)^2 + x(2)^2 + u(0)^2 + u(1)^2\right) \end{aligned}$$

with respect to $(x(0),x(1),x(2))$, and $(u(0),u(1))$ subject to the following constraints

$$\begin{aligned} &x(0) = x \qquad && x(1) = x(0) + u(0) \qquad && x(2) = x(1) + u(1) \\ &u(0) \in [-1,1] && u(1) \in [-1,1] \end{aligned}$$

The constraint $u \in [-1,1]$ is equivalent to two inequality constraints, $u \le 1$ and $-u \le 1$. The first three constraints are equality constraints enforcing satisfaction of the difference equation.

In the second approach, the decision variable is merely $\mathbf{u}$ because the first three constraints are automatically enforced by requiring $\mathbf{x}$ to be a solution of the difference equation. Hence, the optimal control problem becomes minimization with respect to $\mathbf{u} = (u(0)', u(1))'$ of

$$\begin{aligned} V_N(x,\mathbf{u}) &= (1/2)(x^2 + (x+u(0))^2 + (x+u(0)+u(1))^2 + \\ &\qquad u(0)^2 + u(1)^2) \\ &= (3/2)x^2 + \begin{bmatrix} 2x & x \end{bmatrix}\mathbf{u} + (1/2)\mathbf{u}'H\mathbf{u} \end{aligned}$$

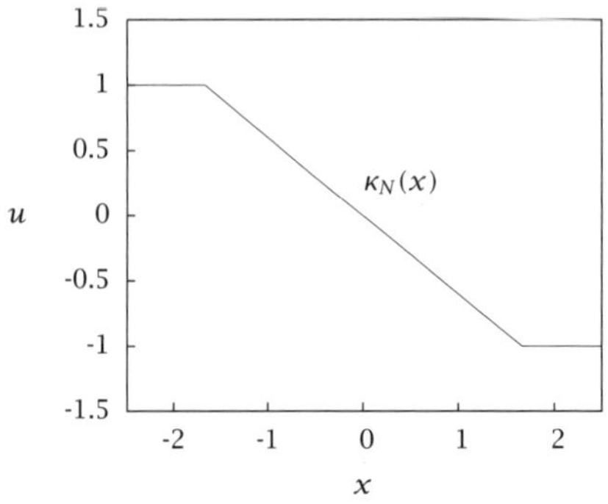

(a) Implicit MPC control law.

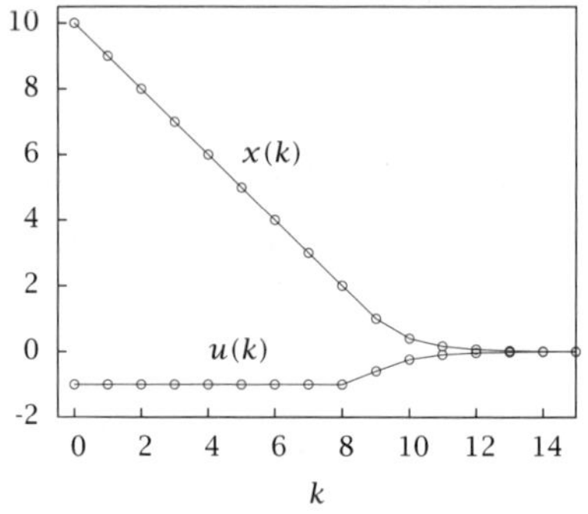

(b) Trajectories of controlled system.

Figure 2.1: Example of MPC.

in which

$$H = \begin{bmatrix} 3 & 1 \\ 1 & 2 \end{bmatrix}$$

subject to the constraint $\mathbf{u} \in \mathcal{U}_N(x)$ where

$$\mathcal{U}_N(x) = \{\mathbf{u} \mid |u(k)| \leq 1 \;\; k = 0, 1\}$$

Because there are no state or terminal constraints, the set $\mathcal{U}_N(x) = \mathcal{U}_N$ for this example does not depend on the parameter x; often it does. Both optimal control problems are quadratic programs.[2] The solution for $x = 10$ is $u^0(1;10) = u^0(2;10) = -1$ so the optimal state trajectory is $x^0(0;10) = 10$, $x^0(1;10) = 9$ and $x^0(2;10) = 8$. The value $V_N^0(10) = 124$. By solving $\mathbb{P}_N(x)$ for every $x \in [-10, 10]$, the optimal control law $\kappa_N(\cdot)$ on this set can be determined, and is shown in Figure 2.1(a). The implicit MPC control law is *time invariant* since the system being controlled, the cost, and the constraints are all time invariant. For our example, the controlled system (the system with MPC) satisfies the difference equation

$$x^+ = x + \kappa_N(x) \qquad \kappa_N(x) = -\text{sat}((3/5)x)$$

and the state and control trajectories for an initial state of $x = 10$ are shown in Figure 2.1(b). It turns out that the origin is exponentially stable for this simple case; often, however, the terminal cost and terminal constraint set have to be carefully chosen to ensure stability. □

[2] A quadratic program is an optimization problem in which the cost is quadratic and the constraint set is polyhedral, i.e., defined by linear inequalities.

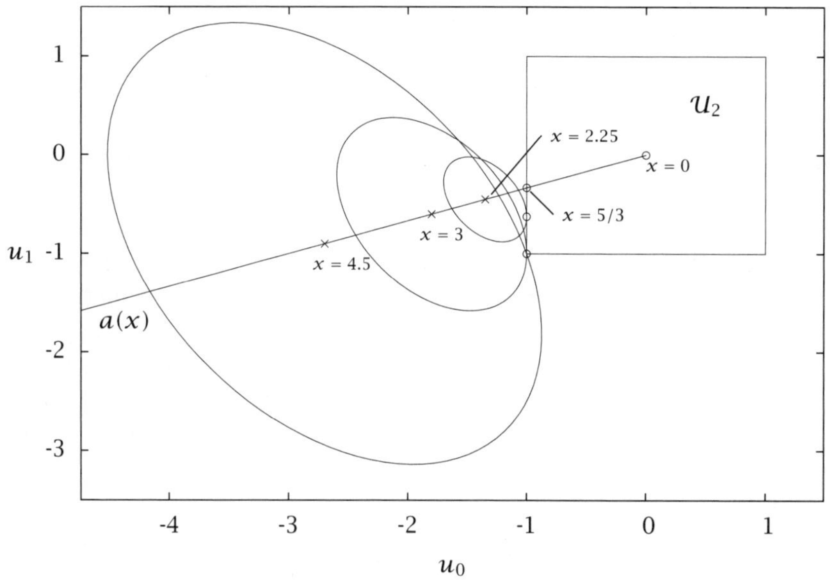

Figure 2.2: Feasible region $\mathcal{U}_2$, elliptical cost contours, and ellipse center, $a(x)$, and constrained minimizers for different values of x.

Example 2.6: Closer inspection of linear quadratic MPC

We revisit the MPC problem discussed in Example 2.5. The objective function is

$$V_N(x, \mathbf{u}) = (1/2)\mathbf{u}'H\mathbf{u} + c(x)'\mathbf{u} + d(x)$$

where $c(x)' = [2\ 1]x$ and $d(x) = (3/2)x^2$. The objective function may be written in the form

$$V_N(x, \mathbf{u}) = (1/2)(\mathbf{u} - a(x))'H(\mathbf{u} - a(x)) + e(x)$$

Expanding the second form shows the two forms are equal if

$$a(x) = -H^{-1}c(x) = K_1 x \qquad K_1 = -(1/5)\begin{bmatrix} 3 \\ 1 \end{bmatrix}$$

and

$$e(x) + (1/2)a(x)'Ha(x) = d(x)$$

Since H is positive definite, $a(x)$ is the unconstrained minimizer of the objective function; indeed $\nabla_{\mathbf{u}} V_N(x, a(x)) = 0$ since

$$\nabla_{\mathbf{u}} V_N(x, \mathbf{u}) = H\mathbf{u} + c(x)$$

The locus of $a(x)$ for $x \geq 0$ is shown in Figure 2.2. Clearly the unconstrained minimizer $a(x) = K_1 x$ is equal to the constrained minimizer $\mathbf{u}^0(x)$ for all x such that $a(x) \in \mathcal{U}_2$ where $\mathcal{U}_2$ is the unit square illustrated in Figure 2.2; since $a(x) = K_1 x$, $a(x) \in \mathcal{U}_2$ for all $x \in \mathbb{X}_1 = [0, x_{c1}]$ where $x_{c1} = 5/3$. For $x > x_{c1}$, the unconstrained minimizer lies outside $\mathcal{U}_2$ as shown in Figure 2.2 for $x = 2.25$, $x = 3$ and $x = 5$. For such x, the constrained minimizer $\mathbf{u}^0(x)$ is a point that lies on the intersection of a level set of the objective function (which is an ellipse) and the boundary of $\mathcal{U}_2$. For $x \in [x_{c1}, x_{c2})$, $\mathbf{u}^0(x)$ lies on the left face of the box $\mathcal{U}_2$ and for $x \geq x_{c2}$, $\mathbf{u}^0(x)$ remains at $(-1, -1)$, the bottom left vertex of $\mathcal{U}_2$.

When $u^0(x)$ lies on the left face of $\mathcal{U}_2$, the gradient $\nabla_{\mathbf{u}} V_N(x, \mathbf{u}^0(x))$ of the objective function is normal to the left face of $\mathcal{U}_2$, i.e., the level set of $V_N^0(\cdot)$ passing through $\mathbf{u}^0(x)$ is tangential to the left face of $\mathcal{U}_2$. The outward normal to $\mathcal{U}_2$ at a point on the left face is $-e_1 = (-1, 0)$ so that at $\mathbf{u} = \mathbf{u}^0(x)$

$$\nabla_{\mathbf{u}} V(x, \mathbf{u}^0(x)) + \lambda(-e_1) = 0$$

for some $\lambda > 0$; this is a standard condition of optimality. Since $\mathbf{u} = [-1 \;\; v]'$ for some $v \in [-1, 1]$ and since $\nabla_{\mathbf{u}} V(x, \mathbf{u}) = H(\mathbf{u} - a(x)) = H\mathbf{u} + c(x)$, the condition of optimality is

$$\begin{bmatrix} 3 & 1 \\ 1 & 2 \end{bmatrix} \begin{bmatrix} -1 \\ v \end{bmatrix} + \begin{bmatrix} 2 \\ 1 \end{bmatrix} x - \begin{bmatrix} \lambda \\ 0 \end{bmatrix} = \begin{bmatrix} 0 \\ 0 \end{bmatrix}$$

or

$$\begin{aligned} -3 + v + 2x - \lambda &= 0 \\ -1 + 2v + x &= 0 \end{aligned}$$

which, when solved, yields $v = (1/2) - (1/2)x$ and $\lambda = -(5/2) + (3/2)x$. Hence,

$$\mathbf{u}^0(x) = b_2 + K_2 x \qquad b_2 = \begin{bmatrix} -1 \\ (1/2) \end{bmatrix} \qquad K_2 = \begin{bmatrix} 0 \\ -(1/2) \end{bmatrix}$$

for all $x \in \mathbb{X}_2 = [x_{c1}, x_{c2}]$ where $x_{c2} = 3$ since $\mathbf{u}^0(x) \in \mathcal{U}_2$ for all x in this range. For all $x \in \mathbb{X}_3 = [x_{c_2}, \infty)$, $\mathbf{u}^0(x) = (-1, -1)'$. Summarizing:

$$\begin{aligned} x \in [0, (5/3)] &\Rightarrow \mathbf{u}^0(x) = K_1 x \\ x \in [(5/3), 3] &\Rightarrow \mathbf{u}^0(x) = K_2 x + b_2 \\ x \in [3, \infty) &\Rightarrow \mathbf{u}^0(x) = b_3 \end{aligned}$$

in which

$$K_1 = \begin{bmatrix} -(3/5) \\ -(1/5) \end{bmatrix} \qquad K_2 = \begin{bmatrix} 0 \\ -(1/2) \end{bmatrix} \qquad b_2 = \begin{bmatrix} -1 \\ (1/2) \end{bmatrix} \qquad b_3 = \begin{bmatrix} -1 \\ -1 \end{bmatrix}$$

The optimal control for $x \leq 0$ may be obtained by symmetry; $\mathbf{u}^0(-x) = -\mathbf{u}^0(x)$ for all $x \geq 0$ so that:

$$\begin{aligned} x \in [0, -(5/3)] &\Rightarrow \mathbf{u}^0(x) = -K_1 x \\ x \in [-(5/3), -3] &\Rightarrow \mathbf{u}^0(x) = -K_2 x - b_2 \\ x \in [-3, -\infty) &\Rightarrow \mathbf{u}^0(x) = -b_3 \end{aligned}$$

It is easily checked that $\mathbf{u}^0(\cdot)$ is continuous and satisfies the constraint for all $x \in \mathbb{R}$. The MPC control law $\kappa_N(\cdot)$ is the first component of $\mathbf{u}^0(\cdot)$ and, therefore, is defined by:

$$\begin{aligned} \kappa_N(x) &= 1 && x \leq -3 \\ \kappa_N(x) &= 1 && x \in [-(5/3), -3] \\ \kappa_N(x) &= -(3/5)x && x \in [-(5/3), (5/3)] \\ \kappa_N(x) &= -1 && x \in [(5/3), 3] \\ \kappa_N(x) &= -1 && x \geq 3 \end{aligned}$$

i.e., $\kappa_N(x) = -\text{sat}((3/5)x)$ which is the saturating control law depicted in Figure 2.1(a). The control law is piecewise affine and the value function piecewise quadratic. The structure of the solution to constrained linear quadratic optimal control problems is explored more fully in Chapter 7. □

As we show in Chapter 3, continuity of the value function is desirable. Unfortunately, this is not true in general; the major difficulty is in establishing that the set-valued function $x \mapsto \mathcal{U}_N(x)$ has certain continuity properties. Continuity of the value function $V_N^0(\cdot)$ and of the implicit control law $\kappa_N(\cdot)$ may be established for a few important cases, however, as is shown by the next result, which assumes satisfaction of our standing assumptions: 2.2 and 2.3 so that the cost function $V_N(\cdot)$ is continuous in $(x, \mathbf{u})$.

Theorem 2.7 (Continuity of value function and control law). *Suppose that Assumptions 2.2 and 2.3 hold.*

(a) Suppose that there are no state constraints so that $\mathbb{X} = \mathbb{X}_f = \mathbb{R}^n$. Then the value function $V_N^0 : \mathcal{X}_N \to \mathbb{R}$ is continuous and $\mathcal{X}_N = \mathbb{R}^n$.

(b) Suppose $f(\cdot)$ is linear ($x^+ = Ax + Bu$) and that the state and control constraints sets $\mathbb{X}$ and $\mathbb{U}$ are polyhedral.[3] Then the value function $V_N^0 : \mathcal{X}_N \to \mathbb{R}$ is continuous.

(c) If, in addition, the solution $\mathbf{u}^0(x)$ of $\mathbb{P}_N(x)$ is unique at each $x \in \mathcal{X}_N$, then the implicit MPC control law $\kappa_N(\cdot)$ is continuous.

The proof of this theorem is given in Section C.3 of Appendix C. The following example, due to Meadows, Henson, Eaton, and Rawlings (1995), shows that there exist nonlinear examples where the value function and implicit control law are not continuous.

Example 2.8: Discontinuous MPC control law

Consider the nonlinear system defined by

$$x_1^+ = x_1 + u$$
$$x_2^+ = x_2 + u^3$$

The control horizon is $N = 3$ and the cost function $V_3(\cdot)$ is defined by

$$V_3(x, \mathbf{u}) := \sum_{k=0}^{2} \ell(x(k), u(k))$$

and the stage cost $\ell(\cdot)$ is defined by

$$\ell(x, u) := |x|^2 + u^2$$

The constraint sets are $\mathbb{X} = \mathbb{R}^2$, $\mathbb{U} = \mathbb{R}$, and $\mathbb{X}_f := \{0\}$, i.e., there are no state and control constraints, and the terminal state must satisfy the constraint $x(3) = 0$. Hence, although there are three control actions, $u(0)$, $u(1)$, and $u(2)$, two must be employed to satisfy the terminal constraint, leaving only one degree of freedom. Choosing $u(0)$ to be the free decision variable automatically constrains $u(1)$ and $u(2)$ to be functions of the initial state x and the first control action $u(0)$. Solving

[3] A set $\mathbb{X}$ is polyhedral if it may be defined as set of linear inequalities, i.e., if it may be expressed in the form $\mathbb{X} = \{x \mid Mx \leq m\}$.

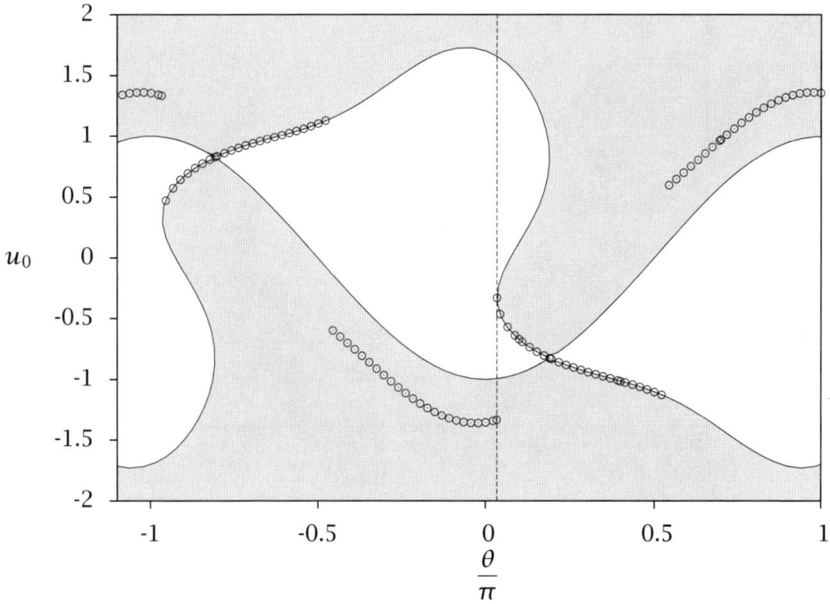

Figure 2.3: First element of control constraint set $\mathcal{U}_3(x)$ (shaded) and control law $\kappa_3(x)$ (circle) versus $x = (\sin(\theta), \cos(\theta))$, $\theta \in [-\pi, \pi]$ on the unit circle for a nonlinear system with terminal constraint.

the equation

$$\begin{aligned} x_1(3) &= x_1 + u(0) + u(1) + u(2) &= 0 \\ x_2(3) &= x_2 + u(0)^3 + u(1)^3 + u(2)^3 &= 0 \end{aligned}$$

for $u(1)$ and $u(2)$ yields

$$\begin{aligned} u(1) &= -x_1/2 - u(0)/2 \pm \sqrt{b} \\ u(2) &= -x_1/2 - u(0)/2 \mp \sqrt{b} \end{aligned}$$

in which

$$b = \frac{3u(0)^3 - 3u(0)^2 x_1 - 3u(0)x_1^2 - x_1^3 + 4x_2}{12(u(0) + x_1)}$$

Clearly a real solution exists only if b is positive, i.e., if both the numerator and denominator in the expression for b have the same sign. The

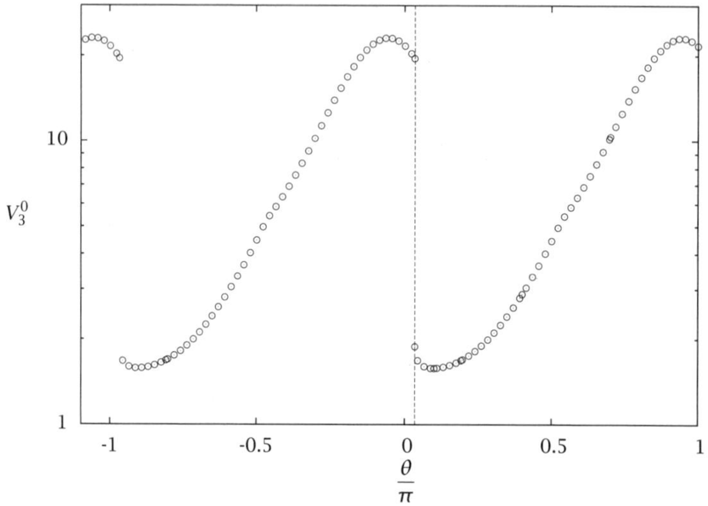

Figure 2.4: Optimal cost $V_3^0(x)$ versus $x = (\sin(\theta), \cos(\theta))$, $\theta \in [-\pi, \pi]$ on the unit circle; the discontinuity in V_3^0 is caused by the discontinuity in $\mathcal{U}_3$ as θ crosses the dashed line in Figure 2.3.

optimal control problem $\mathbb{P}_3(x)$ is defined by

$$V_3^0(x) = \min_{\mathbf{u}}\{V_3(x, \mathbf{u}) \mid \phi(3; x, \mathbf{u}) = 0\}$$

and the implicit MPC control law is $\kappa_3(\cdot)$ where $\kappa_3(x) = u^0(0; x)$, the first element in the minimizing sequence $\mathbf{u}^0(x)$. It can be shown, using analysis presented later in this chapter, that the origin is asymptotically stable for the controlled system $x^+ = f(x, \kappa_N(x))$. That this control law is necessarily discontinuous may be shown as follows. If the control is strictly positive, any trajectory originating in the first quadrant $(x_1, x_2 > 0)$ moves away from the origin. If the control is strictly negative, any control originating in the third quadrant $(x_1, x_2 < 0)$ also moves away from the origin. But the control cannot be zero at any nonzero point lying in the domain of attraction. If it were, this point would be a fixed point for the controlled system, contradicting the fact that it lies in the domain of attraction.

In fact, both the value function $V_3^0(\cdot)$ and the MPC control law $\kappa_3(\cdot)$ are discontinuous. Figures 2.3 and 2.4 show how $\mathcal{U}_3(x)$, $\kappa_3(x)$, and $V_3^0(x)$ vary as $x = (\sin(\theta), \cos(\theta))$ ranges over the unit circle. A further conclusion that can be drawn from this example is that it is possible for the MPC control law to be discontinuous at points where the value function is continuous. □

2.3 Dynamic Programming Solution

We examine next the DP solution of the optimal control problem $\mathbb{P}_N(x)$, not because it provides a practical procedure but because of the insight it provides. DP can rarely be used for constrained and/or nonlinear control problems unless the state dimension n is small. MPC is best regarded as a practical means of implementing the DP solution; for a given state x it provides $V_N^0(x)$ and $\kappa_N(x)$, the value, respectively, of the value function and control law at a *point* x. DP, on the other hand, yields the value function $V_N^0(\cdot)$ and the implicit MPC control law $\kappa_N(\cdot)$.

The optimal control problem $\mathbb{P}_N(x)$ is defined, as before, by (2.8) with the cost function $V_N(\cdot)$ defined by (2.4) and the constraints by (2.5). DP yields an optimal policy $\boldsymbol{\mu}^0 = \{\mu_0^0(\cdot), \mu_1^0(\cdot), \ldots, \mu_{N-1}^0(\cdot)\}$, i.e., a sequence of control laws $\mu_i : \mathcal{X}_i \to \mathbb{U}$, $i = 0, 1, \ldots, N-1$. The domain $\mathcal{X}_i$ of each control law will be defined later. The optimal controlled system is time varying and satisfies

$$x^+ = f(x, \mu_i^0(x)), \ i = 0, 1, \ldots, N-1$$

in contrast with the system using MPC, which is time invariant and satisfies

$$x^+ = f(x, \kappa_N(x)), \ i = 0, 1, \ldots, N-1$$

where $\kappa_N(\cdot) = \mu_0^0(\cdot)$. The optimal control law at time i is $\mu_i^0(\cdot)$ whereas receding horizon control (RHC) uses the time-invariant control law $\kappa_N(\cdot)$ obtained by assuming that at each time t, the terminal time or *horizon* is $t + N$ so that the horizon $t + N$ recedes as t increases. One consequence is that the time-invariant control law $\kappa_N(\cdot)$ is *not* optimal for the problem of controlling $x^+ = f(x, u)$ over the fixed interval $[0, T]$ in such a way as to minimize V_N and satisfy the constraints.

For all $j \in \mathbb{I}_{0:N-1}$, let $V_j(x, \mathbf{u})$, $\mathcal{U}_j(x)$, $\mathbb{P}_j(x)$, and $V_j^0(x)$ be defined, respectively, by (2.4), (2.5), (2.6), and (2.7), with N replaced by j. As shown in Section C.1 of Appendix C, DP solves not only $\mathbb{P}_N(x)$ for all $x \in \mathcal{X}_N$, the domain of $V_N^0(\cdot)$, but also $\mathbb{P}_j(x)$ for all $x \in \mathcal{X}_j$, the domain of $V_j^0(\cdot)$, all $j \in \mathbb{I}_{0:N-1}$. The DP equations are

$$V_j^0(x) = \min_{u \in \mathbb{U}} \{\ell(x,u) + V_{j-1}^0(f(x,u)) \mid f(x,u) \in \mathcal{X}_{j-1}\}, \ \forall x \in \mathcal{X}_j \tag{2.10}$$

$$\kappa_j(x) = \arg\min_{u \in \mathbb{U}} \{\ell(x,u) + V_{j-1}^0(f(x,u)) \mid f(x,u) \in \mathcal{X}_{j-1}\}, \ \forall x \in \mathcal{X}_j \tag{2.11}$$

$$\mathcal{X}_j = \{x \in \mathbb{X} \mid \exists u \in \mathbb{U} \text{ such that } f(x,u) \in \mathcal{X}_{j-1}\} \tag{2.12}$$

for $j = 1, 2, \ldots, N$ (j is *time to go*), with terminal conditions

$$V_0^0(x) = V_f(x) \ \forall x \in \mathcal{X}_0 \qquad \mathcal{X}_0 = \mathbb{X}_f$$

For each j, $V_j^0(x)$ is the optimal cost for problem $\mathbb{P}_j(x)$ if the current state is x, current time is 0 (or i), and the terminal time is j (or $i+j$), and $\mathcal{X}_j$ is its domain; $\mathcal{X}_j$ is also the set of states in $\mathbb{X}$ that can be steered to the terminal set $\mathbb{X}_f$ in j steps by an *admissible* control sequence, i.e., a control sequence that satisfies the control, state, and terminal constraints and, therefore, lies in the set $\mathcal{U}_j(x)$. Hence, for each j

$$\mathcal{X}_j = \{x \in \mathbb{X} \mid \mathcal{U}_j(x) \neq \emptyset\}$$

Definition 2.9 (Feasible preimage of the state). Let $\mathbb{Z} := \mathbb{X} \times \mathbb{U}$. The set-valued function $f_{\mathbb{Z}}^{-1} : \mathbb{X} \to \mathbb{Z}$ is defined by

$$f_{\mathbb{Z}}^{-1}(x) := f^{-1}(x) \cap \mathbb{Z}$$

in which

$$f^{-1}(x) := \{z \in \mathbb{R}^n \times \mathbb{R}^m \mid f(z) = x\}$$

For all $j \geq 0$, let the set $\mathcal{Z}_j \subseteq \mathbb{R}^n \times \mathbb{R}^m$ be defined by

$$\mathcal{Z}_j := f_{\mathbb{Z}}^{-1}(\mathcal{X}_{j-1}) = \{(x,u) \mid f(x,u) \in \mathcal{X}_{j-1}\} \cap \mathbb{Z}$$

The set $\mathcal{X}_j$ may then be expressed as

$$\mathcal{X}_j = \{x \in \mathbb{R}^n \mid \exists u \in \mathbb{R}^m \text{ such that } (x,u) \in \mathcal{Z}_j\}$$

i.e., $\mathcal{X}_j$ is the orthogonal projection of $\mathcal{Z}_j \subseteq \mathbb{R}^n \times \mathbb{R}^m$ onto $\mathbb{R}^n$.

DP yields much more than an optimal control sequence for a given initial state; it yields an optimal feedback *policy* $\boldsymbol{\mu}^0$ or sequence of control laws where

$$\boldsymbol{\mu}^0 := \{\mu_0(\cdot), \mu_1(\cdot), \ldots, \mu_{N-1}(\cdot)\} = \{\kappa_N(\cdot), \kappa_{N-1}(\cdot), \ldots, \kappa_1(\cdot)\}$$

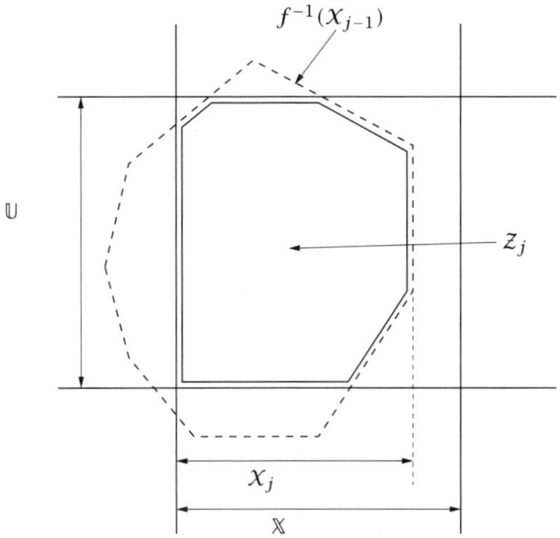

Figure 2.5: The sets Z_j and X_j.

At event (x, i), i.e., at state x at time i, the time to go is $N - i$ and the optimal control is

$$\mu_i^0(x) = \kappa_{N-i}(x)$$

i.e., $\mu_i(\cdot)$ is the control law at time i. Consider an initial *event* $(x, 0)$, i.e., state x at time 0. If the terminal time (horizon) is N, the optimal control for $(x, 0)$ is $\kappa_N(x)$. The successor state, at time 1, is

$$x^+ = f(x, \kappa_N(x))$$

At event $(x^+, 1)$, the time to go to the terminal time is $N - 1$ and the optimal control is $\kappa_{N-1}(x^+) = \kappa_{N-1}(f(x, \kappa_N(x)))$. For a given initial event $(x, 0)$, the optimal policy generates the optimal state and control trajectories $\mathbf{x}^0(x)$ and $\mathbf{u}^0(x)$ that satisfy the difference equations

$$x(0) = x \qquad u(0) = \kappa_N(x) \tag{2.13}$$

$$x(i+1) = f(x(i), u(i)) \qquad u(i) = \kappa_{N-i}(x(i)) \tag{2.14}$$

for $i = 0, 1, \ldots, N - 1$. These state and control trajectories are identical to those obtained, as in MPC, by solving $\mathbb{P}_N(x)$ directly for the particular initial event $(x, 0)$ using a mathematical programming algorithm. Dynamic programming, however, provides a solution for *any* event (x, i) such that $i \in \mathbb{I}_{0:N-1}$ and $x \in X_i$.

Optimal control, in the classic sense of determining a control that minimizes a cost over the interval $[0, T]$, is generally time varying (at event (x, i), $i \in \mathbb{I}_{0:N}$, the optimal control is $\mu_i(x) = \kappa_{N-i}(x)$). Under fairly general conditions, $\mu_i(\cdot) \to \kappa_\infty(\cdot)$ as $N \to \infty$ where $\kappa_\infty(\cdot)$ is the stationary infinite horizon optimal control law. MPC and RHC, on the other hand, employ the time-invariant control $\kappa_N(x)$ for all $i \in \mathbb{I}_{\geq 0}$. Thus the state and control trajectories $\mathbf{x}_{\text{mpc}}(x)$ and $\mathbf{u}_{\text{mpc}}(x)$ generated by MPC for an initial event $(x, 0)$ satisfy the difference equations

$$\begin{aligned} x(0) &= x & u(0) &= \kappa_N(x) \\ x(i+1) &= f(x(i), u(i)) & u(i) &= \kappa_N(x(i)) \end{aligned}$$

and can be seen to differ in general from $\mathbf{x}^0(x)$ and $\mathbf{u}^0(x)$, which satisfy (2.13) and (2.14), and, hence, are *not* optimal for $\mathbb{P}_N(x)$.

Before leaving this section, we obtain some properties of the solution to each partial problem $\mathbb{P}_j(x)$. For this, we require a few definitions.

Definition 2.10 (Positive and control invariant sets).

(a) A set $X \subseteq \mathbb{R}^n$ is positive invariant for $x^+ = f(x)$ if $x \in X$ implies $f(x) \in X$.

(b) A set $X \subseteq \mathbb{R}^n$ is control invariant for $x^+ = f(x, u)$, $u \in \mathbb{U}$, if, for all $x \in X$, there exists a $u \in \mathbb{U}$ such that $f(x, u) \in X$.

We recall from our standing assumptions 2.2 and 2.3 that $f(\cdot)$, $\ell(\cdot)$ and $V_f(\cdot)$ are continuous, that $\mathbb{X}$ and $\mathbb{X}_f$ are closed, $\mathbb{U}$ is compact and that each of these sets contains the origin.

Proposition 2.11 (Existence of solutions to DP recursion). *Suppose Assumptions 2.2 and 2.3 hold. Then*

(a) For all $j \geq 0$, the cost function $V_j(\cdot)$ is continuous in $\mathcal{Z}_j$, and, for each $x \in \mathcal{X}_j$, the control constraint set $\mathcal{U}_j(x)$ is compact and a solution $\mathbf{u}^0(x) \in \mathcal{U}_j(x)$ to $\mathbb{P}_j(x)$ exists.

(b) If $\mathcal{X}_0 := \mathbb{X}_f$ is control invariant for $x^+ = f(x, u)$, $u \in \mathbb{U}$, then, for each $j \in \mathbb{I}_{\geq 0}$, the set $\mathcal{X}_j$ is also control invariant, $\mathcal{X}_j \supseteq \mathcal{X}_{j-1}$, and $0 \in \mathcal{X}_j$. In addition, the set $\mathcal{X}_N$ is positive invariant for $x^+ = f(x, \kappa_N(x))$.

(c) For each $j \geq 0$, the set $\mathcal{X}_j$ is closed.

Proof.

(a) This proof is almost identical to the proof of Proposition 2.4.

(b) By assumption, $\mathcal{X}_0 = \mathbb{X}_f \subseteq \mathbb{X}$ is control invariant. By (2.12)

$$\mathcal{X}_1 = \{x \in \mathbb{X} \mid \exists u \in \mathbb{U} \text{ such that } f(x, u) \in \mathcal{X}_0\}$$

Since $\mathcal{X}_0$ is control invariant for $x^+ = f(x,u), u \in \mathbb{U}$, for every $x \in \mathcal{X}_0$ there exist a $u \in \mathbb{U}$ such that $f(x,u) \in \mathcal{X}_0$ so that $x \in \mathcal{X}_1$. Hence $\mathcal{X}_1 \supseteq \mathcal{X}_0$. Since for every $x \in \mathcal{X}_1$, there exists a $u \in \mathbb{U}$ such that $f(x,u) \in \mathcal{X}_0 \subseteq \mathcal{X}_1$, it follows that $\mathcal{X}_1$ is control invariant for $x^+ = f(x,u)$, $u \in \mathbb{U}$. If for some integer $j \in \mathbb{I}_{\geq 0}$, $\mathcal{X}_{j-1}$ is control invariant for $x^+ = f(x,u)$, it follows by similar reasoning that $\mathcal{X}_j \supseteq \mathcal{X}_{j-1}$ and that $\mathcal{X}_j$ is control invariant. By induction $\mathcal{X}_j$ is control invariant and $\mathcal{X}_j \supseteq \mathcal{X}_{j-1}$ for all $j > 0$. Hence $0 \in \mathcal{X}_j$ for all $j \in \mathbb{I}_{\geq 0}$. That $\mathcal{X}_N$ is positive invariant for $x^+ = f(x, \kappa_N(x))$ follows from (2.11) that $\kappa_N(\cdot)$ steers every $x \in \mathcal{X}_N$ into $\mathcal{X}_{N-1} \supseteq \mathcal{X}_N$.

(c) By Assumption 2.3, $\mathcal{X}_0 = \mathbb{X}_f$ is closed. Suppose, for some $j \in \mathbb{I}_{\geq 1}$, that $\mathcal{X}_{j-1}$ is closed. Then $\mathcal{Z}_j := \{(x,u) \mid f(x,u) \in \mathcal{X}_{j-1}\}$ is closed since $f(\cdot)$ is continuous. To prove that $\mathcal{X}_j$ is closed, take any sequence $\{x_i\}$ in $\mathcal{X}_j$ that converges to, say, $\bar{x}$. For each i, select a $u_i \in \mathbb{U}$ such that $(x_i, u_i) \in \mathcal{Z}_j$. Then, since $\mathbb{U}$ is compact, there exists a subsequence of $\{(x_i, u_i)\}$, indexed by I, such that $x_i \to \bar{x}$ and $u_i \to \bar{u}$ as $i \to \infty, i \in I$. Since $\mathbb{X}$ is closed, $\mathbb{U}$ is compact, and $\mathcal{Z}_j$ is closed, it follows that $\bar{x} \in \mathbb{X}$, $\bar{u} \in \mathbb{U}$ and $(\bar{x}, \bar{u}) \in \mathcal{Z}_j$. Hence $\bar{x} \in \mathcal{X}_j := \{x \in \mathbb{X} \mid \exists u \in \mathbb{U}$ such that $(\mathrm{x}, \mathrm{u}) \in \mathcal{Z}_\mathrm{j}\}$ so that $\mathcal{X}_j$ is closed. By induction $\mathcal{X}_j$ is closed for all $j \in \mathbb{I}_{\geq 0}$. ■

The fact that $\mathcal{X}_N$ is positive invariant for $x^+ = f(x, \kappa_N(x))$ can also be established by observing that $\mathcal{X}_N$ is the set of states x in $\mathbb{X}$ for which there exists a $\mathbf{u}$ that is feasible for $\mathbb{P}_N(x)$, i.e., for which there exists a control $\mathbf{u}$ satisfying the control, state and terminal constraints. It is shown in the next section that for every $x \in \mathcal{X}_N$, there exists a feasible control sequence $\tilde{\mathbf{u}}$ for $\mathbb{P}_N(x^+)$ where $x^+ = f(x, \kappa_N(x))$ is the successor state provided that $\mathbb{X}_f$ is control invariant, i.e., $\mathcal{X}_N$ is positive invariant for $x^+ = f(x, \kappa_N(x))$ if $\mathbb{X}_f$ is control invariant. An important practical consequence is that if $\mathbb{P}_N(x(0))$ can be solved for the initial state $x(0)$, then $\mathbb{P}_N(x(i))$ can be solved for any subsequent state $x(i)$ of the controlled system $x^+ = f(x, \kappa_N(x))$, a property that is sometimes called recursive feasibility. Uncertainty, in the form of additive disturbances, model error or state estimation error, may destroy this important property; techniques to restore this property when uncertainty is present are discussed in Chapter 3.

2.4 Stability

2.4.1 Introduction

To establish stability we employ Lyapunov theorems such as Theorem B.13 in Appendix B. Because we are considering the regulator problem in this chapter, we are concerned with asymptotic or exponential stability of the origin. Hence, we replace $\mathcal{A}$ in Theorem B.13 of Appendix B by $\{0\}$, the set consisting of a single point, the origin. Thus, the origin is asymptotically stable with a region of attraction X for the system $x^+ = f(x)$ if there exist: a Lyapunov function V, a positive invariant set X, two $\mathcal{K}_\infty$ functions $\alpha_1(\cdot)$ and $\alpha_2(\cdot)$, and a positive definite function $\alpha_3(\cdot)$ satisfying

$$V(x) \geq \alpha_1(|x|) \tag{2.15}$$

$$V(x) \leq \alpha_2(|x|) \tag{2.16}$$

$$V(f(x)) \leq V(x) - \alpha_3(|x|) \tag{2.17}$$

for all $x \in X$. Recall that $\alpha : \mathbb{R} \to \mathbb{R}_{\geq 0}$ is a $\mathcal{K}_\infty$ function if it is continuous, strictly increasing, zero at zero, and is unbounded; and α is a positive definite function if it is continuous and positive everywhere except at the origin. Our task in this chapter is to find a function $V(\cdot)$ with these properties for the MPC system $x^+ = f(x, \kappa_N(x))$.

A standard approach to establish stability is to employ the value function of an infinite horizon optimal control problem as a Lyapunov function. This suggests the use of $V_N^0(\cdot)$, the value function for the finite horizon optimal control problem whose solution yields the model predictive controller, as a Lyapunov function. It is simple to show, under mild assumptions on $\ell(\cdot)$, that $V_N^0(\cdot)$ has property (2.15) for all $x \in \mathcal{X}_N$. The value function $V_\infty(\cdot)$ for infinite horizon optimal control problems does satisfy, under mild conditions, $V_\infty^0(f(x, \kappa_\infty(x))) = V_\infty^0(x) - \ell(x, \kappa_\infty(x))$ thereby ensuring satisfaction of property (2.17). Since, as is often pointed out, optimality does not imply stability, this property does not usually hold when the horizon is finite. One of the main tasks of this chapter is show that if the "ingredients" $V_f(\cdot)$, $\ell(\cdot)$, and $\mathbb{X}_f$ of the finite horizon optimal control problem are chosen appropriately, then $V_N^0(f(x, \kappa_N(x))) \leq V_N^0(x) - \ell(x, \kappa_N(x))$ for all x in $\mathcal{X}_N$ enabling property (2.17) to be obtained. Property (2.16), an upper bound on the value function, is more difficult to establish. We show subsequently that the choice of "ingredients" that ensures satisfaction of property (2.17) also ensures satisfaction of property (2.16) but only for all x in $\mathbb{X}_f$ rather than for all $x \in \mathcal{X}_N$. We therefore also address the problem of establishing asymptotic or exponential stability of the origin even if property (2.16) holds only for x in $\mathbb{X}_f$.

We now address a point that we have glossed over. The solution to an optimization problem is not necessarily unique. Thus $\mathbf{u}^0(x)$ and $\kappa_N(x)$ may be set valued; any point in the set $\mathbf{u}^0(x)$ is a solution of $\mathbb{P}_N(x)$. Similarly $\mathbf{x}^0(x)$ is set valued. Uniqueness may be obtained by choosing that element in the set $\mathbf{u}^0(x)$ that has least norm. To avoid expressions such as "let $\mathbf{u}$ be any element of the minimizing set $\mathbf{u}^0(x)$," we shall, in the sequel, use $\mathbf{u}^0(x)$ to denote any sequence in the set of minimizing sequences and use $\kappa_N(x)$ to denote $u^0(0;x)$, the first element of this sequence.

2.4.2 Stabilizing Conditions: No State Constraints

To show as simply as possible that the descent property (2.17) holds if $V_f(\cdot)$ and $\mathbb{X}_f$ are chosen appropriately, we consider first the case when there are no state or terminal constraints, i.e., $\mathbb{X} = \mathbb{X}_f = \mathbb{R}^n$, so that the only constraint is the control constraint. Hence $\mathcal{U}_N(x) = \mathbb{U}^N$, which is independent of x. For this case, $\mathcal{X}_j = \mathbb{R}^n$ for all $j \in \{1, 2, \ldots, N\}$. Let x be any state in $\mathcal{X}_N = \mathbb{R}^n$ at time 0. Then

$$V_N^0(x) = V_N(x, \mathbf{u}^0(x))$$

in which

$$\mathbf{u}^0(x) = \left\{u^0(0;x), u^0(1;x), \ldots, u^0(N-1;x)\right\}$$

is any minimizing control sequence. The resultant optimal state sequence is

$$\mathbf{x}^0(x) = \left\{x^0(0;x), x^0(1;x), \ldots, x^0(N;x)\right\}$$

where $x^0(0;x) = x$ and $x^0(1;x) = x^+$. The successor state to x at time 0 is $x^+ = f(x, \kappa_N(x)) = x^0(1;x)$ at time 1 where $\kappa_N(x) = u^0(0;x)$, and

$$V_N^0(x^+) = V_N(x^+, \mathbf{u}^0(x^+))$$

in which

$$\mathbf{u}^0(x^+) = \left\{u^0(0;x^+), u^0(1;x^+), \ldots, u^0(N-1;x^+)\right\}$$

It is difficult to compare $V_N^0(x)$ and $V_N^0(x^+)$ directly, but

$$V_N^0(x^+) = V_N(x^+, \mathbf{u}^0(x^+)) \leq V_N(x^+, \tilde{\mathbf{u}})$$

where $\tilde{\mathbf{u}}$ is any feasible control sequence for $\mathbb{P}_N(x^+)$, i.e., any control sequence in $\mathbb{U}^N$. To facilitate comparison of $V_N(x^+, \tilde{\mathbf{u}})$ with $V_N^0(x) = V_N(x, \mathbf{u}^0(x))$, we choose

$$\tilde{\mathbf{u}} = \left\{u^0(1;x), \ldots, u^0(N-1;x), u\right\}$$

where u still has to be chosen. Comparing $\tilde{\mathbf{u}}$ with $\mathbf{u}^0(x)$ shows that $\tilde{\mathbf{x}}$, the state sequence due to control sequence $\tilde{\mathbf{u}}$, is

$$\tilde{\mathbf{x}} = \left\{x^0(1;x), x^0(2;x), \ldots, x^0(N;x), f(x^0(N;x), u)\right\}$$

in which $x^0(1;x) = x^+ = f(x, \kappa_N(x))$; since there are no state or terminal constraints, the state sequence $\tilde{\mathbf{x}}$ is clearly feasible if $u \in \mathbb{U}$. Since $\mathbf{x}^0$ coincides with $\tilde{\mathbf{x}}$ and $\mathbf{u}(\cdot)$ coincides with $\tilde{\mathbf{u}}$ for $i = 1, 2, \ldots, N-1$ (but not for $i = N$), a simple calculation yields

$$\begin{aligned} V_N^0(x) &= V_N(x, \mathbf{u}^0(x)) \\ &= \ell(x, \kappa_N(x)) + \sum_{j=1}^{N-1} \ell(x^0(j;x), u^0(j;x)) + V_f(x^0(N;x)) \end{aligned}$$

so that

$$\begin{aligned} V_N(x^+, \tilde{\mathbf{u}}) = V_N^0(x) - \ell(x, \kappa_N(x)) - V_f(x^0(N;x)) + \\ \ell(x^0(N;x), u) + V_f(f(x^0(N;x), u)) \end{aligned}$$

in which $x^+ = f(x, \kappa_N(x))$. Since $V_N^0(x^+) \leq V_N(x^+, \tilde{\mathbf{u}})$, it follows that

$$V_N^0(f(x, \kappa_N(x))) - V_N^0(x) \leq -\ell(x, \kappa_N(x)) \tag{2.18}$$

for all $x \in \mathbb{R}^n$ *provided that* for all $x \in \mathbb{R}^n$, there exists a $u \in \mathbb{U}$ such that

$$V_f(f(x, u)) - V_f(x) + \ell(x, u) \leq 0 \tag{2.19}$$

A function $V_f(\cdot)$ satisfying inequality (2.19) for all $x \in \mathbb{R}^n$ is a *global control-Lyapunov function* (CLF). If $V_f(\cdot)$ is a global CLF, the value function $V_N^0(\cdot)$ has the desired descent property (2.18). Global asymptotic stability of the origin for the system $x^+ = f(x, \kappa_N(x))$ under MPC may be established. If $V_f(\cdot)$ is a global CLF satisfying (2.19), however, there exists a control law $\kappa_f(\cdot)$ satisfying $V_f(f(x, \kappa_f(x))) \leq V_f(x) - \ell(x, \kappa_f(x))$ for all $x \in \mathbb{R}^n$. Global asymptotic stability of the origin for the system $x^+ = f(x, \kappa_f(x))$ may be established. In this case MPC is *not required* to stabilize the system, though it may provide superior performance.

2.4.3 Stabilizing Conditions: Constrained Problems

In this section we consider the case when state and control constraints (2.2) are present. MPC is stabilizing if a global CLF is employed as the terminal cost. A global CLF is seldom available, however, either because the system is nonlinear or because constraints are present. Hence, we must set our sights lower and employ as our terminal cost function $V_f(\cdot)$, a local CLF, one that is defined only on a neighborhood $\mathbb{X}_f$ of the origin where $\mathbb{X}_f \subseteq \mathbb{X}$. A consequent requirement is that the terminal state must be constrained, explicitly or implicitly, to lie in $\mathbb{X}_f$. Our stabilizing condition now takes the form:

Assumption 2.12 (Basic stability assumption).

$$\min_{u \in \mathbb{U}} \{V_f(f(x,u)) + \ell(x,u) \mid f(x,u) \in \mathbb{X}_f\} \leq V_f(x), \quad \forall x \in \mathbb{X}_f$$

This assumption implicitly requires that for each $x \in \mathbb{X}_f$, there exists a $u \in \mathbb{U}$ such that $f(x,u) \in \mathbb{X}_f$, i.e., Assumption 2.12 implies the following assumption.

Assumption 2.13 (Implied invariance assumption). The set $\mathbb{X}_f$ is control invariant for the system $x^+ = f(x,u)$.

Assumptions 2.12 and 2.13 specify properties which, if possessed by the terminal cost function and terminal constraint set, enable us to employ the value function $V_N^0(\cdot)$ for the optimal control problem $\mathbb{P}_N$ as a Lyapunov function. The important descent and monotonicity properties of $V_N^0(\cdot)$ are established in Lemmas 2.14 and 2.15.

Lemma 2.14 (Optimal cost decrease). *Suppose, as usual, that Assumptions 2.2 and 2.3 hold, and that Assumptions 2.12 (and 2.13) hold. Then*

$$V_N^0(f(x,\kappa_N(x))) \leq V_N^0(x) - \ell(x,\kappa_N(x))$$

for all $x \in \mathcal{X}_N$.

Proof. Let x be any point in $\mathcal{X}_N$. Then $V_N^0(x) = V_N(x, \mathbf{u}^0(x))$ where

$$\mathbf{u}^0(x) = \{u^0(0;x), u^0(1;x), \ldots, u^0(N-1;x)\}$$

and $u^0(0;x) = \kappa_N(x)$; the control sequence is feasible for $\mathbb{P}_N(x)$ because it satisfies all control, state, and terminal constraints. The corresponding state sequence is

$$\mathbf{x}^0(x) = \{x^0(0;x), x^0(1;x), \ldots, x^0(N;x)\}$$

where $x^0(0;x) = x$, $x^0(1;x) = f(x, \kappa_N(x))$ and $x^0(N;x) \in \mathbb{X}_f$. At the successor state $x^+ = x^0(1;x)$, we choose, as before, the nonoptimal control sequence $\tilde{\mathbf{u}}$ defined by

$$\tilde{\mathbf{u}} := \{u^0(1;x), \ldots, u^0(N-1;x), u\}$$

where u is still to be chosen. The resultant state sequence is

$$\tilde{\mathbf{x}} = \{x^0(1;x), \ldots, x^0(N;x), f(x^0(N;x), u)\}$$

The control sequence $\tilde{u}(\cdot)$ is feasible, but not necessarily optimal, for $\mathbb{P}_N(x^0(1;x))$ provided that $f(x^0(N;x), u) \in \mathbb{X}_f$. We obtain as before

$$V_N^0(f(x, \kappa_N(x))) \leq V_N^0(x) - \ell(x, \kappa_N(x))$$

provided now that for all $x \in \mathbb{X}_f$, there exists a $u \in \mathbb{U}$ such that

$$V_f(f(x,u)) \leq V_f(x) - \ell(x,u), \text{ and } f(x,u) \in \mathbb{X}_f$$

which is true by Assumptions 2.12 and 2.13. ∎

Lemma 2.14 holds if $\mathbb{U}$ is closed but not necessarily bounded and can be used, with suitable assumptions on $\ell(\cdot)$, to establish asymptotic stability of the origin. The descent property established in Lemma 2.14 may be established also using a monotonicity property of the value function.

2.4.4 Monotonicity of the Value Function

If Assumptions 2.12 (and 2.13) hold, the value function sequence $\{V_j^0(\cdot)\}$ has an interesting monotonicity property, first established for the unconstrained linear quadratic regulator problem, namely, for given x, the value $V_j^0(x)$ decreases as the time to go j increases. We prove this in Lemma 2.15.

Lemma 2.15 (Monotonicity of the value function). *Suppose, as usual, that Assumptions 2.2 and 2.3 hold, and that Assumptions 2.12 (and 2.13) hold. Then*

$$\begin{aligned} V_{j+1}^0(x) &\leq V_j^0(x) && \forall x \in \mathcal{X}_j, \ \forall j \in \mathbb{I}_{0:N-1} \\ V_N^0(x) &\leq V_f(x) && \forall x \in \mathbb{X}_f \end{aligned}$$

Proof. From the DP recursion (2.10)

$$V_1^0(x) = \min_{u \in \mathbb{U}} \{\ell(x,u) + V_0^0(f(x,u)) \mid f(x,u) \in \mathcal{X}_0\}$$

But $V_0^0(\cdot) := V_f(\cdot)$ and $\mathcal{X}_0 := \mathbb{X}_f$. Also, by Assumption 2.12,

$$\min_{u \in \mathbb{U}} \{\ell(x,u) + V_f(f(x,u))\} \leq V_f(x) \qquad \forall x \in \mathbb{X}_f$$

so that

$$V_1^0(x) \leq V_0^0(x) \qquad \forall x \in \mathcal{X}_0$$

Next, suppose that for some $j \geq 1$,

$$V_j^0(x) \leq V_{j-1}^0(x) \qquad \forall x \in \mathcal{X}_{j-1}$$

Then, using the DP equation (2.10)

$$\begin{aligned} V_{j+1}^0(x) - V_j^0(x) = {} & \ell(x, \kappa_{j+1}(x)) + V_j^0(f(x, \kappa_{j+1}(x))) \\ & - \ell(x, \kappa_j(x)) - V_{j-1}^0(f(x, \kappa_j(x))) \qquad \forall x \in \mathcal{X}_j \subseteq \mathcal{X}_{j+1} \end{aligned}$$

Since $\kappa_j(x)$ may *not* be optimal for $\mathbb{P}_{j+1}(x)$ for all $x \in \mathcal{X}_j \subseteq \mathcal{X}_{j+1}$, we have

$$\begin{aligned} V_{j+1}^0(x) - V_j^0(x) \leq {} & \ell(x, \kappa_j(x)) + V_j^0(f(x, \kappa_j(x))) \\ & - \ell(x, \kappa_j(x)) - V_{j-1}^0(f(x, \kappa_j(x))) \qquad \forall x \in \mathcal{X}_j \end{aligned}$$

Also, from (2.12), $x \in \mathcal{X}_j$ implies $f(x, \kappa_j(x)) \in \mathcal{X}_{j-1}$ so that, by assumption, $V_j^0(f(x, \kappa_j(x))) \leq V_{j-1}^0(f(x, \kappa_j(x)))$ for all $x \in \mathcal{X}_j$. Hence

$$V_{j+1}^0(x) \leq V_j^0(x) \qquad \forall x \in \mathcal{X}_j$$

By induction

$$V_{j+1}^0(x) \leq V_j^0(x) \qquad \forall x \in \mathcal{X}_j,\ \forall j \in \{1, 2, \ldots, N-1\}$$

It then follows that $V_N^0(x) \leq V_0^0(x) := V_f(x)$ for all $x \in \mathcal{X}_0 := \mathbb{X}_f$. ∎

Lemma 2.15 also holds if $\mathbb{U}$ is closed but not bounded. The monotonicity property can be used to establish the descent property of $V_N^0(\cdot)$ proved in Lemma 2.14 by noting that

$$\begin{aligned} V_N^0(x) &= \ell(x, \kappa_N(x)) + V_{N-1}^0(f(x, \kappa_N(x))) \\ &= \ell(x, \kappa_N(x)) + V_N^0(f(x, \kappa_N(x))) + \\ &\quad [V_{N-1}^0(f(x, \kappa_N(x))) - V_N^0(f(x, \kappa_N(x)))] \end{aligned}$$

so that using the monotonicity property

$$\begin{aligned} V_N^0(f(x,\kappa_N(x))) &= V_N^0(x) - \ell(x,\kappa_N(x)) + \\ &\quad [V_N^0(f(x,\kappa_N(x))) - V_{N-1}^0(f(x,\kappa_N(x)))] \\ &\leq V_N^0(x) - \ell(x,\kappa_N(x)) \qquad \forall x \in \mathcal{X}_N \end{aligned}$$

which is the desired descent property.

2.4.5 Further Properties of the Value Function $V_N^0(\cdot)$

Lemma 2.14 shows that the value function $V_N^0(\cdot)$ has a descent property that makes it a suitable candidate for a Lyapunov function that may be used to establish stability of the origin for a wide variety of MPC systems. To proceed, we postulate two alternative conditions on the stage cost $\ell(\cdot)$ and terminal cost $V_f(\cdot)$ required to show that $V_N^0(\cdot)$ has the properties given in Appendix B, which are sufficient to establish stability of the origin. Our additional assumption is:

Assumption 2.16 (Bounds on stage and terminal costs).
(a) The stage cost $\ell(\cdot)$ and the terminal cost $V_f(\cdot)$ satisfy

$$\begin{aligned} \ell(x,u) &\geq \alpha_1(|x|) \quad \forall x \in \mathcal{X}_N,\ \forall u \in \mathbb{U} \\ V_f(x) &\leq \alpha_2(|x|) \quad \forall x \in \mathbb{X}_f \end{aligned}$$

in which $\alpha_1(\cdot)$ and $\alpha_2(\cdot)$ are $\mathcal{K}_\infty$ functions, or

(b) The stage cost $\ell(\cdot)$ and the terminal cost $V_f(\cdot)$ satisfy

$$\begin{aligned} \ell(x,u) &\geq c_1|x|^a \quad \forall x \in \mathcal{X}_N,\ \forall u \in \mathbb{U} \\ V_f(x) &\leq c_2|x|^a \quad \forall x \in \mathbb{X}_f \end{aligned}$$

for some $c_1 > 0$, $c_2 > 0$, and $a > 0$.

Note that Assumption 2.16(b) implies 2.16(a) and that both Assumptions 2.16(a) and 2.16(b) are satisfied with $a = 2$ if $\ell(x,u) = (1/2)(x'Qx + u'Ru)$ and Q and R are positive definite. With this extra assumption, $V_N^0(\cdot)$ has the properties summarized in the following result.

Proposition 2.17 (Optimal value function properties).

(a) Suppose that Assumptions 2.2, 2.3, 2.12, 2.13, and 2.16(a) are satisfied. Then there exist $\mathcal{K}_\infty$ functions $\alpha_1(\cdot)$ and $\alpha_2(\cdot)$ such that $V_N^0(\cdot)$ has

the following properties

$$\begin{aligned} V_N^0(x) &\geq \alpha_1(|x|) & \forall x \in \mathcal{X}_N \\ V_N^0(x) &\leq \alpha_2(|x|) & \forall x \in \mathbb{X}_f \\ V_N^0(f(x, \kappa_N(x))) &\leq V_N^0(x) - \alpha_1(|x|) & \forall x \in \mathcal{X}_N \end{aligned}$$

(b) Suppose that Assumptions 2.2, 2.3, 2.12, 2.13, and 2.16(b) are satisfied. Then there exist positive constants c_1, c_2, *and* a *such that* $V_N^0(\cdot)$ *has the following properties*

$$\begin{aligned} V_N^0(x) &\geq c_1|x|^a & \forall x \in \mathcal{X}_N \\ V_N^0(x) &\leq c_2|x|^a & \forall x \in \mathbb{X}_f \\ V_N^0(f(x, \kappa_N(x))) &\leq V_N^0(x) - c_1|x|^a & \forall x \in \mathcal{X}_N \end{aligned}$$

Proof.

(a) The first inequality follows from Assumption 2.16(a) and the fact that $V_N^0(x) \geq \ell(x, \kappa_N(x))$. The second inequality follows from Lemma 2.15 and Assumption 2.16. Finally, the third inequality follows from Lemma 2.14 and Assumption 2.16(a).

(b) The proof of this part is similar to the previous. ■

These properties are almost identical to those required in Theorems B.11 and B.13 in Appendix B to establish asymptotic stability of the origin with a region of attraction $\mathcal{X}_N$. The second property falls short because the upper bound holds for all x in $\mathbb{X}_f$ rather than for all x in $\mathcal{X}_N$. Despite this, asymptotic stability of the origin with a region of attraction $\mathcal{X}_N$ can still be established when $\mathbb{X}_f$ contains the origin in its interior as we show subsequently. Alternatively, the second inequality is sometimes assumed to hold for all $x \in \mathcal{X}_N$, in which case asymptotic stability of the origin can be established using standard theorems in Appendix B; see the subsequent Assumption 2.23. Finally, as we show next, $V_N^0(x) \leq \alpha_2(|x|)$ for all $x \in \mathbb{X}_f$ implies, under some mild assumptions, that $V_N^0(x) \leq \alpha_2(|x|)$ for all $x \in \mathcal{X}_N$ so that, under these assumptions, the results of Appendix B can again be used to prove asymptotic stability of the origin. In the next result, the set X may be $\mathcal{X}_N$, if $\mathcal{X}_N$ is compact, or a sublevel set of $V_N^0(\cdot)$.

Proposition 2.18 (Extension of upper bound to compact set)**.** *Suppose that Assumptions 2.2, 2.3, 2.12, and 2.13 hold, that* $\mathbb{X}_f$ *contains the origin in its interior, and that* $\mathbb{X}_f \subseteq X$ *where* X *is a compact set in* $\mathbb{R}^n$. *If* *there exists a $\mathcal{K}_\infty$ function $\alpha(\cdot)$ such that $V_N^0(x) \leq \alpha(|x|)$ for all $x \in \mathbb{X}_f$, then there exists another $\mathcal{K}_\infty$ function $\beta(\cdot)$ such that $V_N^0(x) \leq \beta(|x|)$ for all $x \in X$.*

Proof. Because the origin lies in the interior of $\mathbb{X}_f$, there exists a $d > 0$ such that $\{x \mid |x| \leq d\} \subset \mathbb{X}_f$. Let $e = \max\{\alpha(|x|) \mid |x| \leq d\} > 0$; then $\alpha(|x|) \geq e$ for all $x \in X \setminus \mathbb{X}_f$. Since X is compact by assumption, $\mathbb{U}$ is compact by Assumption 2.3, and $V_N(\cdot)$ continuous by Proposition 2.4, there exists an upper bound $c > e$ for $V_N(\cdot)$ on $X \times \mathbb{U}^N$ and, hence, for $V_N^0(\cdot)$ on X. Thus $\beta(\cdot) := (c/e)\alpha(\cdot)$ is a $\mathcal{K}_\infty$ function satisfying $\beta(|x|) \geq \alpha(|x|)$ for all x in X and $\beta(|x|) \geq c$ for all $x \in X \setminus \mathbb{X}_f$. Hence $\beta(\cdot)$ is a $\mathcal{K}_\infty$ function satisfying $V_N^0(x) \leq \beta(|x|)$ for all $x \in X$. ■

An immediate consequence of Propositions 2.17 and 2.18 is the following result.

Proposition 2.19 (Lyapunov function on $\mathcal{X}_N$). *Suppose Assumptions 2.2, 2.3, 2.12, 2.13, and 2.16 are satisfied, that $\mathbb{X}_f$ has an interior containing the origin, and that $\mathcal{X}_N$ is bounded. Then, for all $x \in \mathcal{X}_N$*

$$V_N^0(x) \geq \alpha_1(|x|) \tag{2.20}$$

$$V_N^0(x) \leq \alpha_2(|x|) \tag{2.21}$$

$$V_N^0(f(x, \kappa_N(x))) \leq V_N^0(x) - \alpha_1(|x|) \tag{2.22}$$

in which $\alpha_1(\cdot)$ and $\alpha_2(\cdot)$ are $\mathcal{K}_\infty$ functions.

Proof. The result follows directly from Proposition 2.18 since the assumption that the set $\mathcal{X}_N$ is bounded, coupled with the fact that it is closed, as shown in Proposition 2.11, implies that it is compact. ■

Hence, if the hypotheses of Proposition 2.19 are satisfied, Theorems B.11 and B.13 in Appendix B may be used to establish asymptotic stability of the origin in $\mathcal{X}_N$. Sufficient conditions for the boundedness of $\mathcal{X}_N$ are provided by the next result. Recall $f_{\mathbb{Z}}^{-1}(\cdot)$ is given in Definition 2.9.

Proposition 2.20 (Boundedness of $\mathcal{X}_j$). *If either $\mathbb{X}$ is bounded or $\mathbb{X}_f$ is bounded and $f_{\mathbb{Z}}^{-1}(\cdot)$ is bounded on bounded sets, then, for all $j \in \mathbb{I}_{\geq 0}$, $\mathcal{X}_j$ is bounded.*

Proof. That $\mathcal{X}_N$ is bounded if $\mathbb{X}$ is bounded follows immediately from the fact that, by definition, $\mathcal{X}_N \subseteq \mathbb{X}$. Assume then that $\mathbb{X}_f$ is bounded and $f_{\mathbb{Z}}^{-1}(\cdot)$ is bounded on bounded sets. Then the set $\mathcal{Z}_1 = f_{\mathbb{Z}}^{-1}(\mathbb{X}_f)$ is bounded and, hence, so is the set $\mathcal{X}_1$. Suppose, for some $j > 0$, the set $\mathcal{Z}_{j-1}$ is bounded; then its projection onto the set $\mathcal{X}_{j-1}$ is bounded and so is the set $\mathcal{Z}_j = f_{\mathbb{Z}}^{-1}(\mathcal{X}_j)$. Thus the set $\mathcal{X}_j$ is bounded. By induction $\mathcal{X}_j$ is bounded for all $j \in \mathbb{I}_{\geq 0}$. ∎

When $f(\cdot)$ is linear, i.e., $f(x,u) = Ax + Bu$, then $f_{\mathbb{Z}}^{-1}(\cdot)$ is bounded on bounded sets if A is nonsingular. The matrix A is always nonsingular when A and B are obtained by sampling a continuous time system $\dot{x} = A_c x + B_c u$ with u constant between sampling instants. In this case $A = \exp(A_c \Delta)$ and $B = \int_0^\Delta \exp(A_c(\Delta - s))Bds$ so that A is invertible. To show that $f_{\mathbb{Z}}^{-1}(\cdot)$ is bounded on bounded sets, let X be an arbitrary bounded set in $\mathbb{R}^n$ and let x' be an arbitrary point $x' \in X$. Then $f^{-1}(x') = \{(x,u) \mid Ax + Bu = x'\}$. Any (x,u) in $f^{-1}(x')$ satisfies $x = A^{-1}x' - A^{-1}Bu$ so that x lies in the bounded set $A^{-1}X \oplus (-A^{-1}B)$ and u lies in the bounded set $\mathbb{U}$. Hence both $f^{-1}(X)$ and $f_{\mathbb{Z}}^{-1}(X)$ lie in the bounded set $A^{-1}X \oplus (-A^{-1}B)$. A similar result holds for nonlinear systems. If $f(\cdot)$ is obtained by sampling a continuous time system $\dot{x} = f_c(x,u)$ with period Δ and u constant between sampling instants, then $f(\cdot)$ is defined by

$$f(x,u) = x + \int_0^\Delta f_c(x(s), u)ds$$

where $x(s)$ is the solution of $\dot{x} = f_c(x,u)$ at time s if x is the state at time zero and u is the constant input in the interval $[0, \Delta]$.

Proposition 2.21 (Properties of discrete time system). *Suppose that*

(a) $f_c(\cdot)$ is continuous.

(b) There exists a positive constant c such that

$$|f_c(x', u) - f_c(x,u)| \leq c|x' - x| \quad \forall x,\ x' \in \mathbb{R}^n,\ u \in \mathbb{U}$$

Then $f(\cdot)$ and $f_{\mathbb{Z}}^{-1}(\cdot)$ are bounded on bounded sets.

The proof of Proposition 2.21 is discussed in Exercise 2.2. Proposition 2.19 shows that if the terminal constraint set $\mathbb{X}_f$ contains the origin in its interior and if $\mathcal{X}_N$ is bounded, which is often the case, then standard stability theorems, such as Theorems B.11 and B.12 in Appendix B, may be used to establish asymptotic stability of the origin. When $\mathbb{X}_f$ contains the origin in its interior but $\mathcal{X}_N$ is unbounded, asymptotic stability of the origin can still be established using the next result that is a slight generalization of Theorem B.12 in Appendix B.

Theorem 2.22 (Asymptotic stability with unbounded region of attraction). *Suppose $X \subset \mathbb{R}^n$ and $\mathbb{X}_f \subset X$ are positive invariant for the system $x^+ = f(x)$, that $\mathbb{X}_f \subset X$ is closed and contains the origin in its interior, and that there exist a function $V : \mathbb{R}^n \to \mathbb{R}_{\geq 0}$ and two $\mathcal{K}_\infty$ functions $\alpha_1(\cdot)$ and $\alpha_2(\cdot)$ such that*

$$V(x) \geq \alpha_1(|x|) \qquad \forall x \in X \tag{2.23}$$

$$V(x) \leq \alpha_2(|x|) \qquad \forall x \in \mathbb{X}_f \tag{2.24}$$

$$V(f(x)) - V(x) \leq -\alpha_1(|x|) \quad \forall x \in X \tag{2.25}$$

Then the origin is asymptotically stable with a region of attraction X for the system $x^+ = f(x)$.

Proof.

Stability. Because $\mathbb{X}_f$ contains the origin in its interior, there exists a $\delta_1 > 0$ such that $\delta_1 \mathcal{B} \subset \mathbb{X}_f$; here $\mathcal{B}$ denotes the closed unit ball in $\mathbb{R}^n$. Let $\delta \in (0, \delta_1] > 0$ be arbitrary. Let $\phi(i; x)$ denote the solution of $x^+ = f(x)$ at time i if the initial state is x. Suppose that $|x| \leq \delta$ so that $x \in \mathbb{X}_f$. It follows from (2.24) that $V(x) \leq \alpha_2(\delta)$ and from (2.25) that $V(\phi(i; x)) \leq V(x) \leq \alpha_2(\delta)$ for all $i \in \mathbb{I}_{\geq 0}$. From (2.23), $|\phi(i; x)| \leq \alpha_1^{-1}(V(x(i))) \leq (\alpha_1^{-1} \circ \alpha_2)(\delta)$ for all $i \in \mathbb{I}_{\geq 0}$. Hence for all $\varepsilon > 0$, there exists a $\delta > 0$, $\delta := \min\{\delta_1, (\alpha_1^{-1} \circ \alpha_2)^{-1}(\varepsilon)\}$, such that $|x| \leq \delta$ implies that $|\phi(i; x)| \leq \varepsilon$ for all $i \in \mathbb{I}_{\geq 0}$. Stability of the origin is established.

Attractivity. The proof of attractivity is similar to the proof of attractivity in Theorem B.11 of Appendix B. ■

Hence, if we add to the hypotheses of Proposition 2.17 the assumption that $\mathbb{X}_f$ contains the origin in its interior, we can use Theorem 2.22 to establish the asymptotic stability of the origin with a region of attraction $\mathcal{X}_N$ for the system $x^+ = f(x, \kappa_N(x))$.

In situations where $\mathbb{X}_f$ does not have an interior, such as when $\mathbb{X}_f = \{0\}$, we cannot establish an upper bound for $V_N^0(\cdot)$ from Assumptions 2.12 and 2.13, and resort to the following assumption.

Assumption 2.23 (Weak controllability). There exists a $\mathcal{K}_\infty$ function $\alpha(\cdot)$ such that

$$V_N^0(x) \leq \alpha(|x|) \ \forall x \in \mathcal{X}_N$$

Assumption 2.23 is weaker than a controllability assumption though it bounds the cost of steering an initial state x to $\mathbb{X}_f$. It confines attention to those initial states that can be steered to $\mathbb{X}_f$ in N steps while satisfying the control and state constraints, and merely requires that the cost of doing so is not excessive.

2.4.6 Summary

In the sequel we apply the previous results to establish asymptotic or exponential stability of a wide range of MPC systems. To facilitate application, we summarize these results and some of their consequences in the following theorem. Since $\kappa_N(\cdot)$ may be set valued, statements of the form $\kappa_N(x)$ has property A in the sequel should be interpreted as every u in $\kappa_N(x)$ has property A.

Theorem 2.24 (MPC stability).

(a) Suppose that Assumptions 2.2, 2.3, 2.12, 2.13, and 2.16(a) are satisfied and that $\mathcal{X}_N = \mathbb{X}_f = \mathbb{R}^n$ so that $V_f(\cdot)$ is a global CLF. Then the origin is globally asymptotically stable for $x^+ = f(x, \kappa_N(x))$. If, in addition, Assumption 2.16(b) is satisfied, then the origin is globally exponentially stable.

(b) Suppose that Assumptions 2.2, 2.3, 2.12, 2.13, and 2.16(a) are satisfied and that $\mathbb{X}_f$ contains the origin in its interior. Then the origin is asymptotically stable with a region of attraction $\mathcal{X}_N$ for the system $x^+ = f(x, \kappa_N(x))$. If, in addition, Assumption 2.16(b) is satisfied and $\mathcal{X}_N$ is bounded, then the origin is exponentially stable with a region of attraction $\mathcal{X}_N$ for the system $x^+ = f(x, \kappa_N(x))$; if $\mathcal{X}_N$ is unbounded, then the origin is exponentially stable with a region of attraction that is any sublevel set of $V_N^0(\cdot)$.

(c) Suppose that Assumptions 2.2, 2.3, 2.12, 2.13, and 2.23 are satisfied and that $\ell(\cdot)$ satisfies $\ell(x, u) \geq \alpha_1(|x|)$ for all $x \in \mathcal{X}_N$, all $u \in \mathbb{U}$, where $\alpha_1(\cdot)$ is a $\mathcal{K}_\infty$ function. Then the origin is asymptotically stable with a region of attraction $\mathcal{X}_N$ for the system $x^+ = f(x, \kappa_N(x))$. If $\ell(\cdot)$ satisfies $\ell(x, u) \geq c_1|x|^a$ for all $x \in \mathcal{X}_N$, all $u \in \mathbb{U}$, and Assumption 2.23 is satisfied with $\alpha(r) = c_2 r^a$ for some $c_1 > 0$, $c_2 > 0$ and $a > 0$, then the origin is exponentially stable with a region of attraction $\mathcal{X}_N$ for the system $x^+ = f(x, \kappa_N(x))$.

(d) Suppose that Assumptions 2.2, 2.3, 2.12, and 2.13 are satisfied, that $\ell(\cdot)$ satisfies $\ell(x, u) \geq c_1|x|^a + c_1|u|^a$, and that Assumption 2.23 is

satisfied with $\alpha(r) = c_2 r^a$ *for some* $c_1 > 0$, $c_2 > 0$, *and* $a > 0$. *Then* $|\kappa_N(x)| \leq c|x|$ *for all* $x \in \mathcal{X}_N$ *where* $c = (c_2/c_1)^{1/a}$.

Proof.

(a) Since $\mathbb{X}_f = \mathbb{R}^n$, Lemmas 2.14 and 2.15 ensure the existence of $\mathcal{K}_\infty$ functions $\alpha_1(\cdot)$ and $\alpha_2(\cdot)$ such that the value function $V_N^0(\cdot)$ satisfies

$$\begin{aligned} &V_N^0(x) \geq \alpha_1(|x|) \\ &V_N^0(f(x, \kappa_N(x))) \leq V_N^0(x) - \alpha_1(|x|) \\ &V_N^0(x) \leq \alpha_2(|x|) \end{aligned}$$

for all $x \in \mathbb{R}^n$. Asymptotic stability of the origin follows from Theorem B.11 in Appendix B. When Assumption 2.16(b) is satisfied, global exponential stability of the origin follows as in the proof of the next part with $\mathcal{X}_N = \mathbb{R}^n$.

(b) If Assumption 2.16(a) is satisfied, asymptotic stability of the origin follows from Proposition 2.17 and Theorem 2.22. If Assumption 2.16(b) is satisfied and $\mathcal{X}_N$ is bounded, it follows from Propositions 2.18 and 2.19 that there exists c_2 sufficiently large such that the value function satisfies

$$V_N^0(x) \geq c_1 |x|^a \tag{2.26}$$

$$V_N^0(f(x, \kappa_N(x))) \leq V_N^0(x) - c_1 |x|^a \tag{2.27}$$

$$V_N^0(x) \leq c_2 |x|^a \tag{2.28}$$

for all $x \in \mathcal{X}_N$. Consider any initial state $x \in \mathcal{X}_N$, and let $x(i)$ denote the solution at time i of the difference equation $x^+ = f(x, \kappa_N(x))$ with initial condition $x(0) = x$. Since, by Proposition 2.11, $\mathcal{X}_N$ is positive invariant for $x^+ = f(x, \kappa_N(x))$, the entire sequence $\{x(i)\}$ lies in $\mathcal{X}_N$ if the initial state x lies in $\mathcal{X}_N$. Hence $\{x(i)\}$ satisfies

$$V_N^0(x(i+1)) \leq V_N^0(x(i)) - c_1 |x(i)|^a \leq (1 - c_1/c_2) V_N^0(x(i))$$

for all $i \in \mathbb{I}_{\geq 0}$. It follows that

$$V_N^0(x(i)) \leq \gamma^i V_N^0(x(0))$$

for all $i \in \mathbb{I}_{\geq 0}$ in which $\gamma := (1 - c_1/c_2) \in (0, 1)$. Hence

$$|x(i)|^a \leq (1/c_1) V_N^0(x(i)) \leq (1/c_1) \gamma^i V_N^0(x(0)) \leq (c_2/c_1) \gamma^i |x(0)|^a$$

so that

$$|x(i)| \leq c\delta^i |x(0)| \qquad \forall x(0) \in \mathcal{X}_N \quad \forall i \in \mathbb{I}_{\geq 0}$$

in which $c := (c_2/c_1)^{1/a}$ and $\delta := \gamma^{1/a} \in (0,1)$. Since $x(i) \in \mathcal{X}_N$ for all $i \in \mathbb{I}_{\geq 0}$, it follows that the origin is exponentially stable with a region of attraction $\mathcal{X}_N$ for $x^+ = f(x, \kappa_N(x))$. Consider now the case when $\mathcal{X}_N$ is unbounded. It follows from (2.26) that any sublevel set of $V_N^0(\cdot)$ is bounded, and, from (2.27), is positive invariant for $x^+ = f(x, \kappa_N(x))$. The origin is exponentially stable with a region of attraction equal to any sublevel set of $V_N^0(\cdot)$, which follows by similar reasoning for the case when $\mathcal{X}_N$ is bounded by replacing $\mathcal{X}_N$ with the bounded sublevel set of $V_N^0(\cdot)$.

(c) It follows from the proof of Proposition 2.17 and Assumption 2.23 that $V_N^0(\cdot)$ satisfies (2.20)–(2.22) for all $x \in \mathcal{X}_N$. Since $\mathcal{X}_N$ is positive invariant, it follows from Theorem B.13 in Appendix B that the origin is asymptotically stable with a region of attraction $\mathcal{X}_N$ for $x^+ = f(x, \kappa_N(x))$. Suppose now that $\ell(\cdot)$ satisfies $\ell(x,u) \geq c_1|x|^a$ for all $x \in \mathcal{X}_N$, all $u \in \mathbb{U}$ and Assumption 2.23 is satisfied with $\alpha(r) = c_2 r^a$ for some $c_1 > 0$, $c_2 > 0$, and $a > 0$. It follows that $V_N^0(\cdot)$ satisfies (2.26)–(2.28) for all $x \in \mathcal{X}_N$. Exponential stability of the origin for $x^+ = f(x, \kappa_N(x))$ follows by the same reasoning employed in the proof of part (b).

(d) It follows the assumption on $\ell(\cdot)$ and Assumption 2.23 that $c_2|x|^a \geq V_N^0(x) \geq c_1|\kappa_N(x)|^a$ so that $|\kappa_N(x)|^a \leq (c_2/c_1)|x|^a$, which implies $|\kappa_N(x)| \leq (c_2/c_1)^{1/a}|x|$ for all $x \in \mathcal{X}_N$. ■

2.4.7 Controllability and Observability

We have not yet made any assumptions on controllability (stabilizability) or observability (detectability) of the system (2.1) being controlled, which may be puzzling since such assumptions are commonly required in optimal control to, for example, establish existence of a solution to the optimal control problem. The reasons for this omission are that such assumptions are implicitly required, at least locally, for the basic stability Assumption 2.12, and that we restrict attention to $\mathcal{X}_N$, the set of states that can be steered to $\mathbb{X}_f$ in N steps satisfying all constraints.

For example, one version of MPC uses a target set $\mathbb{X}_f = \{0\}$, so that the optimal control problem requires determination of an optimal trajectory terminating at the origin; clearly some assumption on controllability to the origin such as Assumption 2.23 is required. Similarly, if the system being controlled is linear, and the constraints polytopic or polyhedral, a common choice for $\mathbb{X}_f$ is the maximal invariant constraint admissible set for a controlled system where the controller is linear and stabilizing. The terminal constraint set $\mathbb{X}_f$ is then the set $\{x \mid x(i) \in \mathbb{X}, Kx(i) \in \mathbb{U}\}$ where $x(i)$ is the solution at time i of $x^+ = (A + BK)x$, and $u = \kappa_f(x) = Kx$ is a stabilizing control law. Stabilizability of the system being controlled is then required; see Section C.3 of Appendix C for a brief exposition of invariant sets.

Detectability assumptions also are required, mainly in proofs of asymptotic or exponential stability. For example, if the stage cost satisfies $\ell(x,u) = (1/2)(|y|^2 + |u|_R^2)$ where $y = Cx$, the stability proofs commonly establish that $y(k) = Cx(k)$ tends to zero as k tends to infinity. To deduce from this fact that $x(k) \to 0$ requires a detectability assumption on the system $x^+ = f(x,u)$, $y = Cx$. If C is invertible, as we sometimes assume, the system is detectable (since $y(k) \to 0$ implies $x(k) \to 0$).

The requisite assumptions of stabilizability and detectability are made later in the context of discussing specific forms of MPC.

2.4.8 Time-Varying Systems

Most of the control problems discussed in this book are time invariant. Time-varying problems do arise in practice, however, even if the system being controlled is time invariant. One example occurs when an observer or filter is used to estimate the state of the system being controlled since bounds on the state estimation error are often time varying. In the deterministic case, for example, state estimation error decays exponentially to zero. In this section, which may be omitted in the first reading, we show how MPC may be employed for time-varying systems.

The problem. The time-varying nonlinear system is described by

$$x^+ = f(x, u, i)$$

where x is the current state at time i, u the current control, and x^+ the successor state at time $i + 1$. For each integer i, the function $f(\cdot, i)$ is assumed to be continuous. The solution of this system at time k given that the initial state is x at time i is denoted by $\phi(k; x, i, \mathbf{u})$; the solution now depends on both the initial time i and current time k rather than merely on the difference $k - i$ as in the time-invariant case. The cost

$V_N(x, i, \mathbf{u})$ also depends on the initial time i and is defined by

$$V_N(x, i, \mathbf{u}) := \sum_{k=i}^{i+N-1} \ell(x(k), u(k), k) + V_f(x(i+N), i+N)$$

in which $x(k) := \phi(k; x, i, \mathbf{u})$, $\mathbf{u} = \{u(i), u(i+1), \ldots, u(i+N-1)\}$, and the stage cost $\ell(\cdot)$ and terminal cost $V_f(\cdot)$ are time varying. The state and control constraints are also time varying

$$x(k) \in \mathbb{X}(k) \qquad u(k) \in \mathbb{U}(k)$$

for all k. In addition, there is a time-varying terminal constraint

$$x(i+N) \in \mathbb{X}_f(i+N)$$

in which i is the current time. The time-varying optimal control problem at event (x, i) is $\mathbb{P}_N(x, i)$ defined by

$$\mathbb{P}_N(x, i): \quad V_N^0(x, i) = \min\{V_N(x, i, \mathbf{u}) \mid \mathbf{u} \in \mathcal{U}_N(x, i)\}$$

in which $\mathcal{U}_N(x, i)$ is the set of control sequences $\mathbf{u} = \{u(i), u(i+1), \ldots, u(i+N-1)\}$ satisfying the state, control and terminal constraints, i.e.,

$$\mathcal{U}_N(x, i) := \{\mathbf{u} \mid (x, \mathbf{u}) \in \mathbb{Z}_N(i)\}$$

in which, for each i, $\mathbb{Z}_N(i) \subset \mathbb{R}^n \times \mathbb{R}^{Nm}$ is defined by

$$\mathbb{Z}_N(i) := \Big\{(x, \mathbf{u}) \mid u(k) \in \mathbb{U}(k), \quad \phi(k; x, i, \mathbf{u}) \in \mathbb{X}(k), \forall k \in \mathbb{I}_{i,i+N-1},$$
$$\phi(i+N; x, i, \mathbf{u}) \in \mathbb{X}_f(i+N)\Big\}$$

For each time i, the domain of $V_N^0(\cdot, i)$ is $\mathcal{X}_N(i)$ where

$$\begin{aligned}\mathcal{X}_N(i) &:= \{x \in \mathbb{X}(i) \mid \mathcal{U}_N(x, i) \neq \emptyset\} \\ &= \{x \in \mathbb{R}^n \mid \exists \mathbf{u} \text{ such that } (x, \mathbf{u}) \in \mathbb{Z}_N(i)\}\end{aligned}$$

which is the projection of $\mathbb{Z}_N(i)$ onto $\mathbb{R}^n$. Our standing assumptions (2.2 and 2.3) are replaced, in the time-varying case, by

Assumption 2.25 (Continuity of system and cost; time-varying case). The functions $f(\cdot)$, $\ell(\cdot)$, and $V_f(\cdot)$ are continuous; for all $i \in \mathbb{I}_{\geq 0}$, $f(0, 0, i) = 0$, $\ell(0, 0, i) = 0$, and $V_f(0, i) = 0$.

Assumption 2.26 (Properties of constraint sets; time-varying case). For all $i \in \mathbb{I}_{\geq 0}$, $\mathbb{X}(i)$ and $\mathbb{X}_f(i)$ are closed, $\mathbb{X}_f(i) \subset \mathbb{X}(i)$, and $\mathbb{U}(i)$ is compact; each set contains the origin.

Because of the time-varying nature of the problem, we need to extend our definitions of invariance and control invariance.

Definition 2.27 (Time-varying control invariant sets). The sequence of sets $\{X(i) \mid i \in \mathbb{I}_{\geq 0}\}$ is said to be time-varying control invariant for the time-varying system $x^+ = f(x, u, i)$ if, for each $i \in \mathbb{I}_{\geq 0}$, for each $x \in X(i)$, there exists a $u \in \mathbb{U}(i)$ such that $x^+ = f(x, u, i) \in X(i+1)$. The sequence of sets $\{X(i) \mid i \in \mathbb{I}_{\geq 0}\}$ is said to be time-varying positive invariant for the time-varying system $x^+ = f(x, i)$ if, for each $x \in X(i)$, $x^+ = f(x, u, i) \in X(i+1)$.

A sequence of sets $\{X(i)\}$ is a *tube*, and time-varying positive invariance of the sequence is positive invariance of the tube. If (x, i) lies in the tube, i.e., if $x \in X(i)$ for some $i \in \mathbb{I}_{\geq 0}$, then all solutions of $x^+ = f(x, i)$ starting at event (x, i) remain in the tube. The following results, which are analogs of the results for time-invariant systems given previously, are stated without proof.

Proposition 2.28 (Continuous system solution; time-varying case). *Suppose Assumptions 2.25 and 2.26 are satisfied. For each initial time i and final time j, the function* $(x, \mathbf{u}) \mapsto \phi(j; x, i, \mathbf{u})$ *is continuous.*

Proposition 2.29 (Existence of solution to optimal control problem; time-varying case). *Suppose Assumptions 2.25 and 2.26 are satisfied. Then for each time* $i \in \mathbb{I}_{\geq 0}$

(a) The function $(x, \mathbf{u}) \mapsto V_N(x, i, \mathbf{u})$ *is continuous in* $\mathbb{Z}_N(i)$.

(b) For each $x \in \mathcal{X}_N(i)$, *the control constraint set* $\mathcal{U}_N(x, i)$ *is compact.*

(c) For each $x \in \mathcal{X}_N(i)$, *a solution to* $\mathbb{P}_N(x, i)$ *exists.*

(d) $\mathcal{X}_N(i)$ *is closed.*

(e) If $\{\mathbb{X}_f(i)\}$ *is time-varying control invariant for* $x^+ = f(x, u, i)$, *then* $\{\mathcal{X}_N(i)\}$ *is time-varying control invariant for* $x^+ = f(x, u, i)$ *and time-varying positive invariant for* $x^+ = f(x, \kappa_N(x, i), i)$.

(f) $0 \in \mathcal{X}_N(i)$.

Stability. As before, the receding horizon control law $\kappa_N(\cdot)$, which is now time varying, is not necessarily optimal or stabilizing. By choosing the time-varying "ingredients" $V_f(\cdot)$ and $\mathbb{X}_f$ in the optimal control problem appropriately, however, stability can be ensured, as we now show. We replace the stability assumptions (2.12 and 2.13) by their time-varying extension.

Assumption 2.30 (Basic stability assumption; time-varying case). For all $i \in \mathbb{I}_{\geq 0}$, $\min_{u \in \mathbb{U}_i}\{V_f(f(x,u,i), i+1) + \ell(x,u,i) \mid f(x,u,i) \in \mathbb{X}_f(i+1)\} \leq V_f(x,i), \ \forall x \in \mathbb{X}_f(i)$.

This assumption implicitly requires that the sets $\{\mathbb{X}_f(i)\}$ are *time-varying* positive invariant in the following sense.

Assumption 2.31 (Implied invariance assumption; time-varying case). For each $i \in \mathbb{I}_{\geq 0}$ and each $x \in \mathbb{X}_f(i)$, there exists a $u \in \mathbb{U}(i)$ such that $f(x,u,i) \in \mathbb{X}_f(i+1)$.

A direct consequence of Assumption 2.31 is the extension of Lemma 2.14, namely, that the time-varying value function $V_N(\cdot)$ has the descent property that its value at $(f(x, \kappa_N(x,i), i), i+1)$ is less than its value at (x,i) by an amount $\ell(x, \kappa_N(x,i), i)$.

Lemma 2.32 (Optimal cost decrease; time-varying case). *Suppose Assumptions 2.25, 2.26, 2.30 and 2.31 hold. Then,*

$$V_N^0(f(x, \kappa_N(x,i), i), i+1) \leq V_N^0(x,i) - \ell(x, \kappa_N(x,i), i) \tag{2.29}$$

for all $x \in \mathcal{X}_N(i)$, *all* $i \in \mathbb{I}_{\geq 0}$.

Lemma 2.33 (MPC cost is less than terminal cost). *Suppose Assumptions 2.25, 2.26, 2.30 and 2.31 hold. Then,*

$$V_N^0(x,i) \leq V_f(x, i+N) \qquad \forall x \in \mathbb{X}_f(i+N), \quad \forall i \in \mathbb{I}_{\geq 0}$$

The proofs of Lemmas 2.32 and 2.33 are left as Exercises 2.9 and 2.10. Determination of a time-varying terminal cost $V_f(\cdot)$ and time-varying terminal constraint set $\mathbb{X}_f$ is complex. Fortunately there are a few important cases where choice of time-invariant terminal cost and constraint set is possible. The first possibility is $\mathbb{X}_f = \{0\}$ and $V_f(0) = 0$; this choice satisfies Assumptions 2.12 and 2.13, as already demonstrated, as well as Assumptions 2.30 and 2.31. The second possibility arises when $f(\cdot)$ is time invariant, which is the case when output feedback rather than state feedback is employed. In this case, discussed more fully in Chapter 5, time-varying bounds on state estimation error may lead to time-varying constraints even though the underlying system is time invariant. We therefore make the following assumption.

Assumption 2.34 (Bounds on stage and terminal costs; time-varying case).

(a) The terminal cost $V_f(\cdot)$ and terminal constraint set $\mathbb{X}_f$ are time invariant.

(b) The stage cost $\ell(\cdot)$ and the terminal cost $V_f(\cdot)$ satisfy, for all $i \in \mathbb{I}_{\geq 0}$

$$\begin{aligned} \ell(x, u, i) &\geq \alpha_1(|x|) \qquad \forall x \in \mathcal{X}_N(i),\ \forall u \in \mathbb{U}(i) \\ V_f(x) &\leq \alpha_2(|x|) \qquad \forall x \in \mathbb{X}_f \end{aligned}$$

in which $\alpha_1(\cdot)$ and $\alpha_2(\cdot)$ are $\mathcal{K}_\infty$ functions.

Our next result is an analog of Proposition 2.17, and follows fairly simply from Lemmas 2.32 and 2.33 and our assumptions.

Proposition 2.35 (Optimal value function properties; time-varying case). *Suppose Assumptions 2.25, 2.26, 2.30, 2.31, and 2.34 are satisfied. Then there exist two $\mathcal{K}_\infty$ functions $\alpha_1(\cdot)$ and $\alpha_2(\cdot)$ such that, for all $i \in \mathbb{I}_{\geq 0}$*

$$\begin{aligned} V_N^0(x, i) &\geq \alpha_1(|x|) && \forall x \in \mathcal{X}_N(i) \\ V_N^0(x, i) &\leq \alpha_2(|x|) && \forall x \in \mathbb{X}_f \\ V_N^0(f(x, \kappa_N(x, i), i)) &\leq V_N^0(x, i) - \alpha_1(|x|) && \forall x \in \mathcal{X}_N(i) \end{aligned}$$

We can deal with the obstacle posed by the fact that the upper bound on $V_N^0(\cdot)$ holds only in $\mathbb{X}_f$ in much the same way as we did previously for the time-invariant case. For simplicity, however, we invoke instead a uniform controllability assumption.

Assumption 2.36 (Uniform weak controllability). There exists a $\mathcal{K}_\infty$ function $\alpha_1(\cdot)$ such that

$$V_N^0(x, i) \leq \alpha(|x|) \qquad \forall x \in \mathcal{X}_N(i),\ \forall i \in \mathbb{I}_{\geq 0}$$

If Assumptions 2.25, 2.26, 2.30, 2.31, 2.34, and 2.36 are satisfied, it follows from the proof of Proposition 2.35 that, for all $i \in \mathbb{I}_{\geq 0}$

$$V_N^0(x, i) \geq \alpha_1(|x|) \qquad \forall x \in \mathcal{X}_N(i) \tag{2.30}$$

$$V_N^0(x, i) \leq \alpha_2(|x|) \qquad \forall x \in \mathcal{X}_N(i) \tag{2.31}$$

$$V_N^0(f(x, \kappa_N(x, i), i)) \leq V_N^0(x, i) - \alpha_1(|x|) \qquad \forall x \in \mathcal{X}_N(i) \tag{2.32}$$

Since the bounds in inequalities (2.30)–(2.32) hold independently of time $i \in \mathbb{I}_{\geq 0}$, we may employ Theorems B.11 and B.13 of Appendix B with minor modification to obtain the following stability result.

Assumption	Title	Page
2.2	Continuity of system and cost	97
2.3	Properties of constraint sets	97
2.12	Basic stability assumption	115
2.13	Implied invariance assumption	115
2.16	Bounds on stage and terminal costs	118
2.23	Weak controllability	122

Table 2.1: Stability assumptions; time-invariant case.

Assumption	Title	Page
2.25	Continuity of system and cost	127
2.26	Properties of constraint sets	128
2.30	Basic stability assumption	129
2.31	Implied invariance assumption	129
2.34	Bounds on stage and terminal costs	130
2.36	Uniform weak controllability	130

Table 2.2: Stability assumptions; time-varying case.

Theorem 2.37 (MPC stability; time-varying case). *Suppose Assumptions 2.25, 2.26, 2.30, 2.31, 2.34, and 2.36 hold. Then, for each initial time* $i \in \mathbb{I}_{\geq 0}$*, the origin is asymptotically stable with a region of attraction* $\mathcal{X}_N(i)$ *for the time-varying system* $x^+ = f(x, \kappa_N(x, j), j), j \geq i$.

2.5 Examples of MPC

We already have discussed the general principles underlying the design of stabilizing model predictive controllers. The conditions on $\mathbb{X}_f$, $\ell(\cdot)$, and $V_f(\cdot)$ that guarantee stability can be implemented in a variety of ways so that MPC can take many different forms. We present in this section a representative set of examples of MPC and include in these examples the most useful forms for applications. These examples also display the roles of the six main assumptions used to guarantee closed-loop asymptotic stability. These six main assumptions are summarized in Table 2.1 for the time-invariant case and Table 2.2 for the time-varying case. Refering back to this table may prove helpful while reading this section and comparing the various forms of MPC.

One question that is often asked is whether or not the terminal constraint is necessary. Since the conditions given previously are sufficient, necessity cannot be claimed. We discuss this further later. It is evident that the constraint arises because one often has a local, rather than a global, CLF for the system being controlled. In some situations, a global CLF *is* available; in such situations, a terminal constraint is not necessary and the terminal constraint set can be taken to be $\mathbb{R}^n$.

All model predictive controllers determine the control action to be applied to the system being controlled by solving, at each state, an optimal control that is usually constrained. If the constraints in the optimal control problem include hard state constraints, then the feasible region $\mathcal{X}_N$ is a subset of $\mathbb{R}^n$. The analysis given previously shows that if the initial state $x(0)$ lies in $\mathcal{X}_N$, so do all subsequent states, a property known as *recursive feasibility*; this property holds if all the assumptions made in our analysis hold. It is always possible, however, for unanticipated events to cause the state to become infeasible. In this case, the optimal control problem, as stated, cannot be solved, and the controller fails. It is therefore desirable, if this does not conflict with design aims, to employ soft state constraints in place of hard constraints. Otherwise, any implementation of the algorithms described subsequently should be modified to include a feature that enables recovery from faults that causes infeasibility. One remedy is to replace the hard constraints by soft constraints when the current state is infeasible, thereby restoring feasibility, and to revert back to the hard constraints as soon as they are satisfied by the current state.

2.5.1 Unconstrained Systems

For unconstrained systems, $\mathbb{U} = \mathbb{R}^m$ and $\mathbb{X} = \mathbb{R}^n$ so that Assumption 2.3 that postulates $\mathbb{U}$ is compact does not hold.

2.5.1.1 Linear Time-Invariant Systems

Here $f(x,u) = Ax + Bu$ and $\ell(x,u) = (1/2)(|x|_Q^2 + |u|_R^2)$ where $Q > 0$ and $R > 0$. If (A,B) is stabilizable, there exists a stabilizing controller $u = Kx$. Let $A_K := A + BK$, $Q_f := Q + K'RK$ and let $V_f : \mathbb{R}^n \to \mathbb{R}_{\geq 0}$ be defined by $V_f(x) := (1/2)x'P_f x$ where $P_f > 0$ satisfies the Lyapunov equation

$$A_K' P_f A_K + Q_f = P_f$$

Since $V_N^0(x) \geq \ell(x, \kappa_N(x)) \geq (1/2)|x|_Q^2$, it follows that there exist $c_1 > 0$ and $c_2 > 0$ such that

$$V_N^0(x) \geq c_1|x|^2 \qquad V_f(x) \leq c_2|x|^2 \quad \forall x \in \mathbb{R}^n$$

With $f(\cdot)$, $\ell(\cdot)$, and $V_f(\cdot)$ defined this way, problem $\mathbb{P}_N(x)$ is an unconstrained parametric quadratic program[4] of the form $\min_{\mathbf{u}}(1/2)x'Lx + x'M\mathbf{u} + (1/2)\mathbf{u}'N\mathbf{u}$ so that $V_N^0(\cdot)$ is a quadratic function of the parameter x, and $\mathbf{u}^0(\cdot)$ and $\kappa_N(\cdot)$ are linear functions of x. Since

$$V_f(Ax + BKx) - V_f(x) + \ell(x, Kx) = x'[A_K'P_fA_K + Q_f - P_f]x = 0$$

for all $x \in \mathbb{R}^n$,

$$V_f(Ax + BKx) = V_f(x) - \ell(x, Kx) \quad \forall x \in \mathbb{R}^n$$

so that $V_f(\cdot)$ and $\mathbb{X}_f := \mathbb{R}^n$ satisfy Assumptions 2.12 and 2.13 with $u = Kx$; $V_f(\cdot)$ is a global CLF and $\mathcal{X}_N = \mathbb{R}^n$. Hence,

$$V_N^0(f(x, \kappa_N(x))) \leq V_N^0(x) - c_1|x|^2 \quad \forall x \in \mathbb{R}^n$$

It also follows from Lemma 2.15, which does not require the assumption that $\mathbb{U}$ is compact, that

$$V_N^0(x) \leq V_f(x) \leq c_2|x|^2 \quad \forall x \in \mathbb{R}^n$$

Summarizing, we have:

> With these assumptions on $V_f(\cdot)$, $\mathbb{X}_f$, and $\ell(\cdot)$, Assumptions 2.12, 2.13, and 2.16(b) are satisfied and, as shown previously, $V_N^0(\cdot)$ satisfies (2.26)–(2.28). It follows, as shown in Theorem 2.24(a), that the origin is globally exponentially stable for $x^+ = f(x, \kappa_N(x))$.

Since $V_f(\cdot)$ is a global CLF, there exists a simple stabilizing controller, namely $u = Kx$. In this case, there is no motivation to use MPC to obtain a stabilizing controller; standard linear H_2 or H_∞ optimal control theory may be employed to obtain satisfactory control. The situation is different in the time-varying case, which we consider next.

---

[4]An optimization problem is parametric if it takes the form $\min_{\mathbf{u}}\{V_N(x; \mathbf{u}) \mid \mathbf{u} \in \mathcal{U}_N(x)\}$ where $\mathbf{u}$ is the decision variable and x is a parameter; the solution is a function of the parameter x.

2.5.1.2 Linear Time-Varying Systems

Here $f(x, u, i) = A_i x + B_i u$ and $\ell(x, u, i) = (1/2)(|x|^2_{Q_i} + |u|^2_{R_i})$ where $Q_i > 0$ and $R_i > 0$ for $i \in \mathbb{I}_{\geq 0}$. Because of the time-varying nature of the problem, it is impossible to obtain a controller by solving an infinite horizon optimal control problem $\mathbb{P}_\infty(x)$. It is possible, however, to determine for each $x \in \mathcal{X}_N(i)$ and each $i \in \mathbb{I}_{\geq 0}$, the MPC action $\kappa_N(x, i)$. Hence, MPC makes it possible to solve an otherwise intractable problem.

It is difficult to determine a time-varying global CLF satisfying Assumption 2.12 that could serve as the terminal cost function $V_f(\cdot)$, so we impose the condition that $\mathbb{X}_f = \{0\}$ in which case $V_f(\cdot)$ may be chosen arbitrarily; the simplest choice is $V_f(0) = 0$. With this choice, problem $\mathbb{P}_N(x, i)$ is a time-varying unconstrained parametric quadratic program that can be easily solved online either as a parametric program or by DP. The value function is a time-varying quadratic function of the parameter x, and $\kappa_N(\cdot)$ a time-varying linear function of x. The terminal cost function and constraint set satisfy Assumptions 2.30 and 2.31. Our choice of $\ell(\cdot)$ ensures the existence of a $c_1 > 0$ such that $V_N^0(x) \geq \ell(x, \kappa_N(x)) \geq c_1|x|^2$ for all $x \in \mathbb{R}^n$. Because Assumptions 2.30 and 2.31 are satisfied, we can employ Lemma 2.32 to show that $V_N^0(\cdot)$ satisfies the descent property in (2.29). Finally, if we assume that controllability Assumption 2.36 is satisfied, we obtain an upper bound for $V_N^0(\cdot)$. Summarizing, we have:

> With these assumptions on $V_f(\cdot)$ and $\mathbb{X}_f$, Assumptions 2.30, 2.31, and 2.34 are satisfied and if, in addition, Assumption 2.36 is satisfied, then, as shown previously, $V_N^0(\cdot)$ satisfies (2.30)–(2.32). It follows from Theorem 2.37 that, for each initial time i, the origin is asymptotically stable with a region of attraction $\mathcal{X}_N(i)$ for the time-varying system $x^+ = A_j x + B_j \kappa_N(x, j)$, $j \geq i$.

2.5.1.3 Nonlinear Systems

Generally, when the system is nonlinear, albeit unconstrained, it is difficult to obtain a global CLF. We next present two forms of MPC. In the first, which is the simplest, the target set is the origin $\mathbb{X}_f = \{0\}$. In the second, $\mathbb{X}_f$ is a positive invariant ellipsoidal set for the system with linear control based on the linearization of the nonlinear system at the origin. The system to be controlled is

$$x^+ = f(x, u)$$

in which $f(\cdot)$ is continuous. The cost function $V_N(\cdot)$ is defined as before by

$$V_N(x, \mathbf{u}) = \sum_{i=0}^{N-1} \ell(x(i), u(i)) + V_f(x(N))$$

where, for each i, $x(i) := \phi(i; x, \mathbf{u})$, the solution of $x^+ = f(x, u)$ at time i if the initial state is x at time 0 and the control is $\mathbf{u}$. Unless $V_f(\cdot)$ is a global CLF, a terminal constraint set $\mathbb{X}_f$ is required, so the optimal control problem solved online is

$$\mathbb{P}_N(x): \quad V_N^0(x) = \min_{\mathbf{u}}\{V_N(x, \mathbf{u}) \mid \mathbf{u} \in \mathcal{U}_N(x)\}$$

in which, in the absence of state and control constraints,

$$\mathcal{U}_N(x) := \{\mathbf{u} \mid \phi(N; x, \mathbf{u}) \in \mathbb{X}_f\}$$

Problem $\mathbb{P}_N(x)$ is an unconstrained nonlinear parametric program so that global solutions are not usually possible. We ignore this difficulty here and assume in this section that the global solution for any x may be computed online. We address the problem when this is not possible in Section 2.8.

Case 1. $\mathbb{X}_f = \{0\}$, $V_f(0) = 0$. This is the simplest case. As before, we note that Assumptions 2.12 and 2.13 hold if the origin is an equilibrium point, i.e., if $f(0,0) = 0$. If, in addition, we assume that $\ell(\cdot)$ satisfies Assumption 2.16(a), namely that there exists a $\mathcal{K}_\infty$ function $\alpha_1(\cdot)$ such that $\ell(\cdot)$ satisfies $\ell(x, u) \geq \alpha_1(|x|)$ for all $(x, u) \in \mathbb{R}^n \times \mathbb{R}^n$, then, for all $x \in \mathcal{X}_N$

$$V_N^0(x) \geq \ell(x, \kappa_N(x)) \geq \alpha_1(|x|)$$
$$V_N^0(f(x, \kappa_N(x))) - V_f(x) \leq -\ell(x, \kappa_N(x)) \leq -\alpha_1(|x|)$$

where the latter inequality is a consequence of Lemmà 2.14. If we also assume the controllability Assumption 2.23 is satisfied, then

$$V_N^0(x) \leq \alpha(|x|) \ \forall x \in \mathcal{X}_N$$

Summarizing, we have:

If these assumptions on $V_f(\cdot)$, $\mathbb{X}_f$, and $\ell(\cdot)$ hold, and Assumptions 2.2 and 2.3 are satisfied, then Assumptions 2.12, 2.13, and 2.16(a) are satisfied. If, in addition, the controllability Assumption 2.23 is satisfied, then it follows from Theorem 2.24(b) that the origin is asymptotically stable with a region of attraction $\mathcal{X}_N$ for $x^+ = f(x, \kappa_N(x))$.

Case 2. $V_f(x) = (1/2)|x|_P^2$, $\mathbb{X}_f = \{x \mid V_f(x) \leq a\}$. In this case we obtain a terminal cost function $V_f(\cdot)$ and a terminal constraint set $\mathbb{X}_f$ by linearization of the nonlinear system $x^+ = f(x, u)$ at the origin. Hence, for the purpose of this case we assume $f(\cdot)$ and $\ell(\cdot)$ are twice continuously differentiable. Suppose then the linearized system is

$$x^+ = Ax + Bu$$

where $A := f_x(0,0)$ and $B := f_u(0,0)$. We assume that (A, B) is stabilizable and we choose any controller $u = Kx$ such that the origin is globally exponentially stable for the system $x^+ = A_K x$, $A_K := A + BK$, i.e., such that A_K is stable. Suppose also that the stage cost $\ell(\cdot)$ is defined by $\ell(x, u) := (1/2)(|x|_Q^2 + |u|_R^2)$ where Q and R are positive definite; hence $\ell(x, Kx) = (1/2)x'Q^*x$ where $Q^* := (Q + K'RK)$. Let P be defined by the Lyapunov equation

$$A_K' P A_K + 2Q^* = P$$

The reason for the factor 2 will become apparent soon. Since Q^* is positive definite and A_K is stable, P is positive definite. Let the terminal cost function $V_f(\cdot)$ be defined by

$$V_f(x) := (1/2)x'Px$$

Clearly $V_f(\cdot)$ is a global CLF for the linear system $x^+ = Ax + Bu$. Indeed, it follows from its definition that $V_f(\cdot)$ satisfies

$$V_f(A_K x) + x'Q^*x - V_f(x) = 0 \quad \forall x \in \mathbb{R}^n \tag{2.33}$$

Consider now the nonlinear system $x^+ = f(x, u)$ with linear control $u = Kx$. The controlled system satisfies

$$x^+ = f(x, Kx)$$

We wish to show that $V_f(\cdot)$ is a local CLF for $x^+ = f(x, u)$ in some neighborhood of the origin; specifically, we wish to show there exists an $a \in (0, \infty)$ such that

$$V_f(f(x, Kx)) + (1/2)x'Q^*x - V_f(x) \leq 0 \quad \forall x \in W(a) \tag{2.34}$$

where, for all $a > 0$, $W(a) := \text{lev}_a V_f = \{x \mid V_f(x) \leq a\}$ is a sublevel set of V_f. Since P is positive definite, $W(a)$ is an ellipsoid with the origin as its center. Comparing inequality (2.34) with (2.33), we see that (2.34) is satisfied if

$$V_f(f(x,Kx)) - V_f(A_K x) \leq (1/2)x'Q^*x \quad \forall x \in W(a) \tag{2.35}$$

Let $e(\cdot)$ be defined as follows

$$e(x) := f(x,Kx) - A_K x$$

so that

$$V_f(f(x,Kx)) - V_f(A_K x) = (A_K x)'Pe(x) + (1/2)e(x)'Pe(x) \tag{2.36}$$

By definition, $e(0) = f(0,0) - A_K 0 = 0$ and $e_x(x) = f_x(x,Kx) + f_u(x,Kx)K - A_K$. It follows that $e_x(0) = 0$. Since $f(\cdot)$ is twice continuously differentiable, for any $\delta > 0$, there exists a $c_\delta > 0$ such that $|e_{xx}(x)| \leq c_\delta$ for all x in $\delta\mathcal{B}$. From Proposition A.11 in Appendix A,

$$\begin{aligned}|e(x)| &= \left| e(0) + e_x(0)x + \int_0^1 (1-s)x'e_{xx}(sx)x ds \right| \\ &\leq \int_0^1 (1-s)c_\delta |x|^2 ds \leq (1/2)c_\delta |x|^2\end{aligned}$$

for all x in $\delta\mathcal{B}$. From (2.36), we see that there exists an $\varepsilon \in (0,\delta]$ such that (2.35), and, hence, (2.34), is satisfied for all $x \in \varepsilon\mathcal{B}$. Because of our choice of $\ell(\cdot)$, there exists a $c_1 > 0$ such that $V_f(x) \geq \ell(x,Kx) \geq c_1|x|^2$ for all $x \in \mathbb{R}^n$. It follows that $x \in W(a)$ implies $|x| \leq \sqrt{a/c_1}$. We can choose a to satisfy $\sqrt{a/c_1} = \varepsilon$. With this choice, $x \in W(a)$ implies $|x| \leq \varepsilon \leq \delta$, which, in turn, implies (2.34) is satisfied.

We conclude that there exists an $a > 0$ such that $V_f(\cdot)$ and $\mathbb{X}_f := W(a)$ satisfy Assumptions 2.12 and 2.13. For each $x \in \mathbb{X}_f$ there exists a $u = Kx$ such that $V_f(x,u) \leq V_f(x) - \ell(x,u)$ since $\ell(x,Kx) = (1/2)x'Q^*x$. Our assumption that $\ell(x,u) = (1/2)(x'Qx + u'Ru)$ where Q and R are positive definite, and our definition of $V_f(\cdot)$ ensure the existence of positive constants c_1 and c_2 such that $V_N^0(x) \geq c_1|x|^2$ for all $\mathbb{R}^n$, and $V_f(x) \leq c_2|x|^2$ for all $x \in \mathbb{X}_f$ thereby satisfying Assumption 2.16. The set $\mathcal{X}_N = \mathbb{R}^n$ because the optimal control problem $\mathbb{P}_N(x)$ has no state or terminal constraints. Finally, by definition, the set $\mathbb{X}_f$ contains the origin in its interior. Summarizing, we have:

> If these assumptions on $V_f(\cdot)$, $\mathbb{X}_f$, and $\ell(\cdot)$ hold, and Assumptions 2.2 and 2.3 are satisfied, then Assumptions 2.12,

2.13, and 2.16(b) are satisfied, $\mathbb{X}_f$ contains the origin in its interior, and $\mathcal{X}_N = \mathbb{R}^n$. Hence, by Theorem 2.24(b), the origin is globally asymptotically stable for $x^+ = f(x, \kappa_N(x))$. Also, by Theorem 2.24(b), the origin is exponentially stable for $x^+ = f(x, \kappa_N(x))$ with any sublevel set of $V_N^0(\cdot)$ as a region of attraction.

2.5.2 Systems with Control Constraints

Usually, when constraints and/or nonlinearities are present, it is impossible to obtain a *global* CLF to serve as the terminal cost function $V_f(\cdot)$. There are, however, a few special cases where this is possible; we examine two such cases in this section.

2.5.2.1 Linear Stable Systems

The system to be controlled is $x^+ = Ax + Bu$ where A is stable (its eigenvalues lie strictly inside the unit circle) and the control u is subject to the constraint $u \in \mathbb{U}$ where $\mathbb{U}$ is compact and contains the origin in its interior. The stage cost is $\ell(x, u) = (1/2)(x'Qx + u'Ru)$ where Q and R are positive definite. To establish stability of the systems under MPC (or RHC), we wish to obtain a *global* CLF to serve as the terminal cost function $V_f(\cdot)$. This is usually difficult because any linear control law $u = Kx$, say, will transgress the control constraint for x sufficiently large. In other words, it is usually impossible to find a $V_f(\cdot)$ such that there exists a $u \in \mathbb{U}$ satisfying $V_f(Ax + Bu) \leq V_f(x) - \ell(x, u)$ for all x in $\mathbb{R}^n$. Since A is stable, however, it is possible to obtain a Lyapunov function for the autonomous system $x^+ = Ax$ that is a suitable candidate for $V_f(\cdot)$; in fact, for all $Q > 0$, there exists a $P > 0$ such that

$$A'PA + Q = P$$

Let $V_f(\cdot)$ be defined by

$$V_f(x) = (1/2)x'Px$$

With $f(\cdot)$, $\ell(\cdot)$, and $V_f(\cdot)$ defined thus, $\mathbb{P}_N(x)$ is a parametric quadratic problem if the constraint set $\mathbb{U}$ is polyhedral and global solutions may be computed online. The terminal cost function $V_f(\cdot)$ satisfies

$$V_f(Ax) + (1/2)x'Qx - V_f(x) = (1/2)x'(A'PA + Q - P)x = 0$$

for all $x \in \mathbb{X}_f := \mathbb{R}^n$. We see that for all $x \in \mathbb{X}_f$, there exists a u, namely $u = 0$, such that $V_f(Ax + Bu) \leq V_f(x) - \ell(x,u)$; $\ell(x,u) = (1/2)x'Qx$ when $u = 0$. Since there are no state or terminal constraints, $\mathcal{X}_N = \mathbb{R}^n$. It follows that there exist positive constants c_1 and c_2 such that

$$\begin{aligned} V_N^0(x) &\geq c_1|x|^2 \\ V_N^0(f(x, \kappa_N(x))) &\leq V_N^0(x) - c_1|x|^2 \\ V_N^0(x) &\leq c_2|x|^2 \end{aligned}$$

for all $x \in \mathcal{X}_N = \mathbb{R}^n$. Summarizing, we have:

> If these assumptions on $V_f(\cdot)$, $\mathbb{X}_f$, and $\ell(\cdot)$ hold, and Assumption 2.3 is satisfied, then Assumptions 2.12, 2.13, and 2.16(b) are satisfied and $\mathcal{X}_N = \mathbb{X}_f = \mathbb{R}^n$. It follows from Theorem 2.24(a) that the origin is globally, exponentially stable for the controlled system $x^+ = Ax + BK\kappa_N(x)$.

An extension of this approach for unstable A is used in Chapter 6.

2.5.2.2 Neutrally Stable Systems

The system to be controlled is, again, $x^+ = Ax + Bu$, but A is now *neutrally* stable[5] and the control u is subject to the constraint $u \in \mathbb{U}$ where $\mathbb{U}$ is compact, contains the origin in its interior, and has the form

$$\mathbb{U} = \{u \in \mathbb{R}^m \mid u_i \in [a_i, b_i],\ i \in \mathbb{I}_{1:m}\}$$

where $a_i < 0 < b_i$. The linear system $x^+ = Ax$ is therefore Lyapunov stable but not asymptotically stable. We assume that the pair (A, B) is controllable. This problem is much more challenging than the problem considered immediately above since control has to be applied to make the system asymptotically stable, and this control can transgress the control constraints. Any linear control law $u = Kx$, no matter how small K, transgresses the control constraints for large enough x. Recent research, however, has demonstrated the existence of a *global* CLF for $x^+ = Ax + Bu$ where A is neutrally stable; the Lyapunov function is, unusually, based on a *nonlinear* control law of the form $u = \text{sat}(Kx)$ where $\text{sat}(\cdot)$ is the vector saturation function defined by

$$\text{sat}(u) := \begin{bmatrix} \text{sat}(u_1) & \text{sat}(u_2) & \cdots & \text{sat}(u_m) \end{bmatrix}'$$

[5]A linear system is neutrally stable if some of the eigenvalues of A lie on the unit circle and are simple, and the remaining eigenvalues lie within the unit circle.

in which u_i is the ith component of the vector u, and

$$\text{sat}(u_i) := \begin{cases} b_i & u_i \geq b_i \\ u_i & u_i \in [a_i, b_i] \\ a_i & u_i \leq a_i \end{cases}$$

If A is neutrally stable, there exists a $P > 0$ such that

$$A'PA \leq P$$

Note that this is weaker than the corresponding result in the previous section. If κ satisfies

$$\kappa B'PB < I$$

however, then the linear control law $u = Kx$ in which

$$K := -\kappa B'PA$$

globally stabilizes the *unconstrained* system, i.e., the matrix $A_K := A + BK$ is stable. Hence, for all $Q^* > 0$, there exists a positive definite matrix P^* satisfying

$$A_K' P^* A_K + Q^* = P^*$$

Let $\kappa_f(\cdot)$ denote the *nonlinear* control law defined by

$$\kappa_f(x) := \text{sat}(Kx)$$

Then, as shown in Kim, Yoon, Jadbabaie, and Persis (2004), there exists a $\lambda > 0$ such that $V_f(\cdot)$ defined by

$$V_f(x) := (1/2)x'P^*x + \lambda(x'Px)^{3/2}$$

is a *global* CLF for $x^+ = Ax + Bu$, satisfying

$$V_f(Ax + B\kappa_f(x)) - V_f(x) + (1/2)|x|^2_{Q^*} \leq 0 \tag{2.37}$$

for all $x \in \mathbb{R}^n$. Suppose now the optimal control problem defining the receding horizon controller (or model predictive controller) is

$$\mathbb{P}_N(x) : \quad V_N^0(x) = \min_{\mathbf{u}}\{V_N(x, \mathbf{u}) \mid \mathbf{u} \in \mathbb{U}^N\}$$

in which the cost $V_N(\cdot)$ is defined by

$$V_N(x, \mathbf{u}) := \sum_{i=0}^{N-1} \ell(x(i), u(i)) + V_f(x(N))$$

and, for all i, $x(i) = \phi(i; x, \mathbf{u})$, the solution of $x^+ = Ax + Bu$ at time i if the initial state at time 0 is x and the control sequence is $\mathbf{u}$. The stage cost is

$$\ell(x, u) := (1/2)(|x|_Q^2 + |u|_R^2)$$

where Q and R are positive definite and R is diagonal. We wish to ensure that $V_f(\cdot)$, defined previously, and $\mathbb{X}_f := \mathbb{R}^n$ satisfy Assumption 2.12 so that the system with MPC has satisfactory stability properties. But Assumptions 2.12 and 2.13 are satisfied for all $x \in \mathbb{R}^n$ if

$$V_f(Ax + B\kappa_f(x)) - V_f(x) + \ell(x, \kappa_f(x)) \leq 0 \tag{2.38}$$

It follows from (2.37) that Assumption 2.12 is satisfied if

$$(1/2)x'Q^*x \geq \ell(x, \kappa_f(x)) = (1/2)x'Qx + (1/2)\kappa_f(x)'R\kappa_f(x) \tag{2.39}$$

We can achieve this by choosing Q^* appropriately. Suppose

$$Q^* := Q + K'RK$$

Then

$$x'Q^*x = x'Qx + (Kx)'RKx$$

But

$$\kappa_f(x)'R\kappa_f(x) = (\text{sat}(Kx))'R\text{sat}(Kx)$$

Since R is diagonal and positive definite, and since $(\text{sat}(a))^2 \leq a^2$ if a is a scalar, we have

$$\kappa_f(x)'R\kappa_f(x) \leq (\text{sat}(Kx))'R\text{sat}(Kx) \leq (Kx)'RKx \; \forall x \in \mathbb{R}^n$$

It follows that (2.39) and, hence, (2.38) are satisfied. Therefore, with $V_f(\cdot)$ as defined previously, $\mathbb{X}_f := \mathbb{R}^n$, and $\ell(\cdot)$ as defined previously, Assumptions 2.12, 2.13, and 2.16(b) are satisfied. Summarizing, we have:

> If these assumptions on $V_f(\cdot)$, $\mathbb{X}_f$, and $\ell(\cdot)$ hold, and Assumption 2.3 is satisfied, then Assumptions 2.12, 2.13, and 2.16(b) are satisfied, and $\mathcal{X}_N = \mathbb{X}_f = \mathbb{R}^n$. It follows from Theorem 2.24 that the origin is globally exponentially stable for the controlled system $x^+ = Ax + B\kappa_N(x)$.

Because $V_f(\cdot)$ is not quadratic, the optimal control problem $\mathbb{P}_N(x)$ is no longer a quadratic program.

2.5.3 Systems with Control and State Constraints

We turn now to the consideration of systems with control and state constraints. In this situation determination of a global CLF is usually difficult if not impossible. Hence we show how local CLFs may be determined together with an invariant region in which they are valid.

2.5.3.1 Linear Systems

The system to be controlled is $x^+ = Ax + Bu$ where A is not necessarily stable, the control u is subject to the constraint $u \in \mathbb{U}$ where $\mathbb{U}$ is compact and contains the origin in its interior, and the state x is subject to the constraint $x \in \mathbb{X}$ where $\mathbb{X}$ is closed and contains the origin in its interior. The stage cost is $\ell(x, u) = (1/2)(x'Qx + u'Ru)$ where Q and R are positive definite. Because of the constraints, it is difficult to obtain a global CLF. Hence we restrict ourselves to the more modest goal of obtaining a local CLF and proceed as follows. If (A, B) is stabilizable, the solution to the infinite horizon *unconstrained* optimal control problem $\mathbb{P}_\infty^{uc}(x)$ is known; the value function for this problem is $V_\infty^{uc}(x) = (1/2)x'Px$ where P is the unique (in the class of positive semidefinite matrices) solution to the discrete algebraic Riccati equation

$$P = A_K' P A_K + Q^*$$

in which $A_K := A + BK$, $Q^* := Q + K'RK$, and $u = Kx$, in which K is defined by

$$K := -(B'PB + R)^{-1} B'PA'$$

is the optimal controller. The value function $V_\infty^{uc}(\cdot)$ for the infinite horizon unconstrained optimal control problem $\mathbb{P}_\infty^{uc}(x)$ satisfies

$$V_\infty^{uc}(x) = \min_u \{\ell(x, u) + V_\infty^{uc}(Ax + Bu)\} = \ell(x, Kx) + V_\infty^{uc}(A_K x)$$

It is known that P is positive definite. We define the terminal cost $V_f(\cdot)$ by

$$V_f(x) := V_\infty^{uc}(x) = (1/2)x'Px$$

If $\mathbb{X}$ and $\mathbb{U}$ are polyhedral, problem $\mathbb{P}_N(x)$ is a parametric quadratic program that may be solved online using standard software. The terminal cost function $V_f(\cdot)$ satisfies

$$V_f(A_K x) + (1/2)x'Q^*x - V_f(x) \leq 0 \ \forall x \in \mathbb{R}^n$$

The controller $u = Kx$ does not necessarily satisfy the control and state constraints, however. The terminal constraint set $\mathbb{X}_f$ must be chosen with this requirement in mind. We may choose $\mathbb{X}_f$ to be the maximal invariant constraint admissible set for $x^+ = A_K x$; this is the largest set W with respect to inclusion[6] satisfying: (a) $W \subseteq \{x \in \mathbb{X} \mid Kx \in \mathbb{U}\}$, and (b) $x \in W$ implies $x(i) = A_K^i x \in W$ for all $i \geq 0$. Thus $\mathbb{X}_f$, defined this way, is control invariant[7] for $x^+ = Ax + Bu$, $u \in \mathbb{U}$. If the initial state x of the system is in $\mathbb{X}_f$, the controller $u = Kx$ maintains the state in $\mathbb{X}_f$ and satisfies the state and control constraints for all future time ($x(i) = A_K^i x \in \mathbb{X}_f \subset \mathbb{X}$ and $u(i) = Kx(i) \in \mathbb{U}$ for all $i \geq 0$). Hence, with $V_f(\cdot)$, $\mathbb{X}_f$, and $\ell(\cdot)$ as defined previously, Assumptions 2.12, 2.13, and 2.16(b) are satisfied. Summarizing, we have:

> If these assumptions on $V_f(\cdot)$, $\mathbb{X}_f$, and $\ell(\cdot)$ hold, and Assumption 2.3 is satisfied, then Assumptions 2.12, 2.13, and 2.16(b) are satisfied, and $\mathbb{X}_f$ contains the origin in its interior. Hence, by Theorem 2.24, the origin is asymptotically stable with a region of attraction $\mathcal{X}_N$ for the controlled system $x^+ = Ax + B\kappa_N(x)$, and exponentially stable with a region of attraction any sublevel set of $V_N^0(\cdot)$.

It is, of course, not necessary to choose K and $V_f(\cdot)$ as above. Any K such that $A_K = A + BK$ is stable may be chosen, and P may be obtained by solving the Lyapunov equation $A_K' P A_K + Q = P$. With $V_f(x) := (1/2)x'Px$ and $\mathbb{X}_f$ the maximal constraint admissible set for $x^+ = A_K x$, the origin may be shown, as above, to be asymptotically stable with a region of attraction $\mathcal{X}_N$ for $x^+ = Ax + B\kappa_N(x)$, and exponentially stable with a region of attraction any sublevel set of $V_N^0(\cdot)$. The optimal control problem is, again, a quadratic program. The terminal set $\mathbb{X}_f$ may be chosen, as above, to be the maximal invariant constraint admissible set for $x^+ = A_K x$, or it may be chosen to be a suitably small sublevel set of $V_f(\cdot)$; by suitably small, we mean small enough to ensure $\mathbb{X}_f \subseteq \mathbb{X}$ and $K\mathbb{X}_f \subseteq \mathbb{U}$. The set $\mathbb{X}_f$, if chosen this way, is ellipsoidal, a subset of the maximal constraint admissible set, and is positive invariant for $x^+ = A_K x$. The disadvantage of this choice is that $\mathbb{P}_N(x)$ is no longer a quadratic program, though it remains a convex program for which software exists.

---

[6] $W \in \mathcal{W}$ is the largest set in $\mathcal{W}$ with respect to inclusion if $W' \subseteq W$ for any $W' \in \mathcal{W}$.

[7] A set X is positive invariant for $x^+ = f(x)$ if $x \in X$ implies $x^+ = f(x) \in X$; a set X is control invariant for $x^+ = f(x, u)$, $u \in \mathbb{U}$ if $x \in X$ implies the existence of a $u \in \mathbb{U}$ such that $x^+ = f(x, u) \in X$.

The choice $V_f(\cdot) = V_\infty^{uc}(\cdot)$ results in an interesting property of the closed-loop system $x^+ = Ax + B\kappa_N(x)$. Generally, the terminal constraint set $\mathbb{X}_f$ is *not* positive invariant for the controlled system $x^+ = Ax + B\kappa_N(x)$. Thus, in solving $\mathbb{P}_N(x)$ for an initial state $x \in \mathbb{X}_f$, the "predicted" state sequence $\mathbf{x}^0(x) = \{x^0(0;x), x^0(1;x), \ldots, x^0(N;x)\}$ starts and ends in $\mathbb{X}_f$ but does not necessarily remain in $\mathbb{X}_f$. Thus $x^0(0;x) = x \in \mathbb{X}_f$ and $x^0(N;x) \in \mathbb{X}_f$, because of the terminal constraint in the optimal control problem, but, for any $i \in \mathbb{I}_{1:N-1}$, $x^0(i;x)$ may lie outside of $\mathbb{X}_f$. In particular, $x^+ = Ax + B\kappa_N(x) = x^0(1;x)$ may lie outside of $\mathbb{X}_f$; $\mathbb{X}_f$ is *not* necessarily positive invariant for the controlled system $x^+ = Ax + B\kappa_N(x)$.

Consider now the problem $\mathbb{P}_N^{uc}(x)$ defined in the same way as $\mathbb{P}_N(x)$ except that *all* constraints are omitted so that $\mathcal{U}_N(x) = \mathbb{R}^{Nm}$

$$\mathbb{P}_N^{uc}(x): \qquad V_N^{uc}(x) = \min_{\mathbf{u}} V_N(x, \mathbf{u})$$

in which $V_N(\cdot)$ is defined as previously by

$$V_N(x, \mathbf{u}) := \sum_{i=0}^{j-1} \ell(x(i), u(i)) + V_f(x(j))$$

with $V_f(\cdot)$ the value function for the infinite horizon unconstrained optimal control problem, i.e., $V_f(x) := V_\infty^{uc}(x) = (1/2)x'Px$. With these definitions, it follows that

$$V_N^{uc}(x) = V_\infty^{uc}(x) = V_f(x) = (1/2)x'Px$$
$$\kappa_N^{uc}(x) = Kx, \quad K = -(B'PB + R)^{-1}B'PA$$

for all $x \in \mathbb{R}^n$; $u = Kx$ is the optimal controller for the unconstrained infinite horizon problem. But $\mathbb{X}_f$ *is* positive invariant for $x^+ = A_K x$.

We now claim that with $V_f(\cdot)$ chosen to equal to $V_\infty^{uc}(\cdot)$, the terminal constraint set $\mathbb{X}_f$ *is* positive invariant for $x^+ = Ax + B\kappa_N(x)$. We do this by showing that $V_N^0(x) = V_N^{uc}(x) = V_\infty^{uc}(x)$ for all $x \in \mathbb{X}_f$, so that the associated control laws are the same, i.e., $\kappa_N(x) = Kx$. First, because $\mathbb{P}_N^{uc}(x)$ is identical with $\mathbb{P}_N(x)$ except for the absence of all constraints, we have

$$V_N^{uc}(x) = V_f(x) \le V_N^0(x) \quad \forall x \in \mathcal{X}_N \supseteq \mathbb{X}_f$$

Second, from Lemma 2.15,

$$V_N^0(x) \le V_f(x) \quad \forall x \in \mathbb{X}_f$$

Hence $V_N^0(x) = V_N^{uc}(x) = V_f(x)$ for all $x \in \mathbb{X}_f$. That $\kappa_N(x) = Kx$ for all $x \in \mathbb{X}_f$ follows from the uniqueness of the solutions to the problems $\mathbb{P}_N(x)$ and $\mathbb{P}_N^{uc}(x)$. Summarizing, we have:

> If $V_f(\cdot)$ is chosen to be the value function for the unconstrained infinite horizon optimal control problem, if $u = Kx$ is the associated controller, and if $\mathbb{X}_f$ is invariant for $x^+ = A_K x$, then $\mathbb{X}_f$ is also positive invariant for the controlled system $x^+ = Ax + B\kappa_N(x)$. Also $\kappa_N(x) = Kx$ for all $x \in \mathbb{X}_f$.

2.5.3.2 Nonlinear Systems

The system to be controlled is

$$x^+ = f(x, u)$$

in which $f(\cdot)$ is assumed to be twice continuously differentiable. The system is subject to state and control constraints

$$x \in \mathbb{X} \qquad u \in \mathbb{U}$$

in which $\mathbb{X}$ is closed and $\mathbb{U}$ is compact; each set contains the origin in its interior. The cost function is defined by

$$V_N(x, \mathbf{u}) = \sum_{i=0}^{N-1} \ell(x(i), u(i)) + V_f(x(N))$$

in which, for each i, $x(i) := \phi(i; x, \mathbf{u})$, the solution of $x^+ = f(x, u)$ at time i if the initial state is x at time 0 and the control is $\mathbf{u}$. The stage cost $\ell(\cdot)$ is defined by

$$\ell(x, u) := (1/2)(|x|_Q^2 + |u|_R^2)$$

in which Q and R are positive definite. The optimal control problem $\mathbb{P}_N(x)$ is defined by

$$\mathbb{P}_N(x): \quad V_N^0(x) = \min_{\mathbf{u}}\{V_N(x, \mathbf{u}) \mid \mathbf{u} \in \mathcal{U}_N(x)\}$$

in which $\mathcal{U}_N(x)$ is defined by (2.6) and includes the terminal constraint $x(N) = \phi(N; x, \mathbf{u}) \in \mathbb{X}_f$ (in addition to the state and control constraints). Our first task is to choose the ingredients $V_f(\cdot)$ and $\mathbb{X}_f$ of the optimal control problem to ensure asymptotic stability of the origin for the controlled system. We proceed as in Section 2.5.1.3, i.e., we linearize the system at the origin to obtain the linear model

$$x^+ = Ax + Bu$$

in which $A = f_x(0,0)$ and $B = f_u(0,0)$ and assume, as before, that (A, B) is stabilizable. We choose any controller $u = Kx$ such that A_K is stable. Choose $Q^* := (Q + K'RK)$ and let P be defined by the Lyapunov equation

$$A_K' P A_K + 2Q^* = P$$

The terminal cost function $V_f(\cdot)$ is again chosen to be

$$V_f(x) := (1/2)x'Px$$

and $\mathbb{X}_f$ is chosen to be a sublevel set $W(a) := \text{lev}_a V_f := \{x \mid V_f(x) \leq a\}$ for some suitably chosen constant a. As shown in Section 2.5.1.3, under the assumptions made previously, there exists an $a > 0$ such that

$$V_f(f(x, Kx)) + \ell(x, Kx) - V_f(x) \leq 0 \;\forall x \in \mathbb{X}_f := W(a)$$

in which $x^+ = f(x, Kx)$ describes the nonlinear system if the linear controller $u = Kx$ is employed. To take into account the state and control constraints, we reduce a if necessary to satisfy, in addition,

$$\mathbb{X}_f \subseteq \mathbb{X} \qquad K\mathbb{X}_f \subseteq \mathbb{U}$$

With $f(\cdot)$, $\ell(\cdot)$, and $V_f(\cdot)$ defined thus, $\mathbb{P}_N(x)$ is a constrained parametric nonlinear optimization problem for which global solutions cannot necessarily be obtained online; we temporarily ignore this problem. Because $\mathbb{X}_f$ is a sublevel set of $V_f(\cdot)$, it is positive invariant for $x^+ = f(x, Kx)$. It follows that $V_f(\cdot)$ and $\mathbb{X}_f$ satisfy Assumptions 2.12 and 2.13. Summarizing, we have:

> If these assumptions on $V_f(\cdot)$, $\mathbb{X}_f$, and $\ell(\cdot)$ hold, and Assumptions 2.2 and 2.3 are satisfied, then Assumptions 2.12, 2.13, and 2.16(b) are satisfied, and X_f contains the origin in its interior. Hence, by Theorem 2.24(b), the origin is asymptotically stable for $x^+ = f(x, \kappa_N(x))$ in $\mathcal{X}_N$ and exponentially stable for $x^+ = f(x, \kappa_N(x))$ in any sublevel set of $V_N^0(\cdot)$.

Asymptotic stability of the origin in $\mathcal{X}_N$ may also be established when $\mathbb{X}_f := \{0\}$ if Assumption 2.23 is invoked.

2.6 Is a Terminal Constraint Set $\mathbb{X}_f$ Necessary?

While addition of a terminal cost $V_f(\cdot)$ does not materially affect the optimal control problem, addition of a terminal constraint $x(N) \in \mathbb{X}_f$, which is a state constraint, may have a significant effect. In particular, problems with only control constraints are usually easier to solve. So if state constraints are not present or if they are handled by penalty functions (soft constraints), it is highly desirable to avoid the addition of a terminal constraint. Moreover, it is possible to establish continuity of the value function for a range of optimal control problems *if* there are no state constraints; continuity of the value function ensures a degree of robustness (see Chapter 3). It is therefore natural to ask if the terminal constraint can be omitted without affecting stability. There are several answers to this question.

2.6.1 Replacing the Terminal Constraint by a Terminal Cost

A reasonably simple procedure is to replace the terminal constraint $x(N) \in \mathbb{X}_f$ by a terminal cost that is sufficiently large to ensure automatic satisfaction of the terminal constraint.

We assume, as in the examples of MPC discussed in Section 2.5, that the terminal cost function $V_f(\cdot)$, the constraint set $\mathbb{X}_f$, and the stage cost $\ell(\cdot)$ for the optimal control problem $\mathbb{P}_N(x)$ are chosen to satisfy Assumptions 2.12, 2.13 and 2.16 so that there exists a local control law $\kappa_f : \mathbb{X}_f \to \mathbb{U}$ such that $\mathbb{X}_f \subset \{x \in \mathbb{X} \mid \kappa_f(x) \in \mathbb{U}\}$ is positive invariant for $x^+ = f(x, \kappa_f(x))$ and $V_f(f(x, \kappa_f(x))) + \ell(x, \kappa_f(x)) \leq V_f(x)$ for all $x \in \mathbb{X}_f$. We assume that the function $V_f(\cdot)$ is defined on $\mathbb{X}$ even though it possesses the property $V_f(f(x, \kappa_f(x))) + \ell(x, \kappa_f(x)) \leq V_f(x)$ only in $\mathbb{X}_f$. In many cases, even if the system being controlled is nonlinear, $V_f(\cdot)$ is quadratic and positive definite, and $\kappa_f(\cdot)$ is linear. The set $\mathbb{X}_f$ may be chosen to be a sublevel set of $V_f(\cdot)$ so that $\mathbb{X}_f = W(a) := \{x \mid V_f(x) \leq a\}$ for some $a > 0$. We discuss in the sequel a modified form of the optimal control problem $\mathbb{P}_N(x)$ in which the terminal cost $V_f(\cdot)$ is replaced by $\beta V_f(\cdot)$ and the terminal constraint $\mathbb{X}_f$ is omitted, and show that if β is sufficiently large the solution of the modified optimal control problem is such that the optimal terminal state nevertheless lies in $\mathbb{X}_f$ so that terminal constraint is implicitly satisfied.

For all $\beta \geq 1$, let $\mathbb{P}_N^\beta(x)$ denote the modified optimal control problem defined by

$$\hat{V}_N^\beta(x) = \min_{\mathbf{u}} \{V_N^\beta(x, \mathbf{u}) \mid \mathbf{u} \in \hat{\mathcal{U}}_N(x)\}$$

in which the cost function to be minimized is now

$$V_N^\beta(x, \mathbf{u}) := \sum_{i=0}^{N-1} \ell(x(i), u(i)) + \beta V_f(x(N))$$

in which, for all i, $x(i) = \phi(i; x, \mathbf{u})$, the solution at time i of $x^+ = f(x, u)$ when the initial state is x and the control sequence is $\mathbf{u}$. The control constraint set $\hat{\mathcal{U}}_N(x)$ ensures satisfaction of the state and control constraints, but not the terminal constraint, and is defined by

$$\hat{\mathcal{U}}_N(x) := \{\mathbf{u} \mid u(i) \in \mathbb{U},\ x(i) \in \mathbb{X},\ i \in \mathbb{I}_{0,N-1}, x(N) \in \mathbb{X}\}$$

The cost function $V_N^\beta(\cdot)$ with $\beta = 1$ is identical to the cost function $V_N(\cdot)$ employed in the standard problem $\mathbb{P}_N$ considered previously. Let $\hat{\mathcal{X}}_N := \{x \in \mathbb{X} \mid \hat{\mathcal{U}}_N(x) \neq \emptyset\}$ denote the domain of $\hat{V}^\beta(\cdot)$; let $\mathbf{u}^\beta(x)$ denote the solution of $\mathbb{P}_N^\beta(x)$; and let $\mathbf{x}^\beta(x)$ denote the associated optimal state trajectory. Thus

$$\mathbf{u}^\beta(x) = \{u^\beta(0; x), u^\beta(1; x), \ldots, u^\beta(N-1; x)\}$$
$$\mathbf{x}^\beta(x) = \{x^\beta(0; x), x^\beta(1; x), \ldots, x^\beta(N; x)\}$$

where $x^\beta(i; x) := \phi(i; x, \mathbf{u}^\beta(x))$ for all i. The implicit MPC control law is $\kappa_N^\beta(\cdot)$ where $\kappa_N^\beta(x) := u^\beta(0; x)$. Neither $\hat{\mathcal{U}}_N(x)$ nor $\hat{\mathcal{X}}_N$ depend on the parameter β. It can be shown (Exercise 2.11) that *the pair* $(\beta V_f(\cdot), \mathbb{X}_f)$ *satisfies Assumptions 2.12 and 2.13* if $\beta \geq 1$, since these assumptions are satisfied by the pair $(V_f(\cdot), \mathbb{X}_f)$. The absence of the terminal constraint $x(N) \in \mathbb{X}_f$ in problem $\mathbb{P}_N^\beta(x)$, which is otherwise the same as the normal optimal control problem $\mathbb{P}_N(x)$ when $\beta = 1$, ensures that $\hat{V}_N^1(x) \leq V_N^0(x)$ for all $x \in \mathcal{X}_N$ and that $\mathcal{X}_N \subseteq \hat{\mathcal{X}}_N$ where $V_N^0(\cdot)$ is the value function for $\mathbb{P}_N(x)$ and $\mathcal{X}_N$ is the domain of $V_N^0(\cdot)$.

The next task is to show the existence of a $\beta \geq 1$ such that $x^\beta(N; x) = \phi(N; x, \mathbf{u}^\beta) \in \mathbb{X}_f$ for all x in some compact set, also to be determined. To proceed, let the terminal equality constrained optimal problem $\mathbb{P}_N^c(x)$ be defined by

$$V_N^c(x) = \min_{\mathbf{u}}\{J_N(x, \mathbf{u}) \mid \mathbf{u} \in \mathcal{U}_N^c(x)\}$$

in which $J_N(\cdot)$ and $\mathcal{U}_N^c(\cdot)$ are defined by

$$J_N(x, \mathbf{u}) := \sum_{i=0}^{N-1} \ell(x(i), u(i))$$
$$\mathcal{U}_N^c(x) := \hat{\mathcal{U}}_N(x) \cap \{\mathbf{u} \mid \phi(N; x, \mathbf{u}) = 0\}$$

In the definition of $J_N(\cdot)$, $x(i) := \phi(i; x, \mathbf{u})$. Let $\mathbf{u}^c$ denote the solution of $\mathbb{P}_N^c(x)$ and let $\mathcal{X}_N^c := \{x \in \mathbb{X} \mid \mathcal{U}_N^c(x) \neq \emptyset\}$ denote the domain of $V_N^c(\cdot)$. We assume that $\mathcal{X}_N^c$ is compact and has an interior. Clearly $\mathcal{U}_N^c(x) \subseteq \hat{\mathcal{U}}_N^c(x)$ and $\mathcal{X}_N^c \subset \mathcal{X}_N$. We also assume that there exists a $\mathcal{K}_\infty$ function $\alpha^c(\cdot)$ such that

$$V_N^c(x) \leq \alpha^c(|x|)$$

for all $x \in \mathcal{X}_N^c$; this is essentially a controllability assumption. The value function for the modified problem $\mathbb{P}_N^\beta(x)$ satisfies

$$\begin{aligned}\hat{V}_N^\beta(x) &= J_N(x, \mathbf{u}_N^\beta(x)) + \beta V_f(x_N^\beta(N; x)) \\ &\leq J_N(x, \mathbf{u}_N^c(x)) = V_N^c(x) \leq \alpha^c(|x|)\end{aligned}$$

for all $x \in \mathcal{X}_N$ where the first inequality follows from the fact that $\beta V_f(x_N^c(N; x)) = 0$ and $\mathbf{u}_N^c$ is not optimal for $\mathbb{P}_N^\beta(x)$; here $x_N^c(N; x) := \phi(N; x, \mathbf{u}^c(x))$. Hence

$$\beta V_f(x_N^\beta(N; x)) \leq \alpha^c(|x|)$$

for all $x \in \mathcal{X}_N$. Since $\mathcal{X}_N^c$ is compact, there exists a finite β such that $V_f(x_N^\beta(N; x)) \leq a$ for all $x \in \mathcal{X}_N^c$. Hence, there exists a finite $\beta > 1$ such that $x_N^\beta(N; x) \in \mathbb{X}_f$ for all $x \in \mathcal{X}_N^c$.

Suppose then that β is sufficiently large to ensure $x_N^\beta(N; x) \in \mathbb{X}_f$ for all $x \in \mathcal{X}_N^c$. Then the origin is asymptotically or exponentially stable for $x^+ = f(x, \kappa_N^\beta(x))$ with a region of attraction $\mathcal{X}_N^c$.

2.6.2 Omitting the Terminal Constraint

A related procedure is merely to omit the terminal constraint and to require that the initial state lies in a subset of $\mathcal{X}_N$ that is sufficiently small or that N is sufficiently large to ensure that the origin is asymptotically stable for the resultant controller. In either approach, the terminal cost may be modified. Here we examine the first alternative and assume, in the sequel, that $V_f(\cdot)$, $\mathbb{X}_f$ and $\ell(\cdot)$ satisfy Assumptions 2.12, 2.13, and 2.16, and that $\mathbb{X}_f := \{x \mid V_f(x) \leq a\}$ for some $a > 0$. Problem $\mathbb{P}_N^\beta(x)$ and the associated MPC control law $\kappa_N^\beta(\cdot)$ are defined in Section 2.6.1. Limon, Alamo, Salas, and Camacho (2006) show that the origin is asymptotically stable for $x^+ = f(x, \kappa_N^\beta(x))$ and each $\beta \geq 1$, with a region of attraction that depends on the parameter β by establishing the following results.

Lemma 2.38 (Entering the terminal region). *Suppose $\mathbf{u}^\beta(x)$ is optimal for the terminally unconstrained problem $\mathbb{P}_N^\beta(x)$, $\beta \geq 1$, and that $\mathbf{x}^\beta(x)$ is the associated optimal state trajectory. If $x^\beta(N;x) \notin \mathbb{X}_f$, then $x^\beta(i;x) \notin \mathbb{X}_f$ for all $i \in \mathbb{I}_{0:N-1}$.*

Proof. Since, as shown in Exercise 2.11, $\beta V_f(f(x, \kappa_f(x))) \leq \beta V_f(x) - \ell(x, \kappa_f(x))$ and $f(x, \kappa_f(x)) \in \mathbb{X}_f$ for all $x \in \mathbb{X}_f$, all $\beta \geq 1$, it follows that for all $x \in \mathbb{X}_f$ and all $i \in \mathbb{I}_{0:N-1}$

$$\beta V_f(x) \geq \sum_{j=i}^{N-1} \ell(x^f(j;x,i), u^f(j;x,i)) + \beta V_f(x^f(N;x,i)) \geq \hat{V}_{N-i}^\beta(x)$$

in which $x^f(j;x,i)$ is the solution of $x^+ = f(x, \kappa_f(x))$ at time j if the initial state is x at time i, $u^f(j;x,i) = \kappa_f(x^f(j;x,i))$, and $\kappa_f(\cdot)$ is the local control law that satisfies the stability assumptions. The second inequality follows from the fact that the control sequence $\{u^f(j;x,i) \mid i \in \mathbb{I}_{i:N-1}\}$ is feasible for $\mathbb{P}_N^\beta(x)$ if $x \in \mathbb{X}_f$. Suppose contrary to what is to be proved, that there exists a $i \in \mathbb{I}_{0:N-1}$ such that $x^\beta(i;x) \in \mathbb{X}_f$. By the principle of optimality, the control sequence $\{u^\beta(i;x), u^\beta(i+1;x), \ldots, u^\beta(N-1;x)\}$ is optimal for $\mathbb{P}_{N-i}^\beta(x^\beta(i;x))$. Hence

$$\beta V_f(x^\beta(i;x)) \geq \hat{V}_{N-i}^\beta(x^\beta(i;x)) \geq \beta V_f(x^\beta(N;x)) > \beta a$$

since $x^\beta(N;x) \notin \mathbb{X}_f$ contradicting the fact that $x^\beta(i;x) \in \mathbb{X}_f$. This proves the lemma. ■

For all $\beta \geq 1$, let the set Γ_N^β be defined by

$$\Gamma_N^\beta := \{x \mid \hat{V}_N^\beta(x) \leq Nd + \beta a\}$$

We assume in the sequel that there exists a $d > 0$ such $\ell(x, u) \geq d$ for all $x \in \mathbb{X} \setminus \mathbb{X}_f$ and all $u \in \mathbb{U}$. The following result is due to Limon et al. (2006).

Theorem 2.39 (MPC stability; no terminal constraint). *The origin is asymptotically or exponentially stable for the closed-loop system $x^+ = f(x, \kappa_N^\beta(x))$ with a region of attraction Γ_N^β. The set Γ_N^β is positive invariant for $x^+ = f(x, \kappa_N^\beta(x))$.*

Proof. From the Lemma, $x^\beta(N;x) \notin \mathbb{X}_f$ implies $x^\beta(i;x) \notin \mathbb{X}_f$ for all $i \in \mathbb{I}_{0:N}$. This, in turn, implies

$$\hat{V}_N^\beta(x) > Nd + \beta a$$

so that $x \notin \Gamma_N^\beta$. Hence $x \in \Gamma_N^\beta$ implies $x^\beta(N;x) \in \mathbb{X}_f$. It then follows, since $\beta V_f(\cdot)$ and $\mathbb{X}_f$ satisfy Assumptions 2.12 and 2.13, that the origin is asymptotically or exponentially stable for $x^+ = f(x, \kappa_N^\beta(x))$ with a region of attraction Γ_N^β. It also follows that $x \in \Gamma_N^\beta(x)$ implies

$$\hat{V}_N^\beta(x^\beta(1;x)) \leq \hat{V}_N^\beta(x) - \ell(x, \kappa_N^\beta(x)) \leq \hat{V}_N^\beta(x) \leq Nd + \beta$$

so that $x^\beta(1;x) = f(x, \kappa_N^\beta(x)) \in \Gamma_N^\beta$. Hence Γ_N^β is positive invariant for $x^+ = f(x, \kappa_N^\beta(x))$. ■

Limon et al. (2006) then proceed to show that Γ_N^β increases with β or, more precisely, that $\beta_1 \leq \beta_2$ implies that $\Gamma_N^{\beta_1} \subseteq \Gamma_N^{\beta_2}$. They also show that for any x steerable to the interior of $\mathbb{X}_f$ by a feasible control, there exists a β such that $x \in \Gamma_N^\beta$.

An attractive alternative is described by Hu and Linnemann (2002) who merely require that the state and control constraint sets, $\mathbb{X}$ and $\mathbb{U}$ respectively, are closed. Their approach uses, as usual, a terminal cost function $V_f : \mathbb{X}_f \to \mathbb{R}$, a terminal constraint set $\mathbb{X}_f$, and a stage cost $\ell(\cdot)$ that satisfy Assumptions 2.12, 2.13, and 2.16. Let $\mathbb{X}_f$ be a sublevel set of $V_f(\cdot)$ defined by

$$\mathbb{X}_f := \{x \in \mathbb{X} \mid V_f(x) \leq a\}$$

for some $a > 0$. Then the extended function $V_f^e : \mathbb{R}^n \to \mathbb{R}$ is defined by

$$V_f^e(x) := \begin{cases} V_f(x) & x \in \mathbb{X}_f \\ a & x \notin \mathbb{X}_f \end{cases}$$

The function $V_f^e(\cdot)$ is continuous but not continuously differentiable; we show later how the definition may be modified to ensure continuous differentiability, a desirable property for optimization algorithms. The optimization problem $\mathbb{P}_N^e(x)$ solved online is defined by

$$\hat{V}_N^e(x) := \min_{\mathbf{u}}\{V_N^e(x, \mathbf{u}) \mid \mathbf{u} \in \hat{\mathcal{U}}_N(x)\}$$

in which, with $x(N) := \phi(N; x, \mathbf{u})$,

$$V_N^e(x, \mathbf{u}) := J_N(x, \mathbf{u}) + V_f^e(x(N))$$

and $J_N(\cdot)$ and $\hat{\mathcal{U}}_N(x)s$ are defined in Section 2.6.1. Let $\mathbf{u}^e(x)$ denote the solution of $\mathbb{P}_N^e(x)$ and $\mathbf{x}^e(x)$ the associated state trajectory where

$$\mathbf{u}^e(x) = \{u^e(0;x), u^e(1;x), \ldots, u^e(N-1;x)\}$$
$$\mathbf{x}^e(x) = \{x^e(0;x), x^e(1;x), \ldots, x^e(N;x)\}$$

The implicit MPC control law is $\kappa_N^e(\cdot)$ defined by

$$\kappa_N^e(x) := u^e(0;x)$$

We now define a restricted set $\mathcal{X}_N^e$ of initial states by

$$\mathcal{X}_N^e := \{x \mid x^e(N;x) \in \mathbb{X}_f\}$$

Hence, the terminal state of any optimal state trajectory with initial state $x \in \mathcal{X}_N^e$ lies in $\mathbb{X}_f$. It follows, by the usual arguments, that for all $x \in \mathcal{X}_N^e$

$$\hat{V}_N^e(x^+) \leq \hat{V}_N^e(x) - \ell(x, \kappa_N(x))$$

where $x^+ := f(x, \kappa_N^e(x)) = x^e(1;x)$. If $\mathcal{X}_N^e$ is positive invariant for $x^+ = f(x, \kappa_N^e(x))$, the origin is asymptotically stable for the system $x^+ := f(x, \kappa_N^e(x))$ with a region of attraction $\mathcal{X}_N^e$. Note, however, that $x \in \mathcal{X}_N^e$ does not necessarily imply that $x^+ = f(x, \kappa_N^e(x)) \in \mathcal{X}_N^e$. Hu and Linnemann (2002) show that $x \in \mathcal{X}_N^e$ implies

$$V_f^e(x^e(N;x^+)) \leq V_f^e(x^e(N-1;x^+)) - \ell(x^e(N-1;x^+), u^e(N-1;x^+))$$

The proof of this inequality is Exercise 2.12. If $x^e(N-1;x^+) = 0$, then $x^e(N;x^+) = 0 \in \mathcal{X}_N^e$ so that $x^+ \in \mathcal{X}_N^e$. On the other hand, if $x^e(N-1;x^+) \neq 0$, then, from the last inequality, $V_f^e(x^e(N;x^+)) < V_f^e(x^e(N-1;x^+))$. It follows from the definition of $V_f^e(\cdot)$ that $x^e(N;x^+) \in \mathbb{X}_f$, which implies that $x^+ \in \mathcal{X}_N^e$. Hence $\mathcal{X}_N^e$ is positive invariant for $x^+ = f(x, \kappa_N^e(x))$. It follows that the origin is asymptotically stable for $x^+ = f(x, \kappa_N^e(x))$ with a region of attraction $\mathcal{X}_N^e$.

For implementation, it is desirable that $V_f^e(\cdot)$ be continuously differentiable; standard optimization algorithms usually require this property. The essential property that $V_f^e(\cdot)$ should have to ensure asymptotic stability of the origin is that, for any $x \in \mathcal{X}_N^e$, $V_f^e(y) - V_f^e(x) \leq -\ell(x,u)$ for all $u \in \mathbb{U}$ implies that $y \in \mathbb{X}_f$. Suppose, then, that we choose $V_f^e(\cdot)$ to be a continuously differentiable $\mathcal{K}$ function that is equal to $V_f(\cdot)$ in $\mathbb{X}_f$ and is bounded by $a + d/2$ outside $\mathbb{X}_f$ where d is such that $\ell(x,u) \geq d$ for all $x \notin \mathbb{X}_f$, all $u \in \mathbb{U}$. We consider two cases.

(a) Suppose $x \in \mathcal{X}_N^e \setminus \mathbb{X}_f$ and $V_f^e(y) - V_f^e(x) \leq -\ell(x,u)$. Then $V_f^e(y) - V_f^e(x) \leq -\ell(x,u) \leq -d$ for any $u \in \mathbb{U}$. Suppose, contrary to what we wish to prove, that $y \notin \mathbb{X}_f$. The definition of $V_f^e(\cdot)$ implies that $|V_f^e(y) - V_f^e(x)| \leq d/2$, a contradiction. Hence $y \in \mathbb{X}_f$.

(b) Suppose $x \in \mathbb{X}_f$ and $V_f^e(y) - V_f^e(x) \leq -\ell(x,u)$. Then $V_f^e(y) \leq V_f(x) \leq a$ which implies that $y \in \mathbb{X}_f$.

Hence the continuously differentiable version of $V_f^e(\cdot)$ has the essential property stated above so that $\mathcal{X}_N^e$ is positive invariant for $x^+ = f(x, \kappa_N^e(x))$ and the origin is asymptotically stable for $x^+ = f(x, \kappa_N^e(x))$ with a region of attraction $\mathcal{X}_N^e$.

If $V_f(x)$ is equal to the optimal infinite horizon cost for all $x \in \mathbb{X}_f$, then $V_N^e(x)$ is also equal to the optimal infinite horizon cost for all $x \in \mathcal{X}_N^e$.

2.7 Stage Cost $\ell(\cdot)$ not Positive Definite

In the analysis above we assume that the function $(x,u) \mapsto \ell(x,u)$ is positive definite; more precisely, we assume that there exists a $\mathcal{K}_\infty$ function $\alpha_1(\cdot)$ such that $\ell(x,u) \geq \alpha_1(|x|)$ for all (x,u). Often we assume that $\ell(\cdot)$ is quadratic, satisfying $\ell(x,u) = (1/2)(x'Qx + u'Ru)$ where Q and R are positive definite. In this section we consider the case where the stage cost is $\ell(y,u)$ where $y = h(x)$ and the function $h(\cdot)$ is not necessarily invertible. An example is the quadratic stage cost $\ell(y,u) = (1/2)(|y|^2 + u'Ru)$ where $y = Cx$ and C is not invertible; hence the stage cost is $(1/2)(x'Qx + u'Ru)$ where $Q = C'C$ is merely positive semidefinite. Since now $\ell(\cdot)$ does not satisfy $\ell(x,u) \geq \alpha_1(|x|)$ for all (x,u) and some $\mathcal{K}_\infty$ function $\alpha_1(\cdot)$, we have to make an additional assumption in order to establish asymptotic stability of the origin for the closed-loop system. An appropriate assumption is detectability, or input/output-to-state-stability (IOSS) that ensures the state goes to zero as the output and input go to zero. We recall Definition B.42, restated here.

Definition 2.40 (Input/output-to-state stable (IOSS)). The system $x^+ = f(x,u)$, $y = h(x)$ is IOSS if there exist functions $\beta(\cdot) \in \mathcal{KL}$ and $\gamma_1(\cdot)$, $\gamma_2(\cdot) \in \mathcal{K}$ such that for every initial state $x_0 \in \mathbb{R}^n$, every control sequence $\mathbf{u}$, and all $i \geq 0$.

$$|x(i)| \leq \max\{\beta(|x|, i), \gamma_1(\|u\|_{0:i-1}), \gamma_2(\|y\|_{0:i})\}$$

where $x(i) := \phi(i; x, \mathbf{u})$, the solution of $x^+ = f(x,u)$ at time i if the initial state is x and the input sequence is $\mathbf{u}$; $y(i) := h(x(i))$, $\|\mathbf{u}\|_{0:i-1}$ is the max norm of the sequence $\{u(0), u(1), \ldots, u(i-1)\}$ and $\|\mathbf{y}\|_{0:i}$ is the max norm of the sequence $\{y(0), y(1), \ldots, y(i)\}$.

We assume, as usual, that Assumptions 2.2, 2.3, 2.12, and 2.13 are satisfied but in place of Assumption 2.16 we assume that there exists $\mathcal{K}_\infty$ functions $\alpha_1(\cdot)$ and $\alpha_2(\cdot)$ such that

$$\ell(y,u) \geq \alpha_1(|y|) + \alpha_1(|u|) \qquad V_f(x) \leq \alpha_2(|x|)$$

for all (y,u) and all x. We also assume that the system $x^+ = f(x,u)$, $y = h(x)$ is IOSS and that $\mathbb{X}_f$ has an interior. Under these assumptions, the value function $V_N^0(\cdot)$ has the following properties

$$\begin{aligned}
&V_N^0(x) \geq \alpha_1|h(x)| && \forall x \in \mathcal{X}_N \\
&V_N^0(f(x,\kappa_N(x))) \leq V_N^0(x) - \alpha_1(|h(x)|) && \forall x \in \mathcal{X}_N \\
&V_N^0(x) \leq \alpha_2(|x|) && \forall x \in \mathbb{X}_f
\end{aligned}$$

That $V_N^0(f(x,\kappa_N(x))) \leq V_N^0(x) - \ell(h(x),\kappa_N(x))$ follows from the basic stability assumption. The fact that $h(x)$ appears in the first and second inequalities instead of x complicates analysis and makes it necessary to assume the IOSS property. We require the following result:

Proposition 2.41 (Convergence of state under IOSS). *Assume that the system $x^+ = f(x,u)$, $y = h(x)$ is IOSS and that $u(i) \to 0$ and $y(i) \to 0$ as $i \to \infty$. Then $x(i) = \phi(i;x,\mathbf{u}) \to 0$ as $i \to \infty$ for any initial state x.*

This proof of this result is discussed in Exercise 2.16.

Given the IOSS property, one can establish that the origin is attractive for closed-loop system with a region of attraction $\mathcal{X}_N$. For all $x \in \mathcal{X}_N$, all $i \in \mathbb{I}_{\geq 0}$, let $x(i;x) := \phi(i;x,\kappa_N(\cdot))$, the solution at time i of $x^+ = f(x,\kappa_N(x))$ if the initial state is x, $y(i;x) := h(x(i;x))$ and $u(i;x) := \kappa_N(x(i;x))$. It follows from the properties of the value function that, for any initial state $x \in \mathcal{X}_N$, the sequence $\{V_N^0(x(i;x))\}$ is nonincreasing and bounded below by zero, so that $V_N^0(x(i;x)) \to c \geq 0$ as $i \to \infty$. Since $V_N^0(x(i+1)) \leq V_N^0(x(i)) - \ell(x(i;x), y(i;x))$, it follows that $\ell(y(i;x),u(i;x)) \to 0$ and, hence, that $y(i;x) \to 0$ and $u(i;x) \to 0$ as $i \to \infty$. From Proposition 2.41, $x(i;x) \to 0$ as $i \to \infty$ for any initial state $x \in \mathcal{X}_N$.

The stability property also is not difficult to establish. Suppose the initial state x satisfies $|x| \leq \delta$ where δ is small enough to ensure that $\delta\mathcal{B} \subset \mathbb{X}_f$. Then $V_N^0(x) \leq \alpha_2(|x|)$ and, since $\{V_N^0(x(i;x))\}$ is nonincreasing, $V_N^0(x(i;x)) \leq \alpha_2(|x|)$ for all $i \in \mathbb{I}_{\geq 0}$. Since $\alpha_2(|x|) \geq V_N^0(x(i;x)) \geq \ell(y(i;x),u(i;x)) \geq \alpha_1(|y(i;x)|) + \alpha_1(|u(i;x)|)$, it follows that $|y(i;x)| \leq \alpha_3(|x|)$ and $|u(i;x)| \leq \alpha_3(|x|)$ for all $x \in \mathcal{X}_N$, all $i \in \mathbb{I}_{\geq 0}$ where $\alpha_3(\cdot)$ is a $\mathcal{K}$ function defined by $\alpha_3 := \alpha_1^{-1} \circ \alpha_2$, i.e., $\alpha_3(r) = \alpha_1^{-1}(\alpha_2(r))$ for all $r \geq 0$. Hence $x(i;x)$ satisfies

$$|x(i;x)| \leq \max\{\beta(\delta, i), \alpha_3(\delta)\} \leq \max\{\beta(\delta, 1), \alpha_3(\delta)\}$$

for all $x \in \mathcal{X}_N$, all $i \in \mathbb{I}_{\geq 0}$. Thus, for all $\varepsilon > 0$, there exists a $\delta > 0$, such $|x| \leq \delta$ implies $|x(i;x)| \leq \varepsilon$ for all $i \in \mathbb{I}_{\geq 0}$. We have established stability of the origin for $x^+ = f(x, \kappa_N(x))$. Hence the origin is asymptotically stable for the closed-loop system $x^+ = f(x, \kappa_N(x))$ with a region of attraction $\mathcal{X}_N$.

In earlier MPC literature, observability rather than detectability was often employed as the extra assumption required to establish asymptotic stability. Exercise 2.15 discusses this approach.

2.8 Suboptimal MPC

There is a significant practical problem that we have not addressed, namely that if the optimal control problem $\mathbb{P}_N(x)$ solved online is not convex, which is usually the case when the system is nonlinear, the global minimum of $V_N(x, \mathbf{u})$ in $\mathcal{U}_N(x)$ cannot usually be determined. Since we assume, in the stability theory given previously, that the global minimum *is* achieved, we have to consider the impact of this unpalatable fact. It is possible, as we show in this section, to achieve stability *without* requiring globally optimal solutions of $\mathbb{P}_N(x)$. Roughly speaking, all that is required is at state x, a feasible solution $\mathbf{u} \in \mathcal{U}_N(x)$ is found giving a cost $V_N(x, \mathbf{u})$ lower than the cost $V_N(w, \mathbf{v})$ at the previous state w due to the previous control sequence $\mathbf{v} \in \mathcal{U}_N(w)$.

Consider then the usual optimal control problem with the terminal cost $V_f(\cdot)$ and terminal constraint set $\mathbb{X}_f$ satisfying Assumptions 2.12 and 2.13; the state constraint set $\mathbb{X}$ is assumed to be closed and the control constraint set $\mathbb{U}$ to be compact. In addition, we assume that $V_f(\cdot)$ satisfies $V_f(x) \geq \alpha_f(|x|)$ and $V_f(x) \leq \gamma_f(|x|)$ for all $x \in \mathbb{X}_f$ where $\alpha_f(\cdot)$ and $\gamma_f(\cdot)$ are $\mathcal{K}_\infty$ functions. These conditions are satisfied, for example, if $V_f(\cdot)$ is a positive definite quadratic function and $\mathbb{X}_f$ is a sublevel set of $V_f(\cdot)$. The set $\mathbb{X}_f$ is assumed to be a sublevel set of $V_f(\cdot)$, i.e., $\mathbb{X}_f = \{x \mid V_f(x) \leq r\}$ for some $r > 0$. We also make the standard Assumption 2.16(a) that $\ell(x, u) \geq \alpha_1(|x|)$ for all $(x, u) \in \mathbb{X} \times \mathbb{U}$ where $\alpha_1(\cdot)$ is a $\mathcal{K}_\infty$ function, which is satisfied if $\ell(\cdot)$ is a positive definite quadratic function. Let $\mathcal{X}_N$ denote, as before, the set of x for which a control sequence $\mathbf{u}$ exists that satisfies the state, control and terminal constraints, i.e., $\mathcal{X}_N := \{x \in \mathbb{X} \mid \mathcal{U}_N(x) \neq \emptyset\}$ where $\mathcal{U}_N(x)$ is defined by (2.6).

The basic idea behind the suboptimal model predictive controller is simple. Suppose that the current state is x and that $\mathbf{u} = \{u(0), u(1), \ldots, u(N-1)\} \in \mathcal{U}_N(x)$ is a feasible control sequence for $\mathbb{P}_N(x)$. The first element $u(0)$ of $\mathbf{u}$ is applied to the system $x^+ = f(x,u)$. In the absence of uncertainty, the next state is equal to the predicted state $x^+ = f(x, u(0))$. Consider the control sequence $\mathbf{u}^+$ defined by

$$\mathbf{u}^+ = \{u(1), u(2), \ldots, u(N-1), \kappa_f(x(N))\} \tag{2.40}$$

in which $x(N) = \phi(N; x, \mathbf{u})$ and $\kappa_f(\cdot)$ is a local control law with the property that $u = \kappa_f(x)$ satisfies Assumption 2.12 for all $x \in \mathbb{X}_f$. The existence of such a $\kappa_f(\cdot)$, which is usually of the form $\kappa_f(x) = Kx$, is implied by Assumption 2.12. Then, as shown in Section 2.4.3, the control sequence $\mathbf{u}^+ \in \mathcal{U}_N(x)$ satisfies

$$V_N(x^+, \mathbf{u}^+) + \ell(x, u(0)) \leq V_N(x, \mathbf{u}) \tag{2.41}$$

and, hence

$$V_N(x^+, \mathbf{u}^+) \leq V_N(x, \mathbf{u}) - \alpha_1(|x|) \tag{2.42}$$

No optimization is required to get the cost reduction $\ell(x, u(0))$ given by (2.41); in practice the control sequence $\mathbf{u}^+$ can be improved by several iterations of an optimization algorithm. Inequality (2.42) is reminiscent of the inequality $V_N^0(x^+) \leq V_N^0(x) - \ell(x, \kappa_N(x))$ that provides the basis for establishing asymptotic stability of the origin for the controlled systems previously analyzed and suggests that the simple algorithm described previously, which places very low demands on the online optimization algorithm, may also ensure asymptotic stability of the origin. This is almost true. The obstacle to applying standard Lyapunov theory is that there is no obvious Lyapunov function because, at each state x^+, there exist many control sequences $\mathbf{u}^+$ satisfying $V_N(x^+, \mathbf{u}^+) \leq V_N(x, \mathbf{u}) - \alpha_1(|x|)$. The function $(x, \mathbf{u}) \mapsto V_N(x, \mathbf{u})$ is *not* a function of x only and may have many different values for each x; therefore it cannot play the role of the function $V_N^0(x)$ used previously. Moreover, the controller can generate, for a given initial state, many different trajectories, all of which have to be considered.

Global attractivity of the origin in $\mathcal{X}_N$, however, may be established. For all $x(0) \in \mathcal{X}_N$, let $\{(x(0), \mathbf{u}(0)), (x(1), \mathbf{u}(1)), \ldots\}$ denote *any* infinite sequence generated by the controlled system and satisfying, therefore, $V_N(x(i+1), \mathbf{u}(i+1)) \leq V_N(x(i), \mathbf{u}(i)) - \alpha_1(|x(i)|)$ for all i. Then

$\{V_N(x(i),\mathbf{u}(i)) \mid i \in \mathbb{I}_{\geq 0}\}$ is a nonincreasing sequence bounded below by zero. Hence $V_N(x(i),\mathbf{u}(i)) \to V_N^* \geq 0$ so that $V_N(x(i+1),\mathbf{u}(i+1)) - V_N(x(i),\mathbf{u}(i)) \to 0$ as $i \to \infty$. We deduce, from (2.42), that $\alpha_1(|x(i)|) \to 0$ so that $x(i) \to 0$ as $i \to \infty$.

Establishing stability of the origin is more difficult for reasons given previously and requires a minor modification of the controller when the state x is close to the origin. The modification we make to the controller is to require that $\mathbf{u}$ satisfies the following requirement when x lies in $\mathbb{X}_f$

$$V_N(x,\mathbf{u}) \leq V_f(x) \qquad f(x,u(0)) \in \mathbb{X}_f \tag{2.43}$$

where $u(0)$ is the first element in $\mathbf{u}$. Stability of the origin can be established using (2.42), (2.43) and the properties of $V_f(\cdot)$ as shown subsequently. Inequality (2.43) is achieved quite simply by using the control law $u = \kappa_f(x)$ to generate the control u when $x \in \mathbb{X}_f$. Let $\mathbf{x}(x;\kappa_f)$ and $\mathbf{u}(x;\kappa_f)$ denote the state and control sequences generated in this way when the initial state is x; these sequences satisfy

$$x^+ = f(x,\kappa_f(x)) \qquad u = \kappa_f(x)$$

with initial condition $x(0) = x$, so that $x(0;x,\kappa_f) = x$, $x(1;x,\kappa_f) = f(x,\kappa_f(x))$, $x(2;x,\kappa_f) = f(x(1;x,\kappa_f)),\kappa_f(x(1;x,\kappa_f))$, etc. Since Assumption 2.12 is satisfied,

$$V_f(x) \geq \ell(x,\kappa_f(x)) + V_f(f(x,\kappa_f(x)))$$

which, when used iteratively, implies

$$V_f(x) \geq \sum_{i=0}^{N-1} \ell(x(i;x,\kappa_f),\kappa_f(x(i;x,\kappa_f))) + V_f(x(N;x,\kappa_f))$$

Hence, for all $x \in \mathbb{X}_f$

$$\begin{aligned} V_N(x,\mathbf{u}(x;\kappa_f)) &= \sum_{i=0}^{N-1} \ell(x(i;x,\kappa_f),\kappa_f(x(i;x,\kappa_f))) + V_f(x(N;x,\kappa_f)) \\ &\leq V_f(x) \end{aligned}$$

as required. Also, it follows from Assumption 2.12 and the definition of $\kappa_f(\cdot)$ that $x^+ = f(x,u(0)) \in \mathbb{X}_f$ if $x \in \mathbb{X}_f$. Thus the two conditions in (2.43) are satisfied by $\mathbf{u}(x;\kappa_f)$. If desired, $\mathbf{u}(x;\kappa_f)$ may be used for the current control sequence $\mathbf{u}$ or as a "warm start" for an optimization algorithm yielding an improved control sequence. In any case, if (2.43) is satisfied, stability of the origin may be established as follows.

Let $\delta > 0$ be arbitrary but small enough to ensure $\delta\mathcal{B} \subset \mathbb{X}_f$. Suppose the initial state $x(0)$ satisfies $|x(0)| \leq \delta$ so that $x(0) \in \mathbb{X}_f$. As before, let $\{(x(i), \mathbf{u}(i))\}$ denote *any* state-control sequence with initial state $x(0) \in \delta\mathcal{B}$ generated by the suboptimal controller. From (2.42) and (2.43) we deduce that $V_N(x(i), \mathbf{u}(i)) \leq V_N(x(0), \mathbf{u}(0)) \leq V_f(x(0)) \leq \gamma_f(|x(0)|) \leq \gamma_f(\delta)$ for all $i \in \mathbb{I}_{\geq 0}$, all $x(0) \in \delta\mathcal{B}$. It follows from our assumption on $\ell(\cdot)$ that $V_N(x, \mathbf{u}) \geq \alpha_1(|x|)$ for all $x \in \mathcal{X}_N$, all $\mathbf{u} \in \mathbb{U}^N$. Hence

$$\alpha_1(|x(i)|) \leq V_N(x(i), \mathbf{u}(i)) \leq \gamma_f(\delta)$$

so that $|x(i)| \leq (\alpha_1^{-1} \circ \gamma_f)(\delta)$ or all $i \in \mathbb{I}_{\geq 0}$. Hence, for all $\varepsilon > 0$, there exists a $\delta = (\alpha_1^{-1} \circ \gamma_f)^{-1}(\varepsilon) = (\gamma_f^{-1} \circ \alpha_1)(\varepsilon) > 0$ such that $|x(0)| \leq \delta$ implies $|x(i)| \leq \varepsilon$ for all $i \in \mathbb{I}_{\geq 0}$. The origin is, therefore, stable and, hence, asymptotically stable with a region of attraction $\mathcal{X}_N$ for the controlled system.

Suboptimal control algorithm.

Data: Integer N_{iter}.

Input: Current state x, previous state sequence $\mathbf{w} = \{w(0), w(1), \ldots, w(N)\}$, previous control sequence $\mathbf{v} = \{v(0), v(1), \ldots, v(N-1)\}$.

Step 1: If $x \notin \mathbb{X}_f$, use $\{v(1), v(2), \ldots, v(N-1), \kappa_f(w(N))\}$ as a warm start for an optimization algorithm. Perform N_{iter} iterations of the algorithm to obtain an improved control sequence $\mathbf{u} \in \mathcal{U}_N(x)$. Apply control $u = u(0)$ to the system being controlled.

Step 2: If $x \in \mathbb{X}_f$, set $u = \kappa_f(x)$ and apply u to the system being controlled; or perform N_{iter} steps of an optimization algorithm using $\mathbf{u}(x, \kappa_f)$, defined previously, as a warm start to obtain an improved control sequence $\mathbf{u} \in \mathcal{U}_N(x)$ satisfying (2.43) and associated state sequence $\mathbf{w}$.

A nominally stabilizing controller with very low online computational demands may be obtained by merely using the warm starts defined in the algorithm. Improved performance is obtained by using N_{iter} iterations of an optimization algorithm to improve the warm start. It is more important to employ optimization in Step 1 when $x \notin \mathbb{X}_f$ since the warm start when $x \in \mathbb{X}_f$ has good performance if $\kappa_f(\cdot)$ is designed properly.

2.9 Tracking

In preceding sections we were concerned with regulation to the origin and the determination of conditions that ensure stability of the origin for the closed-loop system. In this section we consider the problem of tracking a constant reference signal, i.e., regulation to a set point. Assume that the system to be controlled satisfies

$$x^+ = f(x,u) \quad y = h(x)$$

and is subject to the constraints

$$x \in \mathbb{X} \qquad u \in \mathbb{U}$$

in which $\mathbb{X}$ is closed and $\mathbb{U}$ is compact. If the constant reference signal is r, then we wish to steer the initial state x to a state $\bar{x}$ satisfying $h(\bar{x}) = r$ so that $y = r$.

2.9.1 No Uncertainty

We assume initially that there is no model error and no disturbance. The target state and associated steady-state control are obtained by minimizing $|\bar{u}|^2$ with respect to (x,u) subject to the equality constraints

$$\begin{aligned} x &= f(x,u) \\ r &= h(x) \end{aligned}$$

and the inequality constraints $x \in \mathbb{X}$ and $u \in \mathbb{U}$. We assume that a solution exists and denote the solution by $(\bar{x}(r), \bar{u}(r))$; this notation indicates the dependence of the target state and its associated control on the reference variable r. We require the dimension of r to be less than or equal to m, the dimension of u.

MPC may then be achieved by solving online the optimal control problem $\mathbb{P}_N(x,r)$ defined by

$$V_N^0(x,r) = \min_{\mathbf{u}}\{V_N(x,r,\mathbf{u}) \mid \mathbf{u} \in \mathcal{U}_N(x,r)\}$$

in which the cost function $V_N(\cdot)$ and the constraint set are defined by

$$\begin{aligned} V_N(x,r,\mathbf{u}) &:= \sum_{i=0}^{N-1} \ell(x(i) - \bar{x}(r), u(i) - \bar{u}(r)) + V_f(x,r) \\ \mathcal{U}_N(x,r) &:= \{\mathbf{u} \mid x(i) \in \mathbb{X},\ u(i) \in \mathbb{U}, \forall i \in \mathbb{I}_{0:N-1}; x(N) \in \mathbb{X}_f(r)\} \end{aligned}$$

In these definitions, $x(i) = \phi(i; x, \mathbf{u})$, the solution at time i of $x^+ = f(x,u)$ if the initial state is x and the control sequence is $\mathbf{u}$. Let $\mathbf{u}^0(x,r)$ denote the solution of $\mathbb{P}_N(x,r)$. The MPC control law is $\kappa_N(x,r)$, the first control in the sequence $\mathbf{u}^0(x,r)$. The terminal cost function $V_f(\cdot,r)$ and constraint set $\mathbb{X}_f(r)$ must be chosen to satisfy suitably modified stabilizing conditions. Since both depend on r, the simplest option is to choose a terminal equality constraint so that

$$V_f(\bar{x}(r), r) = 0 \qquad \mathbb{X}_f(r) = \{\bar{x}(r)\} \subset \mathbb{X}$$

If the system is linear, i.e., if $x^+ = Ax + Bu$, an alternative choice is

$$V_f(x,r) = V_f'(x - \bar{x}(r)) \qquad \mathbb{X}_f(r) = \{\bar{x}(r)\} \oplus \mathbb{X}_f' \subset \mathbb{X}$$

in which $V_f'(\cdot)$ and $\mathbb{X}_f'$ are, respectively, the terminal cost function and terminal constraint set derived in Section 2.5.3.1; the reference r must satisfy the constraint that $\mathbb{X}_f(r) \subset \mathbb{X}$. With this choice, $V_f(\bar{x}(r), r) = 0$ and $\mathbb{X}_f(r)$ is control invariant for $x^+ = Ax + Bu$. It is easily shown, with either choice and appropriate assumptions, that the point $\bar{x}(r)$ is asymptotically, or exponentially, stable for the controlled system with a region of attraction $\mathcal{X}_N := \{x \mid \mathcal{U}_N(x) \neq \emptyset\}$.

2.9.2 Offset-Free MPC

If uncertainty is present, in the form of model error or an unknown constant disturbance, the tracking error $y - r$ may converge to a nonzero constant vector, called the *offset*, rather than to the origin. It is possible to ensure zero offset by augmenting the system with a model of the disturbance.

We therefore assume that the system to be controlled satisfies

$$\begin{aligned} x^+ &= f(x,u) \\ d^+ &= d \\ y &= h(x) + d + \nu \end{aligned}$$

in which ν is measurement noise. If we assume, as we do everywhere in this chapter, that the state x is known, then a simple filter may be used to obtain an estimate $\hat{d}$ of the unknown, but constant disturbance d. The filter is described by

$$\hat{d}^+ = \hat{d} + L(y - h(x)) - \hat{d}$$

in which $y - h(x)$ may be regarded as a noisy measurement of d, since $y - h(x) = d + \nu$. The difference equation for the estimation error $\tilde{d} := d - \hat{d}$ is

$$\tilde{d}^+ = A_L \tilde{d} + L\nu$$

in which L is chosen to ensure that $A_L := I - L$ is stable. If there is zero measurement noise, $\tilde{d}(i) \to 0$ exponentially as $i \to \infty$. Since $y - h(x) = d + \nu$, the difference equation for $\hat{d}$ may be written as

$$\hat{d}^+ = \hat{d} + L(\tilde{d} + \nu)$$

Since d is unknown, we have to use $\hat{d}$ for control. Hence, for the purpose of control we employ the difference equations

$$\begin{aligned} x^+ &= f(x, u) \\ \hat{d}^+ &= \hat{d} + L(\tilde{d} + \nu) \end{aligned}$$

If $\hat{d}$ is the current estimate of d, our best estimate of d at any time in the future is also $\hat{d}$. Given the current state $(x, \hat{d})$ of the composite system and the current reference r, we determine the target state and associated control by minimizing $|u|^2$ with respect to (x, u) subject to the equality constraints

$$\begin{aligned} x &= f(x, u) \\ r &= h(x) + \hat{d} \end{aligned}$$

and the inequality constraints $x \in \mathbb{X}$ and $u \in \mathbb{U}$. We assume that a solution to this problem exists and denote the solution by $(\bar{x}(r, \hat{d}), \bar{u}(r, \hat{d}))$.

MPC may then be achieved by solving online the optimal control problem $\mathbb{P}_N(x, r, \hat{d})$ defined by

$$V_N^0(x, r, \hat{d}) = \min_{\mathbf{u}} \{V_N(x, r, \mathbf{u}) \mid \mathbf{u} \in \mathcal{U}_N(x, r, \hat{d})\}$$

in which the cost function $V_N(\cdot)$ and the constraint set are defined by

$$V_N(x, r, \hat{d}, \mathbf{u}) := \sum_{i=0}^{N-1} \ell(x(i) - \bar{x}(r, \hat{d}), u(i) - \bar{u}(r, \hat{d})) + V_f(x, r, \hat{d})$$

$$\mathcal{U}_N(x, r, \hat{d}) := \{\mathbf{u} \mid x(i) \in \mathbb{X},\ u(i) \in \mathbb{U}, \forall i \in \mathbb{I}_{0:N-1}; x(N) \in \mathbb{X}_f(r, \hat{d})\}$$

In these definitions, $x(i) = \phi(i; x, \mathbf{u})$, the solution at time i of $x^+ = f(x, u)$ if the initial state is x and the control sequence is $\mathbf{u}$. Let

$\mathbf{u}^0(x,r,\hat{d})$ denote the solution of $\mathbb{P}_N(x,r,\hat{d})$. The MPC control law is $\kappa_N(x,r,\hat{d})$, the first control in the sequence $\mathbf{u}^0(x,r,\hat{d})$. The terminal cost function $V_f(\cdot,r,\hat{d})$ and constraint set $\mathbb{X}_f(r,\hat{d})$ must be chosen to satisfy suitably modified stabilizing conditions. Since both depend on $(r,\hat{d})$, the simplest option is to choose a terminal equality constraint so that

$$V_f(\bar{x}(r,\hat{d}),r,\hat{d}) = 0 \qquad \mathbb{X}_f(r,\hat{d}) = \{\bar{x}(r,\hat{d})\} \subset \mathbb{X}$$

This constraint is equivalent to requiring that the terminal state is equal to $\bar{x}(r,\hat{d})$ in the optimal control problem $\mathbb{P}_N(x,r,\hat{d})$.

If $\hat{d}$ is constant, standard MPC theory shows, under suitable assumptions, that the constant target state $\bar{x}(r,\hat{d})$ is asymptotically stable for $x^+ = f(x,\kappa_N(x,r,\hat{d}))$ with a region of attraction $\mathcal{X}_N(r,\hat{d}) := \{x \mid \mathcal{U}_N(x,r,\hat{d}) \neq \emptyset\}$. In particular, the state $x(i)$ of the controlled system at time i converges to $\bar{x}(r,\hat{d})$ as $i \to \infty$. We now assume that the disturbance $v(i) \to 0$ and, consequently, that $\hat{d}(i) \to d_s$, $x(i) \to x_s := \bar{x}(r,d_s)$, and $u(i) \to u_s := \bar{u}(r,d_s)$ as $i \to \infty$. Hence $y(i) = h(x(i)) + \hat{d}(i) + v(i) \to h(x_s) + d_s$ as $i \to \infty$. It follows from the difference equations for x and $\hat{d}$ that

$$x_s = f(x_s,u_s) \qquad L(y_s - h(x_s) - d_s) = 0$$

If L is invertible (y and d have the same dimension), it follows that

$$x_s = f(x_s,u_s) \qquad y_s = h(x_s) + d_s$$

But, since $x_s := \bar{x}(r,d_s)$ and $u_s := \bar{u}(r,d_s)$, it follows, by definition, that

$$h(x_s) + d_s = r$$

Hence $y(i) \to y_s = r$ as $i \to \infty$; the offset is asymptotically zero.

If we do not assume that $\hat{d}$ converges to a constant value, however, uncertainty in the evolution of $\hat{d}$ may cause the value function $V_N^0(x,r,\hat{d})$ to increase sufficiently often to destroy stability. Robust output MPC, discussed in Chapter 5, may have to be employed to ensure stability of a set rather than a point.

2.9.3 Unreachable Setpoints

In process control, steady-state optimization is often employed to determine an optimal setpoint, and MPC to steer the state of the system to this setpoint. Because of nonzero process disturbances and discrepancies between the models employed for steady-state optimization and for control, the optimal setpoint may be unreachable. Often an unreachable setpoint is then replaced by a reachable steady-state target that is closest to it. This standard procedure is *suboptimal*, however, and does not minimize tracking error. We show in this section that by defining performance relative to the unreachable setpoint rather than to the closest reachable target, it is possible to achieve improved performance. Standard MPC theory can no longer be used to analyze stability, however, because the value function for the new problem does not necessarily decrease along trajectories of the controlled system. With an infinite horizon, the cost function for the optimal control problem that yields MPC is unbounded.

Suppose the system to be controlled is described by

$$x^+ = Ax + Bu$$

with control constraint $u \in \mathbb{U}$, in which $\mathbb{U}$ is convex and compact. The setpoint pair $(x_{\text{sp}}, u_{\text{sp}})$ is not necessarily reachable. The cost function $V(\cdot)$ for the optimal control problem is

$$V_N(x, \mathbf{u}) := \sum_{i=0}^{N-1} \ell(x(i), u(i))$$

in which $x(i) := \phi(i; x, \mathbf{u})$, the solution of the dynamic system at time i if the initial state at time 0 is x and the control sequence is $\mathbf{u} := \{u(0), u(1), \ldots, u(N-1)\}$. The stage cost $\ell(\cdot)$ is defined to be a quadratic function of the distance from the setpoint

$$\ell(x, u) := (1/2)(|x - x_{\text{sp}}|_Q^2 + |u - u_{\text{sp}}|_R^2)$$

in which Q and R are positive definite. For simplicity of exposition, a terminal constraint $x(N) = x_s$, in which x_s is defined subsequently, is included in the optimal control problem $\mathbb{P}_N(x)$ whose solution yields the model predictive controller; problem $\mathbb{P}_N(x)$ is therefore defined by

$$V_N^0(x) = \min_{\mathbf{u}} \{V_N(x, \mathbf{u}) \mid \mathbf{u} \in \mathcal{U}_N(x)\}$$

in which the control sequence constraint set $\mathcal{U}_N(x)$ is defined by

$$\mathcal{U}_N(x) := \{\mathbf{u} \mid u(i) \in \mathbb{U}, i = 0, 1, \ldots, N-1, \ \phi(N; x, \mathbf{u}) = x_s\}$$

The domain of $V_N^0(\cdot)$, i.e., the set of feasible initial states for $\mathbb{P}_N(x)$, is $\mathcal{X}_N$ defined by

$$\mathcal{X}_N := \{x \mid \mathcal{U}_N(x) \neq \emptyset\}$$

For all $x \in \mathcal{X}_N$, the constraint set $\mathcal{U}_N(x)$ is compact. The set $\mathcal{X}_N$ is thus the set of states that can be steered to x_s in N steps by a control sequence $\mathbf{u}$ that satisfies the control constraint. It follows from its definition that $\mathcal{X}_N$ is closed and is compact if A is invertible. Because $V_N(\cdot)$ is continuous, and $\mathcal{U}_N(x)$ is compact for each $x \in \mathcal{X}_N$, it follows that for each $x \in \mathcal{X}_N$, $\mathbf{u} \mapsto V_N(x, \mathbf{u})$ achieves its minimum, $V_N^0(x)$, in $\mathcal{U}_N(x)$. Let $\mathbf{u}^0(x) = \{u^0(0;x), u^0(1;x), \ldots, u^0(N-1;x)\}$ denote the solution of $\mathbb{P}_N(x)$. Following usual practice, the model predictive control at state x is $\kappa_N(x) := u^0(0;x)$, the first element of the optimal control sequence $\mathbf{u}^0(x)$.

The optimal steady state (x_s, u_s) is defined to be the solution of the optimization problem $\mathbb{P}_s$

$$(x_s, u_s) := \arg\min_{x,u}\{\ell(x,u) \mid x = Ax + Bu,\ u \in \mathbb{U}\}$$

This problem has a solution if $0 \in \mathbb{U}$ since then $(0,0)$ satisfies the constraints. Since $Q, R > 0$, the minimizer (x_s, u_s) is unique. Clearly $\ell(x_s, u_s) > 0$ unless the setpoint $(x_{\text{sp}}, u_{\text{sp}})$ is feasible for $\mathbb{P}_s$; it is this fact that requires a nonstandard method for establishing asymptotic stability of (x_s, u_s). The following theorem is proved in (Rawlings, Bonné, Jørgensen, Venkat, and Jørgensen, 2008)

Theorem 2.42 (MPC stability with unreachable setpoint)**.** *The optimal steady state x_s is asymptotically stable with a region of attraction $\mathcal{X}_N$ for the closed-loop system $x^+ = Ax + B\kappa_N(x)$ using setpoint MPC.*

This paper also discusses relaxing the terminal constraint and using instead a terminal penalty based on a terminal controller.

Example 2.43: Unreachable setpoint MPC

An example is presented to illustrate the advantages of the proposed setpoint tracking MPC (sp-MPC) compared to traditional target tracking MPC (targ-MPC). The regulator cost function for the proposed sp-MPC, is

$$V_N^{\text{sp}}(x, \mathbf{u}) = \frac{1}{2}\sum_{j=0}^{N-1} |x(j) - x_{\text{sp}}|_Q^2 + |u(j) - u_{\text{sp}}|_R^2 + |u(j+1) - u(j)|_S^2$$

Performance measure	targ-MPC (a=targ)	sp-MPC (a=sp)	Δ index (%)
V_u^a	0.016	2.2×10^{-6}	99.98
V_y^a	3.65	1.71	53
V^a	3.67	1.71	54

Table 2.3: Comparison of controller performance for Example 2.43.

in which $Q > 0$, $R, S \geq 0$, at least one of $R, S > 0$, and $x(j) = \phi(j; x, \mathbf{u})$. This system can be put in the standard form defined for terminal constraint MPC by using the augmented state $\tilde{x}(k) := (x(k), u(k-1))$ discussed in Section 1.2.5. The regulator cost function in traditional targ-MPC is

$$V_N^{\text{targ}}(x, \mathbf{u}) = \frac{1}{2} \sum_{j=0}^{N-1} \left| x(j) - x_s \right|_Q^2 + \left| u(j) - u_s \right|_R^2 + \left| u(j+1) - u(j) \right|_S^2$$

The controller performance is assessed using the following three closed-loop control performance measures

$$V_u^a = \frac{1}{2k\Delta} \sum_{j=0}^{k-1} \left| u(j) - u_{\text{sp}} \right|_R^2 + \left| u(j+1) - u(j) \right|_S^2$$

$$V_y^a = \frac{1}{2k\Delta} \sum_{j=0}^{k-1} \left| x(j) - x_{\text{sp}} \right|_Q^2$$

$$V^a = V_u^a + V_y^a \qquad a = (\text{sp}, \text{targ})$$

in which Δ is the process sample time, and $x(j)$ and $u(j)$ are the state and control at time j of the controlled system using either target (a=targ) or setpoint (a=sp) tracking MPC for a specified initial state. For each of the indices defined previously, we define the percentage improvement of sp-MPC compared with targ-MPC by

$$\Delta \text{ index } (\%) = \frac{V^{\text{targ}} - V^{\text{sp}}}{V^{\text{targ}}} \times 100$$

Consider the single-input, single-output system with transfer function

$$G(s) = \frac{-0.2623}{60s^2 + 59.2s + 1}$$

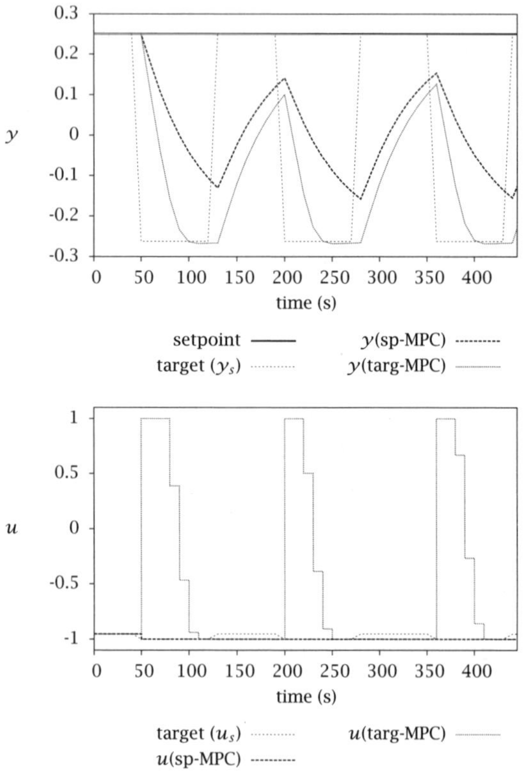

Figure 2.6: Closed-loop performance of sp-MPC and targ-MPC.

sampled with $\Delta = 10$ s. The input u is constrained $|u| \leq 1$. The desired output setpoint is $y_{\text{sp}} = 0.25$, which corresponds to a steady-state input value of -0.953. The regulator parameters are $Q_y = 10, R = 0, S = 1, Q = C'Q_yC + 0.01I_2$. A horizon length of $N = 80$ is used. In time intervals 50–130, 200–270, and 360–430, a state disturbance $d_x = [17.1, 1.77]'$ causes the input to saturate at its lower limit. The output setpoint is unreachable under the influence of this state disturbance. The closed-loop performance of sp-MPC and targ-MPC under the described disturbance scenario are shown in Figure 2.6. The closed-loop performance of the two control formulations are compared in Table 2.3.

In the targ-MPC framework, the controller tries to reject the state disturbance and minimize the deviation from the new steady-state target. This requires a large, undesirable control action that forces the input to move between the upper and lower constraints. The sp-MPC framework, on the other hand, attempts to minimize the deviation from setpoint and subsequently the input just rides the lower input constraint.

The greater cost of control action in targ-MPC is shown by the cost index V_u in Table 2.3. The cost of control action in targ-MPC exceeds that of sp-MPC by nearly 100%. The control in targ-MPC causes the output of the system to move away from the (unreachable) setpoint faster than the corresponding output of sp-MPC. Since the control objective is to be close to the setpoint, this undesirable behavior is eliminated by sp-MPC. □

2.10 Concluding Comments

MPC is an implementation, for practical reasons, of receding horizon control (RHC), in which offline determination of the RHC law $\kappa_N(\cdot)$ is replaced by online determination of its value $\kappa_N(x)$, the control action, at each state x encountered during its operation. Because the optimal control problem that defines the control is a finite horizon problem, neither stability nor optimality of the cost function is achieved by a receding horizon or model predictive controller. This chapter shows how stability may be achieved by adding a terminal cost function and a terminal constraint to the optimal control problem. Adding a terminal cost function adds little or no complexity to the optimal control problem that has to be solved online and usually improves performance. Indeed, the infinite horizon value function $V_\infty^0(\cdot)$ for the constrained problem would be an ideal choice for the terminal penalty because the value function $V_N^0(\cdot)$ for the online optimal control problem would then be equal to $V_\infty^0(\cdot)$ and the controller would inherit the performance advantages of the infinite horizon controller. In addition, the actual trajectories of the controlled system would be precisely equal, in the absence of uncertainty, to those predicted by the online optimizer. Of course, if we knew $V_\infty^0(\cdot)$, the optimal infinite horizon controller $\kappa_\infty(\cdot)$ could be determined and there would be no reason to employ MPC. The infinite horizon cost $V_\infty^0(\cdot)$ is known globally only for special cases, however, such as the linear quadratic unconstrained problem. For more general problems in which constraints and/or nonlinearity are present, its value, or approximate value, in a neighborhood of the setpoint can usually be obtained and the use of this local CLF should, in general, enhance performance. Adding a terminal cost appears to be generally advantageous.

The reason for the terminal constraint is precisely the fact that the terminal penalty is usually merely a local CLF requiring the terminal state to lie in the region where the CLF is valid. Unlike the addition of a terminal penalty, however, addition of a terminal constraint may increase complexity considerably. Because quadratic programs, in which the cost function to be minimized is quadratic and the constraints polyhedral, there is an argument for using polyhedral constraints. Indeed, a potential terminal constraint set for the constrained linear quadratic optimal control problem is the maximal constraint admissible set, which is polyhedral. This set is complex, however, i.e., defined by many linear inequalities, and would appear to be unsuitable for the complex control problems routinely encountered in industry.

A terminal constraint set that is considerably simpler is a suitable sublevel set of the terminal penalty, which is often a simple positive definite quadratic function resulting in a convex terminal constraint set. A disadvantage is that the terminal constraint set is now ellipsoidal rather than polytopic and conventional quadratic programs cannot be employed for the linear quadratic constrained optimal control problem. This does not appear to be a serious disadvantage, however, because the optimal control problem remains convex, so interior point methods may be readily employed.

In the nonlinear case, adding an ellipsoidal terminal constraint set does not appreciably affect the complexity of the optimal control problem. In any case, it is possible to replace the ellipsoidal terminal constraint set by a suitable modification of the terminal penalty as shown in Section 2.6.1. A more serious problem, when the system is nonlinear, is that the optimal control problem is then usually nonconvex so that global solutions, on which many theoretical results are predicated, are usually too difficult to obtain. A method for dealing with this difficulty, which also has the advantage of reducing online complexity, is suboptimal MPC described in Section 2.8.

This chapter also presents some results that contribute to an understanding of the subject but do not provide practical tools. For example, it is useful to know that the domain of attraction for many of the controllers described here is $\mathcal{X}_N$, the set of initial states controllable to the terminal constraint set, but this set cannot usually be computed. The set is, in principle, computable using the DP equations presented in this chapter, and may be computed if the system is linear and the constraints, including the terminal constraint, are polyhedral, provided that the state dimension and the horizon length are suitably small, considerably smaller than in problems routinely encountered in industry. In the nonlinear case, this set cannot usually be computed. Computation difficulties are not resolved if $\mathcal{X}_N$ is replaced by a suitable sublevel set of the value function $V_N^0(\cdot)$. Hence, in practice, both for linear and nonlinear MPC, this set has to be estimated by simulation.

2.11 Notes

MPC has an unusually rich history, making it impossible to summarize here the many contributions that have been made. Here we restrict attention to a subset of this literature that is closely related to the approach adopted in this book. A fuller picture is presented in the review paper (Mayne, Rawlings, Rao, and Scokaert, 2000).

The success of conventional MPC derives from the fact that for deterministic problems (no uncertainty), feedback is not required so the solution to the open-loop optimal control problem solved online for a particular initial state is the same as that obtained by solving the feedback problem using DP, for example. Lee and Markus (1967) pointed out the possibility of MPC in their book on optimal control

> One technique for obtaining a feedback controller synthesis is to measure the current control process state and then compute very rapidly the open-loop control function. The first portion of this function is then used during a short time interval after which a a new measurement of the process state is made and a new open-loop control function is computed for this new measurement. The procedure is then repeated.

Even earlier, Propoi (1963) proposed a form of MPC utilizing linear programming, for the control of linear systems with hard constraints on the control. A big surge in interest in MPC occurred when Richalet, Rault, Testud, and Papon (1978b) advocated its use for process control. A whole series of papers, such as Richalet, Rault, Testud, and Papon (1978a), Cutler and Ramaker (1980), Prett and Gillette (1980), García and Morshedi (1986), and Marquis and Broustail (1988) helped cement its popularity in the process control industries, and MPC soon became the most useful method in modern control technology for control problems with hard constraints with thousands of applications to its credit.

The basic question of stability, an important issue since optimizing a finite horizon cost does not necessarily yield a stabilizing control, was not resolved in this early literature. Early academic research in MPC, reviewed in García, Prett, and Morari (1989), did not employ Lyapunov theory and therefore restricted attention to control of unconstrained linear systems, studying the effect of control and cost horizons on stability. Similar studies appeared in the literature on generalized predictive control (GPC) (Ydstie, 1984; Peterka, 1984; De Keyser and Van Cauwenberghe, 1985; Clarke, Mohtadi, and Tuffs, 1987) that arose to address deficiencies in minimum variance control. Interestingly enough, earlier research on RHC (Kleinman, 1970; Thomas, 1975; Kwon and Pearson, 1977) had shown indirectly that the imposition of a terminal equality constraint in the finite horizon optimal control problem ensured closed-loop stability for linear unconstrained systems. That a terminal equality constraint had an equally beneficial effect for constrained nonlinear discrete time systems was shown by Keerthi and Gilbert (1988) and for constrained nonlinear continuous time systems by Chen and Shaw (1982) and Mayne and Michalska (1990). In each of these papers, Lyapunov stability theory was employed in contrast to the then current literature on MPC and GPC.

The next advance showed that incorporation of a suitable terminal cost and terminal constraint in the finite horizon optimal control problem ensured closed-loop stability; the terminal constraint set is required to be control invariant, and the terminal cost function is required to be a local CLF. Perhaps the earliest proposal in this direction is the brief paper by Sznaier and Damborg (1987) for linear systems with polytopic constraints; in this prescient paper the terminal cost is chosen to be the value function for the *unconstrained* infinite horizon optimal control problem, and the terminal constraint set is the maximal constraint admissible set (Gilbert and Tan, 1991) for the optimal controlled system.[8] A suitable terminal cost and terminal constraint set for constrained nonlinear continuous time systems was proposed in Michalska and Mayne (1993) but in the context of dual mode MPC. In a paper that has had considerable impact, Chen and Allgöwer (1998) showed that similar "ingredients" may be employed to stabilize constrained nonlinear continuous time systems when conventional MPC is employed. Related results were obtained by Parisini and Zoppoli

[8]If the optimal infinite horizon controlled system is described by $x^+ = A_K x$ and if the constraints are $u \in \mathbb{U}$ and $x \in \mathbb{X}$, then the maximal constraint admissible set is $\{x \mid A_K^i x \in \mathbb{X},\ KA_K^i x \in \mathbb{U}\ \forall i \in \mathbb{I}_{\geq 0}\}$.

(1995), and De Nicolao, Magni, and Scattolini (1996).

Stability proofs for the form of MPC proposed, but not analyzed, in Sznaier and Damborg (1987) were finally provided by Chmielewski and Manousiouthakis (1996) and Scokaert and Rawlings (1998). These papers also showed that optimal control for the *infinite* horizon constrained optimal control problem with a specified initial state is achieved if the horizon is chosen sufficiently long. A terminal constraint is not required if a global, rather than a local, CLF is available for use as a terminal cost function. Thus, for the case when the system being controlled is linear and stable, and subject to a convex control constraint, Rawlings and Muske (1993) showed, in a paper that raised considerable interest, that closed-loop stability may be obtained if the terminal constraint is omitted and the infinite horizon cost using zero control is employed as the terminal cost. The resultant terminal cost is a global CLF.

The basic principles ensuring closed-loop stability in these and many other papers including De Nicolao, Magni, and Scattolini (1998), and Mayne (2000) were distilled and formulated as "stability axioms" in the review paper Mayne et al. (2000); they appear as Assumptions 2.12, 2.13 and 2.16 in this chapter. These assumptions provide sufficient conditions for closed-loop stability for a given horizon. There is an alternative literature that shows that closed-loop stability may often be achieved if the horizon is chosen to be sufficiently long. Contributions in this direction include Primbs and Nevistić (2000), Jadbabaie, Yu, and Hauser (2001), as well as Parisini and Zoppoli (1995), Chmielewski and Manousiouthakis (1996), and Scokaert and Rawlings (1998) already mentioned. An advantage of this approach is that it avoids addition of a terminal constraint, although this may be avoided by alternative means as shown in Section 2.6.

2.12 Exercises

Exercise 2.1: Discontinuous MPC

Compute, for Example 2.8, $\mathcal{U}_3(x)$, $V_3^0(x)$ and $\kappa_3(x)$ at a few points on the unit circle.

Exercise 2.2: Boundedness of discrete time model

Complete the proof of Proposition 2.21 by showing that $f(\cdot)$ and $f_{\mathbb{Z}}^{-1}(\cdot)$ are bounded on bounded sets.

Exercise 2.3: Destabilization with state constraints

Consider a state feedback regulation problem with the origin as the setpoint (Muske and Rawlings, 1993). Let the system be

$$A = \begin{bmatrix} 4/3 & -2/3 \\ 1 & 0 \end{bmatrix} \qquad B = \begin{bmatrix} 1 \\ 0 \end{bmatrix} \qquad C = [-2/3 \;\; 1]$$

and the controller objective function tuning matrices be

$$Q = I \qquad R = I \qquad N = 5$$

(a) Plot the unconstrained regulator performance starting from initial condition $x(0) = \begin{bmatrix} 3 & 3 \end{bmatrix}'$.

(b) Add the output constraint $y(k) \leq 0.5$. Plot the response of the constrained regulator (both input and output). Is this regulator stabilizing? Can you modify the tuning parameters Q, R to affect stability as in Section 1.3.4?

(c) Change the output constraint to $y(k) \leq 1 + \epsilon, \epsilon > 0$. Plot the closed-loop response for a variety of ϵ. Are any of these regulators destabilizing?

(d) Set the output constraint back to $y(k) \leq 0.5$ and add the terminal constraint $x(N) = 0$. What is the solution to the regulator problem in this case? Increase the horizon N. Does this problem eventually go away?

Exercise 2.4: Computing the projection of $\mathbb{Z}$ onto $\mathcal{X}_N$

Given a polytope

$$\mathbb{Z} := \{(x, u) \in \mathbb{R}^n \times \mathbb{R}^m \mid Gx + Hu \leq \psi\}$$

write an Octave or MATLAB program to determine $\mathcal{X}$, the projection of $\mathbb{Z}$ onto $\mathbb{R}^n$

$$\mathcal{X} = \{x \in \mathbb{R}^n \mid \exists u \in \mathbb{R}^m \text{ such that } (x, u) \in \mathbb{Z}\}$$

Use algorithms 3.1 and 3.2 in Keerthi and Gilbert (1987).

To check your program, consider a system

$$x^+ = \begin{bmatrix} 1 & 1 \\ 0 & 1 \end{bmatrix} x + \begin{bmatrix} 0 \\ 1 \end{bmatrix} u$$

subject to the constraints $\mathbb{X} = \{x \mid x_1 \leq 2\}$ and $\mathbb{U} = \{u \mid -1 \leq u \leq 1\}$. Consider the MPC problem with $N = 2$, $\mathbf{u} = (u(0), u(1))$, and the set $\mathbb{Z}$ given by

$$\mathbb{Z} = \{(x, \mathbf{u}) \mid x, \phi(1; x, \mathbf{u}), \phi(2; x, \mathbf{u}) \in \mathbb{X} \text{ and } u(0), u(1) \in \mathbb{U}\}$$

Verify that the set

$$\mathcal{X}_2 := \{x \in \mathbb{R}^2 \mid \exists \mathbf{u} \in \mathbb{R}^2 \text{ such that } (x, \mathbf{u}) \in \mathbb{Z}\}$$

is given by

$$\mathcal{X}_2 = \{x \in \mathbb{R}^2 \mid Px \le p\} \qquad P = \begin{bmatrix} 1 & 0 \\ 1 & 1 \\ 1 & 2 \end{bmatrix} \qquad p = \begin{bmatrix} 2 \\ 3 \\ 5 \end{bmatrix}$$

Exercise 2.5: Computing the maximal output admissible set

Write an Octave or MATLAB program to determine the maximal constraint admissible set for the system $x^+ = Fx, y = Hx$ subject to the hard constraint $y \in Y$ in which $Y = \{y \mid Ey \le e\}$. Use algorithm 3.2 in Gilbert and Tan (1991).

To check your program, verify for the system

$$F = \begin{bmatrix} 0.9 & 1 \\ 0 & 0.09 \end{bmatrix} \qquad H = \begin{bmatrix} 1 & 1 \end{bmatrix}$$

subject to the constraint $Y = \{y \mid -1 \le y \le 1\}$, and that the maximal output admissible set is given by

$$O_\infty = \{x \in \mathbb{R}^2 \mid Ax \le b\} \qquad A = \begin{bmatrix} 1 & 1 \\ -1 & 1 \\ 0.9 & 1.01 \\ -0.9 & -1.01 \end{bmatrix} \qquad b = \begin{bmatrix} 1 \\ 1 \\ 1 \\ 1 \end{bmatrix}$$

Show that t^*, the smallest integer t such that $O_t = O_\infty$ satisfies $t^* = 1$.

What happens to t^* as F_{22} increases and approaches 1. What do you conclude for the case $F_{22} \ge 1$?

Exercise 2.6: Terminal constraint and region of attraction

Consider the system

$$x^+ = Ax + Bu$$

subject to the constraints

$$x \in \mathbb{X} \qquad u \in \mathbb{U}$$

in which

$$A = \begin{bmatrix} 2 & 1 \\ 0 & 2 \end{bmatrix} \qquad B = \begin{bmatrix} 1 & 0 \\ 0 & 1 \end{bmatrix}$$

$$\mathbb{X} = \{x \in \mathbb{R}^2 \mid x_1 \le 5\} \qquad \mathbb{U} = \{u \in \mathbb{R}^2 \mid -\mathbf{1} \le u \le \mathbf{1}\}$$

and $\mathbf{1} \in \mathbb{R}^2$ is a vector of ones. The MPC cost function is

$$V_N(x, \mathbf{u}) = \sum_{i=0}^{N-1} \ell(x(i), u(i)) + V_f(x(N))$$

in which

$$\ell(x, u) = (1/2)(|x|_Q^2 + |u|^2) \qquad Q = \begin{bmatrix} \alpha & 0 \\ 0 & \alpha \end{bmatrix}$$

and $V_f(\cdot)$ is the terminal penalty on the final state.

(a) Implement unconstrained MPC with no terminal cost ($V_f(\cdot) = 0$) for a few values of α. Choose a value of α for which the resultant closed loop is unstable. Try $N = 3$.

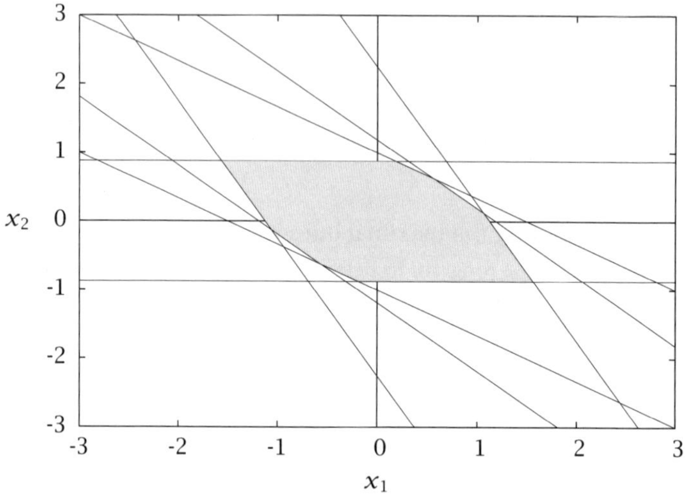

Figure 2.7: Region of attraction (shaded region) for constrained MPC controller of Exercise 2.6.

(b) Implement constrained MPC with no terminal cost or terminal constraint for the value of α obtained in the previous part. Is the resultant closed loop stable or unstable?

(c) Implement constrained MPC with terminal equality constraint $x(N) = 0$ for the same value of α. Find the region of attraction for the constrained MPC controller using the projection algorithm from Exercise 2.4. The result should resemble Figure 2.7.

Exercise 2.7: Infinite horizon cost to go as terminal penalty

Consider the system

$$x^+ = Ax + Bu$$

subject to the constraints

$$x \in \mathbb{X} \qquad u \in \mathbb{U}$$

in which

$$A = \begin{bmatrix} 2 & 1 \\ 0 & 2 \end{bmatrix} \qquad B = \begin{bmatrix} 1 & 0 \\ 0 & 1 \end{bmatrix}$$

and

$$\mathbb{X} = \{x \in \mathbb{R}^2 \mid -5 \le x_1 \le 5\} \qquad \mathbb{U} = \{u \in \mathbb{R}^2 \mid -\mathbf{1} \le u \le \mathbf{1}\}$$

The cost is

$$V_N(x, \mathbf{u}) := \sum_{i=0}^{N-1} \ell(x(i), u(i)) + V_f(x(N))$$

in which

$$\ell(x, u) = (1/2)(|x|_Q^2 + |u|^2) \qquad Q = \begin{bmatrix} \alpha & 0 \\ 0 & \alpha \end{bmatrix}$$

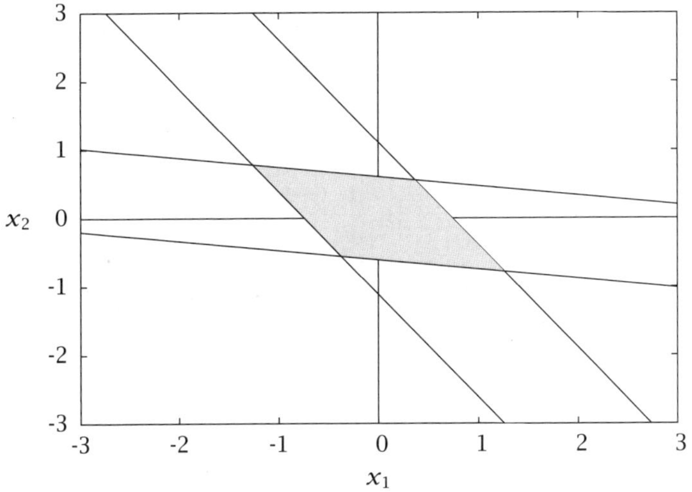

Figure 2.8: The region $\mathbb{X}_f$, in which the unconstrained LQR control law is feasible for Exercise 2.7.

and $V_f(\cdot)$ is the terminal penalty on the final state and $\mathbf{1} \in \mathbb{R}^2$ is a vector of all ones. Use $\alpha = 10^{-5}$ and $N = 3$ and terminal cost $V_f(x) = (1/2)x'\Pi x$ where Π is the solution to the steady-state Riccati equation.

(a) Compute the infinite horizon optimal cost and control law for the unconstrained system.

(b) Find the region $\mathbb{X}_f$, the maximal constraint admissible set using the algorithm in Exercise 2.5 for the system $x^+ = (A + BK)x$ with constraints $x \in \mathbb{X}$ and $Kx \in \mathbb{U}$. You should obtain the region shown in Figure 2.8.

(c) Add a terminal constraint $x(N) \in \mathbb{X}_f$ and implement constrained MPC. Find $\mathcal{X}_N$, the region of attraction for the MPC problem with $V_f(\cdot)$ as the terminal cost and $x(N) \in \mathbb{X}_f$ as the terminal constraint. Contrast it with the region of attraction for the MPC problem in Exercise 2.6 with a terminal constraint $x(N) = 0$.

(d) Estimate $\bar{\mathcal{X}}_N$, the set of initial states for which the MPC control sequence for horizon N is equal to the MPC control sequence for an infinite horizon.
Hint: $x \in \bar{\mathcal{X}}_N$ if and only if $x^0(N;x) \in \text{int}(\mathbb{X}_f)$. Why?

Exercise 2.8: Terminal penalty with and without terminal constraint

Consider the system

$$x^+ = Ax + Bu$$

subject to the constraints

$$x \in \mathbb{X} \qquad u \in \mathbb{U}$$

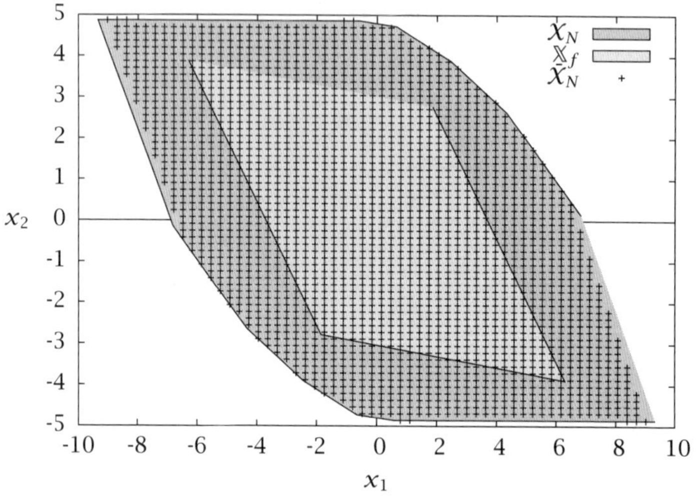

Figure 2.9: The region of attraction for terminal constraint $x(N) \in \mathbb{X}_f$ and terminal penalty $V_f(x) = (1/2)x'\Pi x$ and the estimate of $\bar{\mathcal{X}}_N$ for Exercise 2.8.

in which

$$A = \begin{bmatrix} 2 & 1 \\ 0 & 2 \end{bmatrix} \qquad B = \begin{bmatrix} 1 & 0 \\ 0 & 1 \end{bmatrix}$$

and

$$\mathbb{X} = \{x \in \mathbb{R}^2 \mid -15 \leq x_1 \leq 15\} \qquad \mathbb{U} = \{u \in \mathbb{R}^2 \mid -5 \cdot \mathbf{1} \leq u \leq 5 \cdot \mathbf{1}\}$$

The cost is

$$V_N(x, \mathbf{u}) = \sum_{i=0}^{N-1} \ell(x(i), u(i)) + V_f(x(N))$$

in which

$$\ell(x, u) = (1/2)(|x|_Q^2 + |u|)^2 \qquad Q = \begin{bmatrix} \alpha & 0 \\ 0 & \alpha \end{bmatrix}$$

$V_f(\cdot)$ is the terminal penalty on the final state, and $\mathbf{1} \in \mathbb{R}^2$ is a vector of ones.

Use $\alpha = 10^{-5}$ and $N = 3$ and terminal cost $V_f(x) = (1/2)x'\Pi x$ where $V_f(\cdot)$ is the infinite horizon optimal cost for the unconstrained problem.

(a) Add a terminal constraint $x(N) \in \mathbb{X}_f$, in which $\mathbb{X}_f$ is the maximal constraint admissible set for the system $x^+ = (A + BK)x$ and K is the optimal controller gain for the unconstrained problem. Using the code developed in Exercise 2.7, estimate $\mathcal{X}_N$, the region of attraction for the MPC problem with this terminal constraint and terminal cost. Also estimate $\bar{\mathcal{X}}_N$, the region for which the MPC control sequence for horizon N is equal to the the MPC control sequence for infinite horizon. Your results should resemble Figure 2.9

(b) Remove the terminal constraint and *estimate* the domain of attraction $\hat{\mathcal{X}}_N$ (by simulation). Compare this $\hat{\mathcal{X}}_N$ with $\mathcal{X}_N$ and $\bar{\mathcal{X}}_N$ obtained previously.

(c) Change the terminal cost to $V_f(x) = (3/2)x'\Pi x$ and repeat the previous part.

Exercise 2.9: Decreasing property for the time-varying case

Prove Lemma 2.32.

Exercise 2.10: Terminal cost bound for the time-varying case

Prove Lemma 2.33.

Exercise 2.11: Modification of terminal cost

Refer to Section 2.6.1. Show that the pair $(\beta V_f(\cdot), \mathbb{X}_f)$ satisfies Assumptions 2.12 and 2.13 if $(V_f(\cdot), \mathbb{X}_f)$ satisfies these assumptions, $\beta \geq 1$, and $\ell(\cdot)$ satisfies Assumption 2.16.

Exercise 2.12: Terminal inequality

Refer to Section 2.6.2 where the terms $V_N^e(\cdot)$, $\hat{V}_N^e(\cdot)$, $V_f^e(\cdot)$ and $\mathcal{X}_N^e$ are defined. Prove that the control sequence

$$\{u^e(0;x), u^e(0;x^+), u^e(1;x^+), \ldots, u^e(N-2;x^+)\}$$

is feasible for problem $\mathbb{P}_N^e(x)$. Use this result and the fact that $x^e(N;x)$ lies in $\mathcal{X}_N^e$ to establish that

$$V_f^e(x^e(N;x^+)) \leq V_f^e(x^e(N-1;x^+)) - \ell(x^e(N-1;x^+), u^e(N-1;x^+))$$

in which $x^+ := f(x, u^e(0;x))$.

Exercise 2.13: A Lyapunov theorem for asymptotic stability

Prove the asymptotic stability result for Lyapunov functions.

Theorem 2.44 (Lyapunov theorem for asymptotic stability). *Given the dynamic system*

$$x^+ = f(x) \qquad 0 = f(0)$$

The origin is asymptotically stable if there exist $\mathcal{K}$-functions α, β, γ, and $r > 0$ such that Lyapunov function V satisfies for $x \in r\mathcal{B}$

$$\alpha(|x|) \leq V(x) \leq \beta(|x|)$$
$$V(f(x)) - V(x) \leq -\gamma(|x|)$$

Exercise 2.14: An MPC stability result

Given the following nonlinear model and objective function

$$x^+ = f(x,u), \qquad 0 = f(0,0)$$
$$x(0) = x$$
$$V_N(x, \mathbf{u}) = \sum_{k=0}^{N-1} \ell(x(k), u(k))$$

Consider the terminal constraint MPC regulator

$$\min_{\mathbf{u}} V_N(x, \mathbf{u})$$

subject to

$$x^+ = f(x, u) \qquad x(0) = x \qquad x(N) = 0$$

and denote the first move in the optimal control sequence as $u^0(x)$. Given the closed-loop system

$$x^+ = f(x, u^0(x))$$

(a) Prove that the origin is asymptotically stable for the closed-loop system. State the cost function assumption and controllability assumption required so that the control problem is feasible for some set of defined initial conditions.

(b) What assumptions about the cost function $\ell(x, u)$ are required to strengthen the controller so that the origin is exponentially stable for the closed-loop system? How does the controllability assumption change for this case?

Exercise 2.15: Stability using observability instead of IOSS

Assume that the system $x^+ = f(x, u)$, $y = h(x)$ is ℓ-observable, i.e., there exists a $\alpha \in \mathcal{K}$ and an integer $N_o \geq 0$ such that

$$\sum_{j=0}^{N_o-1} \ell(y(i), u(i)) \geq \alpha(|x|)$$

for all x and all $\mathbf{u}$; here $x(i) := \phi(i; x, \mathbf{u})$ and $y(i) := h(x(i))$. Prove the result given in Section 2.7 that the origin is asymptotically stable for the closed-loop system $x^+ = f(x, \kappa_N(x))$ using the assumption that $x^+ = f(x, u)$, $y = h(x)$ is ℓ-observable rather than IOSS. Assume that $N \geq N_o$.

Exercise 2.16: Input/output-to-state stability (IOSS) and convergence

Prove Proposition 2.41. Hint: consider the solution at time $k + l$ using the state at time k as the initial state.

Exercise 2.17: Equality for quadratic functions

Prove the following result which is useful for analyzing the unreachable setpoint problem.

Lemma 2.45 (An equality for quadratic functions). *Let $\mathbb{X}$ be a nonempty compact subset of $\mathbb{R}^n$, and let $\ell(\cdot)$ be a strictly convex quadratic function on $\mathbb{X}$ defined by $\ell(x) := (1/2)x'Qx + q'x + c, \quad Q > 0$. Consider a sequence $\{x(i) \mid i \in \mathbb{I}_{1:P}\}$ with mean $\bar{x}_P := (1/P)\sum_{i=1}^{P} x(i)$. Then the following holds*

$$\sum_{i=1}^{P} \ell(x(i)) = (1/2)\sum_{i=1}^{P} |x(i) - \bar{x}_P|_Q^2 + P\ell(\bar{x}_P)$$

It follows from this lemma that $\ell(\bar{x}_P) \leq (1/P)\sum_{i=1}^{P} \ell(x(i))$, which is Jensen's inequality for the special case of a quadratic function.

Exercise 2.18: Unreachable setpoint MPC and evolution in a compact set

Prove the following lemma, which is useful for analyzing the stability of MPC with an unreachable setpoint.

Lemma 2.46 (Evolution in a compact set). *Suppose $x(0) = x$ lies in the set $\mathcal{X}_N$. Then the state trajectory $\{x(i)\}$ where, for each i, $x(i) = \phi_f(i;x)$ of the controlled system $x^+ = f(x)$ evolves in a compact set.*

Exercise 2.19: MPC and multivariable, constrained systems

Consider a two-input, two-output process with the following transfer function

$$G(s) = \begin{bmatrix} \dfrac{2}{10s+1} & \dfrac{2}{s+1} \\ \dfrac{1}{s+1} & -\dfrac{4}{s+1} \end{bmatrix}$$

(a) Consider a unit setpoint change in the first output. Choose a reasonable sample time, Δ. Simulate the behavior of an offset-free discrete time MPC controller with $Q = I, S = I$ and large N.

(b) Add the constraint $-1 \le u(k) \le 1$ and simulate the response.

(c) Add the constraint $-0.1 \le \Delta u/\Delta \le 0.1$ and simulate the response.

(d) Add significant noise to both output measurements (make the standard deviation in each output about 0.1). Retune the MPC controller to obtain good performance. Describe which controller parameters you changed and why.

Exercise 2.20: LQR versus LAR

We are now all experts on the linear quadratic regulator (LQR), which employs a linear model and quadratic performance measure. Let's consider the case of a linear model but absolute value performance measure, which we call the linear absolute regulator (LAR)[9]

$$\min_{\mathbf{u}} \sum_{k=0}^{N-1} \left(q\,|x(k)| + r\,|u(k)| \right) + q(N)\,|x(N)|$$

For simplicity consider the following one-step controller, in which u and x are *scalars*

$$\min_{u(0)} V(x(0), u(0)) = |x(1)| + |u(0)|$$

subject to

$$x(1) = Ax(0) + Bu(0)$$

Draw a sketch of $x(1)$ versus $u(0)$ (recall $x(0)$ is a known parameter) and show the x-axis and y-axis intercepts on your plot. Now draw a sketch of $V(x(0), u(0))$ versus $u(0)$ in order to see what kind of optimization problem you are solving. You may want to plot both terms in the objective function individually and then add them together to make your V plot. Label on your plot the places where the cost function V suffers discontinuities in slope. Where is the solution in your sketch? Does it exist for all $A, B, x(0)$? Is it unique for all $A, B, x(0)$?

The motivation for this problem is to change the quadratic program (QP) of the LQR to a linear program (LP) in the LAR, because the computational burden for LPs is often smaller than QPs. The absolute value terms can be converted into linear terms with a standard trick involving slack variables.

[9]Laplace would love us for making this choice, but Gauss would not be happy.

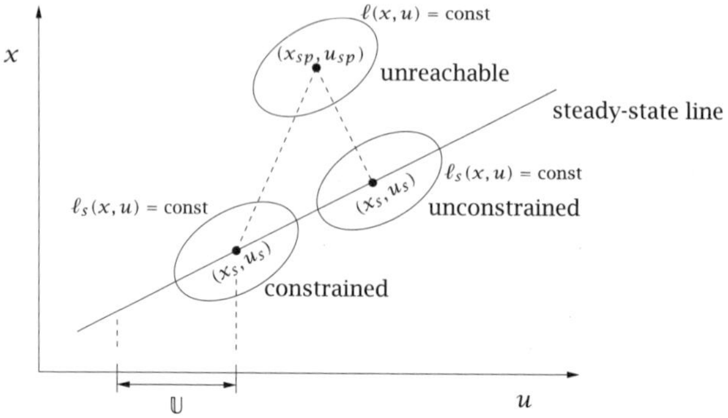

Figure 2.10: Inconsistent setpoint $(x_{\text{sp}}, u_{\text{sp}})$, unreachable stage cost $\ell(x,u)$, and optimal steady states (x_s, u_s), and stage costs $\ell_s(x,u)$ for constrained and unconstrained systems.

Exercise 2.21: Unreachable setpoints in constrained versus unconstrained linear systems

Consider the linear system with input constraint

$$x^+ = Ax + Bu \qquad u \in \mathbb{U}$$

We examine here both unconstrained systems in which $\mathbb{U} = \mathbb{R}^m$ and constrained systems in which $\mathbb{U} \subset \mathbb{R}^m$ is a convex polyhedron. Consider the stage cost defined in terms of setpoints for state and input x_{sp}, u_{sp}

$$\ell(x,u) = (1/2)(|x - x_{\text{sp}}|_Q^2 + |u - u_{\text{sp}}|_R^2)$$

in which we assume for simplicity that $Q, R > 0$. For the setpoint to be unreachable in an unconstrained problem, the setpoint must be *inconsistent*, i.e., not a steady state of the system, or

$$x_{\text{sp}} \neq Ax_{\text{sp}} + Bu_{\text{sp}}$$

Consider also using the stage cost centered at the optimal steady state (x_s, u_s)

$$\ell_s(x,u) = (1/2)(|x - x_s|_Q^2 + |u - u_s|_R^2)$$

The optimal steady state satisfies

$$(x_s, u_s) = \arg\min_{x,u} \ell(x,u)$$

subject to

$$\begin{bmatrix} I - A & -B \end{bmatrix} \begin{bmatrix} x \\ u \end{bmatrix} = 0 \qquad u \in \mathbb{U}$$

Figure 2.10 depicts an inconsistent setpoint, and the optimal steady state for unconstrained and constrained systems.

(a) For unconstrained systems, show that optimizing the cost function with terminal constraint

$$V(x, \mathbf{u}) := \sum_{k=0}^{N-1} \ell(x(k), u(k))$$

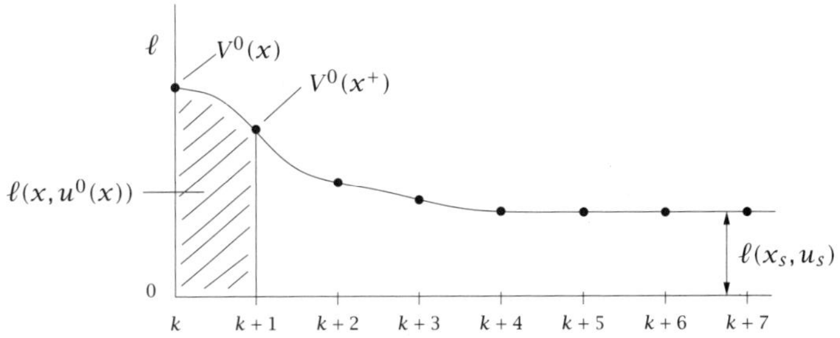

Figure 2.11: Stage cost versus time for the case of unreachable setpoint. The cost $V^0(x(k))$ is the area under the curve to the right of time k.

subject to

$$x^+ = Ax + Bu \qquad x(0) = x \qquad x(N) = x_s$$

gives the same solution as optimizing the cost function

$$V_s(x, \mathbf{u}) := \sum_{k=0}^{N-1} \ell_s(x(k), u(k))$$

subject to the same model constraint, initial condition, and terminal constraint. Therefore, there is no reason to consider the unreachable setpoint problem further for an *unconstrained* linear system. Shifting the stage cost from $\ell(x,u)$ to $\ell_s(x,u)$ provides identical control behavior and is simpler to analyze.

Hint. First define a third stage cost $l(x,u) = \ell(x,u) - \lambda'((I-A)x - Bu)$, and show, for any λ, optimizing with $l(x,u)$ as stage cost is the same as optimizing using $\ell(x,u)$ as stage cost. Then set $\lambda = \lambda_s$, the optimal Lagrange multiplier of the *steady-state* optimization problem.

(b) For *constrained* systems, provide a simple example that shows optimizing the cost function $V(x, \mathbf{u})$ subject to

$$x^+ = Ax + Bu \qquad x(0) = x \qquad x(N) = x_s \qquad u(k) \in \mathbb{U} \quad \text{for all } k \in \mathbb{I}_{0:N-1}$$

does *not* give the same solution as optimizing the cost function $V_s(x, \mathbf{u})$ subject to the same constraints. For *constrained* linear systems, these problems are different and optimizing the unreachable stage cost provides a new design opportunity.

Exercise 2.22: Filing for patent

An excited graduate student shows up at your office. He begins, “Look, I have discovered a great money-making scheme using MPC.” You ask him to tell you about it. “Well,” he says, “you told us in class that the optimal steady state is asymptotically stable even if you use the stage cost measuring distance from the unreachable setpoint, right?” You reply, "Yes, that's what I said." He continues, "OK, well look at this little sketch I drew," and he shows you a picture like Figure 2.11. "So imagine I use the infinite horizon cost function so the open-loop and closed-loop trajectories are identical. If the best steady state is asymptotically stable, then the stage cost asymptotically approaches $\ell(x_s, u_s)$, right?" You reply, "I guess that looks right." He then says, "OK, well if I look at the optimal cost using state x at time k and state x^+ at time $k+1$, by the principle of optimality I get the usual cost decrease"

$$V^0(x^+) \leq V^0(x) - \ell(x, u^0(x)) \tag{2.44}$$

You interrupt, "Wait, these $V^0(\cdot)$ costs are not bounded in this case!" Unfazed, the student replies, "Yeah, I realize that, but this sketch is basically correct regardless. Say we just make the horizon *really long*; then the costs are all finite and this equation becomes closer and closer to being true as we make the horizon longer and longer." You start to feel a little queasy at this point. The student continues, "OK, so if this inequality basically holds, $V^0(x(k))$ is decreasing with k along the closed-loop trajectory, it is bounded below for all k, it converges, and, therefore, $\ell(x(k), u^0(x(k)))$ goes to zero as k goes to ∞." You definitely don't like where this is heading, and the student finishes with, "But $\ell(x, u) = 0$ implies $x = x_{sp}$ and $u = u_{sp}$, and the setpoint is *supposed* to be unreachable. But I have proven that infinite horizon MPC can reach an *unreachable* setpoint. We should patent this!"

How do you respond to this student? Here are some issues to consider.

(a) Does the principle of optimality break down in the unreachable setpoint case?

(b) Are the open-loop and closed-loop trajectories identical in the limit of an infinite horizon controller with an unreachable setpoint?

(c) Does inequality (2.44) hold as $N \to \infty$? If so, how can you put it on solid footing? If not, why not, and with what do you replace it?

(d) Do you file for patent?

Bibliography

C. C. Chen and L. Shaw. On receding horizon control. *Automatica*, 16(3):349–352, 1982.

H. Chen and F. Allgöwer. A quasi-infinite horizon nonlinear model predictive control scheme with guaranteed stability. *Automatica*, 34(10):1205–1217, 1998.

D. Chmielewski and V. Manousiouthakis. On constrained infinite-time linear quadratic optimal control. *Sys. Cont. Let.*, 29:121–129, 1996.

D. W. Clarke, C. Mohtadi, and P. S. Tuffs. Generalized predictive control—Part I. The basic algorithm. *Automatica*, 23(2):137–148, 1987.

C. R. Cutler and B. L. Ramaker. Dynamic matrix control—a computer control algorithm. In *Proceedings of the Joint Automatic Control Conference*, 1980.

R. M. C. De Keyser and A. R. Van Cauwenberghe. Extended prediction self-adaptive control. In H. A. Barker and P. C. Young, editors, *Proceedings of the 7th IFAC Symposium on Identification and System Parameter Estimation*, pages 1255–1260, Oxford, 1985. Pergamon Press.

G. De Nicolao, L. Magni, and R. Scattolini. Stabilizing nonlinear receding horizon control via a nonquadratic penalty. In *Proceedings IMACS Multiconference CESA*, volume 1, pages 185–187, Lille, France, 1996.

G. De Nicolao, L. Magni, and R. Scattolini. Stabilizing receding-horizon control of nonlinear time-varying systems. *IEEE Trans. Auto. Cont.*, 43(7):1030–1036, 1998.

C. E. García and A. M. Morshedi. Quadratic programming solution of dynamic matrix control (QDMC). *Chem. Eng. Commun.*, 46:73–87, 1986.

C. E. García, D. M. Prett, and M. Morari. Model predictive control: Theory and practice—a survey. *Automatica*, 25(3):335–348, 1989.

E. G. Gilbert and K. T. Tan. Linear systems with state and control constraints: The theory and application of maximal output admissible sets. *IEEE Trans. Auto. Cont.*, 36(9):1008–1020, September 1991.

B. Hu and A. Linnemann. Toward infinite-horizon optimality in nonlinear model predictive control. *IEEE Trans. Auto. Cont.*, 47(4):679–682, April 2002.

A. Jadbabaie, J. Yu, and J. Hauser. Unconstrained receding horizon control of nonlinear systems. *IEEE Trans. Auto. Cont.*, 46(5):776–783, 2001.

S. S. Keerthi and E. G. Gilbert. Computation of minimum-time feedback control laws for systems with state-control constraints. *IEEE Trans. Auto. Cont.*, 32: 432–435, 1987.

S. S. Keerthi and E. G. Gilbert. Optimal infinite-horizon feedback laws for a general class of constrained discrete-time systems: Stability and moving-horizon approximations. *J. Optim. Theory Appl.*, 57(2):265–293, May 1988.

J.-S. Kim, T.-W. Yoon, A. Jadbabaie, and C. D. Persis. Input-to-state stability for neutrally stable linear systems subject to control constraints. In *Proceedings of the 43rd IEEE Conference on Decision and Control*, Atlantis, Bahamas, December 2004.

D. L. Kleinman. An easy way to stabilize a linear constant system. *IEEE Trans. Auto. Cont.*, 15(12):692, December 1970.

W. H. Kwon and A. E. Pearson. A modified quadratic cost problem and feedback stabilization of a linear system. *IEEE Trans. Auto. Cont.*, 22(5):838–842, October 1977.

E. B. Lee and L. Markus. *Foundations of Optimal Control Theory*. John Wiley and Sons, New York, 1967.

D. Limon, T. Alamo, F. Salas, and E. F. Camacho. On the stability of MPC without terminal constraint. *IEEE Trans. Auto. Cont.*, 51(5):832–836, May 2006.

P. Marquis and J. P. Broustail. SMOC, a bridge between state space and model predictive controllers: Application to the automation of a hydrotreating unit. In T. J. McAvoy, Y. Arkun, and E. Zafiriou, editors, *Proceedings of the 1988 IFAC Workshop on Model Based Process Control*, pages 37–43, Oxford, 1988. Pergamon Press.

D. Q. Mayne. Nonlinear model predictive control: challenges and opportunities. In F. Allgöwer and A. Zheng, editors, *Nonlinear Model Predictive Control*, pages 23–44. Birkhäuser Verlag, Basel, 2000.

D. Q. Mayne and H. Michalska. Receding horizon control of non-linear systems. *IEEE Trans. Auto. Cont.*, 35(5):814–824, 1990.

D. Q. Mayne, J. B. Rawlings, C. V. Rao, and P. O. M. Scokaert. Constrained model predictive control: Stability and optimality. *Automatica*, 36(6):789–814, 2000.

E. S. Meadows, M. A. Henson, J. W. Eaton, and J. B. Rawlings. Receding horizon control and discontinuous state feedback stabilization. *Int. J. Control*, 62 (5):1217–1229, 1995.

H. Michalska and D. Q. Mayne. Robust receding horizon control of constrained nonlinear systems. *IEEE Trans. Auto. Cont.*, 38(11):1623–1633, 1993.

K. R. Muske and J. B. Rawlings. Model predictive control with linear models. *AIChE J.*, 39(2):262–287, 1993.

T. Parisini and R. Zoppoli. A receding-horizon regulator for nonlinear systems and a neural approximation. *Automatica*, 31(10):1443–1451, 1995.

V. Peterka. Predictor-based self-tuning control. *Automatica*, 20(1):39–50, 1984.

D. M. Prett and R. D. Gillette. Optimization and constrained multivariable control of a catalytic cracking unit. In *Proceedings of the Joint Automatic Control Conference*, pages WP5–C, San Francisco, CA, 1980.

J. A. Primbs and V. Nevistić. Feasibility and stability of constrained finite receding horizon control. *Automatica*, 36:965–971, 2000.

A. I. Propoi. Use of linear programming methods for synthesizing sampled-data automatic systems. *Autom. Rem. Control*, 24(7):837–844, July 1963.

J. B. Rawlings and K. R. Muske. Stability of constrained receding horizon control. *IEEE Trans. Auto. Cont.*, 38(10):1512–1516, October 1993.

J. B. Rawlings, D. Bonné, J. B. Jørgensen, A. N. Venkat, and S. B. Jørgensen. Unreachable setpoints in model predictive control. *IEEE Trans. Auto. Cont.*, 53(9):2209–2215, October 2008.

J. Richalet, A. Rault, J. L. Testud, and J. Papon. Model predictive heuristic control: Applications to industrial processes. *Automatica*, 14:413–428, 1978a.

J. Richalet, A. Rault, J. L. Testud, and J. Papon. Algorithmic control of industrial processes. In *Proceedings of the 4th IFAC Symposium on Identification and System Parameter Estimation*, pages 1119–1167. North-Holland Publishing Company, 1978b.

P. O. M. Scokaert and J. B. Rawlings. Constrained linear quadratic regulation. *IEEE Trans. Auto. Cont.*, 43(8):1163–1169, August 1998.

M. Sznaier and M. J. Damborg. Suboptimal control of linear systems with state and control inequality constraints. In *Proceedings of the 26th Conference on Decision and Control*, pages 761–762, Los Angeles, CA, 1987.

Y. A. Thomas. Linear quadratic optimal estimation and control with receding horizon. *Electron. Lett.*, 11:19–21, January 1975.

B. E. Ydstie. Extended horizon adaptive control. In J. Gertler and L. Keviczky, editors, *Proceedings of the 9th IFAC World Congress*, pages 911–915, Oxford, 1984. Pergamon Press.

3

Robust Model Predictive Control

3.1 Introduction

3.1.1 Types of Uncertainty

Robust control concerns control of systems that are uncertain in some sense so that predicted behavior based on the *nominal* system is not identical to actual behavior. Uncertainty may arise in different ways. The system may have an additive disturbance that is unknown, the state of the system may not be perfectly known, or the model of the system that is used to determine control may be inaccurate.

A system with additive disturbance satisfies the following difference equation

$$x^+ = f(x, u) + w$$

The disturbance w in constrained optimal control problems is usually assumed to be bounded since it is impossible to ensure that a system with unbounded disturbances satisfies the usual state and control constraints. More precisely, we usually assume that w satisfies the constraint $w \in \mathbb{W}$ where $\mathbb{W}$ is a compact subset of $\mathbb{R}^n$ containing the origin.

The situation in which the state is not perfectly measured may be treated in several ways. In the stochastic optimal control literature, where the measured output is $y = Cx + v$ and the disturbance w and measurement noise v are usually assumed to be Gaussian white noise processes, the state or *hyperstate* of the optimal control problem is the conditional density of the state x at time k given prior measurements $\{y(0), y(1), \ldots, y(k-1)\}$. Because this density is usually difficult to compute and use, except in the linear case when it is provided by the Kalman filter, a suboptimal procedure is often adopted. In this suboptimal approach, the state x is replaced by its estimate $\hat{x}$ in a control law determined under the assumption that the state is accessible. This procedure is usually referred to as *certainty equivalence*, a term that was originally employed for the linear quadratic Gaussian (LQG) or similar cases when this procedure did not result in loss of optimality. When $f(\cdot)$ is linear, the evolution of the state estimate $\hat{x}$ may be expressed by a difference equation

$$\hat{x}^+ = f(\hat{x}, u) + \xi$$

in which ξ is the *innovation process*. In controlling $\hat{x}$, we should ensure that the actual state x, which, if the innovation process is bounded, lies in a bounded, possibly time-varying neighborhood of $\hat{x}$, satisfies the constraints of the optimal control problem.

Finally, a system that has parametric uncertainty may be modeled as

$$x^+ = f(x, u, \theta)$$

in which θ represents parameters of the system that are known only to the extent that they belong to a compact set Θ. A much studied example is

$$x^+ = Ax + Bu$$

in which $\theta := (A, B)$ may take any value in $\Theta := \text{co}\{(A_i, B_i) \mid i \in \mathcal{I}\}$ where $\mathcal{I} = \{1, 2, \ldots, I\}$, say, is an index set.

It is possible, of course, for all these types of uncertainty to occur in a single application. In this chapter we focus on the first and third types of uncertainty, namely, additive disturbance and parameter uncertainty. Output MPC, where the controller employs an estimate of the state, rather than the state itself, is treated in Chapter 5.

3.1.2 Feedback Versus Open-Loop Control

It is well known that feedback is required only when uncertainty is present; in the absence of uncertainty, feedback control and open-loop control are equivalent. Indeed, when uncertainty is not present, as for the systems studied in Chapter 2, the optimal control for a given initial state may be computed using either dynamic programming (DP) that provides an optimal control policy or sequence of feedback control laws, or an open-loop optimal control that merely provides a sequence of control actions. A simple example illustrates this fact. Consider the deterministic linear dynamic system defined by

$$x^+ = x + u$$

The optimal control problem, with horizon $N = 3$, is

$$\mathbb{P}_3(x): \qquad V_3^0(x) = \min_{\mathbf{u}_3} V_3(x, \mathbf{u})$$

in which $\mathbf{u} = \{u(0), u(1), u(2)\}$

$$V_3(x, \mathbf{u}) := (1/2) \sum_{i=0}^{2} [(x(i)^2 + u(i)^2)] + (1/2)x(3)^2$$

where, for each i, $x(i) = \phi(i; x, \mathbf{u}) = x + u(0) + u(1) + \ldots + u(i - 1)$, the solution of the difference equation $x^+ = x + u$ at time i if the initial state is $x(0) = x$ and the control (input) sequence is $\mathbf{u} = \{u(0), u(1), u(2)\}$; in matrix operations $\mathbf{u}$ is taken to be the column vector $[u(0), u(1), u(2)]'$. Thus

$$\begin{aligned} V_3(x, \mathbf{u}) &= (1/2)[x^2 + (x + u(0))^2 + (x + u(0) + u(1))^2 + \\ &\qquad (x + u(0) + u(1) + u(2))^2 + u(0)^2 + u(1)^2 + u(2)^2] \\ &= (3/2)x^2 + x \begin{bmatrix} 3 & 2 & 1 \end{bmatrix} \mathbf{u} + (1/2)\mathbf{u}' P_3 \mathbf{u} \end{aligned}$$

in which

$$P_3 = \begin{bmatrix} 4 & 2 & 1 \\ 2 & 3 & 1 \\ 1 & 1 & 2 \end{bmatrix}$$

The vector form of the optimal *open-loop* control sequence for an initial state of x is, therefore,

$$\mathbf{u}^0(x) = -P_3^{-1} \begin{bmatrix} 3 & 2 & 1 \end{bmatrix}' x = -\begin{bmatrix} 0.615 & 0.231 & 0.077 \end{bmatrix}' x$$

The optimal control and state sequences are, therefore,

$$\begin{aligned} \mathbf{u}^0(x) &= \{-0.615x, -0.231x, -0.077x\} \\ \mathbf{x}^0(x) &= \{x, 0.385x, 0.154x, 0.077x\} \end{aligned}$$

To compute the optimal *feedback* control, we use the DP recursions

$$\begin{aligned} V_i^0(x) &= \min_{u \in \mathbb{R}} \{x^2/2 + u^2/2 + V_{i-1}^0(x + u)\} \\ \kappa_i^0(x) &= \arg \min_{u \in \mathbb{R}} \{x^2/2 + u^2/2 + V_{i-1}^0(x + u)\} \end{aligned}$$

with boundary condition

$$V_0^0(x) = (1/2)x^2$$

This procedure gives the value function $V_i^0(\cdot)$ and the optimal control law $\kappa_i^0(\cdot)$ at each i where the subscript i denotes time to go. Solving the DP recursion, for all $x \in \mathbb{R}$, all $i \in \{1, 2, 3\}$, yields

$$\begin{aligned} V_1^0(x) &= (3/4)x^2 & \kappa_1^0(x) &= -(1/2)x \\ V_2^0(x) &= (4/5)x^2 & \kappa_2^0(x) &= -(3/5)x \\ V_3^0(x) &= (21/26)x^2 & \kappa_3^0(x) &= -(8/13)x \end{aligned}$$

Starting at state x at time 0, and applying the optimal control laws iteratively to the *deterministic* system $x^+ = x + u$ (recalling that at time i the optimal control law is $\kappa_{3-i}^0(\cdot)$ since, at time i, $3 - i$ is the time to go) yields

$$\begin{aligned} x^0(0) &= x & u^0(0) &= -(8/13)x \\ x^0(1) &= (5/13)x & u^0(1) &= -(3/13)x \\ x^0(2) &= (2/13)x & u^0(2) &= -(1/13)x \\ x^0(3; x) &= (1/13)x \end{aligned}$$

so that the optimal control and state sequences are, respectively,

$$\begin{aligned} \mathbf{u}^0(x) &= \{-(8/13)x, -(3/13)x, -(1/13)x\} \\ \mathbf{x}^0(x) &= \{x, (5/13)x, (2/13)x, (1/13)x\} \end{aligned}$$

which are identical with the optimal open-loop values computed above.

Consider next an uncertain version of the dynamic system in which uncertainty takes the simple form of an additive disturbance w; the system is defined by

$$x^+ = x + u + w$$

in which the only knowledge of w is that it lies in the compact set $\mathbb{W} := [-1, 1]$. Let $\phi(i; x, \mathbf{u}, \mathbf{w})$ denote the solution of this system at time i if the initial state is x at time 0, and the input and disturbance sequences are, respectively, $\mathbf{u}$ and $\mathbf{w} := \{w(0), w(1), w(2)\}$. The cost now depends on the disturbance sequence — but it also depends, in contrast to the deterministic problem discussed above, on whether the control is open-loop or feedback. To discuss the latter case, we define a feedback policy $\boldsymbol{\mu}$ to be a sequence of control laws

$$\boldsymbol{\mu} := \{\mu_0(\cdot), \mu_1(\cdot), \mu_2(\cdot)\}$$

in which $\mu_i : \mathbb{R} \to \mathbb{R}$, $i = 0, 1, 2$; under policy $\boldsymbol{\mu}$, if the state at time i is x, the control is $\mu_i(x)$. Let $\mathcal{M}$ denote the class of *admissible* policies, for example those policies for which each control law $\mu_i(\cdot)$ is continuous. Then, $\phi(i; x, \boldsymbol{\mu}, \mathbf{w})$ denotes the solution at time $i \in \{0, 1, 2, 3\}$ of the following difference equation

$$x(i+1) = x(i) + \mu_i(x(i)) + w(i) \qquad x(0) = x$$

An open-loop control sequence $\mathbf{u} = \{u(0), u(1), u(2)\}$ is then merely a degenerate policy $\boldsymbol{\mu} = \{\mu_0(\cdot), \mu_1(\cdot), \mu_2(\cdot)\}$ where each control law $\mu_i(\cdot)$ satisfies

$$\mu_i(x) = u(i)$$

for all $x \in \mathbb{R}$ and all $i \in \{0, 1, 2\}$. The cost $V_3(\cdot)$ may now be defined

$$V_3(x, \boldsymbol{\mu}, \mathbf{w}) := (1/2) \sum_{i=0}^{2} [(x(i)^2 + u(i)^2)] + (1/2) x(3)^2$$

where, now, $x(i) = \phi(i; x, \boldsymbol{\mu}, \mathbf{w})$ and $u(i) = \mu_i(x(i))$. Since the disturbance is unpredictable, the value of $\mathbf{w}$ is not known at time 0, so the optimal control problem must "eliminate" it in some meaningful way so that the solution $\boldsymbol{\mu}^0(x)$ does not depend on $\mathbf{w}$. To eliminate $\mathbf{w}$, the optimal control problem $\mathbb{P}_3^*(x)$ is defined by

$$\mathbb{P}_3^*(x): \qquad V_3^0(x) := \inf_{\boldsymbol{\mu} \in \mathcal{M}} J_3(x, \boldsymbol{\mu})$$

in which the cost $J_3(\cdot)$ is defined in such a way that it does not depend on $\mathbf{w}$; inf is used rather than min in this definition since the minimum may not exist. The most popular choice for $J_3(\cdot)$ in the MPC literature is

$$J_3(x, \boldsymbol{\mu}) := \max_{\mathbf{w} \in \mathcal{W}} V_3(x, \boldsymbol{\mu}, \mathbf{w})$$

in which the disturbance $\mathbf{w}$ is assumed to lie in $\mathcal{W}$ a bounded class of admissible disturbance sequences. Alternatively, if the disturbance sequence is random, the cost $J_3(\cdot)$ may be chosen to be

$$J_3(x, \boldsymbol{\mu}) := \mathcal{E} V_3(x, \boldsymbol{\mu}, \mathbf{w})$$

in which $\mathcal{E}$ denotes "expectation" or average, over random disturbance sequences. For our purpose here, we adopt the simple cost

$$J_3(x, \boldsymbol{\mu}) := V_3(x, \boldsymbol{\mu}, \mathbf{0})$$

in which $\mathbf{0} := \{0, 0, 0\}$ is the zero disturbance sequence. In this case, $J_3(x, \boldsymbol{\mu})$ is the nominal cost, i.e., the cost associated with the nominal

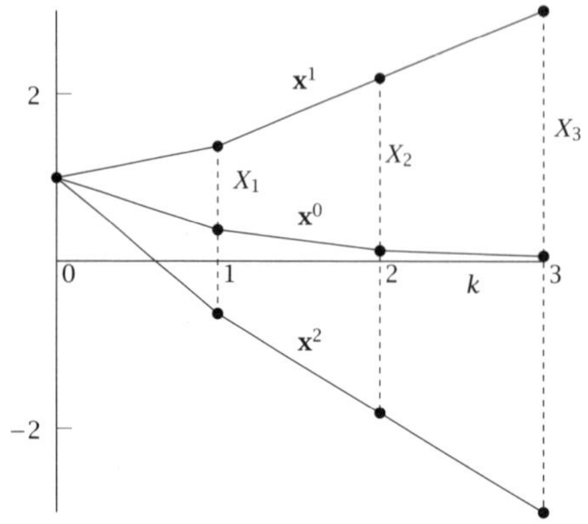

(a) Open-loop trajectories.

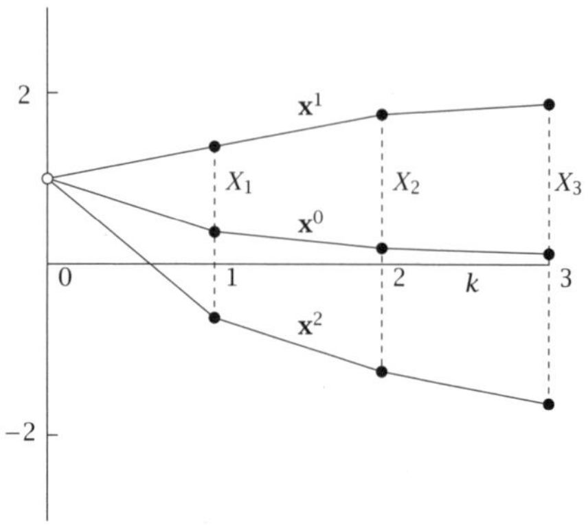

(b) Feedback trajectories.

Figure 3.1: Open-loop and feedback trajectories.

system $x^+ = x + u$ in which the disturbance is neglected. With this cost function, the solution to $\mathbb{P}_3^*(x)$ is the DP solution, obtained previously, to the deterministic *nominal* optimal control problem.

We now compare two solutions to $\mathbb{P}_3(x)$: the open-loop solution in which $\mathcal{M}$ is restricted to be the set of control sequences, and the feedback solution in which $\mathcal{M}$ is the class of admissible policies. The solution to the first problem is the solution to the deterministic problem discussed previously; the optimal control sequence is

$$\mathbf{u}^0(x) = \{-(8/13)x, -(3/13)x, -(1/13)x\}$$

in which x is the initial state at time 0. The solution to the second problem is the sequence of control laws determined previously, also for the deterministic problem, using *dynamic programming*; the optimal policy is $\boldsymbol{\mu}^0 = \{\mu_0^0(\cdot), \mu_1^0(\cdot), \mu_2(\cdot)\}$ where the control laws (functions) $\mu_i(\cdot)$, $i = 0, 1, 2$, are defined by

$$\mu_0^0(x) := \kappa_3^0(x) = -(8/13)x \quad \forall x \in \mathbb{R}$$
$$\mu_1^0(x) := \kappa_2^0(x) = -(3/5)x \quad \forall x \in \mathbb{R}$$
$$\mu_2^0(x) := \kappa_1^0(x) = -(1/2)x \quad \forall x \in \mathbb{R}$$

The two solutions, $\mathbf{u}^0(\cdot)$ and $\boldsymbol{\mu}^0$, when applied to the uncertain system $x^+ = x + u + w$ do *not* yield the same trajectories for all disturbance sequences. This is illustrated in Figure 3.1 for the three disturbance sequences, $\mathbf{w}^0 := \{0, 0, 0\}$, $\mathbf{w}^1 := \{1, 1, 1\}$, and $\mathbf{w}^2 := \{-1, -1, -1\}$;

and initial state $x = 1$ for which the corresponding state trajectories, denoted $\mathbf{x}^0$, $\mathbf{x}^1$, and $\mathbf{x}^2$, are

Open-loop solution.

$$\begin{aligned}
\mathbf{x}^0 &= \{1, (5/13), (2/13), (1/13)\} \\
\mathbf{x}^1 &= \{1, (18/13), (28/13), (40/13)\} \\
\mathbf{x}^2 &= \{1, -(8/13), -(24/13), -(38/13)\}
\end{aligned}$$

Feedback solution.

$$\begin{aligned}
\mathbf{x}^0 &= \{1, (5/13), (2/13), (1/13)\} \\
\mathbf{x}^1 &= \{1, (18/13), (101/65), (231/130)\} \\
\mathbf{x}^2 &= \{1, -(8/13), -(81/65), -(211/130)\}
\end{aligned}$$

Even for the short horizon of 3, the superiority of the feedback solution can be seen although the feedback was designed for the deterministic (nominal) system and therefore did not take the disturbance into account. For the open-loop solution $|x^2(3) - x^1(3)| = 6$, whereas for the feedback case $|x^2(3) - x^1(3)| = 3.4$; the open-loop solution does not restrain the *spread* of the trajectories resulting from the disturbance $\mathbf{w}$. If the horizon length is N, for the open-loop solution, $|x^2(N) - x^1(N)| = 2N$, whereas for the feedback case $|x^2(N) - x^1(N)| \to 3.24$ as $N \to \infty$. The obvious and well-known conclusion is that feedback control is superior to open-loop control when uncertainty is present. Feedback control requires determination of a control *policy*, however, which is a difficult task if nonlinearity and/or constraints are features of the optimal control problem.

3.1.3 Robust MPC

An important feature of conventional, or deterministic, MPC discussed in Chapter 2 is that the solution of the open-loop optimal control problem solved online is identical to that obtained by DP for the given initial state. When uncertainty is present and the state is known or observations of the state are available, feedback control is superior to open-loop control. The optimal control problem solved online must, therefore, permit feedback in order for its solution to coincide with the DP solution. The online optimal control problem with horizon N is $\mathbb{P}_N^*(x)$ in which the decision variable $\boldsymbol{\mu}$ is a sequence of control *laws* rather than $\mathbb{P}_N(x)$ in which the decision variable $\mathbf{u}$ is a sequence of control *actions.* MPC in which the decision variable is a policy has been termed *feedback* MPC to distinguish it from conventional MPC. Both forms of MPC naturally provide feedback control since the control that is implemented depends on the current state x in both cases. But the control that is applied depends on whether the optimal control problem solved is open loop, in which case the decision variable is a control sequence, or feedback, in which case the decision variable is a feedback policy.

In feedback MPC the solution to the optimal control problem $\mathbb{P}_N^*(x)$ is the policy $\boldsymbol{\mu}^0(x) = \{\mu_0^0(\cdot;x), \mu_1^0(\cdot;x), \ldots, \mu_{N-1}^0(\cdot;x)\}$. The constituent control laws are restrictions of those determined by DP and therefore depend on the initial state x as implied by the notation. Thus only the value $u^0(x) = \mu_0(x;x)$ of the control law $\mu_0(\cdot;x)$ at the initial state x need be determined while successive laws need only be determined over a limited range. In the example illustrated in Figure 3.1, $\mu_0(\cdot;x)$ need only be determined at the point $x = 1$, $\mu_1(\cdot;x)$ need only be determined over the interval $[-8/13, 18/13]$, and $\mu_2(\cdot;x)$ over the interval $[-81/65, 101/65]$, whereas in the DP solution these control laws are defined over the infinite interval $(-\infty, \infty)$.

While feedback MPC is superior in the presence of uncertainty, the associated optimal control problem is vastly more complex than the optimal control problem employed in deterministic MPC. The decision variable $\boldsymbol{\mu}$, being a sequence of control laws, is infinite dimensional; each law or function requires, in general, an infinite dimensional grid to specify it. The complexity is comparable to solving the DP equation, so that MPC, which in the deterministic case replaces DP with a solvable open-loop optimization problem, is not easily solved when uncertainty is present. Hence much research effort has been devoted to forms of feedback MPC that sacrifice optimality for simplicity. As in the early days of adaptive control, many different proposals have been made. These proposals for robust MPC are all simpler to implement than the optimal solution provided by DP.

At the current stage of research it is perhaps premature to select a particular approach; we have, nevertheless, selected one approach, *tube-based* MPC that we describe here and in Chapter 5. There is a good reason for our choice. It is well known that standard mathematical optimization algorithms may be used to obtain an optimal open-loop control sequence for an optimal control problem. What is less well known is that there exist algorithms, the second variation algorithms, which provide not only an optimal control sequence but also a *local* time-varying feedback law of the form $u(k) = v(k) + K(k)(x(k) - z(k))$

where $\{v(k)\}$ is the optimal open-loop control sequence and $\{z(k)\}$ the corresponding optimal open-loop state sequence. This policy provides feedback control for states $x(k)$ close to the nominal states $z(k)$. The second variation algorithms are too complex for routine use in MPC because they require computation of the second derivatives with respect to (x, u) of $f(\cdot)$ and $\ell(\cdot)$. When the system is linear, the cost quadratic, and the disturbance additive, however, the optimal control law for the unconstrained infinite horizon case is $u = Kx$. This result may be expressed as a time-varying control law $u(k) = v(k) + K(x(k) - z(k))$ where the state and control sequences $\{z(k)\}$ and $\{v(k)\}$ satisfy the nominal difference equations $z^+ = Az + Bv$, $v = Kz$, i.e., the sequences $\{z(k)\}$ and $\{v(k)\}$ are optimal open-loop solutions for zero disturbance and some initial state. The time-varying control law $u(k) = v(k) + K(x(k) - z(k))$ is clearly optimal in the unconstrained case; it remains optimal for the constrained case in the neighborhood of the nominal trajectory $\{z(k)\}$ if $\{z(k)\}$ and $\{v(k)\}$ lie in the interior of their respective constraint sets.

These comments suggest that a time-varying policy of the form $u(x, k) = v(k) + K(x - z(k))$ might be adequate, at least when $f(\cdot)$ is linear. The nominal control and state sequences, $\{v(k)\}$ and $\{z(k)\}$, respectively, can be determined by solving a standard open-loop optimal control problem of the form usually employed in MPC, and the feedback matrix K can be determined offline. We show that this form of robust MPC has the same order of online complexity as that conventionally used for deterministic systems. It requires a modified form of the online optimal control problem in which the constraints are simply *tightened* to allow for disturbances, thereby constraining the trajectories of the uncertain system to lie in a tube centered on the nominal trajectories. Offline computations are required to determine the modified constraints and the feedback matrix K. We also present, in the last section of this chapter, a modification of this tube-based procedure for nonlinear systems for which a *nonlinear* local feedback policy is required.

A word of caution is necessary. Just as nominal model predictive controllers presented in Chapter 2 may fail in the presence of uncertainty, the controllers presented in this chapter may fail if the actual uncertainty does not satisfy our assumptions, such as when a disturbance that we assume to be bounded exceeds the assumed bounds; the controlled systems are robust only to the specified uncertainties. As always, online fault diagnosis and safe recovery procedures are required to protect the system from unanticipated events.

3.1.4 Tubes

The approach that we adopt is motivated by the following observation. Both open-loop and feedback control generate, in the presence of uncertainty, a *bundle* or *tube* of trajectories, each trajectory in the bundle or tube corresponding to a particular realization of the uncertainty. In Figure 3.1(a), the tube corresponding to $\mathbf{u} = \mathbf{u}^0(x)$ and initial state $x = 1$, is $\{X_0, X_1, X_2, X_3\}$ where $X_0 = \{1\}$; for each i, $X_i = \{\phi(i; x, \mathbf{u}, \mathbf{w}) \mid \mathbf{w} \in \mathcal{W}\}$, the set of states at time i generated by all possible realizations of the disturbance sequence. State constraints must be satisfied by every trajectory in the tube. Control of uncertain systems is best viewed as control of tubes rather than trajectories; the designer chooses, for each initial state, a tube in which all realizations of the state trajectory are controlled to lie. By suitable choice of the tube, satisfaction of state and control constraints may be guaranteed for *every* realization of the disturbance sequence.

Determination of an *exact* tube $\{X_0, X_1, \ldots\}$ corresponding to a given initial state x and policy $\boldsymbol{\mu}$ is difficult even for linear systems, however, and virtually impossible for nonlinear systems. Hence, in the sequel, we show how simple tubes that bound all realizations of the state trajectory may be constructed. For example, for linear systems with convex constraints, a tube $\{X_0, X_1, \ldots,\}$, where for each i, $X_i = \{z(i)\} \oplus Z$, $z(i)$ is the state at time i of a deterministic system, X_i is a polytope, and Z is a positive invariant set, may be designed to bound all realizations of the state trajectory. The exact tube lies inside this simple approximation. Using this construction permits robust model predictive controllers to be designed with not much more computation than that required for deterministic systems.

3.1.5 Difference Inclusion Description of Uncertain Systems

Here we introduce some notation that will be useful in the sequel. A deterministic discrete time system is usually described by a difference equation

$$x^+ = f(x, u) \tag{3.1}$$

We use $\phi(k; x, i, \mathbf{u})$ to denote the solution of (3.1) at time k when the initial state at time i is x and the control sequence is $\mathbf{u} = \{u(0), u(1), \ldots\}$; if the initial time $i = 0$, we write $\phi(k; x, \mathbf{u})$ in place of $\phi(k; (x, 0), \mathbf{u})$.

Similarly, an uncertain system may be described by the difference equation

$$x^+ = f(x, u, w) \tag{3.2}$$

in which the variable w that represents the uncertainty takes values in a specified set $\mathbb{W}$. We use $\phi(k; x, i, \mathbf{u}, \mathbf{w})$ to denote the solution of (3.2) when the initial state at time i is x and the control and disturbance sequences are, respectively, $\mathbf{u} = \{u(0), u(1), \ldots\}$ and $\mathbf{w} = \{w(0), w(1), \ldots\}$. The uncertain system may alternatively be described by a *difference inclusion* of the form

$$x^+ \in F(x, u)$$

in which $F(\cdot)$ is a set-valued map. We use the notation $F : \mathbb{R}^n \times \mathbb{R}^m \rightsquigarrow \mathbb{R}^n$ or[1] $F : \mathbb{R}^n \times \mathbb{R}^m \to 2^{\mathbb{R}^n}$ to denote a function that maps points in $\mathbb{R}^n \times \mathbb{R}^m$ into subsets of $\mathbb{R}^m$. If the uncertain system is described by (3.2), then

$$F(x, u) = f(x, u, \mathbb{W}) := \{f(x, u, w) \mid w \in \mathbb{W}\}$$

If x is the current state, and u the current control, the successor state x^+ lies anywhere in the set $F(x, u)$. If a control policy $\boldsymbol{\mu} := \{\mu_0(\cdot), \mu_1(\cdot), \ldots\}$ is employed, the state evolves according to

$$x^+ \in F(x, \mu_k(x)) \tag{3.3}$$

in which x is the current state, k the current time, and x^+ the successor state at time $k+1$. The system described by (3.3) does not have a single solution for a given initial state; it has a solution for each possible realization $\mathbf{w}$ of the disturbance sequence. We use $S(x, i)$ to denote the set of solutions of (3.3) if the initial state is x at time i. If $\phi(\cdot) \in S(x, i)$ then

$$\phi(t) = \phi(t; x, i, \boldsymbol{\mu}, \mathbf{w})$$

for some admissible disturbance sequence $\mathbf{w}$ where $\phi(t; x, i, \boldsymbol{\mu}, \mathbf{w})$ denotes the solution at time t of

$$x^+ = f(x, \mu_k(x), w)$$

when the initial state is x at time i and the disturbance sequence is $\mathbf{w}$. The policy $\boldsymbol{\mu}$ is defined, as before, to be the sequence $\{\mu_0(\cdot), \mu_1(\cdot), \ldots,$

[1]For any set X, 2^X denotes the set of subsets of X.

$\mu_{N-1}(\cdot)\}$ of control laws. The tube $\mathbf{X} = \{X_0, X_1, \ldots\}$, discussed in Section 3.4, generated when policy $\boldsymbol{\mu}$ is employed, satisfies

$$X_{k+1} = \mathbf{F}(X_k, \mu_k(\cdot)) := \{f(x, \mu_k(x), w) \mid x \in X_k, w \in \mathbb{W}\}$$

in which $\mathbf{F}$ maps sets into sets.

3.2 Nominal (*Inherent*) Robustness

3.2.1 Introduction

Because feedback MPC is complex, it is natural to inquire if nominal MPC, i.e., MPC based on the nominal system ignoring uncertainty, is sufficiently robust to uncertainty. Before proceeding with a detailed analysis, a few comments may be helpful.

MPC uses, as a Lyapunov function, the value function of a parametric optimal control problem. Often the value function is continuous, but this is not necessarily the case, especially if state and/or terminal constraints are present. It is also possible for the value function to be continuous but the associated control law to be discontinuous; this can happen, for example, if the minimizing control is not unique.

It is important to realize that a control law may be stabilizing but not robustly stabilizing; arbitrary perturbations, no matter how small, can destabilize the system. Teel (2004) illustrates this point with the following discontinuous autonomous system ($n = 2$, $x = (x_1, x_2)$)

$$x^+ = f(x) \qquad f(x) = \begin{cases} (0, |x|) & x_1 \neq 0 \\ (0, 0) & \text{otherwise} \end{cases}$$

If the initial state is $x = (1, 1)$, then $\phi(1; x) = (0, \sqrt{2})$ and $\phi(2; x) = (0, 0)$, with similar behavior for other initial states. In fact, all solutions satisfy

$$\phi(k; x) \leq \beta(|x|, k)$$

in which β, defined by

$$\beta(|x|, k) := |x| \max\{2 - k, 0\}$$

is a $\mathcal{KL}$ function, so that the origin is *globally asymptotically stable.* Consider now a perturbed system satisfying

$$x^+ = \begin{bmatrix} \delta \\ |x| + \delta \end{bmatrix}$$

in which $\delta > 0$ is a constant perturbation that causes x_1 to remain strictly positive. If the initial state is $x = \varepsilon(1,1)$, then $x_1(k) = \delta$ for $k \geq 1$, and $x_2(k) > \varepsilon\sqrt{2} + k\delta \to \infty$ as $k \to \infty$, no matter how small δ and ε are. Hence the origin is unstable in the presence of an arbitrarily small perturbation; global asymptotic stability is not a robust property of this system.

This example may appear contrived but, as Teel (2004) points out, it can arise in receding horizon optimal control of a *continuous system*. Consider the following system

$$x^+ = \begin{bmatrix} x_1(1-u) \\ |x|\, u \end{bmatrix}$$

in which the control u is constrained to lie in the set $\mathbb{U} = [-1,1]$. Suppose we choose a horizon length $N = 2$ and choose $\mathbb{X}_f$ to be the origin. If $x_1 \neq 0$, the only feasible control sequence steering x to 0 in two steps is $\mathbf{u} = \{1,0\}$; the resulting state sequence is $\{x, (0,|x|), (0,0)\}$. Since there is only one feasible control sequence, it is also optimal, and $\kappa_2(x) = 1$ for all x such that $x_1 \neq 0$. If $x_1 = 0$, then the only optimal control sequence is $\mathbf{u} = \{0,0\}$ and $\kappa_2(x) = 0$. The resultant closed-loop system satisfies

$$x^+ = f(x) := \begin{bmatrix} x_1(1-\kappa_2(x)) \\ |x|\, \kappa_2(x) \end{bmatrix}$$

in which $\kappa_2(x) = 1$ if $x_1 \neq 0$, and $\kappa_2(x) = 0$ otherwise. Thus

$$f(x) = \begin{cases} (0,|x|) & x_1 \neq 0 \\ (0,0) & \text{otherwise} \end{cases} \tag{3.4}$$

The system $x^+ = f(x)$ is the discontinuous system analyzed previously. Thus, receding horizon optimal control of a continuous system has resulted in a discontinuous system that is globally asymptotically stable but has no robustness.

3.2.2 Difference Inclusion Description of Discontinuous Systems

Consider a discontinuous system

$$x^+ = f(x)$$

in which $f(\cdot)$ is not continuous. An example of such a system occurred in the previous subsection where $f(\cdot)$ satisfies (3.4). Solutions of this system are very sensitive to the value of x_1. An infinitesimal change in x_1 at time 0, say, from 0 can cause a substantial change in the subsequent trajectory resulting, in this example, in a loss of robustness. To design a robust system, one must take into account, in the design process, the system's extreme sensitivity to variations in state. This can be done by *regularizing* the system (Teel, 2004). If $f(\cdot)$ is locally bounded,[2] the *regularization* $x^+ = f(x)$ is defined to be

$$x^+ \in F(x) := \bigcap_{\delta>0} \overline{f(\{x\} \oplus \delta\mathcal{B})}$$

in which $\mathcal{B}$ is the closed unit ball so that $\{x\} \oplus \delta\bar{\mathcal{B}} = \{z \mid |z - x| \leq \delta\}$ and $\overline{A}$ denotes the closure of set A. At points where $f(\cdot)$ is continuous, $F(x) = \{f(x)\}$, i.e., $F(x)$ is the single point $f(x)$. If $f(\cdot)$ is piecewise continuous, e.g., if $f(x) = x$ if $x < 1$ and $f(x) = 2x$ if $x \geq 1$, then $F(x) = \{\lim_{x_i \to x} f(x_i)\}$, the set of all limits of $f(x_i)$ as $x_i \to x$. For our example immediately above, $F(x) = \{x\}$ if $x < 1$ and $F(x) = \{2x\}$ if $x > 1$. When $x = 1$, the limit of $f(x_i)$ as $x_i \to 1$ from below is 1 and the limit of $f(x_i)$ as $x \to 1$ from above is 2, so that $F(1) = \{1, 2\}$. The regularization of $x^+ = f(x)$ where $f(\cdot)$ is defined in (3.4) is $x^+ \in F(x)$ where $F(\cdot)$ is defined by

$$F(x) = \left\{\begin{bmatrix} 0 \\ |x| \end{bmatrix}\right\} \qquad x_1 \neq 0 \tag{3.5}$$

$$F(x) = \left\{\begin{bmatrix} 0 \\ |x| \end{bmatrix}, \begin{bmatrix} 0 \\ 0 \end{bmatrix}\right\} \qquad x_1 = 0 \tag{3.6}$$

If the initial state is $x = (1, 1)$, as before, then the difference inclusion generates the following tube

$$X_0 = \left\{\begin{bmatrix} 1 \\ 1 \end{bmatrix}\right\}, \quad X_1 = \left\{\begin{bmatrix} 0 \\ \sqrt{2} \end{bmatrix}\right\}, \quad X_2 = \left\{\begin{bmatrix} 0 \\ \sqrt{2} \end{bmatrix}, \begin{bmatrix} 0 \\ 0 \end{bmatrix}\right\}, \quad \ldots$$

with $X_k = X_2$ for all $k \geq 2$. The set X_k of possible states clearly does not converge to the origin even though the trajectory generated by the original system does. The regularization reveals that small perturbations can destabilize the system.

3.2.3 When Is Nominal MPC Robust?

The discussion in Section 2.1 shows that nominal MPC is not necessarily robust. It is therefore natural to ask under what conditions nominal

[2] A function $f : \mathbb{R}^p \to \mathbb{R}^n$ is locally bounded if, for every $x \in \mathbb{R}^p$, there exists a neighborhood $\mathcal{N}$ of x and a $c > 0$ such that $|f(z)| \leq c$ for all $z \in \mathcal{N}$.

MPC is robust. To answer this, we have to define robustness precisely. In Appendix B, we define robust stability, and robust asymptotic stability, of a set. We employ this concept later in this chapter in the design of robust model predictive controllers that for a given initial state in the region of attraction, steer *every* realization of the state trajectory to this set. Here, however, we address a slightly different question: when is nominal MPC that steers every trajectory in the region of attraction to the origin robust? Obviously, the disturbance will preclude the controller from steering the state of the perturbed system to the origin; the best that can be hoped for is that the controller will steer the state to some small neighborhood of the origin. Let the nominal (controlled) system be described by $x^+ = f(x)$ where $f(\cdot)$ is not necessarily continuous, and let the perturbed system be described by $x^+ = f(x + e) + w$. Also let $S_\delta(x)$ denote the set of solutions for the perturbed system with initial state x and perturbation sequences $\mathbf{e} := \{e(0), e(1), e(2), \ldots\}$ and $\mathbf{w} := \{w(0), w(1), w(2), \ldots\}$ satisfying $\max\{\|\mathbf{e}\|, \|\mathbf{w}\|\} \leq \delta$ where, for any sequence $\boldsymbol{\nu}$, $\|\boldsymbol{\nu}\|$ denotes the sup norm, $\sup_{k\geq 0} |\nu(k)|$. The definition of robustness that we employ is (Teel, 2004):

Definition 3.1 (Robust global asymptotic stability (GAS)). Let $\mathcal{A}$ be compact, and let $d(x, \mathcal{A}) := \min_a \{|a - x| \mid a \in \mathcal{A}\}$, and $|x|_{\mathcal{A}} := d(x, \mathcal{A})$. The set $\mathcal{A}$ is robust GAS for $x^+ = f(x)$ if there exists a class $\mathcal{KL}$-function $\beta(\cdot)$ such that for each $\varepsilon > 0$ and each compact set C, there exists a $\delta > 0$ such that for each $x \in C$ and each $\phi \in S_\delta(x)$, there holds $|\phi(k; x)|_{\mathcal{A}} \leq \beta(|x|_{\mathcal{A}}, k) + \varepsilon$ for all $k \in \mathbb{I}_{\geq 0}$.

Taking the set $\mathcal{A}$ to be the origin ($\mathcal{A} = \{0\}$) so that $|x|_{\mathcal{A}} = |x|$, we see that if the origin is robustly asymptotically stable for $x^+ = f(x)$, then, for each $\varepsilon > 0$, there exists a $\delta > 0$ such that every trajectory of the perturbed system $x^+ = f(x + e) + w$ with $\max\{\|\mathbf{e}\|, \|\mathbf{w}\|\} \leq \delta$ converges to $\varepsilon\mathcal{B}$ ($\mathcal{B}$ is the closed unit ball); this is the attractivity property. Also, if the initial state x satisfies $|x| \leq \beta^{-1}(\varepsilon, 0)$, then $|\phi(k; x)| \leq \beta(\beta^{-1}(\varepsilon, 0), 0) + \varepsilon = 2\varepsilon$ for all $k \in \mathbb{I}_{\geq 0}$ and for all $\phi \in S_\delta$, which is the Lyapunov stability property. Here the function $\beta^{-1}(\cdot, 0)$ is the inverse of the function $\alpha \mapsto \beta(\alpha, 0)$.

We return to the question: under what conditions is asymptotic stability robust? This is answered by the following important result (Teel, 2004; Kellet and Teel, 2004):

Theorem 3.2 (Lyapunov function and robust GAS). *Suppose $\mathcal{A}$ is com-*

pact and that $f(\cdot)$ is locally bounded.[3] *The set $\mathcal{A}$ is robustly globally asymptotically stable for the system $x^+ = f(x)$ if and only if the system admits a continuous global Lyapunov function for $\mathcal{A}$.*

It is shown in Appendix B that for the system $x^+ = f(x)$, $V : \mathbb{R}^n \to \mathbb{R}_{\geq 0}$ is a global Lyapunov function for set $\mathcal{A}$ if there exist $\mathcal{K}_\infty$ functions $\alpha_1(\cdot)$ and $\alpha_2(\cdot)$, and a continuous positive definite function $\rho(\cdot)$, such that for all $x \in \mathbb{R}^n$

$$\alpha_1(|x|_{\mathcal{A}}) \leq V(x) \leq \alpha_2(|x|_{\mathcal{A}})$$
$$V(f(x)) \leq V(x) - \rho(|x|_{\mathcal{A}})$$

in which $|x|_{\mathcal{A}} := d(x, \mathcal{A})$, the distance of x from the set $\mathcal{A}$. In MPC, the value function of the finite horizon optimal control problem that is solved online is used as a Lyapunov function. In certain cases, such as linear systems with polyhedral constraints, the value function is known to be continuous; see Proposition 7.13. Theorem 3.2, suitably modified because the region of attraction is not global, shows that asymptotic stability is robust, i.e., that asymptotic stability is not destroyed by *small* perturbations.

This result, though important, is limited in its use for applications in that it merely states the existence of a $\delta > 0$ that specifies the permitted magnitude of the perturbations; in practice its value would be required. In the next section we show how the performance of an uncertain system with disturbances of a specified magnitude may be estimated.

Theorem 3.2 characterizes robust stability of the set $\mathcal{A}$ for the system $x^+ = f(x)$ in the sense that it shows robust stability is equivalent to the existence of a continuous global Lyapunov function for the system. It is also possible to characterize robustness of $x^+ = f(x)$ by global asymptotic stability of its regularization $x^+ \in F(x)$. It is shown in Appendix B that for the system $x^+ \in F(x)$, the set $\mathcal{A}$ is *globally asymptotically stable* if there exists a $\mathcal{KL}$-function $\beta(\cdot)$ such that for each $x \in \mathbb{R}^n$ and each $\phi(\cdot) \in S(x)$, i.e., for each solution of $x^+ \in F(x)$ with initial state x, $\phi(k) \leq \beta(|x|, k)$ for all $k \in \mathbb{I}_{\geq 0}$. The following alternative characterization of robust stability of $\mathcal{A}$ for the system $x^+ = f(x)$ appears in (Teel, 2004).

Theorem 3.3 (Robust GAS and regularization). *Suppose $\mathcal{A}$ is compact and that $f(\cdot)$ is locally bounded. The set $\mathcal{A}$ is robust GAS for the system*

[3] A function $f : \mathcal{X} \to \mathcal{Y}$ is locally bounded if, for every $x \in \mathcal{X}$, there exists a neighborhood $\mathcal{N}$ of x such that the set $f(\mathcal{N})$ in $\mathcal{Y}$ is bounded.

$x^+ = f(x)$ if and only if the set $\mathcal{A}$ is globally asymptotically stable for $x^+ \in F(x)$, the regularization of $x^+ = f(x)$.

We saw previously that for $f(\cdot)$ and $F(\cdot)$ defined respectively in (3.4) and (3.6), the origin is not globally asymptotically stable for the regularization $x^+ \in F(x)$ of $x^+ = f(x)$ since not every solution of $x^+ \in F(x)$ converges to the origin. Hence the origin is not robust GAS for the system $x^+ = f(x)$.

3.2.4 Input-to-State Stability

When an uncertain system is nominally asymptotically stable, it is sometimes possible to establish input-to-state stability (ISS) as shown in Section B.6 in Appendix B. We consider the uncertain system described by

$$x^+ = f(x, u, w) \tag{3.7}$$

in which w is a bounded additive disturbance. The constraints that are required to be satisfied are

$$x(i) \in \mathbb{X} \qquad u(i) \in \mathbb{U}$$

for all $i \in \mathbb{I}_{\geq 0} := \{0, 1, 2, \ldots\}$, the set of nonnegative integers. The disturbance w may take any value in the set $\mathbb{W}$. As before, $\mathbf{u}$ denotes the control sequence $\{u(0), u(1), \ldots\}$ and $\mathbf{w}$ the disturbance sequence $\{w(0), w(1), \ldots\}$; $\phi(i; x, \mathbf{u}, \mathbf{w})$ denotes the solution of (3.7) at time i if the initial state is x, and the control and disturbance sequences are, respectively, $\mathbf{u}$ and $\mathbf{w}$.

The *nominal* system is described by

$$x^+ = \bar{f}(x, u) := f(x, u, 0) \tag{3.8}$$

and $\bar{\phi}(i; x, \mathbf{u})$ denotes the solution of the nominal system (3.8) at time i if the initial state is x and the control sequence is $\mathbf{u}$. The *nominal* control problem, defined subsequently, includes, for reasons discussed in Chapter 2, a terminal constraint

$$x(N) \in \mathbb{X}_f$$

The *nominal* optimal control problem is

$$\mathbb{P}_N(x): \qquad V_N^0(x) = \min_{\mathbf{u}}\{V_N(x, \mathbf{u}) \mid \mathbf{u} \in \mathcal{U}_N(x)\}$$

$$\mathbf{u}^0(x) = \arg\min_{\mathbf{u}}\{V_N(x, \mathbf{u}) \mid \mathbf{u} \in \mathcal{U}_N(x)\}$$

in which $\mathbf{u}^0(x) = \{u_0^0(x), u_1^0(x), \ldots, u_{N-1}^0(x)\}$ and the nominal cost $V_N(\cdot)$ is defined by

$$V_N(x, \mathbf{u}) := \sum_{i=0}^{N-1} \ell(x(i), u(i)) + V_f(x(N)) \tag{3.9}$$

In (3.9) and (3.10), $x(i) := \bar{\phi}(i; x, \mathbf{u})$ for all $i \in \mathbb{I}_{0:N-1} = \{0, 1, 2, \ldots, N-1\}$; the set of *admissible* control sequences $\mathcal{U}_N(x)$ is defined by

$$\mathcal{U}_N(x) := \{\mathbf{u} \mid u(i) \in \mathbb{U}, \quad x(i) \in \mathbb{X} \ \forall i \in \mathbb{I}_{0:N-1}, \quad x(N) \in \mathbb{X}_f\} \tag{3.10}$$

which is the set of control sequences such that the nominal system satisfies the control, state, and terminal constraints when the initial state at time 0 is x. Thus, $\mathcal{U}_N(x)$ is the set of feasible controls for the nominal optimal control problem $\mathbb{P}_N(x)$. The set $\mathcal{X}_N \subset \mathbb{R}^n$, defined by

$$\mathcal{X}_N := \{x \in \mathbb{X} \mid \mathcal{U}_N(x) \neq \emptyset\}$$

is the domain of the value function $V_N^0(\cdot)$, i.e., the set of $x \in \mathbb{X}$ for which $\mathbb{P}_N(x)$ has a solution; $\mathcal{X}_N$ is also the domain of the minimizer $\mathbf{u}^0(x)$. The value of the nominal model predictive control at state x is $u^0(0; x)$, the first control in the sequence $\mathbf{u}^0(x)$. Hence the *implicit* nominal MPC control law is $\kappa_N : \mathcal{X}_N \to \mathbb{U}$ defined by

$$\kappa_N(x) = u^0(0; x)$$

We assume, as before, that $\ell(\cdot)$ and $V_f(\cdot)$ are defined by

$$\ell(x, u) := (1/2)(x'Qx + u'Ru) \qquad V_f(x) := (1/2)x'P_f x$$

in which Q, R, and P_f are all positive definite. We also assume that $V_f(\cdot)$ and $\mathbb{X}_f$ satisfy the standard assumption that, for $x \in \mathbb{X}_f$, there exists a $u \in \mathbb{U}$ such that $V_f(\bar{f}(x, u)) \leq V_f(x) - \ell(x, u)$ and that $\mathcal{X}_N$ is compact. Under these assumptions, as shown in Chapter 2, there exist positive constants c_1 and c_2, $c_2 > c_1$, satisfying

$$c_1 |x|^2 \leq V_N^0(x) \leq c_2 |x|^2 \tag{3.11}$$

$$V_N^0(\bar{f}(x, \kappa_N(x))) \leq V_N^0(x) - c_1 |x|^2 \tag{3.12}$$

for all $x \in \mathcal{X}_N$. We also assume:

Assumption 3.4 (Lipschitz continuity of value function). The value function $V_N^0(\cdot)$ is Lipschitz continuous on bounded sets.

Assumption 3.4 is satisfied, as shown in Proposition 7.13, if $f(\cdot)$ is affine, $\ell(\cdot)$ and $V_f(\cdot)$ are quadratic and positive definite, $\mathbb{X}$ is polyhedral, and $\mathbb{X}_f$ and $\mathbb{U}$ are polytopic. Assumption 3.4 is also satisfied, as shown in Theorem C.29, if $V_N(\cdot)$ is Lipschitz continuous on bounded sets, $\mathbb{U}$ is compact, and there are no state constraints, i.e., if $\mathbb{X} = \mathbb{X}_f = \mathbb{R}^n$. It follows from (3.11) and (3.12) that for the nominal system under MPC, the origin is exponentially stable, with a region of attraction $\mathcal{X}_N$; the nominal system under MPC satisfies

$$x^+ = \bar{f}(x, \kappa_N(x)) \tag{3.13}$$

It also follows that there exists a $\gamma \in (0, 1)$ such that

$$V_N^0(\bar{f}(x, \kappa_N(x))) \leq \gamma V_N^0(x)$$

for all $x \in \mathcal{X}_N$ so that $V_N^0(x(i))$ decays exponentially to zero as $i \to \infty$, where $x(i)$ is the state of the controlled system at time i when there is no disturbance. In fact, $V_N^0(x(i)) \leq \gamma^i V_N^0(x(0))$ for all $i \in \mathbb{I}_{\geq 0}$.

We now examine the consequences of applying the nominal model predictive controller $\kappa_N(\cdot)$ to the uncertain system (3.7). The controlled uncertain system satisfies the difference equation

$$x^+ = f(x, \kappa_N(x), w) \tag{3.14}$$

in which w can take any value in $\mathbb{W}$. It is obvious that the state $x(i)$ of the controlled system (3.14) cannot tend to the origin as $i \to \infty$; the best that can be hoped for is that $x(i)$ tends to and remains in some neighborhood of the origin. We shall establish this, if the disturbance w is sufficiently small, using the value function $V_N^0(\cdot)$ of the nominal optimal control problem as an input-to-state stable (ISS) Lyapunov function for the controlled uncertain system (3.14). As before, $V_N^0(\cdot)$ satisfies (3.11) and (3.12). Let

$$R_c := \text{lev}_c V_N^0 = \{x \mid V_N^0(x) \leq c\}$$

be the largest sublevel set of $V_N^0(\cdot)$ contained in $\mathcal{X}_N$; the set R_c is compact. Hence there exists a finite Lipschitz constant d for $V_N^0(\cdot)$ in $R_c \times \mathbb{W}$. Since $R_c \subset \mathcal{X}_N$, the state constraint $x \in \mathbb{X}$ is satisfied everywhere in R_c. Because the uncertain system satisfies (3.14) rather than (3.13), the value function evolves along trajectories of the uncertain system according to

$$V_N^0(f(x, \kappa_N(x), w)) - V_N^0(x) \leq V_N^0(\bar{f}(x, \kappa_N(x))) - V_N^0(x) + d|w|$$

for all $w \in \mathbb{W}$, i.e., according to

$$V_N^0(f(x, \kappa_N(x), w)) \leq \gamma V_N^0(x) + d|w| \tag{3.15}$$

where $\gamma \in (0,1)$. In contrast to the nominal case, the value function does not necessarily decrease along trajectories of the uncertain system; indeed, at the origin ($x = 0$), the value function increases unless $w = 0$. The origin is *not* asymptotically stable for the uncertain system. If $\mathbb{W}$ is sufficiently small, however, a sublevel set $R_b = \{x \mid V_N^0(x) \leq b\} \subset R_c$ of $V_N^0(\cdot)$ satisfying $b < c$ is robust positive invariant for $x^+ = f(x, \kappa_N(x), w)$, $w \in \mathbb{W}$, which we show next. We assume, therefore,

Assumption 3.5 (Restricted disturbances). Let $e := \max_w\{|w| \mid w \in \mathbb{W}\}$; $e \leq (\rho - \gamma)b/d$ for some $\rho \in (\gamma, 1)$.

The first consequence of this assumption is that R_b is robust positive invariant for $x^+ = f(x, \kappa_N(x), w)$, $w \in \mathbb{W}$. Suppose $x \in R_b$ so that $V_N^0(x) \leq b$. Then

$$V_N^0(f(x, \kappa_N(x), w)) \leq \gamma V_N^0(x) + d|w| \leq \gamma b + (\rho - \gamma)b \leq \rho b$$

so that $x^+ \in R_b$ for all $w \in \mathbb{W}$. A second consequence is that R_c is robust positive invariant and that any $x \in R_c \setminus R_b$ is steered by the controller into R_b in finite time since Assumption 3.5 implies $V_N^0(x^+) \leq \rho V_N^0(x)$ for all $x^+ = f(x, \kappa_N(x), w)$, all $x \in R_c \setminus R_b$, all $w \in \mathbb{W}$. Any trajectory with an initial state x in R_c remains in R_c and enters, in finite time, the set R_b where it then remains.

It also follows from (3.11), (3.12) and (3.15) that $V_N^0(\cdot)$ and R_c satisfy Definition B.37 so that $V_N^0(\cdot)$ is an ISS-Lyapunov function in R_c for the uncertain system $x^+ = f(x, \kappa_N(x), w)$, $w \in \mathbb{W}$. By Lemma B.38, the system $x^+ = f(x, \kappa_N(x), w)$, $w \in \mathbb{W}$ is ISS in R_c satisfying, therefore, for some $\beta(\cdot) \in \mathcal{KL}_\infty$, some $\sigma(\cdot) \in \mathcal{K}$,

$$|\phi(i; x, \mathbf{w}_i)| \leq \beta(|x|, i) + \sigma(\|\mathbf{w}_i\|) \leq \beta(|x|, i) + \sigma(e)$$

for all $i \in \mathbb{I}_{\geq 0}$ where $\phi(i; x, \mathbf{w}_i)$ is the solution at time i if the initial state at time 0 is x and the disturbance sequence is $\mathbf{w}_i := \{w(0), w(1), \ldots, w(i-1)\}$.

The next section describes how DP may be used, in principle, to achieve robust receding horizon control (RHC). The purpose of this section is to provide some insight into the problem of robust control;

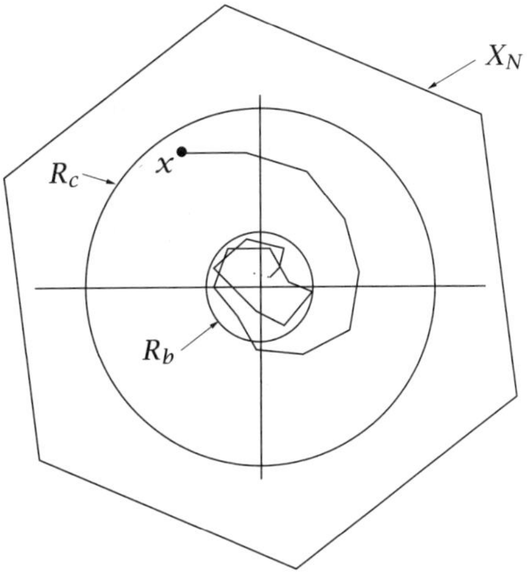

Figure 3.2: The sets $\mathcal{X}_N$, R_b, and R_c.

the section does not show how to obtain robust model predictive controllers that are implementable. Readers whose main concern is implementable robust MPC may prefer to proceed directly to Section 3.4.

3.3 Dynamic Programming Solution

3.3.1 Introduction

In this section we show how robust RHC may be obtained, in principle, using DP. Our concern is to use DP to gain insight. The results we obtain here are not of practical use for complex systems, but reveal the nature of the problem and show what the ideal optimal control problem solved online should be.

In Section 3.2 we examined the inherent robustness of an asymptotically stable system. If uncertainty is present, and it always is, it is preferable to design the controller to be *robust*, i.e., able to cope with some uncertainty. In this section we discuss the design of a robust controller for the system

$$x^+ = f(x, u, w) \tag{3.16}$$

in which a bounded disturbance input w models the uncertainty. The disturbance is assumed to satisfy $w \in \mathbb{W}$ where $\mathbb{W}$ is compact convex, and contains the origin in its interior. The controlled system is required to satisfy the same state and control constraints as above, namely $x \in \mathbb{X}$ and $u \in \mathbb{U}$, as well as a terminal constraint $x(N) \in \mathbb{X}_f$. The solution at time k of (3.16) with control and disturbance sequences $\mathbf{u} = \{u(0), \ldots, u(N-1)\}$ and $\mathbf{w} = \{w(0), \ldots, w(N-1)\}$ if the initial state is x at time 0 is $x(k; x, \mathbf{u}, \mathbf{w})$. Similarly, the solution at time k due to feedback policy $\boldsymbol{\mu}$ and disturbance sequence $\mathbf{w}$ is denoted by $x(k; x, \boldsymbol{\mu}, \mathbf{w})$. As discussed previously, the cost may be taken to be that of the nominal trajectory, or the average, or maximum taken over all possible realizations of the disturbance sequence. Here we employ, as is common in the literature, the maximum over all realizations of the disturbance sequence $\mathbf{w}$, and define the cost due to policy $\boldsymbol{\mu}$ with initial state x to be

$$V_N(x, \boldsymbol{\mu}) := \max_{\mathbf{w}} \{J_N(x, \boldsymbol{\mu}, \mathbf{w}) \mid \mathbf{w} \in \mathcal{W}\} \tag{3.17}$$

in which $\mathcal{W} = \mathbb{W}^N$ is the set of admissible disturbance sequences, and $J_N(x, \boldsymbol{\mu}, \mathbf{w})$ is the cost due to an individual realization $\mathbf{w}$ of the disturbance process and is defined by

$$J_N(x, \boldsymbol{\mu}, \mathbf{w}) := \sum_{i=0}^{N-1} \ell(x(i), u(i), w(i)) + V_f(x(N)) \tag{3.18}$$

in which $\boldsymbol{\mu} = \{u(0), \mu_1(\cdot), \ldots, \mu_{N-1}(\cdot)\}$, $x(i) = \phi(i; x, \boldsymbol{\mu}, \mathbf{w})$, and $u(i) = \mu_i(x(i))$. Let $\mathcal{M}(x)$ denote the set of feedback policies $\boldsymbol{\mu}$ that for a given initial state x satisfy: the state and control constraints, and the terminal constraint for every admissible disturbance sequence $\mathbf{w} \in \mathcal{W}$. The first element $u(0)$ in $\boldsymbol{\mu}$ is a control action rather than a control law because the initial state x is known, whereas future states are uncertain. Thus $\mathcal{M}(x)$ is defined by

$$\begin{aligned} \mathcal{M}(x) := \{\boldsymbol{\mu} \mid\, & u(0) \in \mathbb{U} \\ & \phi(i; x, \boldsymbol{\mu}, \mathbf{w}) \in \mathbb{X},\ \mu_i(\phi(i; x, \boldsymbol{\mu}, \mathbf{w})) \in \mathbb{U} \quad \forall i \in \mathbb{I}_{0:N-1} \\ & \phi(N; x, \boldsymbol{\mu}, \mathbf{w}) \in \mathbb{X}_f \ \forall \mathbf{w} \in \mathcal{W}\} \end{aligned}$$

The robust optimal control problem is

$$\mathbb{P}_N(x): \quad \inf_{\boldsymbol{\mu}} \{V_N(x, \boldsymbol{\mu}) \mid \boldsymbol{\mu} \in \mathcal{M}(x)\} \tag{3.19}$$

The solution to $\mathbb{P}_N(x)$, if it exists, is the policy $\boldsymbol{\mu}^0(x)$

$$\boldsymbol{\mu}^0(x) = \{u^0(0; x), \mu_1^0(\cdot; x), \ldots, \mu_{N-1}^0(\cdot, x)\}$$

and the value function is $V_N^0(x) = V_N(x, \boldsymbol{\mu}^0(x))$. As in conventional MPC, the control applied to the system if the state is x is $u^0(0; x)$, the first element in $\boldsymbol{\mu}^0(x)$; the implicit model predictive feedback control law is $\kappa_N(\cdot)$ defined by

$$\kappa_N(x) := u^0(0; x)$$

3.3.2 Preliminary Results

As in conventional MPC, the value function and implicit control law may, in principle, be obtained by DP. But DP is, in most cases, impossible to use because of its large computational demands. There are, of course, important exceptions such as H_2 and H_∞ optimal control for unconstrained linear systems with quadratic cost functions. DP also can be used for low dimensional constrained optimal control problems when the system is linear, the constraints are affine, and the cost is affine or quadratic. Even when DP is computationally prohibitive, however, it remains a useful tool because of the insight it provides. Because of the cost definition, min-max DP is required. For each $i \in \{0, 1, \ldots, N\}$, let $V_i^0(\cdot)$ and $\kappa_i(\cdot)$ denote, respectively, the partial value function and the optimal solution to the optimal control problem $\mathbb{P}_i$ defined by (3.19) with i replacing N. The DP recursion equations for computing these functions are

$$V_i^0(x) = \min_{u \in \mathbb{U}} \max_{w \in \mathbb{W}} \{\ell(x, u, w) + V_{i-1}^0(f(x, u, w)) \mid f(x, u, \mathbb{W}) \subseteq \mathcal{X}_{i-1}\}$$

$$\kappa_i(x) = (\arg\min_{u \in \mathbb{U}}) \max_{w \in \mathbb{W}} \{\ell(x, u, w) + V_{i-1}^0(f(x, u)) \mid f(x, u, \mathbb{W}) \subseteq \mathcal{X}_{i-1}\}$$

$$\mathcal{X}_i = \{x \in \mathbb{X} \mid \exists\, u \in \mathbb{U} \text{ such that } f(x, u, \mathbb{W}) \subseteq \mathcal{X}_{i-1}\}$$

with boundary conditions

$$V_0^0(x) = V_f(x) \qquad \mathcal{X}_0 = \mathbb{X}_f$$

In these equations, the subscript i denotes the time to go. For each i, $\mathcal{X}_i$ is the domain of $V_i^0(\cdot)$ (and $\kappa_i(\cdot)$) and is therefore the set of states x for which a solution to problem $\mathbb{P}_i(x)$ exists. Thus $\mathcal{X}_i$ is the set of states that can be *robustly* steered by state feedback, i.e., by a policy $\boldsymbol{\mu} \in \mathcal{M}(x)$, to $\mathbb{X}_f$ in i steps or less satisfying all constraints for all disturbance sequences. It follows from these definitions that

$$V_i^0(x) = \max_{w \in \mathbb{W}} \{\ell(x, \kappa_i(x), w) + V_{i-1}^0(f(x, \kappa_i(x), w))\} \tag{3.20}$$

as discussed in Exercise 3.1.

As in the deterministic case studied in Chapter 2, we are interested in obtaining sufficient conditions that ensure that the RHC law $\kappa_N(\cdot)$ is stabilizing. We wish to replace the stabilizing Assumptions 2.12 and 2.13 in Section 2.4.3 of Chapter 2 by conditions appropriate to the robust control problem. The presence of a disturbance requires us to generalize some earlier definitions; we therefore define the terms *robust control invariant* and *robust positive invariant* that generalize our previous definitions of *control invariant* and *positive invariant* respectively.

Definition 3.6 (Robust control invariance). A set $X \subseteq \mathbb{R}^n$ is *robust control invariant* for $x^+ = f(x,u,w)$, $w \in \mathbb{W}$ if, for every $x \in X$, there exists a $u \in \mathbb{U}$ such that $f(x,u,\mathbb{W}) \subseteq X$.

Definition 3.7 (Robust positive invariance). A set X is *robust positive invariant* for $x^+ = f(x,w)$, $w \in \mathbb{W}$ if, for every $x \in X$, $f(x,\mathbb{W}) \subseteq X$.

Stabilizing conditions are imposed on the ingredients $\ell(\cdot), V_f(\cdot)$ and $\mathbb{X}_f$ of the optimal control problem to ensure that the resultant controlled system has desirable stability properties. Our generalization of the stabilizing Assumptions 2.12 and 2.13 that we wish to employ, at least for certain problems, are the following Assumptions 3.8 and 3.9.

Assumption 3.8 (Basic stability assumption; robust case).

(a) For all $x \in \mathbb{X}_f$

$$\min_{u \in \mathbb{U}} \max_{w \in \mathbb{W}} [\Delta V_f + \ell](x,u,w) \leq 0$$

in which $\Delta V_f(x,u,w) = V_f(f(x,u,w)) - V_f(x)$.[4]

(b) $\mathbb{X}_f \subseteq \mathbb{X}$.

Assumption 3.8 implicitly requires that for each $x \in \mathbb{X}_f$, there exists a $u \in \mathbb{U}$ such that $f(x,u,\mathbb{W}) \subseteq \mathbb{X}_f$, i.e., Assumption 3.8 implicitly implies Assumption 3.9.

Assumption 3.9 (Implied stability assumption; robust case). The set $\mathbb{X}_f$ is robust control invariant for $x^+ = f(x,u,w)$, $w \in \mathbb{W}$.

Before proceeding to analyze stability, we should ask if there are any examples that satisfy these conditions. There is at least one important

[4]Generalizing, for any real-valued function $V(\cdot)$, $\Delta V(x,u,w)$ is defined to be $\Delta V(x,u,w) := V(f(x,u,w)) - V(x)$.

example. Assume that $f(x,u,w) = Ax + Bu + Gw$ is linear and the cost is

$$\ell(x,u) = (1/2)(\,|x|_Q^2 + |u|_R^2 - \rho^2\,|w|^2\,) \tag{3.21}$$

in which $Q = C'C$, R is positive definite and $|x|_Q^2$ and $|u|_R^2$ denote, respectively, $x'Qx$ and $u'Ru$. In the absence of constraints, problem $\mathbb{P}_\infty(x)$ becomes a standard infinite horizon, linear quadratic H_∞ optimal control problem. If (A,B,C) has no zeros on the unit circle, which is the case if Q and R are positive definite, the conditions required in Appendix B of Green and Limebeer (1995) for the full information case are satisfied so that there exists a $\tilde{\rho} > 0$ such that a positive definite solution P_f to the associated (generalized) H_∞ algebraic Riccati equation exists for all $\rho > \tilde{\rho}$. Suppose $\rho_f > \tilde{\rho}$ and that P_f is the solution of the H_∞ algebraic Riccati equation, then the associated optimal control and disturbance laws are $u = K_u x$ and $w = K_w x$, respectively, and the matrices $A_f := A + BK_u$ and $A_c := A + BK_u + GK_w$ are both stable. We define the terminal cost function $V_f(\cdot)$ by

$$V_f(x) := (1/2)|x|_{P_f}^2$$

The terminal cost function $V_f(\cdot)$ is the infinite horizon value function, defined globally in $\mathbb{R}^n$ and satisfying $V_f(x) = \max_w\{\ell(x, K_u x, w) + V_f(f(x, K_u x, w))\}$, so that

$$[\Delta V_f + \ell](x, K_u x, w) \leq 0$$

for all (x,w). Hence Assumptions 3.8 and 3.9 are satisfied with $\mathbb{X}_f$ chosen to be any robust positive invariant set for $x^+ = (A + BK_u)x + Gw$, $w \in \mathbb{W}$, that satisfies $\mathbb{X}_f \subseteq \mathbb{X}$ and $K_u\mathbb{X}_f \subseteq \mathbb{U}$, provided such a set exists. Since a positive invariant set for $x^+ = (A + BK_u)x + Gw$ increases with $\mathbb{W}$, and since $\{0\}$ is positive invariant if $\mathbb{W} = \{0\}$, a suitable $\mathbb{X}_f$ exists if $\mathbb{W}$ is sufficiently "small." A similar result can be obtained for a nonlinear system $x^+ = f(x,u,w)$ with $\ell(\cdot)$ defined as in (3.21), provided that $f(\cdot)$ is continuously differentiable, and $A := f_x(0,0,0)$, $B := f_u(0,0,0)$ and $G := f_w(0,0,0)$; see Section 2.5.3.2 of Chapter 2.

Since Assumptions 3.8 and 3.9 appear to be similar to Assumptions 2.12 and 2.13, we would expect to obtain stability results analogous to those obtained in Chapter 2. We do obtain preliminary results that are similar, but the stability properties of the closed-loop system are quite different. Before stating the preliminary results, we note that Assumptions 3.8 and 3.9 imply the existence of a *terminal control law* $\kappa_f : \mathbb{X}_f \to \mathbb{U}$ with the following four properties: (i)

$[\Delta V_f + \ell](x, \kappa_f(x), w) \leq 0$ for all $x \in \mathbb{X}_f$, all $w \in \mathbb{W}$, (ii) $\mathbb{X}_f$ is robust positive invariant for $x^+ = f(x, \kappa_f(x), w)$, (iii) $\mathbb{X}_f \subseteq \mathbb{X}$, and (iv) $\kappa_f(\mathbb{X}_f) \subseteq \mathbb{U}$.

Theorem 3.10 (Recursive feasibility of control policies). *Suppose Assumptions 3.8 and 3.9 hold. Then:*

(a) $\mathcal{X}_N \supseteq \mathcal{X}_{N-1} \supseteq \ldots \supseteq \mathcal{X}_1 \supseteq \mathcal{X}_0 = \mathbb{X}_f$.

(b) $\mathcal{X}_i$ *is robust control invariant for* $x^+ = f(x, u, w)$ $\forall i \in \{0, 1, \ldots, N\}$.

(c) $\mathcal{X}_i$ *is robust positive invariant for* $x^+ = f(x, \kappa_i(x), w) \quad \forall i \in \{0, 1, \ldots, N\}$.

(d) $V_i^0(x) \leq V_{i-1}^0(x) \quad \forall x \in \mathcal{X}_{i-1} \quad \forall i \in \{1, \ldots, N\}$.

(e) $V_N^0(x) \leq V_f(x) \quad \forall\, x \in \mathcal{X}_f$.

(f) $[\Delta V_N^0 + \ell](x, \kappa_N(x), w) \leq [V_N^0 - V_{N-1}^0](f(x, \kappa_N(x), w)) \leq 0$ $\forall (x, w) \in \mathcal{X}_N \times \mathbb{W}$.

(g) For any $x \in \mathcal{X}_N$, $\{\kappa_N(x), \kappa_{N-1}(\cdot), \ldots, \kappa_1(\cdot), \kappa_f(\cdot)\}$ *is a feasible policy for* $\mathbb{P}_{N+1}(x)$, *and, for any* $x \in \mathcal{X}_{N-1}$, $\{\kappa_{N-1}(x), \kappa_{N-2}(\cdot), \ldots, \kappa_1(\cdot), \kappa_f(\cdot)\}$ *is a feasible policy for* $\mathbb{P}_N(x)$.

Proof.

(a)–(c) Suppose, for some i, $\mathcal{X}_i$ is robust control invariant so that any point $x \in \mathcal{X}_i$ can be robustly steered into $\mathcal{X}_i$. By construction, $\mathcal{X}_{i+1}$ is the set of all points x that can be robustly steered into $\mathcal{X}_i$. Hence $\mathcal{X}_{i+1} \supseteq \mathcal{X}_i$ and $\mathcal{X}_{i+1}$ is robust control invariant. But $\mathcal{X}_0 = \mathbb{X}_f$ is robust control invariant. Both (a) and (b) follow by induction. Part (c) follows from (b).

(d) Assume $V_i^0(x) \leq V_{i-1}^0(x)$ for all $x \in \mathcal{X}_{i-1}$. Then from (3.20) we have

$$\begin{aligned}
[V_{i+1}^0 - V_i^0](x) &= \max_{w \in \mathbb{W}} \{\ell(x, \kappa_{i+1}(x), w) + V_i^0(f(x, \kappa_{i+1}(x), w))\} \\
&\quad - \max_{w \in \mathbb{W}} \{\ell(x, \kappa_i(x), w) + V_{i-1}^0(f(x, \kappa_i(x), w))\} \\
&\leq \max_{w \in \mathbb{W}} \{\ell(x, \kappa_i(x), w) + V_i^0(f(x, \kappa_i(x), w))\} \\
&\quad - \max_{w \in \mathbb{W}} \{\ell(x, \kappa_i(x), w) + V_{i-1}^0(f(x, \kappa_i(x), w))\}
\end{aligned}$$

for all $x \in \mathcal{X}_i$ since $\kappa_i(\cdot)$ may *not* be optimal for problem $\mathbb{P}_{i+1}(x)$. We now use the fact that $\max_w\{a(w)\} - \max_w\{b(w)\} \leq \max_w\{a(w) -$

$b(w)\}$, which is discussed in Exercise 3.2, to obtain

$$[V_{i+1}^0 - V_i^0](x) \leq \max_{w \in \mathbb{W}} \{[V_i^0 - V_{i-1}^0](f(x, \kappa_i(x), w))\}$$

for all $(x, w) \in \mathcal{X}_i \times \mathbb{W}$. Also, for all $x \in \mathcal{X}_0 = \mathbb{X}_f$,

$$\begin{aligned}
[V_1^0 - V_0^0](x) &= \min_{u \in \mathbb{U}} \max_{w \in \mathbb{W}} \{\ell(x, u, w) + V_f(f(x, u, w)) - V_f(x)\} \\
&= \min_{u \in \mathbb{U}} \max_{w \in \mathbb{W}} [\Delta V_f + \ell](x, u, w) \\
&\leq 0
\end{aligned}$$

in which the last inequality follows from Assumption 3.8. By induction, $V_i^0(x) \leq V_{i-1}^0(x)\ \forall x \in \mathcal{X}_{i-1},\ \forall i \in \{1, \ldots, N\}$; this is the monotonicity property of the value function for a constrained min-max optimal control problem.

(e) This result is a direct consequence of (a) and (d).

(f) For all $x \in \mathcal{X}_N$, for all $w \in \mathbb{W}$.

$$\begin{aligned}
[\Delta V_N^0 + \ell](x, \kappa_N(x), w) &= V_N^0(f(x, \kappa_N(x), w)) - V_N^0(x) + \ell(x, \kappa_N(x), w) \\
&\leq V_N^0(f(x, \kappa_N(x), w) + \ell(x, \kappa_N(x), w)) \\
&\quad - \ell(x, \kappa_N(x), w) - V_{N-1}^0(f(x, \kappa_N(x), w)) \\
&= [V_N^0 - V_{N-1}^0](f(x, \kappa_N(x), w)) \\
&\leq 0
\end{aligned}$$

in which the last inequality follows from (d) since $f(x, \kappa_N(x), w) \in \mathcal{X}_{N-1}$. The result clearly holds with N replaced by any $i \in \{1, \ldots, N\}$.

(g) Suppose $x \in \mathcal{X}_N$. Then $\boldsymbol{\mu}^0(x) = \{\kappa_N(x), \kappa_{N-1}(\cdot), \ldots, \kappa_1(\cdot)\}$ is a feasible and optimal policy for problem $\mathbb{P}_N(x)$, and steers every trajectory emanating from x into $\mathcal{X}_0 = \mathbb{X}_f$ in N time steps. Because $\mathbb{X}_f$ is positive invariant for $x^+ = f(x, \kappa_f(x), w)$, $w \in \mathbb{W}$, the policy $\{\kappa_N(x), \kappa_{N-1}(\cdot), \ldots, \kappa_1(\cdot), \kappa_f(\cdot)\}$ is feasible for problem $\mathbb{P}_{N+1}(x)$. Similarly, the policy $\{\kappa_{N-1}(x), \kappa_{N-2}(\cdot), \ldots, \kappa_1(\cdot)\}$ is feasible and optimal for problem $\mathbb{P}_{N-1}(x)$, and steers every trajectory emanating from $x \in \mathcal{X}_{N-1}$ into $\mathcal{X}_0 = \mathbb{X}_f$ in $N - 1$ time steps. Therefore the policy $\{\kappa_{N-1}(x), \kappa_{N-2}(\cdot), \ldots, \kappa_1(\cdot), \kappa_f(\cdot)\}$ is feasible for $\mathbb{P}_N(x)$ for any $x \in \mathcal{X}_{N-1}$. ■

3.3.3 Stability of Min-Max Receding Horizon Control

We consider in this subsection the stability properties of min-max RHC for the system $x^+ = f(x, u, w)$ with $\mathbb{P}_N(x)$ defined in (3.19) with

$\ell(x, u, w) := (1/2)(|x|_Q^2 + |u|_R^2) - (\rho^2/2)|w|^2$ and $V_f(x) := (1/2)|x|_{P_f}^2$ where Q, R and P_f are positive definite. In Section 3.3.2, we showed that $\max_{w \in \mathbb{W}}[\Delta V_N^0 + \ell](x, \kappa_N(x), w) \leq 0$ for all $x \in \mathcal{X}_N$ provided that Assumption 3.8 holds. We used the condition $[\Delta V_N^0 + \ell](x, \kappa_N(x), w) \leq 0$ to establish asymptotic stability of the origin for a deterministic system in Chapter 2. Can we do so for the problem considered here? The answer is no; the disturbance w prevents convergence of state trajectories to the origin.

The obstacle appears in theoretical analysis as follows. Our usual conditions for establishing asymptotic stability of the origin for this problem are the existence of a Lyapunov function $V(\cdot)$ satisfying for all $x \in \mathcal{X}_N$

(a) $V(x) \geq \alpha_1(|x|)$

(b) $V(x) \leq \alpha_2(|x|)$

(c) $\max_{w \in \mathbb{W}} \Delta V(x, \kappa_N(x), w) \leq -\alpha_3(|x|)$

in which $\alpha_1(\cdot)$ and $\alpha_2(\cdot)$ are $\mathcal{K}_\infty$ functions and $\alpha_3(\cdot)$ is a positive definite, continuous function.

Choosing $V(\cdot)$ to be the value function $V_N^0(\cdot)$, we see that (a) is satisfied because $V_N^0(x) = \min_{\mu} \max_{\mathbf{w}} J_N(x, \boldsymbol{\mu}, \mathbf{w}) \geq J_N(x, \boldsymbol{\mu}^0(x), \mathbf{0}) \geq \ell(x, \kappa_N(x), 0) \geq (1/2)|x|_Q^2 \geq \alpha_1(|x|)$ for some $\alpha_1(\cdot) \in \mathcal{K}_\infty$ where $\mathbf{0} = \{0, 0, \ldots, 0\}$ is a sequence of zeros and Q is positive definite. Also (b) is satisfied for all $x \in \mathbb{X}_f$ because $V_N^0(x) \leq V_f(x) = (1/2)|x|_{P_f}^2$ where P_f is positive definite, yielding $V_N^0(x) \leq \alpha_2(|x|)$ for all $x \in \mathbb{X}_f$, some $\alpha_2(\cdot) \in \mathcal{K}_\infty$. The region of validity may be extended, as in Chapter 2, to $x \in \mathcal{X}_N$ if $\mathcal{X}_N$ is bounded. The stumbling block is condition (c). We have

$$\Delta V_N^0(x, \kappa_N(x), w) \leq -\ell(x, \kappa_N(x), w)$$

for all $(x, w) \in \mathcal{X}_N \times \mathbb{W}$. Thus $V_N^0(\cdot)$ has the following properties; there exist $\mathcal{K}_\infty$ functions $\alpha_1(\cdot)$ and $\alpha_2(\cdot)$ such that

$$V_N^0(x) \geq \alpha_1(|x|)$$
$$V_N^0(x) \leq \alpha_2(|x|)$$
$$\Delta V_N^0(x, \kappa_N(x), w) \leq -\ell(x, \kappa_N(x), w) \leq -\alpha_1(|x|) + (\rho^2/2)|w|^2$$

for all $(x, w) \in \mathcal{X}_N \times \mathbb{W}$ if $\mathcal{X}_N$ is bounded. The last property, because of the term $(\rho^2/2)|w|^2$, prevents us from establishing asymptotic stability of the origin: the disturbance w prevents convergence of x to the origin. We have to employ alternative notions of stability.

Finite ℓ_2 gain. Suppose Assumptions 3.8 and 3.9 hold. It follows from Theorem 3.10 that

$$\Delta V_N^0(x, \kappa_N(x), w) \leq -\ell(x, \kappa_N(x), w) \tag{3.22}$$

for all $(x, w) \in \mathcal{X}_N \times \mathbb{W}$. Let $\mathbf{x} = \{x(0), x(1), x(2), \ldots\}$, $x(0) = x \in \mathcal{X}_N$, denote any infinite sequence (state trajectory) of the closed-loop system with receding horizon control; $\mathbf{x}$ satisfies

$$x(i+1) = f(x(i), \kappa_N(x(i)), w(i))$$

for some admissible disturbance sequence $\mathbf{w} = \{w(0), w(1), \ldots\}$ in which $w(i) \in \mathbb{W}$ for all i. Using (3.22), which implies $V_N^0(x(i+1)) \leq V_N^0(x(i)) - \ell(x(i), \kappa_N(x(i)), w(i))$ for all i, we deduce that for any positive integer $M > 0$

$$V_N^0(x(M)) \leq V_N^0(x(0)) - \sum_{i=0}^{M-1} \ell(x(i), \kappa_N(x(i)), w(i))$$

If we express $\ell(\cdot)$ in the form

$$\ell(x, u, w) = (1/2)|y|^2 - (\rho^2/2)w^2 \qquad y := \begin{bmatrix} Cx \\ Du \end{bmatrix}$$

in which $Q = C'C$ and $R = D'D$, we obtain

$$\sum_{i=0}^{M-1} |y(i)|^2 \leq \rho^2 \sum_{i=0}^{M-1} |w(i)|^2 + 2V_N^0(x)$$

for any positive integer M. If $\mathbf{w} \in \ell_2$ ($\sum_{i=1}^{\infty} |w(i)|^2 < \infty$), then

$$\sum_{i=0}^{\infty} |y(i)|^2 \leq \rho^2 \sum_{i=0}^{\infty} |w(i)|^2 + 2V_N^0(x)$$

and the closed-loop system $x^+ = f(x, \kappa_N(x), w)$ has finite ℓ_2 gain from w to y.

We showed above that there exist $\mathcal{K}_\infty$ functions $\alpha_1(\cdot)$ and $\alpha_2(\cdot)$ such that

$$V_N^0(x) \geq \alpha_1(|x|)$$
$$V_N^0(x) \leq \alpha_2(|x|)$$
$$\Delta V_N^0(f(x, \kappa_N(x), w)) \leq -\alpha_1(|x|) + (\rho^2/2)|w|^2$$

for all $x \in \mathcal{X}_N$, if $\mathcal{X}_N$ is bounded, and all $w \in \mathbb{W}$. Since $w \mapsto (\rho^2/2)|w|^2$ is a $\mathcal{K}_\infty$ function, the closed-loop system $x^+ = f(x, \kappa_N(x), w)$ is also ISS with w as the input as discussed in Lemma B.38.

Asymptotic stability of the origin. As noted previously, the presence of a bounded disturbance prevents trajectories of the closed-loop system $x^+ = f(x, \kappa_N(x), w)$ from converging to the origin. We show next that asymptotic stability of the origin is possible, however, if the controller $\kappa_N(\cdot)$ is determined on the basis that the disturbance is bounded ($w \in \mathbb{W}$), but the disturbance is either zero or converges to zero as the state tends to the origin.

We showed previously that the value function $V_N^0(\cdot)$ obtained on the basis that $w \in \mathbb{W}$ satisfies

$$V_N^0(x) \geq \alpha_1(|x|)$$
$$V_N^0(x) \leq \alpha_2(|x|)$$
$$\Delta V_N^0(f(x, \kappa_N(x), w)) \leq -\alpha_1(|x|) + (\rho^2/2)|w|^2$$

for all $x \in \mathcal{X}_N$, if $\mathcal{X}_N$ is bounded, all $w \in \mathbb{W}$ where $\alpha_1(\cdot) := (1/2)|x|_Q^2$ and $\alpha_2(\cdot)$ are $\mathcal{K}_\infty$ functions. Since $\Delta V_N^0(x, \kappa_N(x), w)$ is not necessarily negative, the origin is not necessarily asymptotically stable. If, however, the disturbance w is identically zero, then

$$\Delta V_N^0(f(x, \kappa_N(x), w)) \leq -\alpha_1(|x|)$$

for all $x \in \mathcal{X}_N$. This condition, together with the lower and upper bounds on $V_N^0(\cdot)$, is sufficient to establish asymptotic stability of the origin with a domain of attraction $\mathcal{X}_N$.

As the condition $\Delta V_N^0(f(x, \kappa_N(x), w)) \leq -\alpha_1(|x|) + (\rho^2/2)|w|^2$ suggests, however, it is possible for the origin to be asymptotically stable even if some disturbance is present, providing that it decays sufficiently rapidly to zero as the state tends to the origin. We recall that

$$\ell(x, u, w) = (1/2)|x|_Q^2 + (1/2)|u|_R^2 - (\rho^2/2)|w|^2$$

Suppose that w satisfies $(\rho^2/2)|w|^2 \leq |x|_Q^2/4$, or $|w| \leq |x|_Q/(\rho\sqrt{2})$ for all $x \in \mathcal{X}_N$. Then

$$\Delta V_N^0(f(x, \kappa_N(x), w)) \leq -\alpha_1(|x|)/2$$

for all $x \in \mathcal{X}_N$, all $w \in \mathbb{W}$ satisfying $|w| \leq |x|_Q/(\rho\sqrt{2})$. Since $\alpha_1/2$ is a $\mathcal{K}_\infty$ function, asymptotic stability of the origin with a domain of attraction $\mathcal{X}_N$ follows.

Asymptotic stability of an invariant set. In the deterministic case the origin is control invariant since there exists a control, namely $u = 0$, such that $x^+ = f(0,0) = 0$. When bounded disturbances are present, asymptotic stability of the origin must, in general, be replaced by asymptotic stability of an invariant set $\mathcal{O}$ that replaces the origin. Hence, when bounded disturbances are present, we make the following assumption:

Assumption 3.11 (Existence of robust control invariant set).

(a) There exists a compact set $\mathcal{O} \subseteq \mathbb{X}$ that contains the origin and is robust control invariant for $x^+ = f(x,u,w)$ so that for all $x \in \mathcal{O}$ there exists a $u \in \mathbb{U}$ such that $f(x,u,w) \in \mathcal{O}$ for all $w \in \mathbb{W}$.

(b) $\rho = 0$.

Assumption 3.11(a) implies the existence of a control law $\kappa_{\mathcal{O}} : \mathcal{O} \to \mathbb{U}$ such that $\mathcal{O}$ is robust positive invariant for $x^+ = f(x, \kappa_{\mathcal{O}}(x), w)$, i.e., $f(x, \kappa_{\mathcal{O}}(x), w) \in \mathcal{O}$ and $\kappa_{\mathcal{O}}(x) \in \mathbb{U}$ for all $x \in \mathcal{O}$, all $w \in \mathbb{W}$. We assume $\rho = 0$ for simplicity; the term $-\rho^2|w|^2$ in $\ell(\cdot)$ is needed in unconstrained problems to make maximization with respect to the disturbance sequence well defined and is not needed when the constraint $w \in \mathbb{W}$ is present. Accordingly, we replace $\ell(x,u,w)$ by $\ell(x,u)$. Returning to the discussion in Section 3.3.2, we now assume that $\mathcal{O}$ has properties analogous to those of the origin in the deterministic case. Specifically we assume:

Assumption 3.12 (Properties of robust control invariant set).

(a) $\mathcal{O} \subseteq \mathbb{X}_f$.

(b) $V_f(x) = 0$ for all $x \in \mathcal{O}$.

(c) $\ell(x, \kappa_{\mathcal{O}}(x)) = 0$ for all $x \in \mathcal{O}$.

(d) $\kappa_f(x) = \kappa_{\mathcal{O}}(x)$ for all $x \in \mathcal{O}$.

(e) There exists a $\mathcal{K}_\infty$ function $\alpha_1(\cdot)$ such that $\ell(x,u) \geq \alpha_1(|x|_{\mathcal{O}})$ for all $(x,u) \in \mathbb{X} \times \mathbb{U}$.

Since Theorem 3.10 remains true when $\rho = 0$, it is possible to demonstrate, under Assumption 3.12, the assumptions of Section 3.3.2, and the assumption that $\mathcal{X}_N$ is bounded, the existence of $\mathcal{K}_\infty$ functions

$\alpha_1(\cdot)$ and $\alpha_2(\cdot)$ such that

$$V_N^0(x) \geq \alpha_1(|x|_{\mathcal{O}})$$
$$V_N^0(x) \leq \alpha_2(|x|_{\mathcal{O}})$$
$$\Delta V_N^0(f(x, \kappa_N(x), w)) \leq -\alpha_1(|x|_{\mathcal{O}})$$

for all $x \in \mathcal{X}_N, w \in \mathbb{W}$. These bounds differ from those in Proposition 2.19 of Chapter 2 in that $|x|$ is replaced by $|x|_{\mathcal{O}}$ and the term $(\rho^2/2)$ in the last bound is absent. It follows from these bounds that, as shown in Theorem B.23 of Appendix B, the invariant set $\mathcal{O}$ is asymptotically stable for $x^+ = f(x, \kappa_N(x), w)$, $w \in \mathbb{W}$ with a region of attraction $\mathcal{X}_N$.

3.3.4 "Feedback" MPC

The DP solution yields the receding horizon control law $\kappa_N(\cdot)$ but requires extensive computation. In the deterministic case discussed in Chapter 2, $\kappa_N(x)$, the MPC action for a given state x (usually the current state), can be obtained by solving an open-loop optimal control problem. For a given state x, the solutions obtained by DP and by solving the open-loop optimal control problem are identical, in which "open-loop" means the decision variable is the control sequence $\mathbf{u} = \{u(0), u(1), \ldots, u(N-1)\}$. Our first task is to find out if there is a similar relationship when uncertainty is present. DP may again be used to determine the receding horizon control law $\kappa_N(\cdot)$ as shown in Section 3.3. The question arises: does there exist an optimal control problem $\mathbb{P}_N(x)$, parameterized by the state x, the solution of which yields $\kappa_N(x)$, the value of the control law at x? The answer is "yes," but the problem is, unfortunately, no longer an open-loop optimal control problem.

In the deterministic case when $x^+ = f(x, u)$, the decision variable is $\mathbf{u} = \{u(0), u(1), \ldots, u(N-1)\}$, a sequence of control actions, and, if x is the initial state at time 0, a state sequence $\mathbf{x} = \{x(0), x(1), \ldots, x(N)\}$, where $x(0) = x$ and $x(i) = \phi(i; x, \mathbf{u})$, is generated. In the uncertain case when $x^+ = f(x, u, w)$, the decision variable is a control policy $\boldsymbol{\mu} = \{u(0), \mu_1(\cdot)., \ldots, \mu_{N-1}(\cdot)\}$; if x is the initial state, the policy $\boldsymbol{\mu}$ generates a state *tube* $\mathbf{X}(x, \boldsymbol{\mu}) = \{X(0; x), X(1; x, \boldsymbol{\mu}), \ldots, X(N; x, \boldsymbol{\mu})\}$ where $X(0; x) = \{x\}$ and, for all $i \in \mathbb{I}_{\geq 0}$, $X(i; x, \boldsymbol{\mu}) = \{\phi(i; x, \boldsymbol{\mu}, \mathbf{w}) \mid \mathbf{w} \in \mathcal{W}\}$. The tube $\mathbf{X}(x, \boldsymbol{\mu})$ is a bundle of state trajectories, one for each admissible disturbance sequence $\mathbf{w}$; see Figure 3.3. In Figure 3.3(b), the central trajectory corresponds to the disturbance sequence $\mathbf{w} = \{0, 0, 0\}$. The

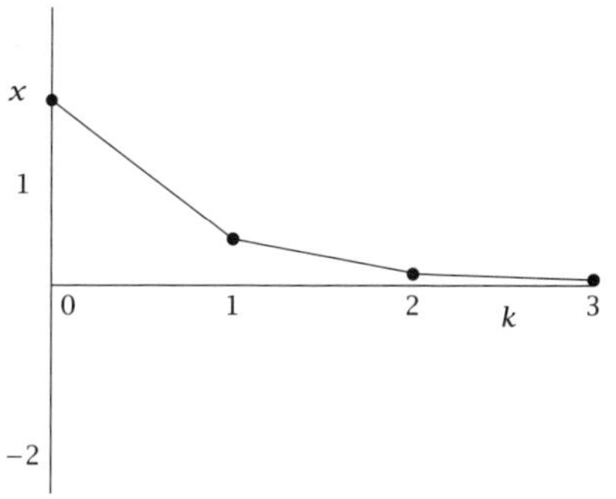

(a) Deterministic case.

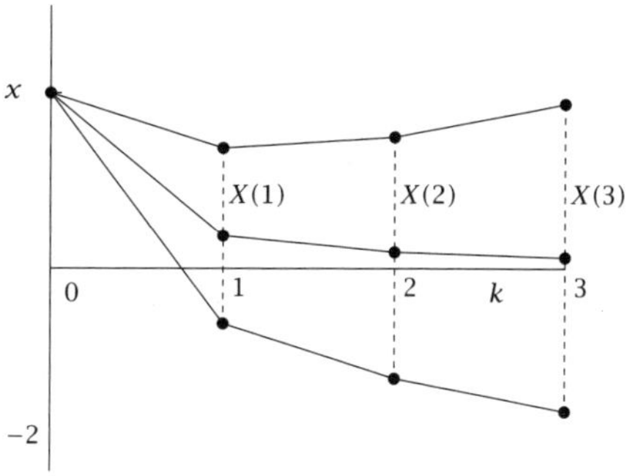

(b) Uncertain case.

Figure 3.3: State trajectory and state tube.

tube **X** may be regarded as the solution of the *set* difference equation

$$X(i+1) = F(X(i), \mu_i(\cdot)) \qquad X(0) = \{x\}$$

in which $F(X, \mu_i(\cdot)) := \{f(x, \mu_i(x), w) \mid x \in X, w \in \mathbb{W}\}$.

If we define $V_N(\cdot)$ and $J_N(\cdot)$ as in (3.17) and (3.18), respectively, then the MPC problem $\mathbb{P}_N(x)$ at state x is, as before

$$\mathbb{P}_N(x): \quad \inf_{\boldsymbol{\mu}} \{V_N(x, \boldsymbol{\mu}) \mid \boldsymbol{\mu} \in \mathcal{M}(x)\}$$

in which $\mathcal{M}(x)$ is the set of feedback policies $\boldsymbol{\mu} = \{u(0), \mu_1(\cdot), \ldots, \mu_{N-1}(\cdot)\}$ that, for a given initial state x, satisfy the state and control constraints $u(0) \in \mathbb{U}$, $\phi(i; x, \boldsymbol{\mu}, \mathbf{w}) \in \mathbb{X}$, $\phi(N; x, \boldsymbol{\mu}, \mathbf{w}) \in \mathbb{X}_f$, and $\mu_i(\phi(i; x, \boldsymbol{\mu}, \mathbf{w})) \in \mathbb{U}$, for all $i \in \{1, \ldots, N-1\}$ and every admissible disturbance sequence $\mathbf{w} \in \mathcal{W}$. This is precisely the problem solved by DP in Section 3.3. So the solution obtained by solving $\mathbb{P}_N(x)$ for the given state x, rather than for every state $x \in \mathcal{X}_N$ as provided by DP, is indeed the DP solution restricted to the sets $X^0(i; x) := \{\phi(i; x, \boldsymbol{\mu}^0(x), \mathbf{w}) \mid \mathbf{w} \in \mathcal{W}\}$, $i \in \{0, 1, \ldots, N\}$. More precisely, the DP solution yields, for each $i \in \{0, 1, \ldots, N\}$, the value function $V_i^0(z)$ and optimal control law $\kappa_i(z)$ for each $z \in \mathcal{X}_i$, whereas the solution to the MPC problem $\mathbb{P}_N(x)$ yields, for each $i \in \{0, 1, \ldots, N\}$, the value function $V_i^0(z)$ and optimal control law $\kappa_i(z)$ for each $z \in X^0(i; x)$.

While it is satisfying to know that one may pose an MPC problem for a given initial state x whose solution is identical to a restriction of the DP solution, this result is of theoretical interest only because, unlike in the deterministic case where the MPC problem is simple enough to solve online, in the uncertain case $\mathbb{P}_N(x)$ is much too complex. One reason for the complexity is that optimization of a bundle of trajectories is required in which each trajectory must satisfy all constraints. A second, even more important, reason is the complexity of the decision variable $\boldsymbol{\mu}$ which is infinite dimensional because it is a sequence of control *laws*.

3.4 Tube-Based Robust MPC

3.4.1 Introduction

To proceed realistically we need to sacrifice optimality for simplicity. Many methods for doing so have been proposed in the literature. We outline next one procedure that achieves this objective and that yields robust MPC by solving online an optimal control problem that has the same order of complexity as that employed for conventional MPC. We simplify the decision variable that, ideally, is a policy by replacing it with a finite-dimensional parameterization that consists of an open-loop control sequence and a simple local feedback controller. In addition, we replace the tube, whose exact determination is difficult, by a simply determined *outer-bounding* tube. The underlying idea is quite simple. We generate the "center" of the tube by using conventional MPC with tighter constraints on the nominal system, and restrict the "size" of the tube by using local feedback that attempts to steer all trajectories of the uncertain system to the central trajectory. The resultant controller may be regarded as a "two degrees of freedom" controller. The local feedback around the nominal trajectory is the inner loop and attenuates disturbances while MPC is used in the outer loop.

In this section we address robust MPC of constrained linear systems. To do so, we make use of some concepts in set algebra. Given two subsets A and B of $\mathbb{R}^n$, we define set addition, set subtraction (sometimes called Minkowski or Pontryagin set subtraction), set multiplication and Hausdorff distance between two sets as follows.

Definition 3.13 (Set algebra and Hausdorff distance).

(a) Set addition: $A \oplus B := \{a + b \mid a \in A, b \in B\}$.

(b) Set subtraction: $A \ominus B := \{x \in \mathbb{R}^n \mid \{x\} \oplus B \subseteq A\}$.

(c) Set multiplication: Let $K \in \mathbb{R}^{m \times n}$. Then $KA := \{Ka \mid a \in A\}$.

(d) The Hausdorff distance $d_H(\cdot)$ between two subsets A and B of $\mathbb{R}^n$

is defined by

$$d_H(A, B) := \max\{\sup_{a \in A} d(a, B), \sup_{b \in B} d(b, A)\}$$

in which $d(x, S)$ denotes the distance of a point $x \in \mathbb{R}^n$ from a set $S \subset \mathbb{R}^n$ and is defined by

$$d(x, S) := \inf_y \{d(x, y) \mid y \in S\} \qquad d(x, y) := |x - y|$$

In these definitions, $\{x\}$ denotes the set consisting of a single point x and $\{x\} \oplus B$ therefore denotes the set $\{x + b \mid b \in B\}$; the set $A \ominus B$ is the largest set C such that $B \oplus C \subseteq A$. A sequence $\{x(i)\}$ is said to converge to a set S if $d(x(i), S) \to 0$ as $i \to \infty$. If $d_H(A, B) \leq \varepsilon$, then the distance of every point $a \in A$ from B is less than or equal to ε and that the distance of every point $b \in B$ from A is less than or equal to ε. We say that the sequence of sets $\{A(i)\}$ converges, in the Hausdorff metric, to the set B if $d_H(A(i), B) \to 0$ as $i \to \infty$.

Our first task is to generate an outer-bounding tube. An excellent background for the following discussion is provided in Kolmanovsky and Gilbert (1998).

3.4.2 Outer-Bounding Tubes for Linear Systems with Additive Disturbances

Consider the following linear system

$$x^+ = Ax + Bu + w$$

in which $w \in \mathbb{W}$, a compact convex subset of $\mathbb{R}^n$ containing the origin. We assume that either $\mathbb{W}$ contains the origin in its interior, or, if not, $w = G\xi$ where $\xi \in \mathbb{R}^p$, $p < n$ lies in the compact convex set Ξ that contains the origin in its interior and (A, G) is controllable. Let $\phi(i; x, \mathbf{u}, \mathbf{w})$ denote the solution of $x^+ = Ax + Bu + w$ at time i if the initial state at time 0 is x, and the control and disturbance sequences are, respectively, $\mathbf{u}$ and $\mathbf{w}$.

Let the nominal system be described by

$$z^+ = Az + Bu$$

and let $\bar{\phi}(i; z, \mathbf{u})$ denote the solution of $z^+ = Az + Bu$ at time i if the initial state at time 0 is z. Then $e := x - z$, the deviation of the actual state x from the nominal state z, satisfies the difference equation

$$e^+ = Ae + w$$

so that

$$e(i) = A^i e(0) + \sum_{j=0}^{i-1} A^j w(j)$$

in which $e(0) = x(0) - z(0)$. If $e(0) = 0$, $e(i) \in S(i)$ where the set $S(i)$ is defined by

$$S(i) := \sum_{j=0}^{i-1} A^j \mathbb{W} = \mathbb{W} \oplus A\mathbb{W} \oplus \ldots \oplus A^{i-1}\mathbb{W}$$

in which $\sum$ and $\oplus$ denote set addition. It follows from our assumptions on $\mathbb{W}$ that $S(i)$ contains the origin in its interior for all $i \geq n$. Let us first consider the tube $\mathbf{X}(x, \mathbf{u})$ generated by the open-loop control sequence $\mathbf{u}$ when $x(0) = z(0) = x$, and $e(0) = 0$. It is easily seen that $\mathbf{X}(x, \mathbf{u}) = \{X(0; x), X(1; x, \mathbf{u}), \ldots, X(N; x, \mathbf{u})\}$ where

$$X(0; x) = \{x\} \qquad X(i; x, \mathbf{u}) := \{\phi(i; x, \mathbf{u}, \mathbf{w}) \mid \mathbf{w} \in \mathcal{W}\} = \{z(i)\} \oplus S(i)$$

and $z(i) = \bar{\phi}(i; x, \mathbf{u})$, the state at time i of the nominal system, is the center of the tube. So it is relatively easy to obtain the exact tube generated by an open-loop control if the system is linear and has a bounded additive disturbance, provided that one can compute the sets $S(i)$. If $\mathbb{W} = G\mathbb{V}$ where $\mathbb{V}$ is convex, then $S(i)$ is convex for all $i \in \mathbb{I}_{\geq 0}$. If, in addition, $\mathbb{V}$ contains the origin in its interior and (A, G) is controllable, then $S(i)$ contains the origin in its interior for all $i \in \mathbb{I}_{\geq n}$.

If A is stable, then, as shown in Kolmanovsky and Gilbert (1998), $S(\infty) := \sum_{j=0}^{\infty} A^j \mathbb{W}$ exists and is positive invariant for $x^+ = Ax + w$, i.e., $x \in S(\infty)$ implies that $Ax + w \in S(\infty)$ for all $w \in \mathbb{W}$; also $S(i) \to S(\infty)$ in the Hausdorff metric as $i \to \infty$. The set $S(\infty)$ is known to be the minimal robust positive invariant set[5] for $x^+ = Ax + w$, $w \in \mathbb{W}$. Also $S(i) \subseteq S(i+1) \subseteq S(\infty)$ for all $i \in \mathbb{I}_{\geq 0}$ so that the tube $\hat{\mathbf{X}}(x, \mathbf{u})$ defined by

$$\hat{\mathbf{X}}(x, \mathbf{u}) := \{\hat{X}(0), \hat{X}(1; x, \mathbf{u}), \ldots, \hat{X}(N; x, \mathbf{u})\}$$

in which

$$\hat{X}(0) = \{x\} \qquad \hat{X}(i; x, \mathbf{u}) = \{z(i)\} \oplus S$$

in which $S = S(\infty)$ is an outer-bounding tube with constant "cross-section" S for the exact tube $\mathbf{X}(x, \mathbf{u})$ ($X(i; x, \mathbf{u}) \subseteq \hat{X}(i; x, \mathbf{u})$ for all $i \in \mathbb{I}_{\geq 0}$). It is sometimes more convenient to use the "constant cross-section" outer-bounding tube $\hat{\mathbf{X}}(x, \mathbf{u})$ in place of the exact tube $\mathbf{X}(x, \mathbf{u})$.

[5]Every other robust positive invariant set X satisfies $X \supseteq S_\infty$.

If we restrict attention to the interval $[0, N]$ as we do in computing the MPC action, then setting $S = S(N)$ yields a less conservative, constrained cross-section, outer-bounding tube for this interval.

While the exact tube $\mathbf{X}(x, \mathbf{u})$, and the outer-bounding tube $\hat{\mathbf{X}}(x, \mathbf{u})$, are easily obtained, their use may be limited for reasons discussed earlier—the sets $S(i)$ may be unnecessarily large simply because an open-loop control sequence rather than a feedback policy was employed to generate the tube. For example, if $\mathbb{W} = [-1, 1]$ and $x^+ = x + u + w$, then $S(i) = (i+1)\mathbb{W}$ increases without bound as time i increases. We must introduce feedback to contain the size of $S(i)$, but wish to do so in a simple way because optimizing over arbitrary policies is prohibitive. The feedback policy we propose is

$$u = v + K(x - z)$$

in which x is the current state of the system $x^+ = Ax + Bu + w$, z is the current state of a nominal system defined below, and v is the current input to the nominal system. With this feedback policy, the state x satisfies the difference equation

$$x^+ = Ax + Bv + BKe + w$$

in which $e := x - z$ is the deviation of the actual state from the nominal state. Let $\phi(i; x, \mathbf{v}, \mathbf{e}, \mathbf{w})$ denote the solution at time i of $x^+ = Ax + Bv + BKe + w$ if its initial state is x at time 0, the control sequence is $\mathbf{v}$, the disturbance sequence is $\mathbf{w}$, and the error sequence is $\mathbf{e}$. The nominal system corresponding to the uncertain system $x^+ = Ax + Bv + BKe + w$ is

$$z^+ = Az + Bv$$

The deviation e now satisfies the difference equation

$$e^+ = A_K e + w \qquad A_K := A + BK$$

which is the same equation used previously except that A, which is possibly unstable, is replaced by A_K, which is stable by design. If K is chosen so that A_K is stable, then the corresponding uncertainty sets $S_K(i)$ defined by

$$S_K(i) := \sum_{j=0}^{i-1} A_K^j \mathbb{W}$$

can be expected to be smaller than the original uncertainty sets $S(i)$, $i \in \mathbb{I}_{\geq 0}$, considerably smaller if A is unstable and i is large, but not necessarily much smaller if A is strongly stable. Our assumptions on $\mathbb{W}$ imply that $S_K(i)$, like $S(i)$, contains the origin in its interior for each i. Since A_K is stable, the set $S_K(\infty) := \sum_{j=0}^{\infty} A_K^j \mathbb{W}$ exists and is positive invariant for $e^+ = A_K e + w$; also, $S_K(i) \to S_K(\infty)$ in the Hausdorff metric as $i \to \infty$. Since K is fixed, the feedback policy $u = K(x - z) + v$ is simply parameterized by the open-loop control sequence $\mathbf{v}$. If $x(0) = z(0) = x$, the tube generated by the feedback policy $\boldsymbol{\mu}$ is $\mathbf{X}(x, \mathbf{v}) = \{X(0), X(1; x, \mathbf{v}), \ldots, X(N; x, \mathbf{v})\}$ where

$$X(0) = \{x\} \qquad X(i; x, \mathbf{v}) := \{\phi_K(i; x, \mathbf{v}, \mathbf{w}) \mid \mathbf{w} \in \mathcal{W}\} = \{z(i)\} \oplus S_K(i)$$

in which $z(i)$ is the solution of the nominal system $z^+ = Az + Bv$ at time i if the initial state is $z(0) = x$, and the control sequence is $\mathbf{v}$. For given initial state x and control sequence $\mathbf{v}$, the solution of $x^+ = Ax + B(v + Ke) + w$ lies in the tube $\mathbf{X}(x, \mathbf{v})$ for every admissible disturbance sequence $\mathbf{w}$. As before, $S_K(i)$ may be replaced by $S_K(\infty)$ to get an outer-bounding tube. If attention is confined to the interval $[0, N]$, $S_K(i)$ may be replaced by $S_K(N)$ to obtain a less conservative outer-bounding tube. If we consider again our previous example, $\mathbb{W} = [-1, 1]$ and $x^+ = x + u + w$, and choose $K = -(1/2)$, then $A_K = 1/2$, $S_K(i) = (1 + 0.5 + \ldots + 0.5^{i-1})\mathbb{W}$, and $S_K(\infty) = 2\mathbb{W} = [-2, 2]$. In contrast, $S(i) \to [-\infty, \infty]$ as $i \to \infty$.

In the preceding discussion, we required $x(0) = z(0)$ so that $e(0) = 0$ in order to ensure $e(i) \in S(i)$ or $e(i) \in S_K(i)$. When A_K is stable, however, it is possible to relax this restriction. This follows from the previously stated fact that $S_K(\infty)$ exists and is robust positive invariant for $e^+ = A_K e + w$, i.e., $e \in S_K(\infty)$ implies $e^+ \in S_K(\infty)$ for all $e^+ \in \{A_K e\} \oplus \mathbb{W}$. Hence, if $e(0) \in S_K(\infty)$, then $e(i) \in S_K(\infty)$ for all $i \in \mathbb{I}_{\geq 0}$, all $\mathbf{w} \in \mathbb{W}^i$.

In tube-based MPC, discussed next, we ensure that $z(i) \to 0$ as $i \to \infty$, so that $x(i)$, which lies in the set $\{z(i)\} \oplus S_K(i)$, converges to the set $S_K(\infty)$ as $i \to \infty$. Even though $S_K(\infty)$ is difficult to compute, this is a useful theoretical property of the controlled system. The controller is required to ensure that state and control constraints are not transgressed. To do this, knowledge of $S_K(\infty)$ is not required. If we know that $e(0) \in S_K(\infty)$ because, for example, $z(0)$ is chosen to satisfy $z(0) = x(0)$, then $x(i)$ lies in $\{z(i)\} \oplus S_K(\infty)$ for all i. All that is then required, in the nominal optimal control problem $\bar{\mathbb{P}}_N$, is knowledge of a set S that is an outer approximation of $S_K(\infty)$. If $x(i)$ lies in $\{z(i)\} \oplus S_K(\infty)$, it certainly lies in $\{z(i)\} \oplus S$. And if $\{z(i)\} \oplus S \subseteq \mathbb{X}$ for all i, then $x(i) \in \{z(i)\} \oplus S_K(\infty)$ certainly satisfies the state constraint

$x(i) \in \mathbb{X}$ for all i and all admissible disturbance sequences. Of course, choosing a large outer-approximating set S results in a degree of conservatism; the choice of S is a tradeoff between simplicity and conservatism. The closer the set S approximates $S_K(\infty)$, the less conservative but more complex S is. If we wish to allow freedom in the choice of $z(0)$, we can choose S to be a robust positive invariant outer approximation of $S_K(\infty)$; then $x(i) \in \{z(i)\} \oplus S$ for all i if $x(0) \in \{z(0)\} \oplus S$.

Consider then the tube $X_\infty(z, i)$ defined by

$$X(z, \mathbf{v}) := \{X_0(z, \mathbf{v}), X_1(z, \mathbf{v}), \ldots, X_N(z, \mathbf{v})\}$$

in which, for each $i \in \{0, 1, \ldots, N\}$,

$$X_i(z, \mathbf{v}) := \{z(i)\} \oplus S \qquad z(i) := \bar{\phi}(i; z, \mathbf{v})$$

and S is an outer approximation of $S_K(\infty)$ ($S_K(\infty) \subseteq S$). It follows from the previous discussion that if $x(0) \in \{z(0)\} \oplus S_K(\infty)$ and S is merely an outer approximation of $S_K(\infty)$ *or* if $x(0) \in \{z(0)\} \oplus S$ where S is a robust positive invariant outer approximation of $S_K(\infty)$ (Raković, Kerrigan, Kouramas, and Mayne, 2005a), then $e(i)$ lies in S for all $i \in \mathbb{I}_{\geq 0}$, and every state trajectory $\{x(i)\}$ of $x^+ = Ax + B(v + Ke) + w$, $w \in \mathbb{W}$. In other words, each trajectory corresponding to an admissible realization of $\mathbf{w}$, lies in the tube $X(z, \mathbf{v})$, as shown in Figure 3.4. An obvious choice for $z(0)$ that ensures $e(0) \in S_K(\infty)$ is $z(0) = x(0)$. Similarly every control trajectory $\{u(i)\}$ of the uncertain system lies in the tube $\{\{v(i)\} \oplus KS_K(\infty)\}$ or in the tube $\{\{v(i)\} \oplus KS\}$.

The fact that the state and control trajectories of the uncertain system lie in known neighborhoods of the state and control trajectories, $\{z(i)\}$ and $\{v(i)\}$ respectively, is the basis for tube-based MPC described subsequently. It follows from this fact that if $\{z(i)\}$ and $\{v(i)\}$ are chosen to satisfy $\{z(i)\} \oplus S_K(\infty) \subseteq \mathbb{X}$ and $\{v(i)\} \oplus KS_K(\infty) \subseteq \mathbb{V}$ for all $i \in \mathbb{I}_{\geq 0}$, then $x(i) \in \mathbb{X}$ and $u(i) \in \mathbb{U}$ for all $i \in \mathbb{I}_{\geq 0}$. Thus $\{z(i)\}$ and $\{v(i)\}$ should be chosen to satisfy the *tighter* constraints $z(i) \in \mathbb{Z}$ and $v(i) \in \mathbb{V}$ for all $i \in \mathbb{I}_{\geq 0}$ where $\mathbb{Z} := X \ominus S$ and $\mathbb{V} := U \ominus KS$ in which $S = S_K(\infty)$ or is an outer approximation of $S_K(\infty)$. If $K = 0$, because A is strongly stable, $\mathbb{X}$ and $\mathbb{V}$ should be chosen to satisfy $\mathbb{Z} = X \ominus S$ and $\mathbb{V} = U$, i.e., there is no need to tighten the constraint on v. It may seem that it is necessary to compute $S_K(\infty)$, or a robust positive invariant outer approximation S, which is known to be difficult, in order to employ this approach. This is not the case, however; we show

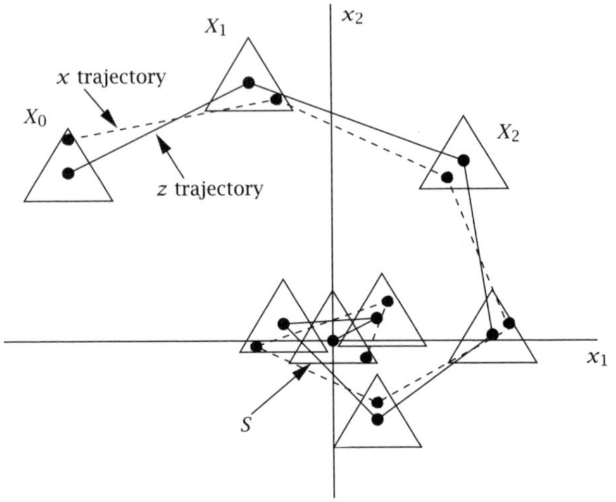

Figure 3.4: Outer-bounding tube $\mathbf{X}(z, \mathbf{v})$; $X_i = \{z(i)\} \oplus S$.

later that the tighter constraint sets $\mathbb{Z}$ and $\mathbb{V}$ may be relatively simply determined.

3.4.3 Tube-Based MPC of Linear Systems with Additive Disturbances

Introduction. Now that we have shown how to contain *all* the trajectories of an uncertain system emanating from the current state within a tube $X(z, \mathbf{v})$ where z is the initial state of the nominal system and $\mathbf{v}$ is an open-loop control sequence, we show how this tool may be used to obtain robust control. We restrict attention in this subsection to constrained linear systems with a bounded additive disturbance. In later sections we consider alternative forms of uncertainty such as parametric uncertainty as well as constrained robust control of constrained nonlinear systems. Our goal is to develop forms of robust MPC that are only marginally more complex than nominal MPC despite the uncertainty.

In this subsection, we discuss first how to formulate an optimal control problem, the solution of which yields a control policy that minimizes a cost, and ensures that the state and control satisfy the given state and control constraints for all admissible bounded disturbances. The basic idea is simple. Choose a trajectory $\{z(i)\}$ for the nominal trajectory depending on the initial state z and the control sequence $\mathbf{v}$, such that each trajectory $\{x(i)\}$ of the system being controlled satisfies the state constraint $x(i) \in \mathbb{X}$ for all $i \in \{0, 1, \ldots, N\}$, and the actual control sequence $\{u(i)\}$ satisfies the control constraint for all $i \in \{0, 1, \ldots, N-1\}$. Recalling that the state satisfies $x(i) \in \{z(i)\} \oplus S$ for all $i \in \{0, 1, \ldots, N\}$ and all admissible disturbance sequences, the state constraint is satisfied if $\{z(i)\} \oplus S \in \mathbb{X}$ or, equivalently, if $z(i) \in \mathbb{X} \ominus S$ for all $i \in \{0, 1, \ldots, N\}$. Similarly the control constraint is satisfied if $u(i) = v(i) + Ke(i) \in \mathbb{U}$ for all $e(i) \in S$ or, equivalently, if $v(i) \in \mathbb{U} \ominus KS$ for all $i \in \{0, 1, \ldots, N-1\}$. These assertions only make sense if the disturbance set $\mathbb{W}$ is sufficiently small to ensure that Assumption 3.14 is satisfied where:

Assumption 3.14 (Restricted disturbances for constraint satisfaction). $S \subset \mathbb{X}$ and $KS \subset \mathbb{U}$.

We suppose Assumption 3.14 holds in the sequel. An assumption like this is not uncommon in robust control; if $\mathbb{W}$ is too large, there is no possibility of satisfying the constraints for all realizations of the disturbance sequence $\mathbf{w}$. Summarizing, the state and control constraints, $x(i) \in \mathbb{X}$ and $u(i) \in \mathbb{U}$, are satisfied at each time i if the time-invariant control law $u = v + K(x - z)$, is employed, and the nominal system $z^+ = Az + Bv$ satisfies the *tighter* constraints

$$z(i) \in \mathbb{Z} := \mathbb{X} \ominus S \tag{3.23}$$

$$v(i) \in \mathbb{V} := \mathbb{U} \ominus KS \tag{3.24}$$

for all relevant i and if, in addition,

$$x(0) \in \{z(0)\} \oplus S \qquad e(0) \in S$$

in which S is robust positive invariant for $e^+ = A_K e + w$, $w \in \mathbb{W}$. Satisfaction of the state constraint at time N, i.e., satisfaction of $x(N) \in \mathbb{X}$, is ensured if the nominal system satisfies the terminal constraint

$$z(N) \in \mathbb{Z}_f \qquad \mathbb{Z}_f \subseteq \mathbb{Z} \tag{3.25}$$

Tube-based robust predictive controller. The first requirement for the simple tube-based model predictive controller is a suitable nominal trajectory. To obtain this, we define a finite horizon optimal control $\bar{\mathbb{P}}_N(z)$ in which z is the current state of the nominal system. The optimal control problem is minimization of a cost function $\bar{V}_N(z, \mathbf{v})$ in which

$$\bar{V}_N(z, \mathbf{v}) := \sum_{k=0}^{N-1} \ell(z(k), v(k)) + V_f(z(N))$$

subject to satisfaction, by the state sequence $\mathbf{z} = \{z(0), z(1), \ldots, z(N)\}$ and the control sequence $\mathbf{v} = \{v(0), v(1), \ldots, v(N-1)\}$, of the nominal difference equation $z^+ = Az + Bv$ and the constraints (3.23)–(3.25). The nominal optimal control problem is, therefore

$$\bar{\mathbb{P}}_N(z): \quad \bar{V}_N^0(z) = \min_{\mathbf{v}}\{\bar{V}_N(z, \mathbf{v}) \mid \mathbf{v} \in \mathcal{V}_N(z)\}$$

in which the constraint set $\mathcal{V}_N(z)$, which depends, as the notation implies, on the parameter z, is defined by

$$\mathcal{V}_N(z) := \{\mathbf{v} \mid v(k) \in \mathbb{V}, \quad \bar{\phi}(k; z, \mathbf{v}) \in \mathbb{Z} \ \forall k \in \{0, 1, \ldots, N-1\}, \\ \bar{\phi}(N; z, \mathbf{v}) \in \mathbb{Z}_f\} \tag{3.26}$$

In (3.26), $\mathbb{Z}_f \subseteq \mathbb{Z}$ is the terminal constraint set. Solution of $\bar{\mathbb{P}}_N(z)$ yields the minimizing control sequence $\mathbf{v}^0(z) = \{v^0(0; z), v^0(1; z), \ldots, v^0(N-1; z)\}$. The model predictive control applied to the nominal system at state z is $v^0(0; z)$, the first control action in the minimizing control sequence. The implicit nominal MPC control law is, therefore, $\bar{\kappa}_N(\cdot)$, defined by

$$\bar{\kappa}_N(z) := v^0(0; z)$$

Let $\mathcal{Z}_N$ denote the domain of $\bar{V}_N^0(\cdot)$, and of $\bar{\kappa}_N(\cdot)$,

$$\mathcal{Z}_N := \{z \in \mathbb{Z} \mid \mathcal{V}_N(z) \neq \emptyset\}$$

We propose to control the uncertain system $x^+ = Ax + Bu + w$ by constraining it to lie in a tube whose center is the solution of the nominal system obtained using the implicit nominal MPC control law $\bar{\kappa}_N(\cdot)$. The control applied to the system being controlled is $u = \kappa_N(x, z)$ in which x is the current state of the system being controlled, z is the current state of the nominal system, and $\kappa_N(\cdot)$ is defined by

$$\kappa_N(x, z) := \bar{\kappa}_N(z) + K(x - z)$$

The composite closed-loop system plus controller therefore satisfy

$$x^+ = Ax + B\kappa_N(x, z) + w \tag{3.27}$$
$$z^+ = Az + B\bar{\kappa}_N(z) \tag{3.28}$$

with initial state (x, x). The center of the tube is the sequence $\mathbf{z} = \{z(0), z(1), \ldots\}$ obtained by solving (3.28) with initial state $z(0) = x$, i.e., for each $i \in \mathbb{I}_{\geq 0}$, $z(i) = \bar{\phi}(i; x, \bar{\kappa}_N(\cdot))$. Since the difference equation

$z^+ = Az + B\bar{\kappa}_N(z)$ is autonomous, the solution $\mathbf{z}$ may be computed beforehand—at least up to a finite number of time steps. The control $u(i)$ applied to the system at time i is, then

$$u(i) = \kappa_N(x(i), z(i)) = v(i) + K[x(i) - z(i)]$$

in which $v(i) = \bar{\kappa}_N(z(i))$. The state sequence $\mathbf{x} = \{x(0), x(1), \ldots\}$ therefore satisfies

$$x(i+1) = Ax(i) + B\kappa_N(x(i), z(i)) + w(i) \qquad x(0) = x$$

To analyze stability of the closed-loop system, we have to consider, since the controller is a dynamic system with state z, the composite system whose state is (x, z) or the equivalent system whose state is (e, z). Since (e, z) and (x, z) are related by an invertible transformation

$$\begin{bmatrix} e \\ z \end{bmatrix} = T \begin{bmatrix} x \\ z \end{bmatrix} \qquad T := \begin{bmatrix} I & -I \\ 0 & I \end{bmatrix}$$

the two systems are equivalent. The composite system whose state is (x, z) satisfies, as shown previously

$$x^+ = Ax + B\kappa_N(x, z) + w \tag{3.29}$$

$$z^+ = Az + B\bar{\kappa}_N(z) \tag{3.30}$$

with initial state $(x(0), z(0)) = (x, x)$ whereas the composite system whose state is (e, z), $e := x - z$, satisfies

$$e^+ = A_K e + w \tag{3.31}$$

$$z^+ = Az + B\bar{\kappa}_N(z) \tag{3.32}$$

with initial state $(e(0), z(0)) = (0, x)$. The latter system is easier to analyze. So one way to proceed is to establish exponential stability of $S_K(\infty) \times \{0\}$ with region of attraction $S_K(\infty) \times \mathcal{Z}_N$ of the composite system described by (3.31) and (3.32); we leave this as Exercise 3.6.

Instead we consider the original system described by (3.29) and (3.30). We know, from the discussion above, that $e(i) \in S_K(\infty) \subseteq S$ and $x(i) \in \{z(i)\} \oplus S_K(\infty) \subseteq \{z(i)\} \oplus S$ for all $k \in \mathbb{I}_{\geq 0}$ if $e(0) \in S_K(\infty)$, and K is such that A_K is stable. Also, we know from Chapter 2 that if the stability Assumptions 2.12 and 2.13 are satisfied for the nominal optimal control problem $\bar{\mathbb{P}}_N(z)$, then the value function $\bar{V}_N^0(\cdot)$ satisfies

$$\bar{V}_N^0(z) \geq \ell(z, \bar{\kappa}_N(z)) \qquad \forall z \in \mathcal{Z}_N \tag{3.33}$$

$$\Delta\bar{V}_N^0(z) \leq -\ell(z, \bar{\kappa}_N(z)) \qquad \forall z \in \mathcal{Z}_N \tag{3.34}$$

$$\bar{V}_N^0(z) \leq V_f(z) \qquad \forall z \in \mathbb{Z}_f \tag{3.35}$$

in which $\Delta\bar{V}_N^0(z) = \bar{V}_N^0(z^+) - \bar{V}_N^0(z)$ with $z^+ = Az + B\bar{\kappa}_N(z)$.

If: (i) $\ell(z,v) = (1/2)|z|_Q^2 + (1/2)|v|_R^2$ in which Q and R are positive definite, (ii) the terminal cost $V_f(z) = (1/2)|z|_{P_f}^2$ in which P_f is positive definite, (iii) Assumption 3.14 holds, and (iv) the terminal cost $V_f(\cdot)$ and terminal constraint set $\mathbb{Z}_f$ satisfy the stability Assumptions 2.12 and 2.13, and (v) $\mathcal{Z}_N$ is compact, then there exist constants c_1 and c_2 such that (3.33)–(3.35) become

$$\bar{V}_N^0(z) \geq c_1|z|^2 \qquad \forall z \in \mathcal{Z}_N \tag{3.36}$$

$$\Delta\bar{V}_N^0(z) \leq -c_1|z|^2 \qquad \forall z \in \mathcal{Z}_N \tag{3.37}$$

$$\bar{V}_N^0(z) \leq c_2|z|^2 \qquad \forall z \in \mathcal{Z}_N \tag{3.38}$$

Hence the origin is exponentially stable for the nominal system $z^+ = Az + B\bar{\kappa}_N(z)$ with a region of attraction $\mathcal{Z}_N$, i.e., there exists a $c > 0$ and a $\gamma \in (0,1)$ such that $|z(i)| \leq c|z(0)|\gamma^i$ for all $i \in \mathbb{I}_{\geq 0}$. Since $x(i) = z(i) + e(i)$ where $e(i) \in S_K(\infty)$

$$|x(i)|_{S_K(\infty)} = d(z(i) + e(i), S_K(\infty)) \leq d(z(i) + e(i), e(i)) = |z(i)|$$

Hence, for all $i \in \mathbb{I}_{\geq 0}$,

$$|x(i)|_{S_K(\infty)} \leq c|z(0)|\gamma^i$$

Let $\mathcal{A} \subset \mathbb{R}^n \times \mathbb{R}^n$ be defined as follows

$$\mathcal{A} := S_K(\infty) \times \{0\}$$

so that

$$|(x,z)|_{\mathcal{A}} = |x|_{S_K(\infty)} + |z|$$

It follows from the previous discussion that the state (x,z) of the composite system satisfies

$$|(x(i),z(i))|_{\mathcal{A}} = |x(i)|_{S_K(\infty)} + |z(i)| \leq 2c|z(0)|\gamma^i \leq 2c\gamma^i|(x(0),z(0))|_{\mathcal{A}}$$

for all $i \in \mathbb{I}_{\geq 0}$ since $|z(0)| \leq |x(0)|_{S_K(\infty)} + |z(0)| = |(x(0),z(0))|_{\mathcal{A}}$. We have proved:

Proposition 3.15 (Exponential stability of tube-based MPC). *The set $\mathcal{A} := S_K(\infty) \times \{0\}$ is exponentially stable with a region of attraction $(\mathcal{Z}_N \oplus S_K(\infty)) \times \mathcal{Z}_N$ for the composite system* (3.29) *and* (3.30).

Proposition 3.15 remains true if $S_K(\infty)$ is replaced by S where $S \supset S_K(\infty)$ is robust positive invariant for $e^+ = A_K e + w$, $w \in \mathbb{W}$. The tube-based model predictive controller is formally described by the following algorithm in which i denotes current time.

Tube-based model predictive controller.

Initialization: At time $i = 0$, set $x = z = x(0)$ in which $x(0)$ is the current state.

Step 1 (Compute control): At time i and current state (x, z), solve the nominal optimal control problem $\bar{\mathbb{P}}_N(z)$ to obtain the nominal control action $v = \bar{\kappa}_N(z)$ and the control action $u = v + K(x - z)$.

Step 2 (Check): If $\bar{\mathbb{P}}_N(z)$ is infeasible, adopt safety/recovery procedure.

Step 3 (Apply control): Apply the control u to the system being controlled.

Step 4 (Update): Measure the successor state x^+ of the system being controlled and compute the successor state $z^+ = f(z, v)$ of the nominal system.

Step 5: Set $(x, z) = (x^+, z^+)$, set $i = i + 1$, and go to Step 1.

In this algorithm, $\bar{\kappa}_N(z)$ is, of course, the first element in the control sequence $\mathbf{v}^0(z)$ obtained by solving the nominal optimal control problem $\bar{\mathbb{P}}_N(z)$. Step 2, the check step, is not activated if the assumptions made previously are satisfied and, therefore, is ignored in our analysis.

Computation of $\mathbb{Z}$ and $\mathbb{V}$. To implement the tube-based controller, we need inner approximations $\mathbb{Z}$ and $\mathbb{V}$ to be, respectively, the sets $\hat{\mathbb{Z}} := \mathbb{X} \ominus S_K(\infty)$ and $\hat{\mathbb{V}} := \mathbb{U} \ominus K S_K(\infty)$; computation of the set $S_K(\infty)$, a difficult task, is not necessary. Suppose we have a single state constraint

$$y := c'x \leq d$$

so that $\mathbb{X} = \{x \in \mathbb{R}^n \mid c'x \leq d\}$. Then, since, for all $i \in \mathbb{I}_{\geq 0}$, $x(i) = z(i) + e(i)$ where $e(i) \in S_K(\infty)$ if $e(0) \in S_K(\infty)$, it follows that $c'x(i) \leq d$ if

$$c'z(i) \leq d - \max\{c'e \mid e \in S_K(\infty)\}$$

Let ϕ_∞ be defined as follows

$$\phi_\infty := \max_e \{c'e \mid e \in S_K(\infty)\}$$

Hence

$$\hat{\mathbb{Z}} = \{z \in \mathbb{R}^n \mid c'z \leq d - \phi_\infty\}$$

is a suitable constraint for the nominal system, i.e., $z \in \hat{\mathbb{Z}}$ implies $c'x = c'z + c'e \leq d$ or $x \in \mathbb{X}$ for all $e \in S_K(\infty)$. To obtain $\hat{\mathbb{Z}}$, we need to compute ϕ_∞. But computing ϕ_∞ requires solving an infinite dimensional optimization problem, which is impractical. We can obtain an inner approximation to $\hat{\mathbb{Z}}$, which is all we need to implement robust MPC, by computing an upper bound to ϕ_∞. We now show how this may be done (Raković, Kerrigan, Kouramas, and Mayne, 2003). We require the following assumption.

Assumption 3.16 (Compact convex disturbance set). The compact convex set $\mathbb{W}$ contains the origin in its interior.

For each $i \in \mathbb{I}_{\geq 0}$, let ϕ_i be defined as follows

$$\phi_i := \max_e \{c'e \mid e \in S_K(i)\}$$

It can be shown that

$$\phi_N = \max_{\{w_i\}} \{c' \sum_{i=0}^{N-1} A_K^i w_i \mid w_i \in \mathbb{W}\}$$

and that

$$\phi_\infty = \max_{\{w_i\}} \{c' \sum_{i=0}^{\infty} A_K^i w_i \mid w_i \in \mathbb{W}\}$$

Suppose now we choose the feedback matrix K and the horizon N so that

$$A_K^N w \in \alpha \mathbb{W} \ \forall w \in \mathbb{W}$$

where $\alpha \in (0, 1)$. Because A_K is stable and $\mathbb{W}$ contains the origin in its interior, this choice is always possible. It follows from the definitions of ϕ_∞ and ϕ_N that

$$\begin{aligned}
\phi_\infty &= \phi_N + \max_{\{w_i\}} \{c' \sum_{i=N}^{\infty} A_K^i w_i \mid w_i \in \mathbb{W}\} \\
&= \phi_N + \max_{\{w_i\}} \{c'(A_K^N w_0 + A_K A_K^N w_1 + A_K^2 A_K^N w_2 + \ldots) \mid w_i \in \mathbb{W}\} \\
&\leq \phi_N + \max_{\{w_i\}} \{c'(\alpha w_0 + A_K \alpha w_1 + A_K^2 \alpha w_2 + \ldots) \mid w_i \in \mathbb{W}\}
\end{aligned}$$

where the last line follows from the fact that $A_K^N w \in \alpha W$ if $w \in \mathbb{W}$. It follows that

$$\phi_\infty \leq \phi_N + \alpha \phi_\infty$$

or

$$\phi_\infty \leq (1 - \alpha)^{-1} \phi_N$$

Hence an upper bound for ϕ_∞ may be obtained by determining ϕ_N, i.e., by solving a linear program. The constant $(1-\alpha)^{-1}$ may be made as close as desired to 1 by choosing α suitably small. The set $\mathbb{Z}$ defined by

$$\mathbb{Z} := \{z \in \mathbb{R}^n \mid c'z \leq d - (1-\alpha)^{-1}\phi_N\} \subseteq \hat{\mathbb{Z}}$$

is a suitable constraint set for the robust controller. If there are several state constraints

$$y_j := c_j' x \leq d_j \;\forall j \in \mathcal{J}$$

and K and N are chosen as previously to satisfy $A_K^N w \in \alpha\mathbb{W}$ for all $w \in \mathbb{W}$ and some $\alpha \in (0,1)$, then a suitable constraint set for the controller is the set

$$\mathbb{Z} := \{z \in \mathbb{R}^n \mid c_j' z \leq d_j - (1-\alpha)^{-1}\phi_N^j, \;\forall j \in \mathcal{J}\} \subseteq \hat{\mathbb{Z}}$$

in which, for each $j \in \mathcal{J}$,

$$\phi_N^j := \max_{\{w_i\}}\{c_j' e \mid e \in S_K(i)\} = \max_{\{w_i\}}\{c_j' \sum_{i=0}^{N-1} A_K w_i \mid w_i \in \mathbb{W}\}$$

A similar procedure may be used to obtain a suitable constraint set $\mathbb{V} \subseteq \hat{\mathbb{V}} = \mathbb{U} \ominus KS_K(\infty)$. Suppose $\mathbb{U}$ is described by

$$\mathbb{U} := \{u \in \mathbb{R}^m \mid a_j' u \leq b_j \;\forall j \in \mathcal{I}\}$$

If K and N are chosen as above, then a suitable constraint set $\mathbb{V}$ for the nominal system is

$$\mathbb{V} := \{v \in \mathbb{R}^m \mid a_j' v \leq b_j - (1-\alpha)^{-1}\theta_N^j, \; j \in \mathcal{I}\}$$

in which, for each $j \in \mathcal{I}$,

$$\theta_N^j := \max_{\{w_i\}}\{a_j' K e \mid e \in S_K(i)\} = \max_{\{w_i\}}\{a_j' K \sum_{i=0}^{N-1} A_K w_i \mid w_i \in \mathbb{W}\}$$

Critique. A feature of the robust controller that may appear strange is the fact that the nominal state trajectory $\{z(i)\}$ is completely independent of the state trajectory $\{x(i)\}$ of the uncertain system. Although the control $u = \kappa_N(x,z)$ applied to the uncertain system depends on the state of both systems, the control $v = \bar{\kappa}_N(z)$ applied to the nominal system depends only on the state z of the nominal system. This feature arises because we are considering a very specific problem: determination of a control that steers an uncertain linear system robustly from a known initial state x to the neighborhood $\{x_f\} \oplus S_K(\infty)$ of a desired final state x_f; $x_f = 0$ in the previous analysis. More generally, the target state x_f and a slowly varying external disturbance d will vary with time, and the control $u = \kappa_N(x, z, x_f, d)$ will depend on these variables.

A form of feedback from x to v and, hence, to z is easily added. Step 1 in the controller algorithm presented previously may be changed as follows.

Step 1 (Compute control): At time i and current state (x, z), solve the nominal optimal control problems $\bar{\mathbb{P}}_N(x)$ and $\bar{\mathbb{P}}_N(z)$ to obtain $\bar{\kappa}_N(z)$ and $\bar{\kappa}_N(x)$. If $\bar{V}_N^0(x) \leq \bar{V}_N^0(z)$ and $x \in \mathbb{Z}$, set $z = x$ and $u = v = \bar{\kappa}_N(x)$. Otherwise set $v = \bar{\kappa}_N(z)$ and $u = v + K(x - z)$.

Since the modified controller produces, at state (x, z), either a nominal cost $\bar{V}_N^0(z)$ or $\bar{V}_N^0(x) \leq \bar{V}_N^0(z)$ where $x \in \mathbb{Z}$ becomes the updated value of z, the analysis and conclusions above remain valid. The modification provides improved performance. From an alternative viewpoint, the modified controller may be regarded as an improved version of nominal MPC in which the nominal control $\bar{\kappa}_N(x)$ is replaced by a safe control $\bar{\kappa}_N(z)$ if $\bar{\kappa}_N(x)$ does not lead to a cost reduction because of the disturbance w.

As pointed out previously, the nominal controller $\bar{\kappa}_N(\cdot)$ steers the nominal state z to the desired final state, the origin in our analysis, while the feedback controller K keeps the state x of the uncertain system close to the nominal state z. Hence the feedback controller K should be chosen to reduce the effect of the additive disturbance; its choice depends, therefore, on the nature of the disturbance as shown in the examples in Section 3.6 that illustrate the fact that the feedback control $u = v + K(x - z)$ may even have a higher sampling rate than the nominal control $v = \bar{\kappa}_N(z)$ in order to attenuate more effectively high frequency disturbances.

3.4.4 Improved Tube-Based MPC of Linear Systems with Additive Disturbances

In this section we describe a version of the tube-based model predictive controller that has pleasing theoretical properties and that does not require computation of a nominal trajectory. It is, however, more difficult to implement since it requires knowledge of $S_K(\infty)$ or of a robust positive invariant outer approximation S. This section should therefore be omitted by readers interested only in easily implementable controllers.

We omitted, in Section 3.4.3, to make use of an additional degree of freedom available to the controller, namely $z(0)$, the initial state of the nominal system. Previously we set $z(0) = x(0) = x$. It follows from the discussion at the end of Section 3.4.3 that every trajectory of the system $x^+ = Ax + Bv + BKe + w$ emanating from an initial state x lies in the tube $X(z, \mathbf{v})$, provided that the initial state x of the closed-loop system and the initial state z of the nominal system satisfy

$$x \in \{z\} \oplus S$$

in which S is either $S_K(\infty)$ or a robust positive invariant set for $e^+ = A_K e + w$ that is an outer approximation of $S_K(\infty)$. So, we may optimize the choice of the initial state z of the nominal system, provided we satisfy the constraints $x \in \{z\} \oplus S$ and $z \in \mathbb{Z}$. But we can go further. We can optimize the choice of z at every time i because, if the current state x of the closed-loop system and the current state z of the nominal system satisfy $x \in \{z\} \oplus S$, and the input to the system being controlled is $v + K(x - z)$ where v is the input to the nominal system, then the subsequent states x^+ and z^+ satisfy $x^+ \in \{z^+\} \oplus S$. To this end, we define a new finite horizon optimal control problem $\mathbb{P}_N^*(x)$, to be solved online, that reduces the cost $\bar{V}_N^0(z)$ obtained in Section 3.4.3

$$\begin{aligned}\mathbb{P}_N^*(x): \quad V_N^*(x) &= \min_z \{\bar{V}_N^0(z) \mid x \in \{z\} \oplus S,\ z \in \mathbb{Z}\} \\ &= \min_{\mathbf{v},z} \{\bar{V}_N(z, \mathbf{v}) \mid \mathbf{v} \in \mathcal{V}_N(z),\ x \in \{z\} \oplus S,\ z \in \mathbb{Z}\}\end{aligned}$$

Because of the extra freedom provided by varying z, the domain of the value function $V_N^*(\cdot)$ is $\mathcal{X}_N := \mathcal{Z}_N \oplus S$ where $\mathcal{Z}_N$ is the domain of $\bar{V}_N^0(\cdot)$. The solution to problem $\mathbb{P}_N^*(x)$ is $z^*(x)$ and $\mathbf{v}^*(x) = \mathbf{v}^0(z^*(x))$; optimizing with respect to z means that z in $\bar{\mathbb{P}}_N(z)$ is replaced by $z^*(x)$. It follows that

$$V_N^*(x) = \bar{V}_N^0(z^*(x)) \tag{3.39}$$

for all $x \in \mathcal{X}_N$. The control applied to the system $x^+ = Ax + Bu + w$ at state x is $\kappa_N(x)$ defined by

$$\kappa_N(x) := \kappa_N^*(x) + K(x - z^*(x))$$

in which $\kappa_N^*(x) = \bar{\kappa}_N(z^*(x))$ is the first element in the sequence $\mathbf{v}^*(x) = \mathbf{v}^0(z^*(x))$. The main change from the simple tube-based model predictive controller is that z is replaced by $z^*(x)$. A theoretical advantage is that the applied control $\kappa_N(x)$ depends only on the current state x and not on the composite state (x, z) as in the simple controller.

It follows from (3.36), (3.37), (3.38) and (3.39) that the value function $V_N^*(\cdot)$ satisfies

$$V_N^*(x) = \bar{V}_N^0(z^*(x)) \geq c_1|z^*(x)|^2 \tag{3.40}$$

$$V_N^*(x) = \bar{V}_N^0(z^*(x)) \leq c_2|z^*(x)|^2 \tag{3.41}$$

$$\Delta V_N^*(x,w) \leq \Delta\bar{V}_N^0(z^*(x)) \leq -c_1|z^*(x)|^2 \tag{3.42}$$

for all $(x,w) \in \mathcal{X}_N \times \mathbb{W}$ in which the last line follows from the fact that

$$\begin{aligned} \Delta V_N^*(x,w) &:= V_N^*(x^+) - V_N^*(x) = \bar{V}_N^0(z^*(x^+)) - \bar{V}_N^0(z^*(x)) \\ &\leq \bar{V}_N^0((z^*(x))^+) - \bar{V}_N^0(z^*(x)) = \Delta\bar{V}_N^0(z^*(x)) \leq -c_1|z^*(x)|^2 \end{aligned}$$

with $x^+ = Ax + B\kappa_N(x) + w$ and $(z^*(x))^+ = Az^*(x) + B\bar{\kappa}_N(z^*(x))$. Next we note that

$$V_N^*(x) = 0 \ \forall x \in S$$

This equality follows from the fact that for all $x \in S$, the constraint $x \in \{z\} \oplus S$ in problem $\mathbb{P}_N^*(x)$ is satisfied by $z = 0$ since $0 \in S$. Because $\bar{V}_N^0(0) = 0$, it follows that $V_N^*(x) = \bar{V}_N^0(z^*(x)) \leq \bar{V}_N^0(0) = 0$; since $V_N^*(x) \geq 0$ we deduce that $V_N^*(x) = 0$. It also follows that $z^*(x) = 0$ for all $x \in S$ so that $z^*(x)$ is a "measure" of how far x is from the set S.

For each $i \in \mathbb{I}_{\geq 0}$, let $x(i) := \phi(i; x(0), \kappa_N(\cdot), \mathbf{w})$, the solution of $x^+ = Ax + B\kappa_N(x) + w$ at time i if the initial state at time 0 is $x(0)$. It follows from (3.40)–(3.42) that $V_N^*(x(i)) \leq \gamma^i V_N^*(x(0))$ where $\gamma := 1 - c_1/c_2 \in (0,1)$. Hence there exist $c > 0$ and $\delta = \sqrt{\gamma}$ such that

$$|z^*(x(i))| \leq c\delta^i |z^*(x(0))| \tag{3.43}$$

for all $i \in \mathbb{I}_{\geq 0}$. For all i, $x(i) = z^*(x(i)) + e(i)$ where $e(i) \in S$ so that $|x(i)|_S = d(z^*(x(i)) + e(i), S) \leq d(z^*(x(i)) + e(i), e(i)) = |z^*(x(i))|$. In fact, though this is harder to show, $d(\{z\} \oplus S, S) = |z|$. Hence

$$|x(i)|_S \leq |z^*(x(i))| \leq c|z^*(x(0))|\delta^i$$

so that $x(i)$ converges robustly exponentially fast to S but S is not necessarily robustly exponentially stable for $x^+ = Ax + B\kappa_N(x) + w$, $w \in \mathbb{W}$.

We define the sets $X(i)$ for $i \in \mathbb{I}_{\geq 0}$ by

$$X(i) := \{z^*(x(i))\} \oplus S \tag{3.44}$$

The Hausdorff distance between $X(i)$ and S satisfies

$$d_H(X(i), S) = |z^*(x(i))| \le c\delta^i |z^*(x(0))| = c\delta^i d_H(X(0), S)$$

for all $i \in \mathbb{I}_{\ge 0}$. Exercise 3.4 shows that $d_H(\{z\} \oplus S, S) = |z|$. We have therefore proved the following.

Proposition 3.17 (Exponential stability of tube-based MPC without nominal trajectory). *The set S is exponentially stable with a region of attraction $\mathcal{Z}_N \oplus S$ for the set difference equation*

$$X^+ = F(X, \mathbb{W})$$

in which $F : 2^{\mathbb{X}} \to 2^{\mathbb{X}}$ is defined by

$$F(X) := \{Ax + B\kappa_N(x) + w \mid x \in X, w \in \mathbb{W}\}$$

Robust exponential stability of S for $X^+ = F(X, \mathbb{W})$ is not as strong as robust exponential stability of S for $x^+ = Ax + B\kappa_N(x) + w$, $w \in \mathbb{W}$. To establish the latter, we would have to show that for some $c > 0$ and all $i \in \mathbb{I}_{\ge 0}$, $|x(i)|_S \le c\delta^i |x(0)|_S$. Instead we have merely shown that $|x(i)|_S \le c\delta^i |z^*(x(0))|$.

3.5 Tube-Based MPC of Linear Systems with Parametric Uncertainty

Introduction. Section 3.4 shows how it is possible to construct bounding tubes and, consequently, tube-based model predictive controllers when the uncertainty in the system takes the form of a bounded additive disturbance w. For this kind of uncertainty, the tube has a constant cross-section S or a cross-section S_k that increases with time k and converges to S.

Here we consider a different form of uncertainty, parametric uncertainty in linear constrained systems. More specifically, we consider here robust control of the system

$$x^+ = Ax + Bu$$

in which the parameter $p := (A, B)$ can, at any time, take any value in the convex set $\mathcal{P}$ defined by

$$\mathcal{P} := \text{co}\{(A_j, B_j) \mid j \in \mathcal{J}\}$$

in which $\mathcal{J} := \{1, 2, \ldots, J\}$. We make the following assumption.

Assumption 3.18 (Quadratic stabilizability)**.** The system $x^+ = Ax + Bu$ is quadratically stabilizable, i.e., there exists a positive definite function $V_f : x \mapsto (1/2)x'P_f x$, a feedback control law $u = Kx$, and a positive constant ε such that

$$V_f((A + BK)x) - V_f(x) \leq -\varepsilon |x|^2 \tag{3.45}$$

for all $x \in \mathbb{R}^n$ and all $p = (A, B) \in \{(A_j, B_j) \mid j \in \mathcal{J}\}$. The origin is globally exponentially stable for $x^+ = A_K x := (A + BK)x$ for all $(A, B) \in \{(A_j, B_j) \mid j \in \mathcal{J}\}$.

The feedback matrix K and the positive definite matrix P_f may be determined using linear matrix inequalities. Because $\mathcal{P}$ is convex and $V_f(\cdot)$ is strictly convex, (3.45) is satisfied for all $x \in \mathbb{R}^n$ and all $(A, B) \in \mathcal{P}$. The system is subject to the same constraints as before

$$x \in \mathbb{X} \qquad u \in \mathbb{U}$$

in which $\mathbb{X}$ and $\mathbb{U}$ are assumed, for simplicity, to be compact and polytopic; each set contains the origin in its interior. We define the nominal system to be

$$z^+ = \bar{A}z + \bar{B}v$$

in which

$$\bar{A} := (1/J) \sum_{j=1}^{J} A_j \qquad \bar{B} := (1/J) \sum_{j=1}^{J} B_j$$

The origin is globally exponentially stable for $x^+ = \bar{A}_K x := (\bar{A} + \bar{B}K)x$. The difference equation $x^+ = Ax + Bu$ of the system being controlled may be expressed in the form

$$x^+ = \bar{A}x + \bar{B}u + w$$

in which the disturbance[6] $w = w(x, u, p)$ is defined by

$$w := (A - \bar{A})x + (B - \bar{B})u$$

Hence, the disturbance w lies in the set $\mathbb{W}$ defined by

$$\mathbb{W} := \{(A - \bar{A})x + (B - \bar{B})u \mid (A, B) \in \mathcal{P},\ (x, u) \in \mathbb{X} \times \mathbb{U}\}$$

Clearly $\mathbb{W}$ is polytopic. The state and control constraint sets, $\mathbb{Z}$ and $\mathbb{V}$ for the nominal optimal control problem, defined in Section 3.4, whose

[6]The controller "regards" w as a disturbance and "assumes" that the system being controlled is $x^+ = \bar{A}x + \bar{B}u + w$.

solution yields implicitly the nominal control law $\bar{\kappa}_N(\cdot)$ are chosen to satisfy

$$\mathbb{Z} \oplus S_K(\infty) \subset \mathbb{X} \qquad \mathbb{V} \oplus KS_K(\infty) \subset \mathbb{U}$$

in which, as before,

$$S_K(\infty) := \sum_{i=0}^{\infty} (\bar{A}_K)^i \mathbb{W}$$

The origin is exponentially stable for the nominal system $z^+ = \bar{A}z + \bar{B}\bar{\kappa}_N(z)$ with a region of attraction $\mathcal{Z}_N$. We know from Section 3.4 that the control law $\kappa_N(x,z) = \bar{\kappa}_N(z) + K(x-z)$ results in satisfaction of the state and control constraints $x \in \mathbb{X}$ and $u \in \mathbb{U}$ for all admissible disturbance sequences provided that the initial state $(x(0), z(0))$ of the composite satisfies $(x(0), z(0)) \in \mathcal{M}_N := \{(x,z) \mid x \in \{z\} \oplus S_K(\infty), z \in \mathcal{Z}_N\} \subseteq \mathbb{X} \times \mathcal{Z}_N$. The set $S_K(\infty) \times \{0\}$ is robustly exponentially stable for the composite controlled system with a region of attraction $\mathcal{M}_N$.

Unlike the robust control problem studied in Section 3.4, the disturbance w now depends on x and u. In the sequel, we make use of the fact that $w \to 0$ uniformly in $p \in \mathcal{P}$ as $(x,u) \to 0$ to prove, under some assumptions, that the origin is robustly asymptotically stable for the composite system $x^+ = \bar{A}x + \bar{B}\kappa_N(x,z) + w$, $z^+ = \bar{A}z + \bar{B}\bar{\kappa}_N(z)$ with a region of attraction $\mathcal{M}_N$. We choose $\kappa_N(x,z)$ as in Section 3.4.3, to ensure that the origin is exponentially stable for $z^+ = \bar{A}z + \bar{B}\bar{\kappa}_N(z)$ with a region of attraction $\mathcal{Z}_N$ and $\kappa_N(x,z) := \bar{\kappa}_N(z) + K(x-z)$. The approach we adopt to establish that the origin is robustly asymptotically stable for the composite system may be summarized as follows. We consider, for the purpose of analysis, two sequences of nested sets $\{\mathbb{X}_i \mid i \in \mathbb{I}_{\geq 0}\}$ and $\{\mathbb{U}_i \mid i \in \mathbb{I}_{\geq 0}\}$ where, for each i, $\mathbb{X}_i := (1/2)^i \mathbb{X}$ and $\mathbb{U}_i := (1/2)^i \mathbb{U}$. For all $(x,u) \in \mathbb{X}_i \times \mathbb{U}_i$, all i, $w \in \mathbb{W}_i := (1/2)^i \mathbb{W}$. Clearly $\mathbb{W}_i$ is polytopic for all $i \in \mathbb{I}_{\geq 0}$. Let $S_0 := S_K(\infty)$ and, for each $i \in \mathbb{I}_{\geq 0}$, let $S_i \subset \mathbb{R}^n$ be defined as follows

$$S_i := (1/2)^i S_0$$

Clearly $S_i = \sum_{j=0}^{\infty} (\bar{A}_K)^j \mathbb{W}_i$ for each i. For each i, the set S_i is robust positive invariant for $e^+ = \bar{A}_K e + w$, $w \in \mathbb{W}_i$. We now make the assumption:

Assumption 3.19 (Restricted parameter uncertainty). The set $\mathcal{P}$ is sufficiently small to ensure that $\mathbb{W} = \mathbb{W}_0$ satisfies

$$S_0 \subset (1/4)\mathbb{X}_0 \qquad KS_0 \subset (1/4)\mathbb{U}_0$$

where $\mathbb{X}_0 = \mathbb{X}$ and $\mathbb{U}_0 = \mathbb{U}$.

Consequently $S_i \subset (1/2)\mathbb{X}_{i+1}$ and $KS_i \subset (1/2)U_{i+1}$ for all $i \in \mathbb{I}_{\geq 0}$. Consider now the solution $(x(i), z(i))$ at time i of the composite system $x^+ = \bar{A}x + \bar{B}\kappa_N(x,z) + w$, $z^+ = \bar{A}z + \bar{B}\bar{\kappa}_N(z)$ at time i if the initial state is $(x,z) \in \mathcal{M}_N$ so that $x \in \mathbb{X}_0$ and $e := x - z \in S_0$. Hence $e(i) \in S_0$ for all $i \geq 0$. Since $x(i) = z(i) + e(i)$ and $u(i) = v(i) + Ke(i)$, $z(i) \to 0$ and $v(i) \to 0$ as $i \to \infty$ and $e(i) \in S_0$ for all i, it follows that there exists a finite time i_0 such that $x(i) = z(i) + e(i) \in 2S_0 \subset \mathbb{X}_1$ and $u(i) = v(i) + Ke(i) \in 2KS_0 \subset \mathbb{U}_1$ for all $i \geq i_0$ and every admissible disturbance sequence. Thus, for all $i \geq i_0$, $w(i) \in \mathbb{W}_1$ so that $e(i) \to S_1$ as $i \to \infty$. Since $z(i) \to 0$, $v(i) \to 0$ and $e(i) \to S_1$ as $i \to \infty$, there exists a finite time $i_1 > i_0$ such that $x(i) = z(i) + e(i) \in 2S_1 \subset \mathbb{X}_2$ and $u(i) = v(i) + Ke(i) \in 2KS_1 \subset \mathbb{U}_2$ for all $i \geq i_i$ and every admissible disturbance sequence. Proceeding similarly, we deduce that for all $j \in \mathbb{I}_{\geq 0}$, there exists a finite time i_j such that $x(i) \in \mathbb{X}_{j+1}$ and $u(i) \in \mathbb{U}_{j+1}$ for all $i \geq i_j$ and every admissible disturbance sequence. Hence the initial state (x,z) is robustly steered to the origin. Since (x,z) is an arbitrary point in $\mathcal{M}_N$, the origin is robustly attractive for the composite system with a region of attraction $\mathcal{M}_N$.

To prove stability of the origin for the system

$$\begin{aligned} x^+ &= \bar{A}x + \bar{B}\kappa_N(x,z) + w \\ z^+ &= \bar{A}z + \bar{B}\bar{\kappa}_N(z) \end{aligned}$$

we consider the equivalent system,

$$\begin{aligned} e^+ &= \bar{A}_K e + w \\ z^+ &= \bar{A}z + \bar{B}\bar{\kappa}_N(z) \end{aligned}$$

The disturbance w lies in a set that gets smaller as (x,u) approaches the origin; indeed $x \in \epsilon\mathbb{X}$ and $u \in \epsilon\mathbb{U}$ implies $w \in \epsilon\mathbb{W}$. The states (x,z) and (e,z) of these two composite systems are related by

$$\begin{bmatrix} e \\ z \end{bmatrix} = T \begin{bmatrix} x \\ z \end{bmatrix}, \quad T := \begin{bmatrix} I & -I \\ 0 & I \end{bmatrix}$$

since $e := x - z$. Since T is invertible, the two composite systems are equivalent and stability for one system implies stability for the other. We assume that the value function $\bar{V}_N^0(\cdot)$ for the nominal optimal control problem has the usual properties:

$$\begin{aligned} \bar{V}_N^0(z) &\geq c_1|z|^2 + c_1|\bar{\kappa}_N(z)|^2 \\ \bar{V}_N^0(f(z, \bar{\kappa}_N(z))) &\leq \bar{V}_N^0(z) - c_1|z|^2 \\ \bar{V}_N^0(z) &\leq c_2|z|^2 \end{aligned}$$

for all $z \in \mathcal{Z}_N$; these properties arise when the stage cost is quadratic and positive definite, $\mathcal{Z}_N$ is bounded, and an appropriate terminal cost and constraint set are employed. The first inequality, which is a minor extension of the inequality normally employed, follows from the definition of $\bar{V}_N^0(\cdot)$ and of $\ell(\cdot)$. It follows from these conditions that there exists a $c > 0$ such that

$$|\bar{\kappa}_N(z)| \leq c|z|$$

for all $z \in \mathcal{Z}_N$. For all $\alpha \geq 0$, let $\text{lev}_\alpha V$ denote the sublevel set of $\bar{V}_N^0(\cdot)$ defined by

$$\text{lev}_\alpha V := \{z \in \mathcal{Z}_N \mid \bar{V}_N^0(z) \leq \alpha\}$$

and let $S := S_K(\infty)$; S is robust positive invariant for $e^+ = A_K e + w$, $w \in \mathbb{W}$ and, from Assumption 3.19, $S \subset (1/4)\mathbb{X}$ and $KS \subset (1/4)\mathbb{U}$.

We show below that, for all $\epsilon \in (0, 3/4]$, there exists a $\delta > 0$ such that $(z(0), e(0)) \leq \delta$ implies $z(i) \in c(3/4)\mathbb{X}$ and $e(i) \in \epsilon S$ for all $i \in \mathbb{I}_{\geq 0}$ thereby establishing robust stability of the origin for the composite system. The upper limit of $3/4$ on ϵ is not a limitation since the analysis shows that, for every $\epsilon \geq 3/4$, there exists a $\delta > 0$ such that $(z(0), e(0)) \leq \delta$ implies $z(i) \in \epsilon^* \mathbb{X}$ and $e(i) \in \epsilon^* S$ for all $i \in \mathbb{I}_{\geq 0}$ where $\epsilon^* = 3/4 \leq \epsilon$.

Let $\epsilon \in (0, 3/4]$ be arbitrary. From the properties of $\bar{V}_N^0(\cdot)$ and $\bar{\kappa}_N(\cdot)$, we may deduce the existence of an $\alpha > 0$ such that $\text{lev}_\alpha V \subseteq \epsilon(3/4)\mathbb{X}$ and $\bar{\kappa}_N(\text{lev}_\alpha V) \subseteq \epsilon(3/4)\mathbb{U}$. Hence there exists a $\delta \in (0, \epsilon)$ such that $\delta\mathcal{B} \subseteq \text{lev}_\alpha V \cap \epsilon S$ so that $|(z(0), e(0))| < \delta$ implies $z(0) \in \text{lev}_\alpha V$ and $e(0) \in \epsilon S$.

Suppose next that $z(i) \in \text{lev}_\alpha V$ and $e(i) \in \epsilon S$. Then $x(i) = z(i) + e(i) \in \epsilon(3/4)\mathbb{X} \oplus \epsilon(1/4)\mathbb{X} = \epsilon\mathbb{X}$. Similarly, $u(i) = \bar{\kappa}_N(z(i)) + Ke(i) \subseteq \bar{\kappa}_N(\text{lev}_\alpha V) \oplus \epsilon KS \subseteq \epsilon(3/4)\mathbb{U} \oplus \epsilon(1/4)\mathbb{U} = \epsilon\mathbb{U}$. Hence $w(i) \in \epsilon\mathbb{W}$. Since $\text{lev}_\alpha V$ is positive invariant for $z^+ = Az + B\bar{\kappa}_N(z)$, it follows that $z(i+1) \in \text{lev}_\alpha V$. Since ϵS is robust positive invariant for $e^+ = A_K e + w$, $w \in \epsilon\mathbb{W}$, it follows that $e(i+1) \in \epsilon S$. By induction, $z(i) \in \epsilon\mathbb{X}$ and $e(i) \in \epsilon S$ for all $i \in \mathbb{I}_{\geq 0}$. We have proved:

Proposition 3.20 (Asymptotic stability of tube-based MPC). *The origin is asymptotically stable with a region of attraction $\mathcal{M}_N$ for the composite controlled system.*

This result shows that the standard tube-based model predictive system has a degree of robustness against parametric uncertainty, provided that we can bound the disturbance w resulting from model error so that it lies in some compact set $\mathbb{W}_0$ that is sufficiently small. The controller described previously is conservative since the nominal system is designed on the basis that the disturbance w lies in $\mathbb{W}_0$. A less conservative design would exploit the fact that $w \in \mathbb{W}_i = (1/2)^i \mathbb{W}_0$ when $(x, u) \in (1/2)^i(\mathbb{X} \times \mathbb{U})$.

3.6 Tube-Based MPC of Nonlinear Systems

Satisfactory control in the presence of uncertainty requires feedback. As shown in Section 3.3.4, MPC of uncertain systems ideally requires optimization over control policies rather than control sequences, resulting in an optimal control problem that is usually impossibly complex. Practicality demands simplification; hence, in tube-based MPC of constrained linear systems we replace the control policy $\boldsymbol{\mu} = \{\mu_0(\cdot), \mu_1(\cdot), \ldots, \mu_{N-1}(\cdot)\}$, in which each element $\mu_i(\cdot)$ is an arbitrary function, by the simpler policy $\boldsymbol{\mu}$ in which each element has the simple form $\mu_i(x) = v(i) + K(x - z(i))$ in which $v(i)$ and $z(i)$, the control and state of the nominal system at time i, are determined using conventional MPC. The feedback gain K, which defines the local control law, is determined offline; it can be chosen so that all possible trajectories of the uncertain system lie in a tube centered on the nominal trajectory $\{z(0), z(1), \ldots\}$. The "cross-section" of the tube is a constant set S so that every possible state of the uncertain system at time i lies in the set $\{z(i)\} \oplus S$. This enables the nominal trajectory to be determined using MPC, to ensure that all possible trajectories of the uncertain system satisfy the state and control constraints, and that all trajectories converge to an invariant set centered on the origin.

It would be desirable to extend this methodology to the control of constrained nonlinear systems, but we face some formidable challenges. It is possible to define a nominal system and, as shown in Chapter 2, to determine, using MPC with "tightened" constraints, a nominal trajectory that can serve as the center of a tube. But it seems to be prohibitively difficult to determine a local control law that steers all trajectories of the uncertain system toward the nominal trajectory, and of a set centered on the nominal trajectory in which these trajectories can be guaranteed to lie.

We overcome these difficulties by employing *two* model predictive controllers. The first uses MPC with tightened constraints to determine, as before, a nominal trajectory; the second, the ancillary controller, uses MPC to steer the state of the uncertain system toward the nominal trajectory. We avoid the difficult task of determining, a priori, a local control law by employing MPC that merely determines a suitable control action for the current state.

The system to be controlled is described by a nonlinear difference equation

$$x^+ = f(x, u) + w \tag{3.46}$$

in which the additive disturbance is assumed to lie in the compact set $\mathbb{W}$ that contains the origin. The state x and the control u are required to satisfy the constraints

$$x \in \mathbb{X} \qquad u \in \mathbb{U}$$

The solution of (3.46) at time i if the initial state at time 0 is x and the control is generated by policy $\boldsymbol{\mu}$ is $\phi(i; x, \boldsymbol{\mu}, \mathbf{w})$, in which $\mathbf{w}$ denotes, as usual, the disturbance sequence $\{w(0), w(1), \ldots\}$. Similarly, $\phi(i; x, \kappa, \mathbf{w})$ denotes the solution of (3.46) at time i if the initial state at time 0 is x and the control is generated by a time invariant control law $\kappa(\cdot)$.

The nominal system is obtained by neglecting the disturbance w and is therefore described by

$$z^+ = f(z, v)$$

Its solution at time i if its initial state is z is denoted by $\bar{\phi}(i; z, \mathbf{v})$, in which $\mathbf{v} := \{v(0), v(1), \ldots\}$ is the nominal control sequence. The deviation between the actual and nominal state is $e := x - z$ and satisfies

$$e^+ = f(x, u) - f(z, v) + w$$

Because $f(\cdot)$ is nonlinear, this difference equation cannot be simplified as in the linear case where e^+ is independent of x and z, depending only on their difference $e = x - z$ and on w.

3.6.1 The Central Path

The central path is a feasible trajectory for the nominal system that is sufficiently far from the boundaries of the original constraints to enable the ancillary controller for the uncertain system to satisfy these constraints. It is generated by the solution to a nominal optimal control problem $\bar{\mathbb{P}}_N(z)$ where z is the state of the nominal system. The cost function $\bar{V}_N(\cdot)$ for the nominal optimal control problem is defined by

$$\bar{V}_N(z, \mathbf{v}) := \sum_{k=0}^{N-1} \ell(z(k), v(k)) + V_f(z(N)) \tag{3.47}$$

in which $z(k) = \bar{\phi}(k; z, \mathbf{v})$ and z is the initial state. The function $\ell(\cdot)$ is defined by

$$\ell(z, v) := (1/2)(|z|_Q^2 + |v|_R^2)$$

in which Q and R are positive definite, $|z|_Q^2 := z^T Q z$, and $|v|_R^2 := v^T Q v$. We impose the following state and control constraints on the nominal system

$$z \in \mathbb{Z} \qquad v \in \mathbb{V}$$

in which $\mathbb{Z} \subset \mathbb{X}$ and $\mathbb{V} \subset \mathbb{U}$. The choice of $\mathbb{Z}$ and $\mathbb{V}$ is more difficult than in the linear case because it is difficult to bound the deviation $e = x - z$ of the state x of the uncertain system from the state z of the nominal system. We assume that these two constraint sets are compact. The terminal cost function $V_f(\cdot)$ together with the terminal constraint set $\mathbb{Z}_f \subseteq \mathbb{X}$ for the nominal system are chosen as described in Chapter 2 and Section 3.4 to satisfy the usual "stability axioms." The state and control constraints, and the terminal constraint $z(N) \in \mathbb{Z}_f$ impose a parametric constraint $\mathbf{v} \in \mathcal{V}_N(z)$ on the nominal control sequence in which $\mathcal{V}_N(z)$ is defined by

$$\mathcal{V}_N(z) := \{\mathbf{v} \mid v(k) \in \mathbb{V}, \quad \bar{\phi}(k; z, \mathbf{v}) \in \mathbb{Z} \quad \forall k \in \mathbb{I}_{0:N-1}, \\ \bar{\phi}(N; z, \mathbf{v}) \in \mathbb{Z}_f\}$$

For each z, the set $\mathcal{V}_N(z)$ is compact; it is bounded because of the assumptions on $\mathbb{V}$, and closed because of the continuity of $\bar{\phi}(\cdot)$. The nominal optimal control problem $\bar{\mathbb{P}}_N(z)$ is defined by

$$\bar{\mathbb{P}}_N(z): \quad \bar{V}_N^0(z) = \min_{\mathbf{v}}\{\bar{V}_N(z, \mathbf{v}) \mid \mathbf{v} \in \mathcal{V}_N(z)\}$$

A solution exists if z is feasible for $\bar{\mathbb{P}}_N(z)$ because $\bar{V}_N(\cdot)$ is continuous and $\mathcal{V}_N(z)$ is compact. Let $\mathcal{Z}_N := \{z \mid \mathcal{V}_N(z) \neq \emptyset\}$ denote the domain of $\bar{V}_N^0(z)$, the set of feasible states for $\bar{\mathbb{P}}_N(z)$. By virtue of our assumptions, the set $\mathcal{Z}_N$ is bounded. The solution of $\bar{\mathbb{P}}_N(z)$ is the minimizing control sequence

$$\mathbf{v}^0(z) = \{v^0(0; z), v^0(1; z), \ldots, v^0(N-1; z)\}$$

which we assume is unique, and the associated optimal state sequence is

$$\mathbf{z}^0(z) = \{z, z^0(1; z), \ldots, z^0(N; z)\}$$

The first element $v^0(0; z)$ of $\mathbf{v}^0(z)$ is the control that is applied in MPC. The implicit MPC control law is, therefore, $\bar{\kappa}_N(\cdot)$ defined by

$$\bar{\kappa}_N(z) := v^0(0; z)$$

The nominal system under MPC satisfies

$$z^+ = f(z, \bar{\kappa}_N(z))$$

The central path that defines the ancillary control problem defined in the next subsection consists of the state trajectory

$$\mathbf{z}^*(z) := \{z^*(0; z), z^*(1; z), \ldots\}$$

and the control trajectory

$$\mathbf{v}^*(z) := \{v^*(0; z), v^*(1; z), \ldots\}$$

in which z is the initial state of the nominal system. These trajectories are the solutions of the controlled nominal system described by

$$z^+ = f(z, \bar{\kappa}_N(z))$$

so that for all i

$$z^*(i; z) = \bar{\phi}(i; z, \bar{\kappa}_N) \qquad v^*(i; z) = \bar{\kappa}_N(z^*(i; z)) \tag{3.48}$$

If the terminal cost function $V_f(\cdot)$ and terminal constraint set $\mathbb{Z}_f$ are chosen to satisfy the usual stability assumptions, which we assume to be the case, and $\mathcal{Z}_N$ is bounded, there exist $c_1 > 0$ and $\bar{c}_2 > c_1$ such that

$$\begin{aligned}
\bar{V}_N^0(z) &\geq c_1 |z|^2 \\
\bar{V}_N^0(z) &\leq \bar{c}_2 |z|^2 \\
\Delta \bar{V}_N^0(z) &\leq -c_1 |z|^2
\end{aligned}$$

for all $z \in \mathcal{Z}_N$ in which

$$\Delta \bar{V}_N^0(z) := \bar{V}_N^0(f(z, \bar{\kappa}_N(z))) - \bar{V}_N^0(z)$$

It follows that the origin is exponentially stable with a region of attraction $\mathcal{Z}_N$ for the system $z^+ = f(z, \bar{\kappa}_N(z))$. The state of the controlled nominal system converges to the origin exponentially fast.

3.6.2 Ancillary Controller

The purpose of the ancillary controller is to maintain the state of the uncertain system $x^+ = f(x, u) + w$ close to the trajectory of the nominal system $z^+ = f(z, \bar{\kappa}_N(z))$. The ancillary controller replaces the controller $u = v + K(x - z)$ employed in the linear case. To obtain u in the nonlinear control, we determine a control sequence that minimizes the cost of the deviation between the trajectories of the two systems, $x^+ = f(x,u)$ and $z^+ = f(z, \bar{\kappa}_N(z))$, with initial states x and z, respectively, and choose u to be the first element of this sequence. If the optimal control problem is properly posed, the resultant control u steers the state of the deterministic system $x^+ = f(x,u)$ toward the nominal trajectory, and, hence, as in the linear case, tends to keep the trajectory of the uncertain system $x^+ = f(x,u) + w$ close to the nominal trajectory.

The ancillary controller is, therefore, based on the composite system

$$x^+ = f(x,u) \tag{3.49}$$

$$z^+ = f(z, \bar{\kappa}_N(z)) \tag{3.50}$$

The cost $V_N(x,z,\mathbf{u})$ that measures the distance between the trajectories of these two systems is defined by

$$V_N(x,z,\mathbf{u}) := \sum_{i=0}^{N-1} \ell(x(i) - z^*(i;z), u(i) - v^*(i;z)) \tag{3.51}$$

in which, for each i, $x(i) := \bar{\phi}(i;x,\mathbf{u})$ is the solution of (3.49) at time i if the initial state is x and the control input sequence is $\mathbf{u}$; $z^*(i;z)$ and $v^*(i;z)$ are defined in (3.48). For the purpose of analysis it is convenient to suppose that the entire infinite sequences $\mathbf{z}^*(z)$ and $\mathbf{v}^*(z)$ have been precalculated. In practice, apart from initialization, generation of the sequences used in (3.51) require only one solution of $\mathbb{P}_N$ at each iteration. It is not necessary for the cost function $\ell(\cdot)$ in (3.51) to be the same function as in (3.47) that defines the cost for the nominal controller. Indeed, as we show subsequently, it is not even necessary for the ancillary controller to have the same sample time as the nominal controller. The ancillary control problem is the minimization of $V_N(x,z,\mathbf{u})$ with respect to $\mathbf{u}$ subject to merely one state constraint, the terminal equality constraint $x(N) = z^*(N;z)$. The tube-based controller implicitly satisfies the state and input constraints. The terminal constraint is chosen for simplicity to ensure stability. Hence, the ancillary control problem $\mathbb{P}_N(x,z)$ is defined by

$$V_N^0(x,z) = \min_{\mathbf{u}}\{V_N(x,z,\mathbf{u}) \mid \mathbf{u} \in \mathcal{U}_N(x,z)\}$$

$$\mathcal{U}_N(x,z) := \{\mathbf{u} \in \mathbb{U}^N \mid \bar{\phi}(N;x,\mathbf{u}) = z^*(N;z)\}$$

in which $\mathcal{U}_N(x,z)$ is the constraint set. For each (x,z), the set $\mathcal{U}_N(x,z)$ is compact. There is no terminal cost and the terminal constraint set is the single state $z^*(N;z) = \bar{\phi}(N;z,\bar{\kappa}_N(\cdot))$

$$\mathbb{X}_f(z) = \{z^*(N;z)\}$$

For each $z \in \mathcal{Z}_N$, the domain of the value function $V_N^0(\cdot, z)$ and of the minimizer is the set $X_N(z)$ defined by

$$X_N(z) := \{x \in \mathbb{X} \mid \mathcal{U}_N(x,z) \neq \emptyset\}$$

For each $z \in \mathcal{Z}_N$, the set $X_N(z)$ is bounded. For future reference, let the set $\mathcal{M}_N \subset \mathbb{R}^n \times \mathbb{R}^n$ be defined by

$$\mathcal{M}_N := \{(x,z) \mid z \in \mathcal{Z}_N,\ x \in X_N(z)\}$$

The set $\mathcal{M}_N$ is bounded. For any $(x,z) \in \mathcal{M}_N$, the minimizing control sequence is $\mathbf{u}^0(x,z) = \{u^0(0;x,z), u^0(1;x,z), \ldots, u^0(N-1;x,z)\}$, and the control applied to the system is $u^0(0;x,z)$, the first element in this sequence. The corresponding optimal state sequence is $\mathbf{x}^0(x,z) = \{x, x^0(1;x,z), \ldots, x^0(N;x,z)\}$. The implicit ancillary control law is, therefore, $\kappa_N(\cdot)$ defined by

$$\kappa_N(x,z) := u^0(0;x,z)$$

The composite uncertain system then satisfies

$$x^+ = f(x, \kappa_N(x,z)) + w \tag{3.52}$$
$$z^+ = f(x, \bar{\kappa}_N(z)) \tag{3.53}$$

If $x = z$, then, as is easily verified, $V_N^0(x,z) = 0$ and

$$u^0(i;x,z) = v^*(i;z),\ i = 0,1,\ldots,N-1$$

so that the control and state trajectories of the two systems (3.49) and (3.50) are identical. In particular

$$\kappa_N(z,z) = \bar{\kappa}_N(z)$$

If some controllability assumptions are satisfied, the value function $V_N^0(\cdot)$ has properties analogous to those of $\bar{V}_N^0(\cdot)$, except that the bounds are $\mathcal{K}_\infty$ functions of $x - z$ rather than of x

$$V_N^0(x,z) \geq c_1|x-z|^2 \tag{3.54}$$
$$V_N^0(x,z) \leq c_2|x-z|^2 \tag{3.55}$$
$$\Delta V_N^0(x,z) \leq -c_1|x-z|^2 \tag{3.56}$$

for all $(x, z) \in \mathcal{M}_N$ in which, now

$$\Delta V_N^0(x, z) := V_N^0(f(z, \kappa_N(x, z)), f(x, \bar{\kappa}_N(z))) - V_N^0(x, z)$$

Note that $\Delta V_N^0(x, z)$ is the change in the value as x changes to $x^+ = f(x, \kappa_N(x, z))$ *and* z changes to $z^+ = f(x, \bar{\kappa}_N(z))$; the effect of the disturbance w is ignored in this expression. It follows from (3.54)-(3.56) that

$$V_N^0(f(x, \kappa_N(x, z)), f(z, \bar{\kappa}_N(z))) \leq \gamma V_N^0(x, z)$$

in which $\gamma := 1 - c_1/c_2 \in (0, 1)$ and, hence, that the origin is exponentially stable with a region of attraction $\mathcal{M}_N$ for the composite *deterministic* system $x^+ = f(x, \kappa_N(x, z))$, $z^+ = f(z, \bar{\kappa}_N(z))$. This property is sufficient to bound $e = x - z$ for the composite uncertain system $x^+ = f(x, \kappa_N(x, z)) + w$, $z^+ = f(z, \bar{\kappa}_N(z))$ and allows us, as shown subsequently, to determine suitable tightened constraint sets $\mathbb{Z}$ and $\mathbb{V}$. Assuming these sets have been determined, a robust MPC algorithm for nonlinear systems can be proposed; we do this next.

3.6.3 Controller Algorithm

Suppose $\mathbb{Z}$ and $\mathbb{V}$ have been chosen. In the following algorithm, $\mathbf{v}^*$ denotes the control sequence $\{v^*(0), v^*(1), \ldots, v^*(N-1)\}$, and $\mathbf{z}^*$ denotes the state sequence $\{z^*(0), z^*(1), \ldots, z^*(N)\}$. The controller algorithm is:

Robust control algorithm.

Initialization: At time 0, set $i = 0$, $x = x(0)$, and $z = x$. Solve $\bar{\mathbb{P}}_N$ for N time steps to obtain the nominal closed-loop state and control sequences $\mathbf{v}^* = \mathbf{v}^*(z) = \{v^*(0; z), v^*(1; z), \ldots, v^*(N-1; z)\}$ and $\mathbf{z}^* = \mathbf{z}^*(z) = \{z^*(0; z), z^*(1; z), \ldots, z^*(N; z)\}$, and set $u = \bar{\kappa}_N(z) = v^*(0; z)$.[7]

Step 1 (Compute control): At time i, compute $u = \kappa_N(x, z)$ by solving $\mathbb{P}_N(x, z)$.

Step 2 (Control): Apply u to the system being controlled.

Step 3 (Update x): Set $x = x^+$ where $x^+ = f(x, u) + w$ is the successor state.

[7]Recall $z^*(0; z) = z$ and $v^*(0; z) = \bar{\kappa}_N(z)$.

Step 4 (Update z, $\mathbf{v}^*$, and $\mathbf{z}^*$): Compute $v^* = \bar{\kappa}_N(z^*(N))$ and $z^* = f(z^*(N), v^*)$ by solving $\mathbb{P}_N(z^*(N))$. Set $z = z^*(1)$. Set $\mathbf{v}^* = \{v^*(1), \ldots, v^*(N-1), v^*\}$ and set $\mathbf{z}^* = \{z^*(1), \ldots, z^*(N), z^*\}$.

Step 5 (Repeat): Set $i = i + 1$. Go to Step 1.

A check step may be incorporated as done previously to safeguard against unanticipated events.

3.6.4 Analysis

Because of the nonlinearity and the terminal equality constraint in problem $\mathbb{P}_N(x, z)$, analysis is technical and requires use of the implicit function theorem. Full details appear on the website `www.che.wisc.edu/~jbraw/mpc` and the references cited there. Here we give an outline of the analysis. Ideally we would like, as in the linear case, to have a constant set S such that given $z(i)$, the state of the nominal system at time i, we could assert that the state $x(i)$ of the uncertain system lies in $\{z(i)\} \oplus S$. Instead, as we show subsequently, for each state $z(i)$ of the nominal system, the state $x(i)$ of the uncertain system lies, for some $d > 0$, in the set $S_d(z(i))$ where the set-valued $S_d(\cdot)$ is defined, for all $z \in \mathcal{Z}_N$

$$S_d(z) := \{x \in \mathbb{R}^n \mid V_N^0(x, z) \leq d\}$$

The set $S_d(z)$ is a sublevel set of the function $x \mapsto V_N^0(x, z)$. Since $S_0(z) = \{z\}$, the set $S_d(z)$ is a neighborhood of z. The set $S_d(z)$, that varies with z, replaces the set $\{z\} \oplus S$ employed in Section 3.4 because of the following important property that holds under certain controllability and differentiability assumptions:

Proposition 3.21 (Existence of tubes for nonlinear systems). *There exists a $d > 0$ such that if the state (x, z) of the composite system* (3.52) *and* (3.53) *lies in $\mathcal{M}_N$ and satisfies $x \in S_d(z)$, then the successor state (x^+, z^+) satisfies $x^+ \in S_d(z^+)$, i.e.,*

$$x^+ = f(x, \kappa_N(x, z)) + w \in S_d(z^+) \qquad z^+ = f(z, \bar{\kappa}_N(z))$$

for all w satisfying $|w| \leq (1 - \gamma)d/k(z)$ where $k(z)$ is a local Lipschitz constant for $x \mapsto V_N^0(x, z)$.

If $w \in \mathbb{W}$ implies $|w| \leq (1 - \gamma)d/k$ where k is an upper bound for $k(z)$ in $\mathcal{Z}_N$, then every solution of the system $x^+ = f(x, \kappa_N(x, z)) + w$, $w \in \mathbb{W}$ lies in the tube $\mathbf{S} := \{S_d(z), S_d(z^*(1; z)), S_d(z^*(2; z)), \ldots\}$ for

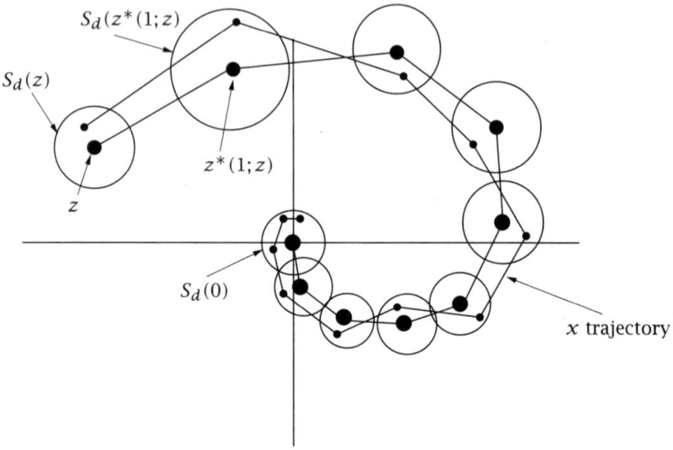

Figure 3.5: Tube for a nonlinear system.

all disturbance sequences $\{w(i)\}$ satisfying $w(i) \in \mathbb{W}$ for all $i \in \mathbb{I}_{\geq 0}$. Figure 3.5 illustrates this result and the fact that the cross-section of the tube varies with the state of the nominal system.

3.6.5 Choosing $\mathbb{Z}$ and $\mathbb{V}$

The tightened constraint sets $\mathbb{Z}$ and $\mathbb{V}$ may, in principle, be computed. Suppose there exists a single state constraint $c'x \leq e$. The tightened state constraint set is $\mathbb{Z} := \{x \mid c'x \leq f\}$ where $f < e$. Assuming that the constant d is known, the tightened state constraint set is suitable provided that

$$\phi(z) := \max_x \{c'x \mid x \in S_d(z)\} \leq e$$

for all $z \in \mathbb{Z} \cap \mathcal{Z}_N$, i.e., for all $z \in \mathcal{Z}_N$ satisfying $c'z \leq f$. In practice, $\phi(z)$ could be computed for a finite number of representative points in $\mathbb{Z} \cap \mathcal{Z}_N$. Since $S_d(z) := \{x \mid V_N^0(x,z) \leq d\}$, $\phi(z)$ may be computed using

$$-\phi(z) = \min_x \{-c'x \mid V_N^0(x,z) \leq d\} = \min_{(x,\mathbf{u})} \{-c'x \mid V_N(x,z,\mathbf{u}) \leq d\}$$

Other state constraints may be similarly treated. The tightened control constraint set also may be computed.

An alternative is the following. If, as is often the case even for nonlinear systems, the sets $\mathbb{X}$ and $\mathbb{U}$ are polyhedral, we may choose tightened constraint sets $\mathbb{Z} = \alpha\mathbb{X}$ and $\mathbb{V} = \beta\mathbb{V}$ where $\alpha,\ \beta \in (0,1)$

by a simple modification of the defining inequalities. If, for example, $\mathbb{X} = \{x \mid Ax \leq a\}$, then $\alpha\mathbb{X} = \{x \mid Ax \leq \alpha a\}$. This choice may be tested by Monte Carlo simulation of the controlled system. If constraints are violated in the simulation, α and β may be reduced; if the constraints are too conservative, α and β may be increased. For each choice of α and β, the controller provides a degree of robustness that can be adjusted by modifying the "tuning" parameters α and β.

Example 3.22: Robust control of an exothermic reaction

Consider the control of a continuous-stirred-tank reactor. We use a model derived in Hicks and Ray (1971) and modified by Kameswaran and Biegler (2006). The reactor is described by the second-order differential equation

$$\begin{aligned}
\dot{x}_1 &= (1/\theta)(1 - x_1) - kx_1 \exp(-M/x_2) \\
\dot{x}_2 &= (1/\theta)(x_f - x_2) - kx_1 \exp(-M/x_2) - \alpha u(x_2 - x_c) + w
\end{aligned}$$

in which x_1 is the product concentration, x_2 is the temperature, and u is the coolant flowrate. The model parameters are $\theta = 20$, $k = 300$, $M = 5$, $x_f = 0.3947$, $x_c = 0.3816$, and $\alpha = 0.117$. The state, control and disturbance constraint sets are

$$\begin{aligned}
\mathbb{X} &= \{x \in \mathbb{R}^2 \mid x_1 \in [0, 2],\ x_2 \in [0, 2]\} \\
\mathbb{U} &= \{u \in \mathbb{R} \mid u \in [0, 2]\} \\
\mathbb{W} &= \{w \in \mathbb{R} \mid w \in [-0.001, 0.001]\}
\end{aligned}$$

The controller is required to steer the system from a locally stable steady state $x(0) = (0.9831, 0.3918)$ at time 0, to a locally unstable steady state $z_e = (0.2632, 0.6519)$. Because the desired terminal state is z_e rather than the origin, the stage cost $\ell(z, v)$ is replaced by $\ell(z - z_e, v - v_e)$ where $\ell(z, v) := (1/2)(|z|^2 + v^2)$ and (z_e, v_e) is an equilibrium pair satisfying $z_e = f(z_e, v_e)$; the terminal constraint set $\mathbb{Z}_f$ is chosen to be $\{z_e\}$. The constraint sets for the nominal control problem are $\mathbb{Z} = \mathbb{X}$ and $\mathbb{V} = [0.02, 2]$. Since the state constraints are not activated, there is no need to tighten $\mathbb{X}$. The disturbance is chosen to be $w(t) = A\sin(\omega t)$ where A and ω are independent uniformly distributed random variables, taking values in the sets $[0, 0.001]$ and $[0, 1]$, respectively. The horizon length is $N = 40$ and the sample time is $\Delta = 3$ giving a horizon time of 120. The ancillary controller uses $\ell_a(x, u) = (1/2)(|x|^2 + u^2)$ and the same horizon and sample time.

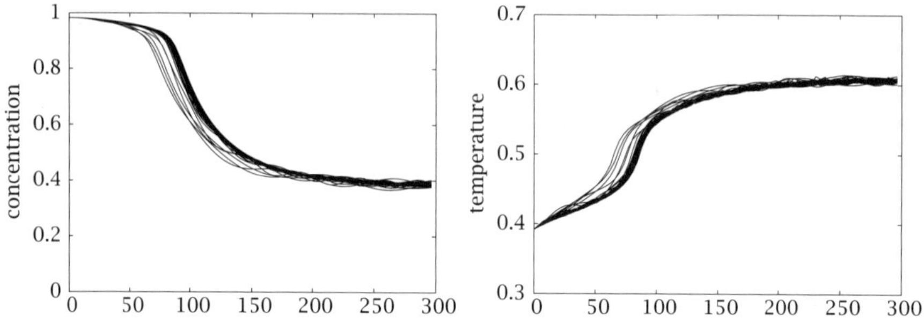

(a) Standard MPC.

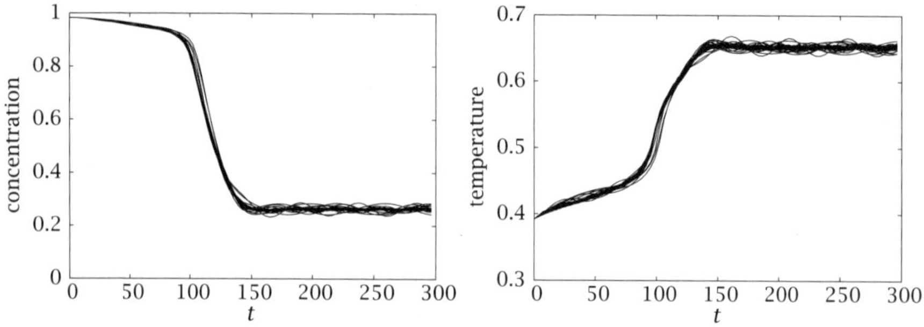

(b) Tube-based MPC.

Figure 3.6: Comparison of 100 realizations of standard and tube-based MPC for the chemical reactor example.

For comparison, a standard MPC controller using the same stage cost $\ell(\cdot)$, and the same terminal constraint set $\mathbb{Z}_f$ employed in the central path controller is simulated. Figure 3.6(a) illustrates the performance standard MPC, and Figure 3.6(b) the performance of tube-based MPC for 100 realizations of the disturbance sequence. Tube-based MPC, as expected, has a smaller spread of trajectories than is the case for standard MPC. Because each controller has the same stage cost and terminal constraint, the spread of trajectories in the steady-state phase when $z(t) = z_e$ is the same for the two controllers. Because the control constraint set for the tube-based central controller is tighter than that for the standard controller, however, the tube-based controller is somewhat slower than the standard controller.

The ancillary controller may be tuned to reduce more effectively

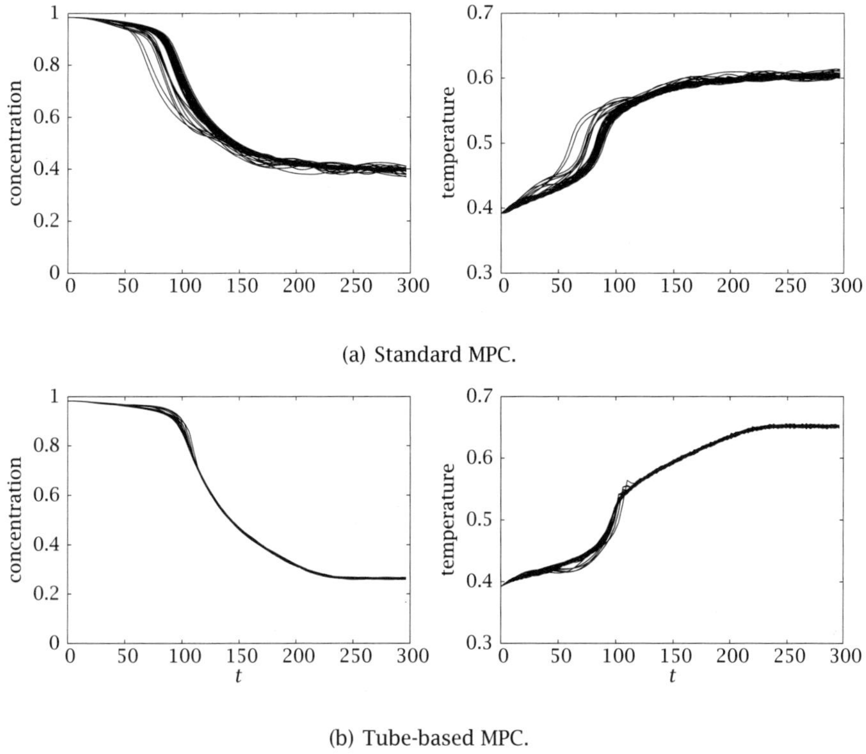

Figure 3.7: Comparison of standard and tube-based MPC with an aggressive ancillary controller.

the spread of trajectories due to the external disturbance. It can be said that the main purpose of the central controller is to steer the system from one equilibrium state to another, while the purpose of the ancillary controller is to reduce the effect of the disturbance. These different objectives may require different stage costs. Our next simulation compares the performance of the standard and tube-based MPC when a more "aggressive" stage cost is employed for the ancillary controller. Figure 3.7 shows the performance of these two controllers when the nominal and standard MPC controller employ $\ell(z - z_e, v - v_e)$ with $\ell(z, v) := (1/2)|z|^2 + 5v^2$ and the ancillary controller employs $\ell_a(x, u) = 50|x|^2 + (1/20)u^2$. The tube-based MPC controller reduces the spread of the trajectories during both the transient *and* the steady state phases.

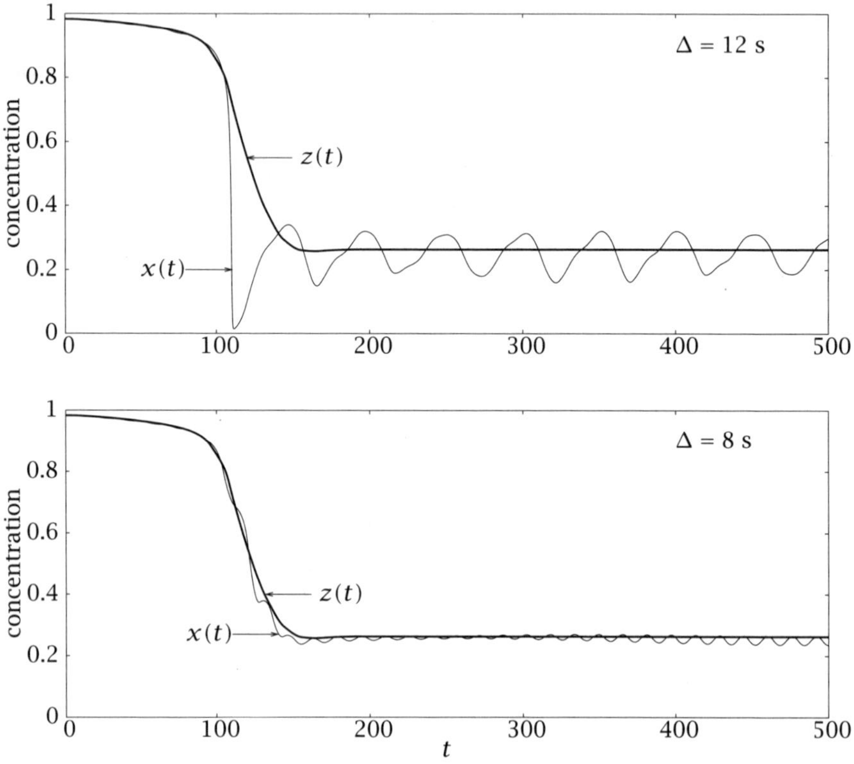

Figure 3.8: Concentration versus time for the ancillary controller for sample time 12 (top) and 8 (bottom).

It is also possible to tune the sample time of the ancillary controller. This feature may be useful when the disturbance frequency lies outside the pass band of the central path (nominal) controller. Figure 3.8 shows how concentration varies with time when the disturbance is $w(t) = 0.002\sin(0.4t)$, the sample time of the central path controller is 12 whereas the sample time of the ancillary controller is 12 (top figure) and 8 (bottom figure). The central path controller employs $\ell(z - z_e, v - v_e)$, and the ancillary controller employs $\ell(x, u)$ where $\ell(x, u) := (1/2)(|x|^2 + u^2)$. The ancillary controller with the smaller sample time is more effective in rejecting the disturbance. $\square$

3.7 Notes

There is now a considerable volume of research on robust MPC; for a review of the literature up to 2000 see Mayne, Rawlings, Rao, and Scokaert (2000). Early literature examines robustness of nominal MPC under perturbations in Scokaert, Rawlings, and Meadows (1997), and robustness under model uncertainty in De Nicolao, Magni, and Scattolini (1996), and Magni and Sepulchre (1997). Sufficient conditions for robust stability of nominal MPC with modeling error are provided in Santos and Biegler (1999). Teel (2004) provides an excellent discussion of the interplay between nominal robustness and continuity of the Lyapunov function, and also presents some illuminating examples of nonrobust MPC.

The limitations of nominal MPC when uncertainty is present motivated the introduction of feedback, or closed-loop, MPC in which the decision variable is a *policy*, i.e., a sequence of control laws, rather than a sequence of control actions (Mayne, 1995; Kothare, Balakrishnan, and Morari, 1996; Mayne, 1997; Lee and Yu, 1997; Scokaert and Mayne, 1998). With this formulation, the implicit MPC control law can be the same as the receding horizon control law obtained by DP. See Section 3.3.4 and papers such as Magni, De Nicolao, Scattolini, and Allgöwer (2003), where a H_∞ MPC control law is obtained. But such results are *conceptual* because the decision variable is infinite dimensional. Hence practical controllers employ suboptimal policies that are finitely parameterized, an extreme example being nominal MPC. To avoid constraint violation, suboptimal MPC often requires tightening of the constraints in the optimal control problem solved online (Michalska and Mayne, 1993; Chisci, Rossiter, and Zappa, 2001; Mayne and Langson, 2001). Of particular interest is the demonstration in Marruedo, Álamo, and Camacho (2002) that using a sequence of nested constraint sets yields input-to-state stability of nominal MPC if the disturbance is sufficiently small. This procedure was extended in Grimm, Messina, Tuna, and Teel (2007), who do not require the value function to be continuous and do not require the terminal cost to be a control-Lyapunov function. The robust suboptimal controllers discussed in this chapter employ the concept of tubes introduced in the pioneering papers by Bertsekas and Rhodes (1971a,b), and developed for continuous time systems by Aubin (1991) and Khurzhanski and Valyi (1997) and, for linear systems, use a control parameterization proposed by Rossiter, Kouvaritakis, and Rice (1998). Robust positive invariant sets are employed to construct tubes as shown in (Chisci et al., 2001) and (Mayne and Langson, 2001). Useful references are the surveys by Blanchini (1999) and Kolmanovsky and Gilbert (1995), as well as the recent book by Blanchini and Miani (2008). Kolmanovsky and Gilbert (1995) provide an extensive coverage of the theory and computation of minimal and maximal robust (disturbance) invariant sets. The computation of approximations to robust invariant sets that are themselves invariant is discussed in a series of papers by Raković and colleagues (Raković et al., 2003, 2005a; Raković, Mayne, Kerrigan, and Kouramas, 2005b; Kouramas, Raković, Kerrigan, Allwright, and Mayne, 2005). The tube-based controllers described previously are based on the papers (Langson, Chryssochoos, Raković, and Mayne, 2004; Mayne, Seron, and Raković, 2005).

Because robust MPC is still an active area of research, other methods for achieving robustness have been proposed. Diehl, Bock, and Kostina (2006) simplify the robust nonlinear MPC problem by using linearization, also employed in (Nagy and Braatz, 2004), and present some efficient numerical procedures to determine an approximately optimal control sequence. Goulart, Kerrigan, and Maciejowski (2006) propose a control that is an affine function of current and past states; the decision variables are the associated parameters. This method subsumes the tube-based controllers described in this chapter and has the advantage that a separate nominal trajectory is not required. A disadvantage is the increased complexity of the decision variable, although an efficient computational procedure that reduces computational time per iteration from $O(N^6)$ to $O(N^3)$ has been developed by Goulart, Kerrigan, and Ralph (2008).

Considerable attention has recently been given to input-to-state stability of uncertain systems. Thus Limon, Alamo, Raimondo, de la Peña, Bravo, and Camacho (2008) present the theory of input-to-state stability as a unifying framework for robust MPC, generalizes the tube-based MPC described in (Langson et al., 2004), and extends existing results on min-max MPC. Another example of research in this vein is the paper by Lazar, de la Peña, Hemeels, and Alamo (2008) that utilizes input-to-state practical stability to establish robust stability of feedback min-max MPC. A different approach is described by Angeli, Casavola, Franże, and Mosca (2008) where it is shown how to construct, for each time i, an ellipsoidal inner approximation $\mathcal{E}_i$ to the set $\mathcal{T}_i$ of states that can be robustly steered in i steps to a robust control invariant set $\mathcal{T}$. All that is required from the online controller is the determination of the minimum i such that the current state x lies in $\mathcal{X}_i$ and a control that steers $x \in \mathcal{X}_i$ into the set $\mathcal{X}_{i-1} \subset \mathcal{X}_i$.

3.8 Exercises

Exercise 3.1: Removing the outer min in a min-max problem

Show that $V_i^0 : \mathcal{X}_i \to \mathbb{R}$ and $\kappa_i : \mathcal{X}_i \to \mathbb{U}$ defined by

$$V_i^0(x) = \min_{u\in\mathbb{U}} \max_{w\in\mathbb{W}} \{\ell(x,u,w) + V_{i-1}^0(f(x,u,w)) \mid f(x,u,\mathbb{W}) \subset \mathcal{X}_{i-1}\}$$

$$\kappa_i(x) = \arg\min_{u\in\mathbb{U}} \max_{w\in\mathbb{W}} \{\ell(x,u,w) + V_{i-1}^0(f(x,u,w)) \mid f(x,u,\mathbb{W}) \subset \mathcal{X}_{i-1}\}$$

$$\mathcal{X}_i = \{x \in \mathbb{X} \mid \exists u \in \mathbb{U} \text{ such that } f(x,u,\mathbb{W}) \subset \mathcal{X}_{i-1}\}$$

satisfy

$$V_i^0(x) = \max_{w\in\mathbb{W}} \{\ell(x,\kappa_i(x),w) + V_{i-1}^0(f(x,\kappa_i(x),w))\}$$

Exercise 3.2: Maximizing a difference

Prove the claim used in the proof of Theorem 3.10 that

$$\max_w\{a(w)\} - \max_w\{b(w)\} \leq \max_w\{a(w) - b(w)\}$$

Also show the following minimization version

$$\min_w\{a(w)\} - \min_w\{b(w)\} \geq \min_w\{a(w) - b(w)\}$$

Exercise 3.3: Equivalent constraints

Assuming that S is a polytope and, therefore, defined by linear inequalities, show that the constraint $x \in \{z\} \oplus S$ (on z for given x) may be expressed as $Bz \leq b + Bx$, i.e., z must lie in a polytope. If S is symmetric ($x \in S$ implies $-x \in S$), show that $x \in \{z\} \oplus S$ is equivalent to $z \in \{x\} \oplus S$.

Exercise 3.4: Hausdorff distance between translated sets

Prove that the Hausdorff distance between two sets $\{x\} \oplus S$ and S, where S is a compact subset of $\mathbb{R}^n$ and x and y are points in $\mathbb{R}^n$, is $|x - y|$.

Exercise 3.5: Exponential convergence of $X(i)$

Prove that the sequence of sets $\{X(i)\}$ defined in (3.44) by $X(i) := \{z^*(x(i))\} \oplus S$ converges exponentially to the set S.

Exercise 3.6: Exponential stability of composite system

Show that the set $S_K(\infty) \times \{0\}$ is exponentially stable with a region of attraction $S_K(\infty) \times Z_N$ for the composite system described by (3.31) and (3.32).

Exercise 3.7: Simulating a robust MPC controller

This exercise explores robust MPC for linear systems with an additive bounded disturbance

$$x^+ = Ax + Bu + w$$

The first task, using the tube-based controller described in Section 3.4.3 is to determine state and control constraint sets $\mathbb{Z}$ and $\mathbb{V}$ such that if the nominal system $z^+ = Az + Bv$ satisfies $z \in \mathbb{Z}$ and $v \in \mathbb{V}$, then the actual system $x^+ = Ax + Bu + w$ with $u = v + K(x - z)$ where K is such that $A + BK$ is strictly stable, satisfies the constraints $x \in \mathbb{X}$ and $u \in \mathbb{U}$.

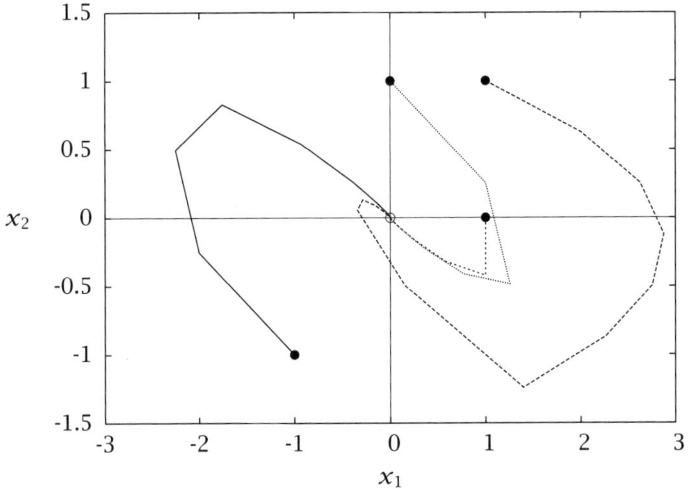

Figure 3.9: Closed-loop robust MPC state evolution with $|w| \leq 0.1$ from four different x_0.

(a) To get started, consider the scalar system

$$x^+ = x + u + w$$

with constraint sets $\mathbb{X} = \{x \mid x \leq 2\}$, $\mathbb{U} = \{u \mid |u| \leq 1\}$ and $\mathbb{W} = \{w \mid |w| \leq 0.1\}$. Choose $K = -(1/2)$ so that $A_K = 1/2$. Determine $\mathbb{Z}$ and $\mathbb{V}$ so that if the nominal system $z^+ = z + v$ satisfies $z \in \mathbb{Z}$ and $v \in \mathbb{V}$, the uncertain system $x^+ = Ax + Bu + w$, $u = v + K(x - z)$ satisfies $x \in \mathbb{X}$, $u \in \mathbb{U}$.

(b) Repeat part (a) for the following uncertain system

$$x^+ = \begin{bmatrix} 1 & 1 \\ 0 & 1 \end{bmatrix} x + \begin{bmatrix} 0 \\ 1 \end{bmatrix} u + w$$

with the constraint sets $\mathbb{X} = \{x \in \mathbb{R}^2 \mid x_1 \leq 2\}$, $\mathbb{U} = \{u \in \mathbb{R} \mid |u| \leq 1\}$ and $\mathbb{W} = [-0.1, 0.1]$. Choose $K = \begin{bmatrix} -0.4 & -1.2 \end{bmatrix}$.

(c) Determine a model predictive controller for the nominal system and constraint sets $\mathbb{Z}$ and $\mathbb{V}$ used in (b).

(d) Implement robust MPC for the uncertain system and simulate the closed loop system for a few initial states and a few disturbance sequences for each initial state. The phase plot for initial states $[-1, -1]$, $[1, 1]$, $[1, 0]$ and $[0, 1]$ should resemble Figure 3.9.

Bibliography

D. Angeli, A. Casavola, G. Franže, and E. Mosca. An ellipsoidal off-line MPC scheme for uncertain polytopic discrete-time systems. *Automatica*, 44: 3113–3119, 2008.

J. P. Aubin. *Viability Theory*. Systems & Control: Foundations & Applications. Birkhauser, Boston, Basel, Berlin, 1991.

D. P. Bertsekas and I. B. Rhodes. Recursive state estimation for a set-membership description of uncertainty. *IEEE Trans. Auto. Cont.*, 16:117–128, 1971a.

D. P. Bertsekas and I. B. Rhodes. On the minimax reachability of target sets and target tubes. *Automatica*, 7(2):233–247, 1971b.

F. Blanchini. Set invariance in control. *Automatica*, 35:1747–1767, 1999.

F. Blanchini and S. Miani. *Set-Theoretic methods in Control*. Systems & Control: Foundations and Applications. Birkhäuser, 2008.

L. Chisci, J. A. Rossiter, and G. Zappa. Systems with persistent disturbances: predictive control with restricted constraints. *Automatica*, 37(7):1019–1028, 2001.

G. De Nicolao, L. Magni, and R. Scattolini. Robust predictive control of systems with uncertain impulse response. *Automatica*, 32(10):1475–1479, 1996.

M. Diehl, H. G. Bock, and E. Kostina. An approximation technique for robust nonlinear optimization. *Math. Prog.*, 107:213–230, 2006. Series B.

P. J. Goulart, E. C. Kerrigan, and J. M. Maciejowski. Optimization over state feedback policies for robust control with constraints. *Automatica*, 42:523–533, 2006.

P. J. Goulart, E. C. Kerrigan, and D. Ralph. Efficient robust optimization for robust control with constraints. *Math. Prog.*, 114(1):115–147, July 2008.

M. Green and D. J. N. Limebeer. *Linear Robust Control*. Prentice-Hall, Englewood Cliffs, New Jersey 07632, 1995.

G. Grimm, M. J. Messina, S. E. Tuna, and A. R. Teel. Nominally robust model predictive control with state constraints. *IEEE Trans. Auto. Cont.*, 52(10): 1856–1870, October 2007.

G. A. Hicks and W. H. Ray. Approximation methods for optimal control synthesis. *Can. J. Chem. Eng.*, 49:522–528, August 1971.

S. Kameswaran and L. T. Biegler. Simultaneous dynamic optimization strategies: Recent advances and challenges. *Comput. Chem. Eng.*, 30:1560–1575, September 2006.

C. M. Kellet and A. R. Teel. Discrete-time asymptotic controllability implies smooth control-Lyapunov function. *Sys. Cont. Let.*, 52:349–359, 2004.

A. B. Khurzhanski and I. Valyi. *Ellipsoidal-valued dynamics for estimation and control.* Systems & Control: Foundations & Applications. Birkhauser, Boston, Basel, Berlin, 1997.

I. Kolmanovsky and E. G. Gilbert. Maximal output admissible sets for discrete-time systems with disturbance inputs. In *Proceedings of the American Control Conference*, Seattle, June 1995.

I. Kolmanovsky and E. G. Gilbert. Theory and computation of disturbance invariant sets for discrete-time linear systems. *Math. Probl. Eng.*, 4(4):317–367, 1998.

M. V. Kothare, V. Balakrishnan, and M. Morari. Robust constrained model predictive control using linear matrix inequalities. *Automatica*, 32(10):1361–1379, 1996.

K. I. Kouramas, S. V. Raković, E. C. Kerrigan, J. C. Allwright, and D. Q. Mayne. On the minimal robust positively invariant set for linear difference inclusions. In *Proceedings of the 44th IEEE Conference on Decision and Control and European Control Conference ECC 2005*, pages 2296–2301, Sevilla, Spain, December 2005.

W. Langson, I. Chryssochoos, S. V. Raković, and D. Q. Mayne. Robust model predictive control using tubes. *Automatica*, 40:125–133, January 2004.

M. Lazar, D. M. de la Peña, W. P. M. H. Hemeels, and T. Alamo. On input-to-state stability of min-max nonlinear model predictive control. *Sys. Cont. Let.*, 57: 39–48, 2008.

J. H. Lee and Z. Yu. Worst-case formulations of model predictive control for systems with bounded parameters. *Automatica*, 33(5):763–781, 1997.

D. Limon, T. Alamo, D. M. Raimondo, D. M. de la Peña, J. M. Bravo, and E. F. Camacho. Input-to-state stability: an unifying framework for robust model predictive control. In L. Magni, D. M. Raimondo, and F. Allgöwer, editors, *International Workshop on Assessment and Future Directions of Nonlinear Model Predictive Control*, Pavia, Italy, September 2008.

L. Magni and R. Sepulchre. Stability margins of nonlinear receding-horizon control via inverse optimality. *Sys. Cont. Let.*, 32:241–245, 1997.

L. Magni, G. De Nicolao, R. Scattolini, and F. Allgöwer. Robust model predictive control for nonlinear discrete-time systems. *Int. J. Robust and Nonlinear Control*, 13:229–246, 2003.

D. L. Marruedo, T. Álamo, and E. F. Camacho. Stability analysis of systems with bounded additive uncertainties based on invariant sets: stability and feasibility of MPC. In *Proceedings of the American Control Conference*, pages 364–369, Anchorage, Alaska, May 2002.

D. Q. Mayne. Optimization in model based control. In *Proceedings of the IFAC Symposium Dynamics and Control of Chemical Reactors, Distillation Columns and Batch Processes*, pages 229–242, Helsingor, Denmark, June 1995.

D. Q. Mayne. Nonlinear model predictive control: An assessment. In J. C. Kantor, C. E. García, and B. Carnahan, editors, *Proceedings of Chemical Process Control - V*, pages 217–231. CACHE, AIChE, 1997.

D. Q. Mayne and W. Langson. Robustifying model predictive control of constrained linear systems. *Electron. Lett.*, 37(23):1422–1423, 2001.

D. Q. Mayne, J. B. Rawlings, C. V. Rao, and P. O. M. Scokaert. Constrained model predictive control: Stability and optimality. *Automatica*, 36(6):789–814, 2000.

D. Q. Mayne, M. M. Seron, and S. V. Raković. Robust model predictive control of constrained linear systems with bounded disturbances. *Automatica*, 41 (2):219–224, February 2005.

H. Michalska and D. Q. Mayne. Robust receding horizon control of constrained nonlinear systems. *IEEE Trans. Auto. Cont.*, 38(11):1623–1633, 1993.

Z. Nagy and R. Braatz. Open-loop and closed-loop robust optimal control of batch processes using distributional and worst-case analysis. *J. Proc. Cont.*, pages 411–422, 2004.

S. V. Raković, E. C. Kerrigan, K. I. Kouramas, and D. Q. Mayne. Approximation of the minimal robustly positively invariant set for discrete-time LTI systems with persistent state disturbances. In *Proceedings 42nd IEEE Conference on Decision and Control*, volume 4, pages 3917–3918, Maui, Hawaii, USA, December 2003.

S. V. Raković, E. C. Kerrigan, K. I. Kouramas, and D. Q. Mayne. Invariant approximations of the minimal robustly positively invariant sets. *IEEE Trans. Auto. Cont.*, 50(3):406–410, 2005a.

S. V. Raković, D. Q. Mayne, E. C. Kerrigan, and K. I. Kouramas. Optimized robust control invariant sets for constrained linear discrete-time systems. In *Proceedings of 16th IFAC World Congress on Automatic Control*, Prague, Czechoslavakia, 2005b.

J. A. Rossiter, B. Kouvaritakis, and M. J. Rice. A numerically robust state-space approach to stable-predictive control strategies. *Automatica*, 34(1):65–73, 1998.

L. O. Santos and L. T. Biegler. A tool to analyze robust stability for model predictive control. *J. Proc. Cont.*, 9:233–245, 1999.

P. O. M. Scokaert and D. Q. Mayne. Min-max feedback model predictive control for constrained linear systems. *IEEE Trans. Auto. Cont.*, 43(8):1136–1142, August 1998.

P. O. M. Scokaert, J. B. Rawlings, and E. S. Meadows. Discrete-time stability with perturbations: Application to model predictive control. *Automatica*, 33(3): 463–470, 1997.

A. R. Teel. Discrete time receding horizon control: is the stability robust. In Marcia S. de Queiroz, Michael Malisoff, and Peter Wolenski, editors, *Optimal control, stabilization and nonsmooth analysis*, volume 301 of *Lecture notes in control and information sciences*, pages 3–28. Springer, 2004.

4

State Estimation

4.1 Introduction

We now turn to the general problem of estimating the state of a noisy dynamic system given noisy measurements. We assume that the system generating the measurements is given by

$$\begin{aligned} x^+ &= f(x, w) \\ y &= h(x) + v \end{aligned} \tag{4.1}$$

in which the process disturbance, w, measurement disturbance, v, and system initial state, $x(0)$, are independent random variables with stationary probability densities. One of our main purposes is to provide a state estimate to the MPC regulator as part of a feedback control system, in which case the model changes to $x^+ = f(x, u, w)$ with both process disturbance w and control input u. But state estimation is a general technique that is often used in monitoring applications without any feedback control. In Chapter 5, we discuss the combined use of state estimation with MPC regulation. In this chapter we consider state estimation as an independent subject. For notational convenience, we often neglect the control input u as part of the system model in this chapter.

4.2 Full Information Estimation

Of all the estimators considered in this chapter, full information estimation will prove to have the best theoretical properties in terms of stability and optimality. Unfortunately, it will also prove to be computationally intractable except for the simplest cases, such as a linear system model. Its value therefore lies in clearly defining what is *desirable* in a state estimator. One method for practical estimator design

	System variable	Decision variable	Optimal decision
state	x	χ	$\hat{x}$
process disturbance	w	ω	$\hat{w}$
measured output	y	η	$\hat{y}$
measurement disturbance	v	ν	$\hat{v}$

Table 4.1: System and state estimator variables.

therefore is to come as close as possible to the properties of full information estimation while maintaining a tractable online computation. This design philosophy leads directly to moving horizon estimation (MHE).

First we define some notation necessary to distinguish the system variables from the estimator variables. We have already introduced the system variables (x, w, y, v). In the estimator optimization problem, these have corresponding decision variables, which we denote $(\chi, \omega, \eta, \nu)$. The *optimal* decision variables are denoted $(\hat{x}, \hat{w}, \hat{y}, \hat{v})$ and these optimal decisions are the estimates provided by the state estimator. This notation is summarized in Table 4.1. Next we summarize the relationships between these variables

$$\begin{aligned} x^+ &= f(x, w) & y &= h(x) + v \\ \chi^+ &= f(\chi, \omega) & y &= h(\chi) + \nu \\ \hat{x}^+ &= f(\hat{x}, \hat{w}) & y &= h(\hat{x}) + \hat{v} \end{aligned}$$

Notice that it is always the system measurement y that appears in the second column of equations. We can also define the decision variable output, $\eta = h(\chi)$, but notice that ν measures the fitting error, $\nu = y - h(\chi)$, and we must use the system measurement y and not η in this relationship. Therefore, we do not satisfy a relationship like $\eta = h(\chi) + \nu$, but rather

$$\begin{aligned} y &= h(\chi) + \nu & \eta &= h(\chi) \\ y &= h(\hat{x}) + \hat{v} & \hat{y} &= h(\hat{x}) \end{aligned}$$

We begin with a reasonably general definition of the full information estimator that produces an estimator that is *stable*, which we also shall define subsequently. The full information objective function is

$$V_T(\chi(0), \boldsymbol{\omega}) = \ell_x(\chi(0) - \overline{x}_0) + \sum_{i=0}^{T-1} \ell_i(\omega(i), \nu(i)) \tag{4.2}$$

subject to

$$\chi^+ = f(\chi, \omega) \qquad y = h(\chi) + \nu$$

in which T is the current time, $y(i)$ is the measurement at time i, and $\overline{x}_0$ is the prior information on the initial state.[1] Because $\nu = y - h(\chi)$ is the error in fitting the measurement y, $\ell_i(\omega, \nu)$ costs the model disturbance and the fitting error. These are the two error sources we reconcile in all state estimation problems.

The full information estimator is then defined as the solution to

$$\min_{\chi(0), \boldsymbol{\omega}} V_T(\chi(0), \boldsymbol{\omega}) \tag{4.3}$$

We denote the solution as $\hat{x}(0|T)$, $\hat{w}(i|T), 0 \leq i \leq T-1$, $T \geq 1$, and the optimal cost as V_T^0. We also use $\hat{x}(T) := \hat{x}(T|T)$ to simplify the notation. The optimal solution and cost also depend on the measurement sequence $\mathbf{y}$, and the prior $\overline{x}_0$, but this dependency is made explicit only when necessary. The choice of stage costs $\ell_x(\cdot)$ and $\ell_i(\cdot)$ is made after we define the class of disturbances affecting the system.

The next order of business is to decide what class of systems to consider if the goal is to obtain a stable state estimator. A standard choice in most nonlinear estimation literature is to assume system observability. The drawback with this choice is that it is overly restrictive for even linear systems. As discussed in Chapter 1, for linear systems we require only detectability for stable estimation (Exercise 1.33). We therefore start instead with an assumption of detectability that is appropriate for nonlinear systems. First we require the definition of i-IOSS (Sontag and Wang, 1997)

Definition 4.1 (i-IOSS). The system $x^+ = f(x, w), y = h(x)$ is *incrementally input/output-to-state stable* (i-IOSS) if there exists some $\beta(\cdot) \in \mathcal{KL}$ and $\gamma_1(\cdot), \gamma_2(\cdot) \in \mathcal{K}$ such that for every two initial states z_1 and

[1] Notice we have dropped the final measurement $y(T)$ compared to the problem considered in Chapter 1 to formulate the prediction form rather than the filtering form of the state estimation problem. This change is purely for notational convenience, and all results developed in this chapter can also be expressed in the filtering form of MHE.

z_2, and any two disturbance sequences $\mathbf{w}_1$ and $\mathbf{w}_2$

$$|x(k; z_1, \mathbf{w}_1) - x(k; z_2, \mathbf{w}_2)| \leq \beta(|z_1 - z_2|, k) + \gamma_1(\|\mathbf{w}_1 - \mathbf{w}_2\|_{0:k-1}) + \gamma_2(\|\mathbf{y}_{z_1,\mathbf{w}_1} - \mathbf{y}_{z_2,\mathbf{w}_2}\|_{0:k})$$

The notation $x(k; x_0, \mathbf{w})$ denotes the solution to $x^+ = f(x, w)$ satisfying initial condition $x(0) = x_0$ with disturbance sequence $\mathbf{w} = \{w(0), w(1), \ldots\}$. We also require the system with an "initial" condition at a time k_1 other than $k_1 = 0$, and use the notation $x(k; x_1, k_1, \mathbf{w})$ to denote the solution to $x^+ = f(x, w)$ satisfying the condition $x(k_1) = x_1$ with disturbance sequence $\mathbf{w} = \{w(0), w(1), \ldots\}$.

One of the most important and useful implications of the i-IOSS property is the following proposition.

Proposition 4.2 (Convergence of state under i-IOSS). *If system $x^+ = f(x, w), y = h(x)$ is i-IOSS, $w_1(k) \to w_2(k)$ and $y_1(k) \to y_2(k)$ as $k \to \infty$, then*

$$x(k; z_1, \mathbf{w}_1) \to x(k; z_2, \mathbf{w}_2) \qquad \text{for all } z_1, z_2$$

The proof of this proposition is discussed in Exercise 4.3.

The class of disturbances (w, v) affecting the system is defined next. Often we assume these are random variables with stationary probability densities, and often zero-mean normal densities. When we wish to establish estimator stability, however, we wish to show that if the disturbances affecting the measurement converge to zero, then the estimate error also converges to zero. So here we restrict attention to *convergent* disturbances.

Assumption 4.3 (Convergent disturbances). The sequence $(w(k), v(k))$ for $k \in \mathbb{I}_{\geq 0}$ are bounded and converge to zero as $k \to \infty$.

Remark 4.4 (Summable disturbances). If the disturbances satisfy Assumption 4.3, then there exists a $\mathcal{K}$-function $\gamma_w(\cdot)$ such that the disturbances are summable

$$\sum_{i=0}^{\infty} \gamma_w(|(w(i), v(i))|) \quad \text{is bounded}$$

See Sontag (1998b, Proposition 7) for a statement and proof of this result.[2]

[2] This result is also useful in establishing the converse Lyapunov function theorem for asymptotic stability as discussed in Exercise B.4 of Appendix B.

Given this class of disturbances, the estimator stage cost is chosen to satisfy the following property.

Assumption 4.5 (Positive definite stage cost)**.** The initial state cost and stage costs are continuous, bounded, and positive definite. The costs satisfy the following inequalities for all $x \in \mathbb{R}^n$, $w \in \mathbb{R}^g$, and v in $\mathbb{R}^p$

$$\underline{\gamma}_x(|x|) \leq \ell_x(x) \qquad \leq \gamma_x(|x|) \tag{4.4}$$

$$\underline{\gamma}_w(|(w,v)|) \leq \ell_i(w,v) \leq \gamma_w(|(w,v)|) \qquad i \geq 0 \tag{4.5}$$

in which $\underline{\gamma}_x, \underline{\gamma}_w, \gamma_x, \gamma_w \in \mathcal{K}_\infty$ and γ_w is defined in Remark 4.4.

Notice that if we change the class of disturbances affecting the system, we may also have to change the stage cost in the state estimator to satisfy $\ell_i(w,v) \leq \gamma_w(|(w,v)|)$ in (4.5). The standard stage cost is the quadratic function, but slowly decaying disturbances in the data require "stronger" than quadratic stage costs to ensure summability. An interaction between anticipated disturbances affecting the system and choice of stage cost in the state estimator is hardly surprising, but Remark 4.4 and Assumption 4.5 make the requirements explicit.

Next we define estimator stability. Again, because the system is nonlinear, we must define stability of a solution. Consider the zero estimate error solution for all $k \geq 0$. This solution arises when the system's initial state is equal to the estimator's prior and there are zero disturbances, $x_0 = \overline{x}_0$, $(w(i), v(i)) = 0$ all $i \geq 0$. In this case, the optimal solution to the full information problem is $\hat{x}(0|T) = \overline{x}_0$ and $\hat{w}(i|T) = 0$ for all $0 \leq i \leq T$, $T \geq 1$, which also gives perfect agreement of estimate and measurement $h(\hat{x}(i|T)) = y(i)$ for $0 \leq i \leq T$, $T \geq 1$. The perturbation to this solution are: the system's initial state (distance from $\overline{x}_0$), and the process and measurement disturbances. We next define stability properties so that *asymptotic stability* considers the case $x_0 \neq \overline{x}_0$ with zero disturbances, and *robust stability* considers the case in which $(w(i), v(i)) \neq 0$.

Definition 4.6 (Global asymptotic stability)**.** The estimate is based on the *noise-free* measurement $\mathbf{y} = h(\mathbf{x}(x_0, 0))$. The estimate is (nominally) globally asymptotically stable (GAS) if there exists a $\mathcal{KL}$-function $\beta(\cdot)$ such that for all $x_0, \overline{x}_0$ and $k \in \mathbb{I}_{\geq 0}$

$$|x(k; x_0, 0) - \hat{x}(k)| \leq \beta(|x_0 - \overline{x}_0|, k)$$

It bears mentioning that the standard definition of estimator stability for *linear systems* is consistent with Definition 4.6.

Definition 4.7 (Robust global asymptotic stability)**.** The estimate is based on the *noisy* measurement $\mathbf{y} = h(\mathbf{x}(x_0, \mathbf{w})) + \mathbf{v}$. The estimate is robustly GAS if for all x_0 and $\overline{x}_0$, and $(\mathbf{w}, \mathbf{v})$ satisfying Assumption 4.3, the following hold.

(a) The estimate converges to the state; as $k \to \infty$

$$\hat{x}(k) \to x(k; x_0, \mathbf{w})$$

(b) For every $\varepsilon > 0$ there exists $\delta > 0$ such that

$$\gamma_x(|x_0 - \overline{x}_0|) + \sum_{i=0}^{\infty} \gamma_w\big(\,|(w(i), v(i))|\,\big) \leq \delta \tag{4.6}$$

implies $|x(k; x_0, \mathbf{w}) - \hat{x}(k)| \leq \varepsilon$ for all $k \in \mathbb{I}_{\geq 0}$.

The first part of the definition ensures that converging disturbances lead to converging estimates. The second part provides a bound on the transient estimate error given a bound on the disturbances. Note also that robust GAS implies GAS (see also Exercise 4.9). With the pieces in place, we can state the main result of this section.

Theorem 4.8 (Robust GAS of full information estimates)**.** *Given an i-IOSS (detectable) system and measurement sequence generated by* (4.1) *with disturbances satisfying Assumption 4.3, then the full information estimate with stage cost satisfying Assumption 4.5 is robustly GAS.*

Proof.

(a) First we establish that the full information cost is bounded for all $T \geq 1$ including $T = \infty$. Consider a candidate set of decision variables

$$\chi(0) = x_0 \qquad \omega(i) = w(i) \quad 0 \leq i \leq T-1$$

The full information cost for this choice is

$$V_T(\chi(0), \boldsymbol{\omega}) = \ell_x(x_0 - \overline{x}_0) + \sum_{i=0}^{T-1} \ell_i(w(i), v(i))$$

From Remark 4.4, the sum is bounded for all T including the limit $T = \infty$. Therefore, let V_∞ be an upper bound for the right-hand side. The optimal cost exists for all $T \geq 0$ because V_T is a continuous function and goes to infinity as any of its arguments goes to infinity due to the lower bounds in Assumption 4.5. Next we show that the optimal cost sequence converges. Evaluate the cost at time $T-1$ using the optimal solution from time T. We have that

$$V_{T-1}(\hat{x}(0|T), \hat{\mathbf{w}}_T) = V_T^0 - \ell_T(\hat{w}(T|T), \hat{v}(T|T))$$

Optimization at time $T-1$ can only improve the cost giving

$$V_T^0 \geq V_{T-1}^0 + \ell_T(\hat{w}(T|T), \hat{v}(T|T))$$

and we see that the optimal sequence $\{V_T^0\}$ is nondecreasing and bounded above by V_∞. Therefore the sequence converges and the convergence implies

$$\ell_T(\hat{w}(T|T), \hat{v}(T|T)) \to 0$$

as $T \to \infty$. The lower bound in (4.5) then gives that $\hat{v}(T) = y(T) - h(\hat{x}(T|T)) \to 0$ and $\hat{w}(T|T) \to 0$ as $T \to \infty$. Since the measurement satisfies $y = h(x) + v$, and $v(T)$ converges to zero, we have that

$$h(x(T)) - h(\hat{x}(T|T)) \to 0 \qquad \hat{w}(T|T) \to 0 \qquad T \to \infty$$

Because the system is i-IOSS, we have the following inequality for all x_0, $\hat{x}(0|k)$, $\mathbf{w}$, $\hat{\mathbf{w}}_k$, and $k \geq 0$,

$$\begin{aligned} \left| x(k; x_0, \mathbf{w}) - x(k; \hat{x}(0|k), \hat{\mathbf{w}}_k) \right| \leq \beta(\left| x_0 - \hat{x}(0|k) \right|, k) + \\ \gamma_1\left(\left\| \mathbf{w} - \hat{\mathbf{w}}_k \right\|_{0:k-1} \right) + \gamma_2\left(\left\| h(\mathbf{x}) - h(\hat{\mathbf{x}}_k) \right\|_{0:k} \right) \end{aligned} \tag{4.7}$$

Since $w(k)$ converges to zero, $w(k) - \hat{w}(k)$ converges to zero, and $h(x(k)) - h(\hat{x}(k))$ converges to zero. From Proposition 4.2 we conclude that $\left| x(k; x_0, \mathbf{w}) - x(k; \hat{x}(0|k), \hat{\mathbf{w}}_k) \right|$ converges to zero. Since the state estimate is $\hat{x}(k) := x(k; \hat{x}(0|k), \hat{\mathbf{w}}_k)$ and the state is $x(k) = x(k; x_0, \mathbf{w})$, we have that

$$\hat{x}(k) \to x(k) \qquad k \to \infty$$

and the estimate converges to the system state. This establishes part (a) of the robust GAS definition.[3]

(b) Assume that (4.6) holds for some arbitrary $\delta > 0$. This gives immediately an upper bound on the optimal full information cost function for all T, $0 \leq T \leq \infty$, i.e, $V_\infty = \delta$. We then have the following bounds on the initial state estimate for all $k \geq 0$, and the initial state

$$\underline{\gamma}_x(|\hat{x}(0|k) - \overline{x}_0|) \leq \delta \qquad \gamma_x(|x_0 - \overline{x}_0|) \leq \delta$$

---

[3]It is not difficult to extend this argument to conclude $\hat{x}(i|k) \to x(i; x_0, \mathbf{w})$ as $k \to \infty$ for $k - N \leq i \leq k$ and any finite $N \geq 0$.

These two imply a bound on the initial estimate error, $|x_0 - \hat{x}(0|k)| \leq \underline{\gamma}_x^{-1}(\delta) + \gamma_{\overline{x}}^{-1}(\delta)$. The process disturbance bounds are for all $k \geq 0$, $0 \leq i \leq k$

$$\underline{\gamma}_w(|\hat{w}(i|k)|) \leq \delta \qquad \gamma_w(|w(k)|) \leq \delta$$

and we have that $|w(i) - \hat{w}(i|k)| \leq \underline{\gamma}_w^{-1}(\delta) + \gamma_{\overline{w}}^{-1}(\delta)$. A similar argument gives for the measurement disturbance $|v(i) - \hat{v}(i|k)| \leq \underline{\gamma}_w^{-1}(\delta) + \gamma_{\overline{w}}^{-1}(\delta)$. Since $-(v(i) - \hat{v}(i|k)) = h(x(i)) - h(\hat{x}(i|k))$, we have that

$$|h(x(i)) - h(\hat{x}(i|k))| \leq \underline{\gamma}_w^{-1}(\delta) + \gamma_{\overline{w}}^{-1}(\delta)$$

We substitute these bounds in (4.7) and obtain for all $k \geq 0$

$$|x(k) - \hat{x}(k)| \leq \overline{\beta}(\underline{\gamma}_x^{-1}(\delta) + \gamma_{\overline{x}}^{-1}(\delta)) + (\gamma_1 + \gamma_2)(\underline{\gamma}_w^{-1}(\delta) + \gamma_{\overline{w}}^{-1}(\delta))$$

in which $\overline{\beta}(s) := \beta(s, 0)$ is a $\mathcal{K}$-function. Finally we choose δ such that the right-hand side is less than ε, which is possible since the right-hand side defines a $\mathcal{K}$-function, which goes to zero with δ. This gives for all $k \geq 0$

$$|x(k) - \hat{x}(k)| \leq \varepsilon$$

and part (b) of the robust GAS definition is established. ■

4.2.1 State Estimation as Optimal Control of Estimate Error

Given the many structural similarities between estimation and regulation, the reader may wonder why the stability analysis of the full information estimator presented in the previous section looks rather different than the zero-state regulator stability analysis presented in Chapter 2. To provide some insight into essential *differences*, as well as similarities, between estimation and regulation, consider again the estimation problem in the simplest possible setting with a linear time invariant model and Gaussian noise

$$\begin{aligned} x^+ &= Ax + Gw \qquad & w &\sim N(0, Q) \\ y &= Cx + v \qquad & v &\sim N(0, R) \end{aligned} \tag{4.8}$$

and random initial state $x(0) \sim N(\overline{x}(0), P^-(0))$. In full information estimation, we define the objective function

$$V_T(\chi(0), \boldsymbol{\omega}) = \frac{1}{2}\left(|\chi(0) - \overline{x}(0)|^2_{(P^-(0))^{-1}} + \sum_{i=0}^{T-1} |\omega(i)|^2_{Q^{-1}} + |\nu(i)|^2_{R^{-1}} \right)$$

subject to $\chi^+ = A\chi + G\omega$, $y = C\chi + v$. Denote the solution to this optimization as

$$(\hat{x}(0|T), \hat{\mathbf{w}}_T) = \arg\min_{\chi(0),\boldsymbol{\omega}} V_T(\chi(0), \boldsymbol{\omega})$$

and the trajectory of state estimates comes from the model $\hat{x}(i+1|T) = A\hat{x}(i|T) + G\hat{w}(i|T)$. We define estimate error as $\tilde{x}(i|T) = x(i) - \hat{x}(i|T)$ for $0 \leq i \leq T-1$, $T \geq 1$.

Because the system is *linear*, the estimator is stable if and only if it is stable with zero process and measurement disturbances. So analyzing stability is equivalent to the following simpler question. If noise-free data are provided to the estimator, $(w(i), v(i)) = 0$ for all $i \geq 0$ in (4.8), is the estimate error asymptotically stable as $T \to \infty$ for all x_0? We next make this statement precise. First we note that the noise-free measurement satisfies $y(i) - C\hat{x}(i|T) = C\tilde{x}(i|T), 0 \leq i \leq T$ and the initial condition term can be written in estimate error as $\hat{x}(0) - \overline{x}(0) = -(\tilde{x}(0) - a)$ in which $a = x(0) - \overline{x}(0)$. For the noise-free measurement we can therefore rewrite the cost function as

$$V_T(a, \tilde{x}(0), \mathbf{w}) = \frac{1}{2}\left(\left|\tilde{x}(0) - a\right|^2_{(P^-(0))^{-1}} + \sum_{i=0}^{T-1} \left|C\tilde{x}(i)\right|^2_{R^{-1}} + \left|w(i)\right|^2_{Q^{-1}} \right) \tag{4.9}$$

in which we list explicitly the dependence of the cost function on parameter a. For estimation we solve

$$\min_{\tilde{x}(0),\mathbf{w}} V_T(a, \tilde{x}(0), \mathbf{w}) \tag{4.10}$$

subject to $\tilde{x}^+ = A\tilde{x} + Gw$. Now consider problem (4.10) as an optimal control problem using w as manipulated variable and minimizing an objective that measures size of estimate error $\tilde{x}$ and control w. We denote the optimal solution as $\tilde{x}^0(0; a)$ and $\mathbf{w}^0(a)$. Substituting these into the model equation gives optimal estimate error $\tilde{x}^0(j|T; a), 0 \leq j \leq T, 0 \leq T$. Parameter a denotes how far $x(0)$, the system's initial state generating the measurement, is from $\overline{x}(0)$, the prior. If we are lucky and $a = 0$, the optimal solution is $(\tilde{x}^0, \mathbf{w}^0) = 0$, and we achieve zero cost in V_T^0 and zero estimate error $\tilde{x}(j|T)$ at all time in the trajectory $0 \leq j \leq T$ for all time $T \geq 1$. The stability analysis in estimation is to show that the origin for $\tilde{x}$ is asymptotically stable. In other words, we wish to show there exists a KL-function β such that $\left|\tilde{x}^0(T; a)\right| \leq \beta(|a|, T)$ for all $T \in \mathbb{I}_{\geq 0}$.

We note the following differences between standard regulation and the estimation problem (4.10). First we see that (4.10) is slightly nonstandard because it contains an extra decision variable, the initial state, and an extra term in the cost function, (4.9). Indeed, without this extra term, the regulator could choose $\widetilde{x}(0) = 0$ to zero the estimate error immediately, choose $\mathbf{w} = 0$, and achieve zero cost in $V_T^0(a)$ for all a. The nonstandard regulator allows $\widetilde{x}(0)$ to be manipulated as a decision variable, but penalizes its distance from a. Next we look at the stability question. The stability analysis is to show there exists KL-function β such that $\left|\widetilde{x}^0(T;a)\right| \leq \beta(|a|, T)$ for all $T \in \mathbb{I}_{\geq 0}$. Here convergence is a question about the terminal state in a sequence of *different* optimal control problems with increasing horizon length T. That is also not the standard regulator convergence question, which asks how the state trajectory evolves using the optimal control law. In standard regulation, we inject the optimal first input and ask whether we are successfully moving the system to the origin as time increases. In estimation, we do not inject anything into the system; we are provided more information as time increases and ask whether our explanation of the data is improving (terminal estimate error is decreasing) as time increases.

Because stability is framed around the behavior of the terminal state, we would not choose *backward* dynamic programming (DP) to solve (4.10), as in standard regulation. We do not seek the optimal first control move as a function of a known initial state. Rather we seek the optimal terminal state $\widetilde{x}^0(T;a)$ as a function of the parameter a appearing in the cost function. This problem is better handled by *forward* DP as discussed in Sections 1.3.2 and 1.4.3 of Chapter 1 when solving the full information state estimation problem. Exercise 4.12 discusses how to solve (4.10); we obtain the following recursion for the optimal terminal state

$$\widetilde{x}^0(k+1;a) = (A - \widetilde{L}(k)C)\,\widetilde{x}^0(k;a) \tag{4.11}$$

for $k \geq 0$. The initial condition for the recursion is $\widetilde{x}^0(0;a) = a$. The time-varying gains $\widetilde{L}(k)$ and associated cost matrices $P^-(k)$ required are

$$\begin{aligned} P^-(k+1) &= GQG' + AP^-(k)A' \\ &\qquad - AP^-(k)C'(CP^-(k)C' + R)^{-1}CP^-(k)A \end{aligned} \tag{4.12}$$

$$\widetilde{L}(k) = AP^-(k)C'(CP^-(k)C' + R)^{-1} \tag{4.13}$$

in which $P^-(0)$ is specified in the problem. As expected, these are the standard estimator recursions developed in Chapter 1. Asymptotic stability of the estimate error can be established by showing that $V(k, \widetilde{x}) := (1/2)\widetilde{x}' P(k)^{-1}\widetilde{x}$ is a Lyapunov function for (4.11) (Jazwinski, 1970, Theorem 7.4). Notice that this Lyapunov function is *not* the optimal cost of (4.10) as in a standard regulation problem. The optimal cost of (4.10), $V_T^0(a)$, is an *increasing* function of T rather than a decreasing function of T as required for a Lyapunov function. Also note that the argument used in Jazwinski (1970) to establish that $V(k, x)$ is a Lyapunov function for the *linear* system is *more complicated* than the argument used in Section 4.2 to prove stability of full information estimation for the *nonlinear* system. Although one can find Lyapunov functions valid for estimation, they do not have the same simple connection to optimal cost functions as in standard regulation problems, even in the linear, unconstrained case. Stability arguments based instead on properties of $V_T^0(a)$ are simpler and more easily adapted to cover new situations arising in research problems. If a Lyapunov function is required for further analysis, a converse theorem guarantees its existence.

4.2.2 Duality of Linear Estimation and Regulation

For linear systems, the estimate error $\widetilde{x}$ in full information and state x in regulation to the origin display an interesting duality that we summarize briefly here. Consider the following steady-state estimation and infinite horizon regulation problems.

Estimator problem.

$$\begin{aligned} x(k+1) &= Ax(k) + Gw(k) \\ y(k) &= Cx(k) + v(k) \end{aligned}$$

$$R > 0 \qquad Q > 0 \qquad (A, C) \text{ detectable} \qquad (A, G) \text{ stabilizable}$$

$$\widetilde{x}(k+1) = \left(A - \widetilde{L}C\right)\widetilde{x}(k)$$

Regulator problem.

$$\begin{aligned} x(k+1) &= Ax(k) + Bu(k) \\ y(k) &= Cx(k) \end{aligned}$$

$$R > 0 \qquad Q > 0 \qquad (A, B) \text{ stabilizable} \qquad (A, C) \text{ detectable}$$

$$x(k+1) = (A + BK)x(k)$$

Regulator	Estimator
A	A'
B	C'
C	G'
k	$l = N - k$
$\Pi(k)$	$P^-(l)$
$\Pi(k-1)$	$P^-(l+1)$
Π	P^-
Q	Q
R	R
P_f	$P^-(0)$
K	$-\widetilde{L}'$
$A + BK$	$(A - \widetilde{L}C)'$
x	$\widetilde{x}'$

Regulator	Estimator
$R > 0, \quad Q > 0$	$R > 0, \quad Q > 0$
(A, B) stabilizable	(A, C) detectable
(A, C) detectable	(A, G) stabilizable

Table 4.2: Duality variables and stability conditions for linear quadratic regulation and least squares estimation.

In Appendix A, we derive the dual dynamic system following the approach in Callier and Desoer (1991), and obtain the duality variables in regulation and estimation listed in Table 4.2.

We also have the following result connecting controllability of the original system and observability of the dual system

Lemma 4.9 (Duality of controllability and observability). *(A, B) is controllable (stabilizable) if and only if (A', B') is observable (detectable).*

This result can be established directly using the Hautus lemma and is left as an exercise. This lemma and the duality variables allows us to translate stability conditions for infinite horizon regulation problems into stability conditions for full information estimation problems and vice versa. For example, the following is a basic theorem covering convergence of Riccati equations in the form that is useful in establishing exponential stability of regulation as discussed in Chapter 1.

Theorem 4.10 (Riccati iteration and regulator stability). *Given (A, B) stabilizable, (A, C) detectable, $Q > 0$, $R > 0$, $P_f \geq 0$, and the discrete*

Riccati equation

$$\Pi(k-1) = C'QC + A'\Pi(k)A - \\ A'\Pi(k)B(B'\Pi(k)B + R)^{-1}B'\Pi(k)A, \quad k = N, \ldots, 1$$
$$\Pi(N) = P_f$$

Then

(a) There exists $\Pi \geq 0$ *such that for every* $P_f \geq 0$

$$\lim_{k \to -\infty} \Pi(k) = \Pi$$

and Π *is the unique solution of the steady-state Riccati equation*

$$\Pi = C'QC + A'\Pi A - A'\Pi B(B'\Pi B + R)^{-1}B'\Pi A$$

among the class of positive semidefinite matrices.

(b) The matrix $A + BK$ *in which*

$$K = -(B'\Pi B + R)^{-1}B'\Pi A$$

is a stable matrix.

Bertsekas (1987, pp.59–64) provides a proof for a slightly different version of this theorem. Exercise 4.13 explores translating this theorem into the form that is useful for establishing exponential convergence of full information estimation.

4.3 Moving Horizon Estimation

As displayed in Figure 1.4 of Chapter 1, in MHE we consider only the N most recent measurements, $\mathbf{y}_N(T) = \{y(T-N), y(T-N+1), \ldots y(T-1)\}$. For $T > N$, the MHE objective function is given by

$$\hat{V}_T(\chi(T-N), \boldsymbol{\omega}) = \Gamma_{T-N}(\chi(T-N)) + \sum_{i=T-N}^{T-1} \ell_i(\omega(i), \nu(i))$$

subject to $\chi^+ = f(\chi, \omega)$, $y = h(\chi) + \nu$. The MHE problem is defined to be

$$\min_{\chi(T-N), \boldsymbol{\omega}} \hat{V}_T(\chi(T-N), \boldsymbol{\omega}) \tag{4.14}$$

in which $\boldsymbol{\omega} = \{\omega(T-N), \ldots, \omega(T-1)\}$. The designer chooses the prior weighting $\Gamma_k(\cdot)$ for $k > N$. Until the data horizon is full, i.e., for times $k \leq N$, we generally *define* the MHE problem to be the full information problem.

4.3.1 Zero Prior Weighting

Here we discount the early data completely and choose $\Gamma_i(\cdot) = 0$ for all $i \geq N$. Because it discounts the past data completely, this form of MHE must be able to asymptotically reconstruct the state using only the most recent N measurements. The first issue is establishing existence of the solution. Unlike the full information problem, in which the positive definite initial penalty guarantees that the optimization takes place over a bounded (compact) set, here there is zero initial penalty. So we must restrict the system further than i-IOSS to ensure solution existence. We show next that observability is sufficient for this purpose.

Definition 4.11 (Observability). The system $x^+ = f(x, w), y = h(x)$ is *observable* if there exist finite $N_o \in \mathbb{I}_{\geq 1}$, $\gamma_1(\cdot), \gamma_2(\cdot) \in \mathcal{K}$ such that for every two initial states z_1 and z_2, and any two disturbance sequences $\mathbf{w}_1, \mathbf{w}_2$, and all $k \geq N_o$

$$|z_1 - z_2| \leq \gamma_1(\|\mathbf{w}_1 - \mathbf{w}_2\|_{0:k-1}) + \gamma_2(\|\mathbf{y}_{z_1,\mathbf{w}_1} - \mathbf{y}_{z_2,\mathbf{w}_2}\|_{0:k})$$

At any time $T \geq N$ consider decision variables $\chi(T-N) = x(T-N)$ and $\omega(i) = w(i)$ for $T - N \leq i \leq T - 1$. For these decision variables the cost function has the value

$$\hat{V}_T(\chi(T-N), \boldsymbol{\omega}) = \sum_{i=T-N}^{T-1} \ell_i(w(i), v(i)) \tag{4.15}$$

which is less than V_∞ defined in the full information problem. Observability then ensures that for all $k \geq N \geq N_o$

$$|x(k-N) - \hat{x}(k-N|k)| \leq \gamma_2(\|\mathbf{v}\|_{k-N:k})$$

Since $v(k)$ is bounded for all $k \geq 0$ by Assumption 4.3, observability has bounded the distance between the initial estimate in the horizon and the system state for all $k \geq N$. That along with continuity of $\hat{V}_T(\chi, \boldsymbol{\omega})$ ensures existence of the solution to the MHE problem by the Weierstrass theorem (Proposition A.7). But the solution does not have to be unique.

We show next that final-state observability (FSO) is the natural system requirement for MHE with zero prior weighting to provide stability and convergence.

Definition 4.12 (Final-state observability). The system $x^+ = f(x, w)$, $y = h(x)$ is *final-state observable* (FSO) if there exist finite $N_o \in \mathbb{I}_{\geq 1}$,

$\overline{\gamma}_1(\cdot)$, $\overline{\gamma}_2(\cdot) \in \mathcal{K}$ such that for every two initial states z_1 and z_2, and any two disturbance sequences $\mathbf{w}_1, \mathbf{w}_2$, and all $k \geq N_o$

$$|x(k; z_1, \mathbf{w}_1) - x(k; z_2, \mathbf{w}_2)| \leq \overline{\gamma}_1\big(\, \|\mathbf{w}_1 - \mathbf{w}_2\|_{0:k-1} \,\big) + \\ \overline{\gamma}_2\big(\, \|\mathbf{y}_{z_1,\mathbf{w}_1} - \mathbf{y}_{z_2,\mathbf{w}_2}\|_{0:k} \,\big)$$

Notice that FSO is not the same as observability. It is weaker than observability and stronger than i-IOSS (detectability) as discussed in Exercise 4.11. Consider two equal disturbance sequences, $\mathbf{w}_1 = \mathbf{w}_2$, and two equal measurement sequences $\mathbf{y}_1 = \mathbf{y}_2$. FSO implies that for every pair z_1 and z_2, $x(N_o; z_1, \mathbf{w}_1) = x(N_o; z_2, \mathbf{w}_1)$; we know the *final* states at time $k = N_o$ are equal. FSO does not imply that the *initial* states are equal as required by observability. We can of course add the nonnegative term $\beta(|z_1 - z_2|, k)$ to the right-hand side of the FSO inequality and obtain the i-IOSS inequality, so FSO implies i-IOSS. Exercise 4.11 treats observability, FSO, and detectability of the linear time-invariant system, which can be summarized compactly in terms of the eigenvalues of the partitioned state transition matrix corresponding to the unobservable modes.

Next we show that the MHE cost function converges to zero as $T \to \infty$ for all x_0 and converging disturbances $(w(i), v(i))$. Since $(w(i), v(i))$ converges to zero, (4.15) implies that $\hat{V}_T$ converges to zero as $T \to \infty$. The optimal cost at T, $\hat{V}_T^0$, is bounded above by $\hat{V}_T$ so $\hat{V}_T^0$ also converges to zero. The optimal cost is

$$\hat{V}_T^0 = \sum_{i=T-N}^{T-1} \ell_i(\hat{w}(i|T), y(i) - \hat{x}(i|T))$$

in which $(\hat{x}(i|T), \hat{w}(i|T))$ are the optimal decisions for $T - N \leq i \leq T - 1$, $T \geq N$. Since $\hat{V}_T^0$ converges to zero, we have

$$y(i) - h(\hat{x}(i|T)) \to 0 \qquad \hat{w}(i|T) \to 0$$

as $T \to \infty$. Since $y = h(x) + v$ and $v(i)$ converges to zero, and $w(i)$ converges to zero, we also have

$$h(x(i)) - h(\hat{x}(i|T)) \to 0 \qquad w(i) - \hat{w}(i|T) \to 0 \tag{4.16}$$

for $T - N \leq i \leq T - 1$, $T \geq N$.

We have the following theorem for this estimator.

Theorem 4.13 (Robust GAS of MHE with zero prior weighting). *Consider an observable system and measurement sequence generated by* (4.1) *with disturbances satisfying Assumption 4.3. The MHE estimate with zero prior weighting, $N \geq N_o$, and stage cost satisfying* (4.5)*, is robustly GAS.*

Proof. We establish the two parts of Definition 4.7

(a) Consider the system to be at state $x(k-N)$ at time $k-N$ and subject to disturbances $\mathbf{w}_k = \{w(k-N), \ldots w(k-1)\}$. At time k, the estimator has initial state $\hat{x}(k-N|k)$ and disturbance sequence $\hat{w}_k$. We have that $x(k; x(k-N), k-N, \mathbf{w}_k) = x(k)$ and $x(k; \hat{x}(k-N|k), k-N, \hat{\mathbf{w}}_k) = \hat{x}(k)$, and the FSO property gives for $k \geq N \geq N_o$

$$|x(k) - \hat{x}(k)| \leq \\ \overline{\gamma}_1\left(\|\mathbf{w}_k - \hat{\mathbf{w}}_k\|_{k-N:k-1}\right) + \overline{\gamma}_2\left(\|h(\mathbf{x}_k) - h(\hat{\mathbf{x}}_k)\|_{k-N:k}\right) \quad (4.17)$$

By (4.16) the right-hand side converges to zero as $k \to \infty$, which gives

$$\hat{x}(k) \to x(k)$$

as $k \to \infty$ for all x_0 and measurement sequence generated by (4.1) with disturbances satisfying Assumption 4.3.

(b) For $k \leq N$, MHE is equivalent to full information estimation, and Theorem 4.8 applies. So we consider $k > N$. Assume (4.6) holds for some $\delta > 0$. This implies

$$\|\mathbf{w}_k - \hat{\mathbf{w}}_k\|_{k-N:k-1} \leq \underline{\gamma}_w^{-1}(\delta) + \gamma_{\overline{w}}^{-1}(\delta)$$
$$\|h(\mathbf{x}_k) - h(\hat{\mathbf{x}}_k)\|_{k-N:k} \leq \underline{\gamma}_w^{-1}(\delta) + \gamma_{\overline{w}}^{-1}(\delta)$$

Using these bounds in (4.17) gives for $k > N$

$$|x(k) - \hat{x}(k)| \leq (\overline{\gamma}_1 + \overline{\gamma}_2)(\underline{\gamma}_w^{-1}(\delta) + \gamma_{\overline{w}}^{-1}(\delta))$$

Choose an $\varepsilon > 0$. Since the right-hand side defines a $\mathcal{K}$-function, we can choose $\delta > 0$ small enough to meet the bound $|x(k) - \hat{x}(k)| \leq \varepsilon$ for all $k > N$. Coupled with Theorem 4.8 to cover $k \leq N$, we have established part (b) of robust GAS. ∎

4.3.2 Nonzero Prior Weighting

The two drawbacks of zero prior weighting are: the system had to be assumed *observable* rather than detectable to ensure existence of the solution to the MHE problem; and a large horizon N may be required to obtain performance comparable to full information estimation. We address these two disadvantages by using nonzero prior weighting. To get started, we use forward DP, as we did in Chapter 1 for the unconstrained linear case, to decompose the full information problem exactly into the MHE problem (4.14) in which $\Gamma(\cdot)$ is chosen as arrival cost.

Definition 4.14 (Full information arrival cost). The full information arrival cost is defined as

$$Z_T(p) = \min_{\chi(0),\boldsymbol{\omega}} V_T(\chi(0), \boldsymbol{\omega}) \tag{4.18}$$

subject to

$$\chi^+ = f(\chi, \omega) \qquad y = h(\chi) + v \qquad \chi(T; \chi(0), \boldsymbol{\omega}) = p$$

We have the following equivalence.

Lemma 4.15 (MHE and full information estimation). *The MHE problem* (4.14) *is equivalent to the full information problem* (4.3) *for the choice* $\Gamma_k(\cdot) = Z_k(\cdot)$ *for all* $k > N$ *and* $N \geq 1$.

The proof is left as an exercise. This lemma is the essential insight provided by the DP recursion. But notice that evaluating arrival cost in (4.18) has the same computational complexity as solving a full information problem. So next we generate an MHE problem that has simpler computational requirements, but retains the excellent stability properties of full information estimation. We proceed as follows.

Definition 4.16 (MHE arrival cost). The MHE arrival cost $\hat{Z}(\cdot)$ is defined for $T > N$ as

$$\begin{aligned} \hat{Z}_T(p) &= \min_{z,\boldsymbol{\omega}} \hat{V}_T(z, \boldsymbol{\omega}) \\ &= \min_{z,\boldsymbol{\omega}} \Gamma_{T-N}(z) + \sum_{i=T-N}^{T-1} \ell_i(\omega(i), v(i)) \end{aligned} \tag{4.19}$$

subject to

$$\chi^+ = f(\chi, \omega) \qquad y = h(\chi) + v \qquad \chi(T; z, T-N, \boldsymbol{\omega}) = p$$

For $T \leq N$ we usually define the MHE problem to be the full information problem, so $\hat{Z}_T(\cdot) = Z_T(\cdot)$ and $\hat{V}_T^0 = V_T^0$. Notice from the second equality in the definition that the MHE arrival cost at T is defined in terms of the prior weighting at time $T - N$.

We next show that choosing a prior weighting that *underbounds* the MHE arrival cost is the key sufficient condition for stability and convergence of MHE.

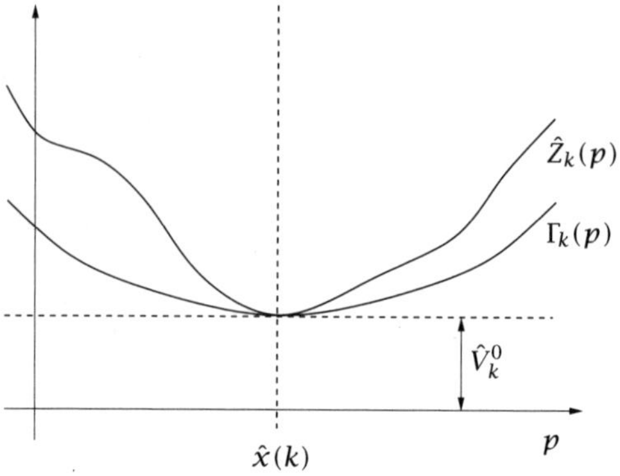

Figure 4.1: MHE arrival cost $\hat{Z}_k(p)$, underbounding prior weighting $\Gamma_k(p)$, and MHE optimal value $\hat{V}_k^0$; for all p and $k > N$, $\hat{Z}_k(p) \geq \Gamma_k(p) \geq \hat{V}_k^0$, and $\hat{Z}_k(\hat{x}(k)) = \Gamma_k(\hat{x}(k)) = \hat{V}_k^0$.

Assumption 4.17 (Prior weighting). We assume that $\Gamma_k(\cdot)$ is continuous and satisfies the following inequalities for all $k > N$

(a) Upper bound

$$\Gamma_k(p) \leq \hat{Z}_k(p) = \min_{z,\boldsymbol{\omega}} \Gamma_{k-N}(z) + \sum_{i=k-N}^{k-1} \ell_i(\omega(i), \nu(i)) \tag{4.20}$$

subject to $\chi^+ = f(\chi, \omega), y = h(\chi) + \nu, \chi(k; z, k-N, \boldsymbol{\omega}) = p$.

(b) Lower bound

$$\Gamma_k(p) \geq \hat{V}_k^0 + \underline{\gamma}_p(|p - \hat{x}(k)|) \tag{4.21}$$

in which $\underline{\gamma}_p \in \mathcal{K}_\infty$.

This assumption is depicted in Figure 4.1.

To establish convergence of the MHE estimates, it will prove useful to have an upper bound for the MHE optimal cost. Next we establish the stronger result that the MHE arrival cost is bounded above by the full information arrival cost as stated in the following proposition.

Proposition 4.18 (Arrival cost of full information greater than MHE).

$$\hat{Z}_T(\cdot) \leq Z_T(\cdot) \qquad T \geq 1 \tag{4.22}$$

Proof. We know this result holds for $T \in \mathbb{I}_{1:N}$ because MHE is equivalent to full information for these T. Next we show that the inequality at T

implies the inequality at $T+N$. Indeed, we have by the definition of the arrival costs

$$\hat{Z}_{T+N}(p) = \min_{z,\omega} \Gamma_T(z) + \sum_{i=T}^{T+N-1} \ell_i(\omega(i), \nu(i))$$

$$Z_{T+N}(p) = \min_{z,\omega} Z_T(z) + \sum_{i=T}^{T+N-1} \ell_i(\omega(i), \nu(i))$$

in which both optimizations are subject to the same constraints $\chi^+ = f(\chi, \omega)$, $y = h(\chi) + \nu$, $\chi(k; z, k-N, \boldsymbol{\omega}) = p$. From (4.20) $\Gamma_T(\cdot) \leq \hat{Z}_T(\cdot)$, and $\hat{Z}_T(\cdot) \leq Z_T(\cdot)$ by assumption. Together these imply the optimal values satisfy $\hat{Z}_{T+N}(p) \leq Z_{T+N}(p)$ for all p, and we have established $\hat{Z}_{T+N}(\cdot) \leq Z_{T+N}(\cdot)$. Therefore we have extended (4.22) from $T \in \mathbb{I}_{1:N}$ to $T \in \mathbb{I}_{1:2N}$. Continuing this recursion establishes (4.22) for $T \in \mathbb{I}_{\geq 1}$. ∎

Given (4.22) we also have the analogous inequality for the optimal costs of MHE and full information

$$\hat{V}_T^0 \leq V_T^0 \qquad T \geq 1 \tag{4.23}$$

Assumption 4.19 (MHE detectable system). We say a system $x^+ = f(x, w)$, $y = h(x)$ is *MHE detectable* if the system augmented with an extra disturbance w_2

$$x^+ = f(x, w_1) + w_2 \qquad y = h(x)$$

is i-IOSS with respect to the augmented disturbance (w_1, w_2).

Note that MHE detectable is stronger than i-IOSS (detectable) but weaker than observable and FSO. See also Exercise 4.10.

Theorem 4.20 (Robust GAS of MHE). *Consider an MHE detectable system and measurement sequence generated by* (4.1) *with disturbances satisfying Assumption 4.3. The MHE estimate defined by* (4.14) *using the prior weighting function* $\Gamma_k(\cdot)$ *satisfying Assumption 4.17 and stage cost satisfying Assumption 4.5 is robustly GAS.*

Proof. The MHE solution exists for $T \leq N$ by the existence of the full information solution, so we consider $T > N$. For disturbances satisfying Assumption 4.3, we established in the proof of Theorem 4.8 for the full information problem that $V_T^0 \leq V_\infty$ for all $T \geq 1$ including $T = \infty$. From Proposition 4.18 and (4.23), we have that the MHE optimal cost also has the upper bound $\hat{V}_T^0 \leq V_\infty$ for all $T \geq 1$ including $T = \infty$. Since we have assumed $f(\cdot)$ and $h(\cdot)$ are continuous, $\Gamma_i(\cdot)$ is continuous for $i > N$, and $\ell_i(\cdot)$ is continuous for all $i \geq 0$, the MHE cost function $\hat{V}_T(\cdot)$ is continuous for $T > N$. The lower bound on Γ_i for $i > N$ and ℓ_i for all $i \geq 0$ imply that for $T > N$, $\hat{V}_T(\chi(T-N), \boldsymbol{\omega})$ goes to infinity as either $\chi(T-N)$ or $\boldsymbol{\omega}$ goes to infinity. Therefore the MHE optimization takes place over a bounded, closed set for $T > N$, and the the solution exists by the Weierstrass theorem.

(a) Consider the solution to the MHE problem at time T, $(\hat{x}(T-N|T), \hat{\mathbf{w}}_T)$. We have that

$$\hat{V}_T^0 = \Gamma_{T-N}(\hat{x}(T-N|T)) + \sum_{i=T-N}^{T-1} \ell_i(\hat{w}(i|T), \hat{v}(i|T))$$

From (4.21) we have

$$\Gamma_{T-N}(\hat{x}(T-N|T)) \geq \hat{V}_{T-N}^0 + \underline{\gamma}_p(|\hat{x}(T-N|T) - \hat{x}(T-N|T-N)|)$$

Using this inequality in the previous equation we have

$$\hat{V}_T^0 \geq \hat{V}_{T-N}^0 + \underline{\gamma}_p(|\hat{x}(T-N|T) - \hat{x}(T-N|T-N)|) + \sum_{i=T-N}^{T-1} \ell_i(\hat{w}(i|T), \hat{v}(i|T)) \tag{4.24}$$

and we have established that the sequence $\{\hat{V}_{T+iN}^0\}$ is a nondecreasing sequence in $i = 1, 2, \ldots$ for any fixed $T \geq 1$. Since $\hat{V}_k^0$ is bounded above for all $k \geq 1$, the sequence $\hat{V}_{T+iN}^0$ converges as $i \to \infty$ for any $T \geq 1$. This convergence gives as $T \to \infty$

$$\underline{\gamma}_p(|\hat{x}(T-N|T) - \hat{x}(T-N|T-N)|) \to 0 \tag{4.25}$$

$$\sum_{i=T-N}^{T-1} \ell_i(\hat{w}(i|T), \hat{v}(i|T)) \to 0 \tag{4.26}$$

Next we create a single estimate sequence by concatenating MHE sequences from times $N, 2N, 3N, \ldots$. This gives the state sequence and corresponding $\overline{\mathbf{w}}_1$ and $\overline{\mathbf{w}}_2$ sequences listed in the following table so that

$$\overline{x}^+ = f(\overline{x}, \overline{w}_1) + \overline{w}_2 \quad \text{for } k \geq 0 \qquad y = h(\overline{x}) + \overline{v}$$

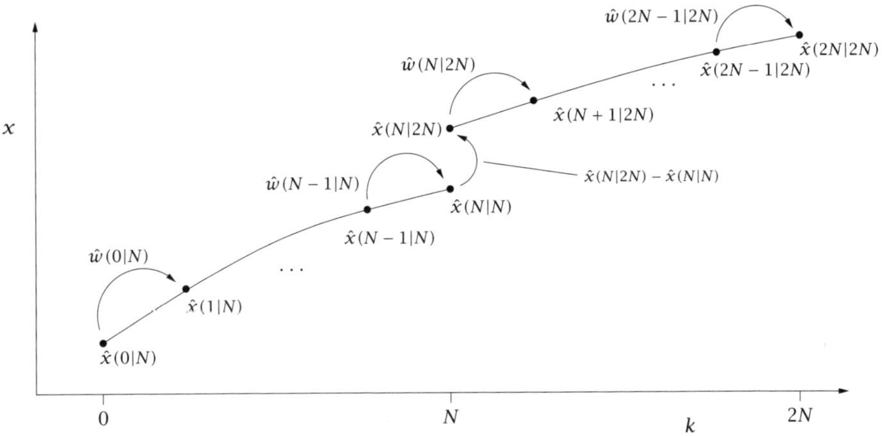

Figure 4.2: Concatenating two MHE sequences to create a single state estimate sequence from time 0 to $2N$.

$\overline{\mathbf{x}}$	$\overline{\mathbf{w}}_1$	$\overline{\mathbf{w}}_2$	$\overline{\mathbf{v}}$
$\hat{x}(0\|N)$	$\hat{w}(0\|N)$	0	$\hat{v}(0\|N)$
$\hat{x}(1\|N)$	$\hat{w}(1\|N)$	0	$\hat{v}(1\|N)$
...	...	...	...
$\hat{x}(N-1\|N)$	$\hat{w}(N-1\|N)$	$\hat{x}(N\|2N) - \hat{x}(N\|N)$	$\hat{v}(N-1\|N)$
$\hat{x}(N\|2N)$	$\hat{w}(N\|2N)$	0	$\hat{v}(N\|2N)$
$\hat{x}(N+1\|2N)$	$\hat{w}(N+1\|2N)$	0	$\hat{v}(N+1\|2N)$
...	...	...	...
$\hat{x}(2N-1\|2N)$	$\hat{w}(2N-1\|2N)$	$\hat{x}(2N\|3N) - \hat{x}(2N\|2N)$	$\hat{v}(2N-1\|2N)$
$\hat{x}(2N\|3N)$	$\hat{w}(2N\|3N)$	0	$\hat{v}(2N\|3N)$
$\hat{x}(2N+1\|3N)$	$\hat{w}(2N+1\|3N)$	0	$\hat{v}(2N+1\|3N)$
...	...	...	...

Notice that every N rows in the array, there is a nonzero entry in the $\mathbf{w}_2$ column. That disturbance is required to move from one MHE sequence to the next as shown in Figure 4.2. But (4.25) implies that $\overline{w}_2(k) \to 0$ as integer $k \to \infty$, and (4.26) implies that $\overline{w}_1(k) \to 0$ as $k \to \infty$. Therefore $|(w_1(k), 0) - (\overline{w}_1(k), \overline{w}_2(k))| \to 0$ as $k \to \infty$. We also have from (4.26) that $h(x(k)) - h(\overline{x}(k)) = \overline{v}(k) - v(k) \to 0$ as $k \to \infty$. Next we apply the MHE-detectability assumption to the $\mathbf{x}$ and $\overline{\mathbf{x}}$ sequences, to obtain the inequality

$$|x(k) - \overline{x}(k)| \leq \beta(|x(0) - \hat{x}(0|N)|, k) + \gamma_1(\,\|(\mathbf{w}_1, 0) - (\overline{\mathbf{w}}_1, \overline{\mathbf{w}}_2)\|_{0:k-1}) + \gamma_2(\,\|h(\mathbf{x}) - h(\overline{\mathbf{x}})\|_{0:k}) \quad (4.27)$$

From Proposition 4.2 we conclude that $\overline{x}(k) \to x(k)$ as $k \to \infty$. Therefore we have that $\hat{x}(iN+j|(i+1)N) \to x(iN+j)$ for all $j = 0, 1, \ldots, N-1$ as integer $i \to \infty$. Note that only $\hat{x}(iN|iN)$ is missing from this argument. But $x(iN) = f(x(iN-1), w(iN-1))$ and $\hat{x}(iN|iN) = f(\hat{x}(iN-1|iN), \hat{w}(iN-1|iN))$. Since $\hat{x}(iN-1|iN) \to x(iN-1)$, $\hat{w}(iN-1|iN) \to w(iN-1)$, and $f(\cdot)$ is continuous, we have that $\hat{x}(iN|iN) \to x(iN)$ as well. We can repeat this concatenation construction using the MHE sequences $N+j, 2N+j, 3N+j, \ldots$ for $j = 1, \ldots, N-1$ to conclude that $\hat{x}(k) \to x(k)$ as $k \to \infty$, and convergence is established.

(b) As previously, assume the following holds for some $\delta > 0$

$$\gamma_x(|x_0 - \overline{x}_0|) + \sum_{i=0}^{\infty} \gamma_w(\,|(w(i), v(i))|\,) \leq \delta$$

We wish to show that for every $\varepsilon > 0$ there exists $\delta > 0$ such that this equation implies $|x(k; x_0, \mathbf{w}) - \hat{x}(k)| \leq \varepsilon$ for all $k \geq 0$. We know such a δ exists for $k \leq N$ from Theorem 4.8. We therefore consider $k > N$. The optimal MHE cost is bounded above by the optimal full information, which is bounded above by δ, $\hat{V}_T^0 \leq V_T^0 \leq \delta$ for $T \geq 0$. So we have using $T = N$,

$$\underline{\gamma}_x(|\hat{x}(0|N) - \overline{x}_0|) \leq \delta \qquad \gamma_x(|x_0 - \overline{x}_0|) \leq \delta$$

which gives $|x(0) - \hat{x}(0|N)| \leq (\underline{\gamma}_x^{-1} + \gamma_x^{-1})(\delta)$. From (4.24) and the fact that $\hat{V}_T^0 \leq \delta$, we know that

$$\underline{\gamma}_p(\,|\hat{x}((i+1)N|iN) - \hat{x}(iN|iN)|\,) \leq \delta$$
$$|\hat{x}((i+1)N|iN) - \hat{x}(iN|iN)| \leq \underline{\gamma}_p^{-1}(\delta)$$

which implies $|\overline{w}_2(k)| \leq \underline{\gamma}_p^{-1}(\delta)$ for all $k \geq 0$. Examining the terms in the $\overline{\mathbf{w}}_1$ column, we conclude as before that $|w(k) - \overline{w}_1(k)| \leq (\underline{\gamma}_p^{-1} + \underline{\gamma}_w^{-1})(\delta)$ for all $k \geq 0$. The $\overline{\mathbf{v}}$ column gives the bound

$$|v(k) - \overline{v}(k)| = |h(x(k)) - h(\overline{x}(k))| \leq (\underline{\gamma}_w^{-1} + \gamma_w^{-1})(\delta)$$

We also have the bounds

$$\begin{aligned}
\|(\mathbf{w}_1, 0) - (\overline{\mathbf{w}}_1, \overline{\mathbf{w}}_2)\| &= \max_{k \geq 0} |(w_1(k), 0) - (\overline{w}_1(k), \overline{w}_2(k))| \\
&= \max_{k \geq 0} |(w_1(k) - \overline{w}_1(k), -\overline{w}_2(k))| \\
&\leq \max_{k \geq 0} |w_1(k) - \overline{w}_1(k)| + |\overline{w}_2(k)| \\
&\leq (2\underline{\gamma}_p^{-1} + \underline{\gamma}_w^{-1})(\delta)
\end{aligned}$$

Substituting these into (4.27) gives

$$|x(k) - \overline{x}(k)| \leq \overline{\beta}((\underline{\gamma}_x^{-1} + \gamma_{\overline{x}}^{-1})(\delta)) + \gamma_1((2\underline{\gamma}_p^{-1} + \underline{\gamma}_w^{-1})(\delta)) + \gamma_2((\underline{\gamma}_w^{-1} + \gamma_{\overline{w}}^{-1})(\delta))$$

Recall $\overline{\beta}(s) := \beta(s, 0)$, which is a $\mathcal{K}$-function, and the right-hand side therefore defines a $\mathcal{K}$-function, so we can make $|x(k) - \overline{x}(k)|$ as small as desired for all $k > N$. This gives a bound for $|x(iN) - \hat{x}(iN + j|iN)|$ for all $i \geq 1$ and j satisfying $0 \leq j \leq N - 1$. Next we use the continuity of $f(\cdot)$ to make $|x(iN) - \hat{x}(iN|iN)|$ small for all $i \geq 0$. Finally we repeat the concatenation construction using the MHE sequences $N + j, 2N + j, 3N + j, \ldots$ for $j = 1, \ldots, N - 1$ to make $|x(k) - \hat{x}(k)|$ as small as desired for all $k > N$, and part (b) of the robust GAS definition is established. ■

Satisfying the prior weighting *inequality* (4.20) is computationally less complex than satisfying the equality (4.16) required in the MHE arrival cost recursion, as we show subsequently in the constrained, linear case. But for the general nonlinear case, ensuring satisfaction of even (4.20) remains a key technical challenge for MHE research.

4.3.3 Constrained Estimation

Constraints in estimation may be a useful way to add information to the estimation problem. We may wish to enforce physically known facts such as: concentrations of impurities, although small, must be nonnegative, fluxes of mass and energy must have the correct sign given temperature and concentration gradients, and so on. Unlike the regulator, the estimator has no way to enforce these constraints on the *system*. Therefore, it is important that any constraints imposed on the estimator are satisfied by the system generating the measurements. Otherwise we may prevent convergence of the estimated state to the system state. For this reason, care should be used in adding constraints to estimation problems.

Because we have posed state estimation as an optimization problem, it is straightforward to add constraints to the formulation. We assume that the system generating the data satisfy the following constraints.

Assumption 4.21 (Estimator constraint sets).

(a) For all $k \in \mathbb{I}_{\geq 0}$, the sets $\mathbb{W}_k$, $\mathbb{X}_k$, and $\mathbb{V}_k$ are nonempty and closed, and $\mathbb{W}_k$ and $\mathbb{V}_k$ contain the origin.

(b) For all $k \in \mathbb{I}_{\geq 0}$, the disturbances and state satisfy

$$x(k) \in \mathbb{X}_k \qquad w(k) \in \mathbb{W}_k \qquad v(k) \in \mathbb{V}_k$$

(c) The prior satisfies $\overline{x}_0 \in \mathbb{X}_0$.

Constrained full information. The constrained full information estimation objective function is

$$V_T(\chi(0), \boldsymbol{\omega}) = \ell_x(\chi(0) - \overline{x}_0) + \sum_{i=0}^{T-1} \ell_i(\omega(i), \nu(i)) \tag{4.28}$$

subject to

$$\chi^+ = f(\chi, \omega) \qquad y = h(\chi) + \nu$$
$$\chi(i) \in \mathbb{X}_i \qquad \omega(i) \in \mathbb{W}_i \qquad \nu(i) \in \mathbb{V}_i \qquad i \in \mathbb{I}_{0:T-1}$$

The constrained full information problem is

$$\min_{\chi(0), \boldsymbol{\omega}} V_T(\chi(0), \boldsymbol{\omega}) \tag{4.29}$$

Theorem 4.22 (Robust GAS of constrained full information). *Consider an i-IOSS (detectable) system and measurement sequence generated by* (4.1) *with constrained, convergent disturbances satisfying Assumptions 4.3 and 4.21. The constrained full information estimator* (4.29) *with stage cost satisfying Assumption 4.5 is robustly GAS.*

Constrained MHE. The constrained moving horizon estimation objective function is

$$\hat{V}_T(\chi(T-N), \boldsymbol{\omega}) = \Gamma_{T-N}(\chi(T-N)) + \sum_{i=T-N}^{T-1} \ell_i(\omega(i), \nu(i)) \tag{4.30}$$

subject to

$$\chi^+ = f(\chi, \omega) \qquad y = h(\chi) + \nu$$
$$\chi(i) \in \mathbb{X}_i \qquad \omega(i) \in \mathbb{W}_i \qquad \nu(i) \in \mathbb{V}_i \qquad i \in \mathbb{I}_{T-N:T-1}$$

The constrained MHE is given by the solution to the following problem

$$\min_{\chi(T-N), \boldsymbol{\omega}} \hat{V}_T(\chi(T-N), \boldsymbol{\omega}) \tag{4.31}$$

Theorem 4.23 (Robust GAS of constrained MHE). *Consider an MHE detectable system and measurement sequence generated by* (4.1) *with convergent, constrained disturbances satisfying Assumptions 4.3 and 4.21. The constrained MHE estimator* (4.31) *using the prior weighting function $\Gamma_k(\cdot)$ satisfying Assumption 4.17 and stage cost satisfying Assumption 4.5 is robustly GAS.*

Because the *system* satisfies the state and disturbance constraints due to Assumption 4.21, both full information and MHE optimization problems are feasible at all times. Therefore the proofs of Theorems 4.22 and 4.23 closely follow the proofs of their respective unconstrained versions, Theorems 4.8 and 4.20, and are omitted.

4.3.4 Smoothing and Filtering Update

We next focus on *constrained linear systems*

$$x^+ = Ax + Gw \qquad y = Cx + v \tag{4.32}$$

We proceed to strengthen several results of the previous sections for this special case. First, the i-IOSS assumption of full information estimation and the MHE detectability assumption both reduce to the assumption that (A, C) *is detectable* in this case. We usually choose a constant quadratic function for the estimator stage cost for all $i \in \mathbb{I}_{\geq 0}$

$$\ell_i(w, v) = (1/2)(|w|^2_{Q^{-1}} + |v|^2_{R^{-1}}) \qquad Q, R > 0 \tag{4.33}$$

In the unconstrained linear problem, we can of course find the full information arrival cost exactly; it is

$$Z_k(z) = V_k^0 + (1/2)\,|z - \hat{x}(k)|_{(P^-(k))^{-1}} \qquad k \geq 0$$

in which $P^-(k)$ satisfies the recursion (4.12) and $\hat{x}(k)$ is the full information estimate at time k. We use this quadratic function for the MHE prior weighting.

Assumption 4.24 (Prior weighting for linear system).

$$\Gamma_k(z) = \hat{V}_k^0 + (1/2)\,|z - \hat{x}(k)|_{(P^-(k))^{-1}} \qquad k > N \tag{4.34}$$

in which $\hat{V}_k^0$ is the optimal MHE cost at time k.

Because the unconstrained arrival cost is available, we usually choose it to be the prior weighting in MHE, $\Gamma_k(\cdot) = Z_k(\cdot)$, $k \geq 0$. This choice implies robust GAS of the MHE estimator also for the *constrained case* as we next demonstrate. To ensure the form of the estimation problem to be solved online is a quadratic program, we specialize the constraint sets to be polyhedral regions.

Assumption 4.25 (Polyhedral constraint sets). For all $k \in \mathbb{I}_{\geq 0}$, the sets $\mathbb{W}_k$, $\mathbb{X}_k$, and $\mathbb{V}_k$ in Assumption 4.21 are nonempty, closed polyhedral regions containing the origin.

Corollary 4.26 (Robust GAS of constrained MHE). *Consider a detectable linear system and measurement sequence generated by* (4.32) *with convergent, constrained disturbances satisfying Assumptions 4.3 and 4.25. The constrained MHE estimator* (4.31) *using prior weighting function satisfying* (4.34) *and stage cost satisfying* (4.33) *is robustly GAS.*

This corollary follows as a special case of Theorem 4.23.

The MHE approach discussed to this point uses at all time $T > N$ the MHE estimate $\hat{x}(T-N)$ and prior weighting function $\Gamma_{T-N}(\cdot)$ derived from the unconstrained arrival cost as shown in (4.34). We call this approach a "filtering update" because the prior weight at time T is derived from the solution of the MHE "filtering problem" at time $T-N$, i.e., the estimate of $\hat{x}(T-N) := \hat{x}(T-N|T-N)$ given measurements up to time $T-N-1$. For implementation, this choice requires storage of a window of N prior filtering estimates to be used in the prior weighting functions as time progresses.

Next we describe a "smoothing update" that can be used instead. In the smoothing update we wish to use $\hat{x}(T-N|T-1)$ (instead of $\hat{x}(T-N|T-N)$) for the prior and wish to find an appropriate prior weighting based on this choice. For the linear *unconstrained* problem we can find an exact prior weighting that gives an equivalence to the full information problem. When constraints are added to the problem, however, the smoothing update provides a different MHE than the filtering update. Like the filtering prior, the smoothing prior weighting does give an underbound for the constrained full information problem, and therefore maintains the excellent stability properties of MHE with the filtering update. As mentioned previously the unconstrained full information arrival cost is given by

$$Z_{T-N}(z) = V^0_{T-N} + (1/2)\,|z - \hat{x}(T-N)|^2_{(P^-(T-N))^{-1}} \qquad T > N \quad (4.35)$$

in which $\hat{x}(T-N)$ is the optimal estimate for the unconstrained full information problem. Next we consider using $\hat{x}(T-N|T-2)$ in place of

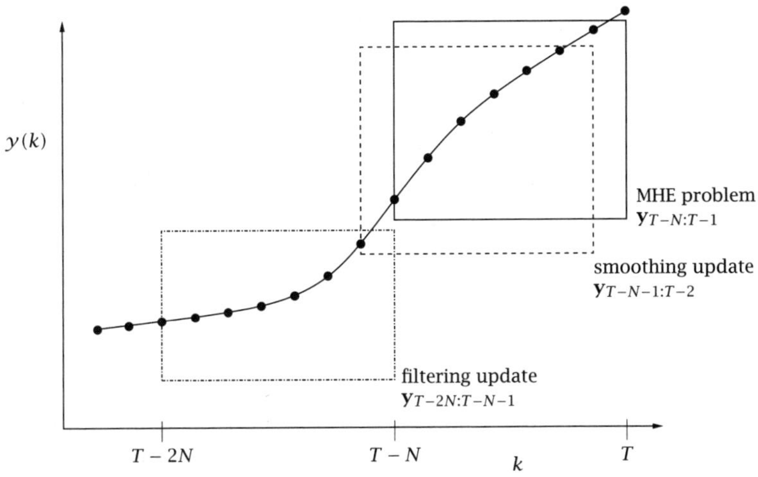

Figure 4.3: Smoothing update.

$\hat{x}(T-N) := \hat{x}(T-N|T-1)$. We might guess that the proper weight for this prior estimate would be the smoothed covariance $P(T-N|T-2)$ instead of $P^-(T-N) := P(T-N|T-1)$, and that guess is correct, but not complete. Notice that the smoothed prior $\hat{x}(T-N|T-2)$ is influenced by the measurements $\mathbf{y}_{0:T-2}$. But the sum of stage costs in the MHE problem at time T depends on measurements $\mathbf{y}_{T-N:T-1}$, so we have to adjust the prior weighting so we do not double count the data $\mathbf{y}_{T-N:T-2}$. The correct prior weighting for the smoothing update has been derived by Rao, Rawlings, and Lee (2001), which we summarize next. The following notation is useful; for any square matrix R and integer $k \geq 1$, define $\text{diag}_k(R)$ to be the following

$$\text{diag}_k(R) := \underbrace{\begin{bmatrix} R & & & \\ & R & & \\ & & \ddots & \\ & & & R \end{bmatrix}}_{k \text{ times}} \qquad \mathcal{O}_k = \begin{bmatrix} 0 & & & & \\ & C & & & \\ & CA & C & & \\ & \vdots & \vdots & \ddots & \\ & CA^{k-2} & CA^{k-3} & \cdots & C \end{bmatrix}$$

$$W_k = \text{diag}_k(R) + \mathcal{O}_k(\text{diag}_k(Q))\mathcal{O}_k'$$

We require the smoothed covariance $P(T-N|T-2)$, which we can obtain from the following recursion (Rauch, Tung, and Striebel, 1965;

Bryson and Ho, 1975)

$$P(k|T) = P(k) +$$
$$P(k)A'(P^-(k+1))^{-1}\Big(P(k+1|T) - P^-(k+1)\Big)(P^-(k+1))^{-1}AP(k)$$

We iterate this equation backwards $N-1$ times starting from the known value $P(T-1|T-2) := P^-(T-1)$ to obtain $P(T-N|T-2)$. The smoothing arrival cost is then given by

$$\widetilde{Z}_{T-N}(z) = \hat{V}^0_{T-1} + (1/2)\,|z - \hat{x}(T-N|T-2)|^2_{(P(T-N|T-2))^{-1}}$$
$$- (1/2)\,|\mathbf{y}_{T-N:T-2} - \mathcal{O}_{N-1}z|^2_{(W_{N-1})^{-1}} \qquad T > N$$

See Rao et al. (2001) and Rao (2000, pp.80–93) for a derivation that shows $\widetilde{Z}_T(\cdot) = Z_T(\cdot)$ for $T > N$.[4] Examining this alternative expression for arrival cost we see that the second term accounts for the use of the smoothed covariance and the smoothed estimate, and the third term subtracts the effect of the measurements that have been double counted in the MHE objective as well as the smoothed prior estimate. Setting the prior weighting $\Gamma_{T-N}(\cdot) = Z_{T-N}(\cdot)$ from (4.35) or $\Gamma_{T-N}(\cdot) = \widetilde{Z}_{T-N}(\cdot)$ from (4.4) give the same results as the Kalman filter for the unconstrained linear problem. But the two arrival costs are approximations of the true arrival cost and give different results once constraints are added to the problem or we use a nonlinear system model. Since the unconstrained arrival cost $\widetilde{Z}_k(\cdot)$ is also an underbound for the constrained arrival cost, MHE based on the smoothing update also provides a robustly GAS estimator for constrained linear systems satisfying the conditions of Theorem 4.26.

4.4 Extended Kalman Filtering

The extended Kalman filter (EKF) generates estimates for *nonlinear* systems by first linearizing the nonlinear system, and then applying the linear Kalman filter equations to the linearized system. The approach can be summarized in a recursion similar in structure to the Kalman filter (Stengel, 1994, pp.387–388)

$$\hat{x}^-(k+1) = f(\hat{x}(k), u(k))$$
$$P^-(k+1) = \overline{A}(k)P(k)\overline{A}(k)' + \overline{h}(k)Q\overline{h}(k)'$$
$$\hat{x}^-(0) = \overline{x}_0 \qquad P^-(0) = Q_0$$

[4]Note that Rao et al. (2001) and Rao (2000) contain some minor typos in the smoothed covariance recursion and the formula for W_k.

The mean and covariance after measurement are given by

$$\hat{x}(k) = \hat{x}^-(k) + L(k)(y(k) - h(\hat{x}^-(k)))$$
$$L(k) = P^-(k)\overline{C}(k)'(R + \overline{C}(k)P^-(k)\overline{C}(k)')^{-1}$$
$$P(k) = P^-(k) - L(k)\overline{C}(k)P^-(k)$$

in which the following linearizations are made

$$\overline{A}(k) = \frac{\partial f(x,u)}{\partial x} \quad \overline{C}(k) = \frac{\partial h(x)}{\partial x}$$

and all partial derivatives are evaluated at $\hat{x}(k)$ and $u(k)$, and $\overline{h}(k) = h(\hat{x}(k), u(k))$. The densities of w, v and x_0 are assumed to be normal. Many variations on this theme have been proposed, such as the iterated EKF and the second-order EKF (Gelb, 1974, 190–192). Of the nonlinear filtering methods, the EKF method has received the most attention due to its relative simplicity and demonstrated effectiveness in handling some nonlinear systems. Examples of implementations include estimation for the production of silicon/germanium alloy films (Middlebrooks and Rawlings, 2006), polymerization reactions (Prasad, Schley, Russo, and Bequette, 2002), and fermentation processes (Gudi, Shah, and Gray, 1994). The EKF is at best an *ad hoc* solution to a difficult problem, however, and hence there exist many pitfalls to the practical implementation of EKFs (see, for example, (Wilson, Agarwal, and Rippin, 1998)). These problems include the inability to accurately incorporate physical state constraints and the naive use of linearization of the nonlinear model.

Until recently, few properties regarding the stability and convergence of the EKF have been established. Recent research shows bounded estimation error and exponential convergence for the continuous and discrete EKF forms given observability, small initial estimation error, small noise terms, and no model error (Reif, Günther, Yaz, and Unbehauen, 1999; Reif and Unbehauen, 1999; Reif, Günther, Yaz, and Unbehauen, 2000). Depending on the system, however, the bounds on initial estimation error and noise terms may be unrealistic. Also, initial estimation error may result in bounded estimate error but not exponential convergence, as illustrated by Chaves and Sontag (2002).

Julier and Uhlmann (2004a) summarize the status of the EKF as follows.

> The extended Kalman filter is probably the most widely used estimation algorithm for nonlinear systems. However, more

> than 35 years of experience in the estimation community has shown that it is difficult to implement, difficult to tune, and only reliable for systems that are almost linear on the time scale of the updates.

We seem to be making a transition from a previous era in which new approaches to nonlinear filtering were criticized as overly complex because "the EKF works," to a new era in which researchers are demonstrating ever simpler examples in which the EKF fails completely. The unscented Kalman filter is one of the methods developed specifically to overcome the problems caused by the naive linearization used in the EKF.

4.5 Unscented Kalman Filtering

The linearization of the nonlinear model at the current state estimate may not accurately represent the dynamics of the nonlinear system behavior even for one sample time. In the EKF prediction step, the mean propagates through the full nonlinear model, but the covariance propagates through the linearization. The resulting error is sufficient to throw off the correction step and the filter can diverge even with a perfect model. The unscented Kalman filter (UKF) avoids this linearization at a single point by sampling the nonlinear response at several points. The points are called sigma points, and their locations and weights are chosen to satisfy the given starting mean and covariance (Julier and Uhlmann, 2004a,b).[5] Given $\hat{x}$ and P, choose sample points, z^i, and weights, w^i, such that

$$\hat{x} = \sum_i w^i z^i \qquad P = \sum_i w^i (z^i - \hat{x})(z^i - \hat{x})'$$

Similarly, given $w \sim N(0, Q)$ and $v \sim N(0, R)$, choose sample points n^i for w and m^i for v. Each of the sigma points is propagated forward at each sample time using the nonlinear system model. The locations and weights of the transformed points then update the mean and covariance

$$z^i(k+1) = f(z^i(k), u(k)) + G(z^i(k), u(k))n^i(k)$$
$$\eta^i = h(z^i) + m^i \quad \text{all } i$$

[5]Note that this idea is fundamentally different than the idea of particle filtering, which is discussed subsequently. The sigma points are chosen deterministically, for example as points on a selected covariance contour ellipse or a simplex. The particle filtering points are chosen by random sampling.

From these we compute the forecast step

$$\hat{x}^- = \sum_i w^i z^i \qquad \hat{y}^- = \sum_i w^i \eta^i$$
$$P^- = \sum_i w^i (z^i - \hat{x}^-)(z^i - \hat{x}^-)'$$

After measurement, the EKF correction step is applied after first expressing this step in terms of the covariances of the innovation and state prediction. The output error is given as $\tilde{y} := y - \hat{y}^-$. We next rewrite the Kalman filter update as

$$\hat{x} = \hat{x}^- + L(y - \hat{y}^-)$$
$$L = \underbrace{\mathcal{E}((x - \hat{x}^-)\tilde{y}')}_{P^-C'} \; \underbrace{\mathcal{E}(\tilde{y}\tilde{y}')^{-1}}_{(R+CP^-C')^{-1}}$$
$$P = P^- - L \underbrace{\mathcal{E}((x - \hat{x}^-)\tilde{y}')'}_{CP^-}$$

in which we approximate the two expectations with the sigma point samples

$$\mathcal{E}((x - \hat{x}^-)\tilde{y}') \approx \sum_i w^i (z^i - \hat{x}^-)(\eta^i - \hat{y}^-)'$$
$$\mathcal{E}(\tilde{y}\tilde{y}') \approx \sum_i w^i (\eta^i - \hat{y}^-)(\eta^i - \hat{y}^-)'$$

See Julier, Uhlmann, and Durrant-Whyte (2000); Julier and Uhlmann (2004a); van der Merwe, Doucet, de Freitas, and Wan (2000) for more details on the algorithm. An added benefit of the UKF approach is that the partial derivatives $\partial f(x,u)/\partial x, \partial h(x)/\partial x$ are not required. See also (Nørgaard, Poulsen, and Ravn, 2000) for other derivative-free nonlinear filters of comparable accuracy to the UKF. See (Lefebvre, Bruyninckx, and De Schutter, 2002; Julier and Uhlmann, 2002) for an interpretation of the UKF as a use of statistical linear regression.

The UKF has been tested in a variety of simulation examples taken from different application fields including aircraft attitude estimation, tracking and ballistics, and communication systems. In the chemical process control field, Romanenko and coworkers have compared the EKF and UKF on a strongly nonlinear exothermic chemical reactor (Romanenko and Castro, 2004), and a pH system (Romanenko, Santos, and Afonso, 2004). The reactor has nonlinear dynamics and a linear measurement model, i.e., a subset of states is measured. In this case, the

UKF performs significantly better than the EKF when the process noise is large. The pH system has linear dynamics but a strongly nonlinear measurement, i.e., the pH measurement. In this case, the authors show a modest improvement in the UKF over the EKF.

4.6 Interlude: EKF, UKF, and MHE Comparison

One nice feature enjoyed by the EKF and UKF formulations is the recursive update equations. One-step recursions are computationally efficient, which may be critical in online applications with short sample times. The MHE computational burden may be reduced by shortening the length of the moving horizon, N. But use of short horizons may produce inaccurate estimates, especially after an unmodeled disturbance. This unfortunate behavior is the result of the system nonlinearity. As we saw in Sections 1.4.3–1.4.4, for *linear systems*, the full information problem and the MHE problem are identical to a one-step recursion using the appropriate state penalty coming from the filtering Riccati equation. Losing the equivalence of a one-step recursion to full information or a finite moving horizon problem brings into question whether the one-step recursion can provide equivalent estimator performance. We show in the following example that the EKF and the UKF do not provide estimator performance comparable to MHE.

Example 4.27: EKF and UKF

Consider the following set of reversible reactions taking place in a well-stirred, isothermal, gas-phase batch reactor

$$\mathrm{A} \underset{k_{-1}}{\overset{k_1}{\rightleftharpoons}} \mathrm{B} + \mathrm{C} \qquad 2\mathrm{B} \underset{k_{-2}}{\overset{k_2}{\rightleftharpoons}} \mathrm{C}$$

The material balance for the reactor is

$$\frac{d}{dt}\begin{bmatrix} c_A \\ c_B \\ c_C \end{bmatrix} = \begin{bmatrix} -1 & 0 \\ 1 & -2 \\ 1 & 1 \end{bmatrix} \begin{bmatrix} k_1 c_A - k_{-1} c_B c_C \\ k_2 c_B^2 - k_{-2} c_C \end{bmatrix}$$

$$\frac{dx}{dt} = f_c(x)$$

with states and measurement

$$x = \begin{bmatrix} c_A & c_B & c_C \end{bmatrix}' \qquad y = RT \begin{bmatrix} 1 & 1 & 1 \end{bmatrix} x$$

in which c_j denotes the concentration of species j in mol/L, R is the gas constant, and T is the reactor temperature in K. The measurement is the reactor pressure in atm, and we use the ideal gas law to model the pressure. The model is nonlinear because of the two second-order reactions. We model the system plus disturbances with the following discrete time model

$$\begin{aligned} x^+ &= f(x) + w \\ y &= Cx + v \end{aligned}$$

in which f is the solution of the ODEs over the sample time, Δ, i.e, if $s(t, x_0)$ is the solution of $dx/dt = f_c(x)$ with initial condition $x(0) = x_0$ at $t = 0$, then $f(x) = s(\Delta, x)$. The state and measurement disturbances, w and v, are assumed to be zero-mean independent normals with constant covariances Q and R. The following parameter values are used in the simulations

$$RT = 32.84 \, \text{mol} \cdot \text{atm/L}$$

$$\Delta = 0.25 \quad k_1 = 0.5 \quad k_{-1} = 0.05 \quad k_2 = 0.2 \quad k_{-2} = 0.01$$

$$C = \begin{bmatrix} 1 & 1 & 1 \end{bmatrix} RT \qquad P(0) = (0.5)^2 I \qquad Q = (0.001)^2 I \qquad R = (0.25)^2$$

$$\overline{x}(0) = \begin{bmatrix} 1 \\ 0 \\ 4 \end{bmatrix} \qquad x(0) = \begin{bmatrix} 0.5 \\ 0.05 \\ 0 \end{bmatrix}$$

The prior density for the initial state, $N(\overline{x}(0), P(0))$, is deliberately chosen to poorly represent the actual initial state to model a large initial disturbance to the system. We wish to examine how the different estimators recover from this large unmodeled disturbance.

Solution

Figure 4.4 (top) shows a typical EKF performance for these conditions. Note that the EKF cannot reconstruct the state for this system and that the estimates converge to incorrect steady states displaying negative concentrations of A and B. For some realizations of the noise sequences, the EKF may converge to the correct steady state. Even for these cases, however, negative concentration estimates still occur during the transient, which correspond to physically impossible states. Figure 4.4 (bottom) presents typical results for the clipped EKF, in which negative values of the filtered estimates are set to zero. Note that although the estimates converge to the system states, this estimator gives pressure

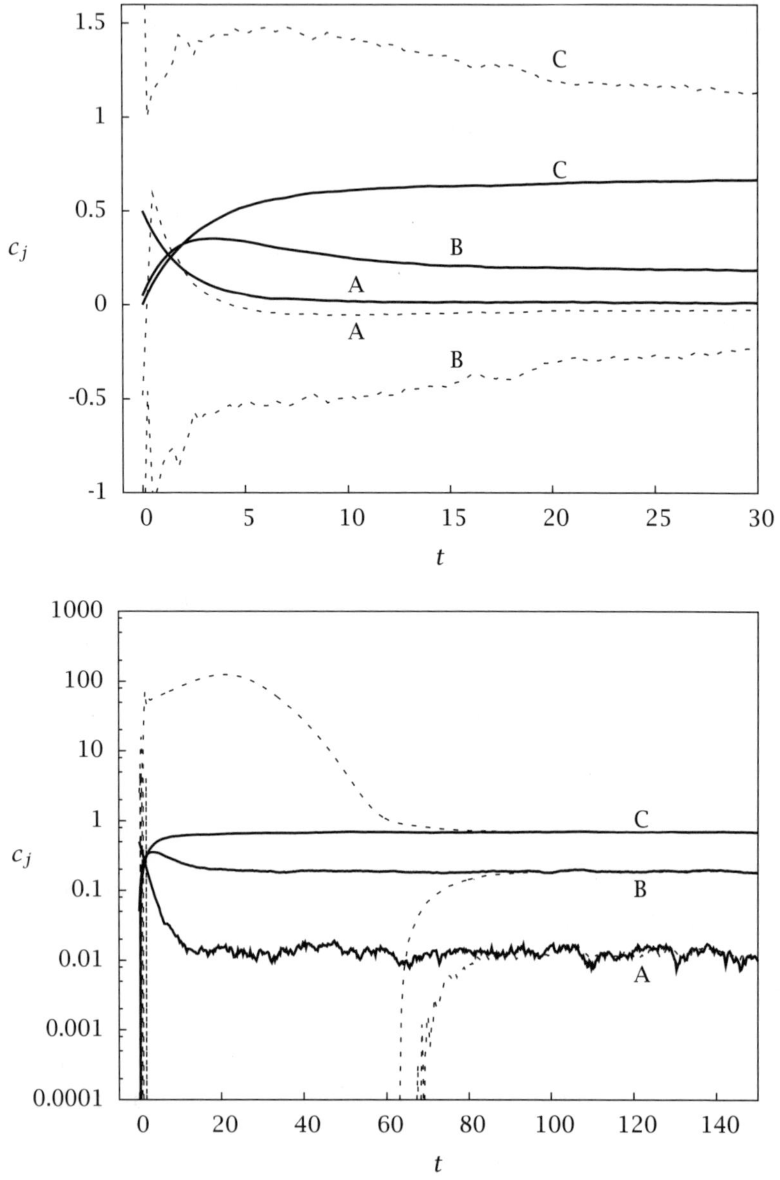

Figure 4.4: Evolution of the state (solid line) and EKF state estimate (dashed line). Top plot shows negative concentration estimates with the standard EKF. Bottom plot shows large estimate errors and slow convergence with the clipped EKF.

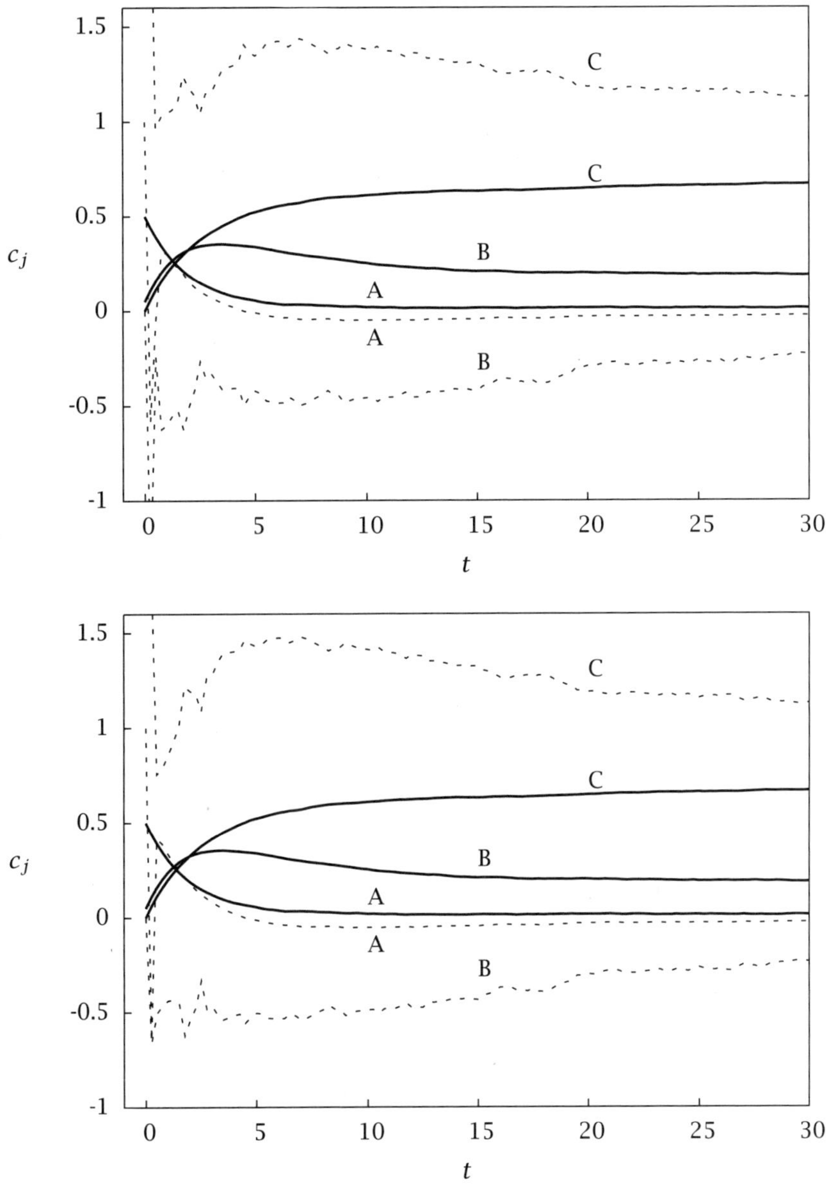

Figure 4.5: Evolution of the state (solid line) and UKF state estimate (dashed line). Top plot shows negative concentration estimates with the standard UKF. Bottom plot shows similar problems even if constraint scaling is applied.

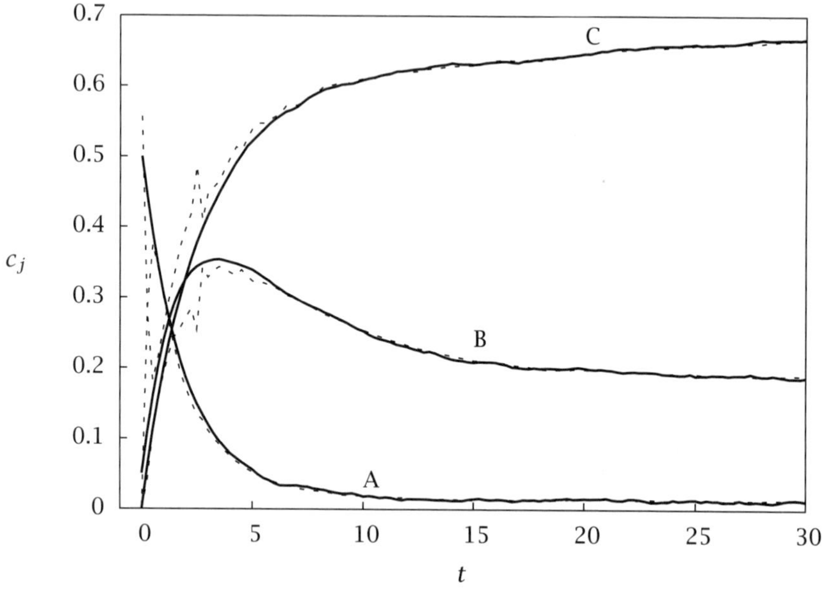

Figure 4.6: Evolution of the state (solid line) and MHE state estimate (dashed line).

estimates that are two orders of magnitude larger than the measured pressure before convergence is achieved.

The standard UKF achieves results similar to the EKF as shown in Figure 4.5 (top). Vachhani, Narasimhan, and Rengaswamy (2006) have proposed a modification to the UKF to handle constrained systems. In this approach, the sigma points that violate the constraints are scaled back to the feasible region boundaries and the sigma point weights are modified accordingly. If this constrained version of the UKF is applied to this case study, the estimates do not significantly improve as shown in Figure 4.5 (bottom). The UKF formulations used here are based on the algorithm presented by Vachhani et al. (2006, Sections 3 and 4) with the tuning parameter κ set to $\kappa = 1$. Adjusting this parameter using other suggestions from the literature (Julier and Uhlmann, 1997; Qu and Hahn, 2009; Kandepu, Imsland, and Foss, 2008) and trial and error does not substantially improve the UKF estimator performance. Better performance is obtained in this example if the sigma points that violate the constraints are simply saturated rather than rescaled to the feasible region boundaries. This form of clipping still does not prevent the occurrence of negative concentrations in this example, however. Negative concentration estimates are not avoided by either scaling or clipping of the sigma points. As a solution to this problem, the use of constrained optimization for the sigma points is proposed (Vachhani et al., 2006; Teixeira, Tôrres, Aguirre, and Bernstein, 2008). If one is willing to perform online optimization, however, MHE with a short horizon is likely to provide more accurate estimates at similar computational cost compared to approaches based on optimizing the locations of the sigma points.

Finally, Figure 4.6 presents typical results of applying constrained MHE to this example. For this simulation we choose $N = 10$ and the smoothing update for the arrival cost approximation. Note that MHE recovers well from the poor initial prior. Comparable performance is obtained if the filtering update is used instead of the smoothing update to approximate the arrival cost. The MHE estimates are also insensitive to the choice of horizon length N for this example. □

The EKF, UKF, and all one-step recursive estimation methods, suffer from the "short horizon syndrome" by *design*. One can try to reduce the harmful effects of a short horizon through tuning various other parameters in the estimator, but the basic problem remains. Large initial state errors lead to inaccurate estimation and potential estimator divergence. The one-step recursions such as the EKF and UKF can be viewed as one extreme in the choice between speed and accuracy in that only a single measurement is considered at each sample. That is similar to an MHE problem in which the user chooses $N = 1$. Situations in which $N = 1$ lead to poor MHE performance often lead to unreliable EKF and UKF performance as well.

4.7 Particle Filtering

Particle filtering is a different approach to the state estimation problem in which statistical sampling is used to approximate the evolution of the conditional density of the state given measurements (Handschin and Mayne, 1969). This method also handles nonlinear dynamic models and can address nonnormally distributed random disturbances to the state and measurement.

Sampled density. Consider a smooth probability density, $p(x)$. In particle filtering we find it convenient to represent this smooth density

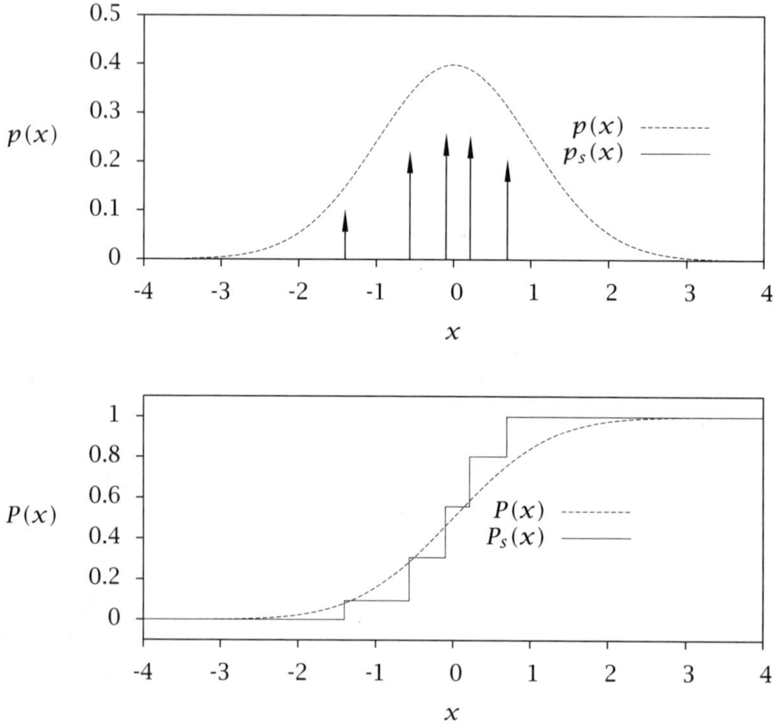

Figure 4.7: Top: exact density $p(x)$ and a sampled density $p_s(x)$ with five samples for $\xi \sim N(0, 1)$. Bottom: corresponding exact $P(x)$ and sampled $P_s(x)$ cumulative distributions.

as a weighted, sampled density, $p_s(x)$

$$p(x) \approx p_s(x) := \sum_{i=1}^{s} w_i \delta(x - x_i)$$

in which $x_i, i = 1, \ldots s$ are the samples, w_i are the weights. As an example, the top of Figure 4.7 displays a normally distributed scalar random variable represented by a sampled density with five samples. The sampled density is a series of impulses at the sample locations x_i. In this example, the weights w_i are the values of $p(x_i)$, normalized to sum to unity. It may seem strange to represent a well-behaved function like $p(x)$ with such a "rough" function like $p_s(x)$, but we will see the advantages shortly. Sometimes we may wish to study convergence of a sampled density to the original density as the number of samples becomes large. To define convergence of this representation of the probability distribution, we refer to the corresponding cumulative distribution rather than the density. From integration, the sampled cumulative distribution is

$$P_s(x) = \sum_{i \in \mathbb{I}_x} w_i \qquad \mathbb{I}_x = \{i | x_i \leq x\}$$

The bottom of Figure 4.7 shows the corresponding cumulative sampled distribution for the sampled density with five samples. The cumulative sampled distribution is a staircase function with steps of size w_i at the sample locations x_i. We can then measure convergence of $P_s(x)$ to $P(x)$ as $s \to \infty$ in any convenient function norm. We delay further discussion of convergence until Section 4.7.2 in which we present some of the methods for choosing the samples and the weights.

In the sequel, we mostly drop the subscript s on sampled densities and cumulative distributions when it is clear from context that we are referring to this type of representation of a probability distribution. We can conveniently calculate the expectation of any function of a random variable having a sampled density by direct integration to obtain

$$\begin{aligned}
\mathcal{E}(f(\xi)) &= \int p_s(x) f(x) dx \\
&= \int \sum_i w_i \delta(x - x_i) f(x) dx \\
&= \sum_i w_i f(x_i)
\end{aligned}$$

For example, we often wish to evaluate the mean of the sampled density, which is

$$\mathcal{E}(\xi) = \sum_i w_i x_i$$

The convenience of integrating the sampled density is one of its main attractive features. If we create a new function (not necessarily a density) by multiplication of $p(x)$ by another function $g(x)$

$$\overline{p}(x) = g(x) p(x)$$

we can easily obtain the sampled function $\overline{p}$. We simply adjust the weights and leave the samples where they are

$$\begin{aligned}
\overline{p}(x) &= g(x)p(x) \\
&= \sum_i g(x) w_i \delta(x - x_i) \\
&= \sum_i w_i g(x_i) \delta(x - x_i) \\
\overline{p}(x) &= \sum_i \overline{w}_i \delta(x - x_i) \qquad \overline{w}_i = w_i g(x_i)
\end{aligned} \tag{4.36}$$

4.7.1 The Sampled Density of a Transformed Random Variable

Given the random variable ξ, assume we have a sampled density for its density $p_\xi(x)$

$$p_\xi(x) = \sum_{i=1}^{s} w_i \delta(x - x_i)$$

Define a new random variable η by an invertible, possibly nonlinear transformation

$$\eta = f(\xi) \qquad \xi = f^{-1}(\eta)$$

We wish to find a sampled density for the random variable η, $p_\eta(y)$. Denote the sampled density for η as

$$p_\eta(y) = \sum_{i=1}^{s} \overline{w}_i \delta(y - y_i)$$

We wish to find formulas for $\overline{w}_i$ and y_i in terms of w_i, x_i and f. We proceed as in the development of equation (A.30) in Appendix A. We wish to have an equivalence for every function $g(x)$

$$\begin{aligned}
\mathcal{E}_{p_\xi}(g(\xi)) &= \mathcal{E}_{p_\eta}(g(f^{-1}(\eta))) \\
\int p_\xi(x) g(x) dx &= \int p_\eta(y) g(f^{-1}(y)) dy \qquad \text{for all } g(\cdot)
\end{aligned}$$

Using the sampled densities on both sides of the equation

$$\sum_i w_i g(x_i) = \sum_i \overline{w}_i g(f^{-1}(y_i))$$

One solution to this equation that holds for every g is the simple choice

$$y_i = f(x_i) \qquad \overline{w}_i = w_i \tag{4.37}$$

We see that for the transformed sampled density, we transform the samples and use the weights of the original density.

Example 4.28: Sampled density of the lognormal

The random variable η is distributed as a lognormal if its logarithm is distributed as a normal. Let $\xi \sim N(0, P)$ and consider the transformation

$$\eta = e^{\xi} \qquad \xi = \log(\eta) \quad \eta > 0$$

Represent p_ξ as a sampled density, use (4.37) to find a sampled density of p_η, and plot histograms of the two sampled densities. Compare the sampled density of p_η to the lognormal density. The two densities are given by

$$\begin{aligned} p_\xi(x) &= \frac{1}{\sqrt{2\pi P}} e^{-x^2/2P} \\ p_\eta(y) &= \frac{1}{y\sqrt{2\pi P}} e^{-\log^2(y)/2P} \end{aligned}$$

Solution

First we take samples x_i from $N(0,1)$ for ξ. Figure 4.8 shows the histogram of the sampled density for 5000 samples. Next we compute $y_i = e_i^x$ to generate the samples of η. The histogram of this sampled density is shown in Figure 4.9. Notice the good agreement between the sampled density and the lognormal density, which is shown as the continuous curve in Figure 4.9. □

Noninvertible transformations. Next consider η to be a noninvertible transformation of ξ

$$\eta = f(\xi) \qquad f \text{ not invertible}$$

Let ξ's sampled density be given by $\{x_i, w_i\}$. The sampled density $\{f(x_i), w_i\}$ remains a valid sampled density for η, which we show next

$$\xi \sim \{x_i, w_i\} \qquad \eta \sim \{f(x_i), w_i\}$$

We wish to show that

$$\mathcal{E}_{p_\eta}(g(\eta)) = \mathcal{E}_{p_\xi}(g(f(\xi))) \qquad \text{for all } g(\cdot)$$

Taking the expectations

$$\int p_\eta(y) g(y) dy = \int p_\xi(x) g(f(x)) dx$$

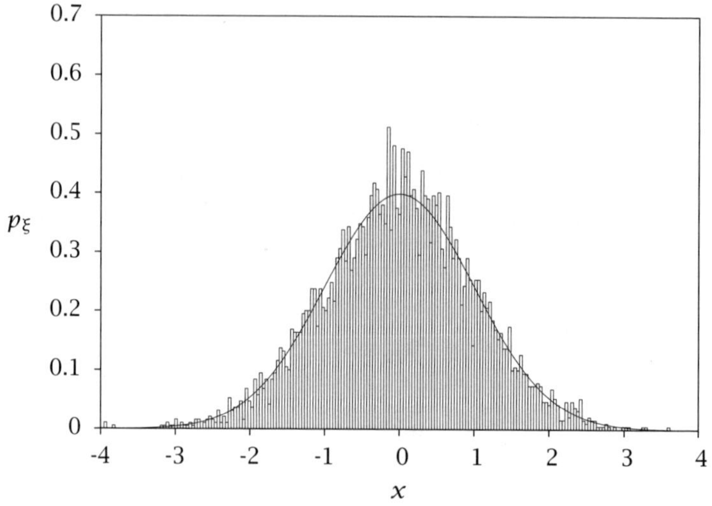

Figure 4.8: Sampled and exact probability densities for $\xi \sim N(0,1)$; 5000 samples.

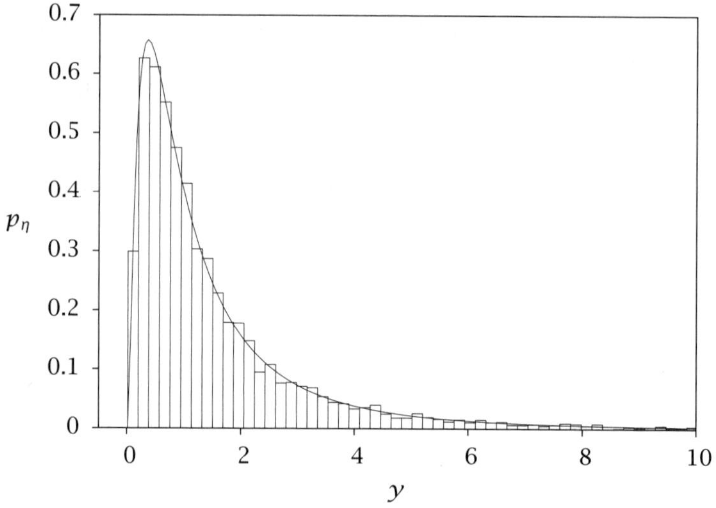

Figure 4.9: Sampled and exact probability densities for nonlinear transformation $\eta = e^\xi$; 5000 samples. The exact density of η is the lognormal, shown as the continuous curve.

Letting η's sampled density be $\{y_i, \overline{w}_i\}$, and using ξ's sampled density give

$$\sum_{i=1}^{s} \overline{w}_i g(y_i) = \sum_{i=1}^{s} w_i g(f(x_i))$$

and setting $\overline{w}_i = w_i, y_i = f(x_i), i = 1, \ldots, s$ achieves equality for all $g(\cdot)$, and we have established the result. The difference between the noninvertible and invertible cases is that we do not have a method to obtain samples of ξ from samples of η in the noninvertible case. We can transform the sampled density in only one direction, from p_ξ to p_η.

4.7.2 Sampling and Importance Sampling

Consider a random variable ξ with a smooth probability density $p(x)$. Assume one is able to draw samples x_i of ξ with probability

$$p_{\text{sa}}(x_i) = p(x_i) \tag{4.38}$$

in which $p_{\text{sa}}(x_i)$ denotes the probability of drawing a sample with value x_i. In this case, if one draws s samples, a sampled density for ξ is given by

$$p_s(x) = \sum_i w_i \delta(x - x_i) \qquad w_i = 1/s, \quad i = 1, \ldots, s \tag{4.39}$$

and the weights are all equal to $1/s$.

Convergence of sampled densities. It is instructive to examine how a typical sampled density converges with sample size to the density from which the samples are drawn. Consider a set of s samples. When drawing multiple samples of a density, we assume the samples are mutually independent

$$p_{\text{sa}}(x_1, x_2, \ldots, x_s) = p_{\text{sa}}(x_1) p_{\text{sa}}(x_2) \cdots p_{\text{sa}}(x_s)$$

We denote the cumulative distribution of the sampled density as

$$P_s(x; s) = \sum_{i \in \mathbb{I}_x} w_i \qquad \mathbb{I}_x = \{i | x_i \leq x\}$$

in which the second argument s is included to indicate P_s's dependence on the sample size. The value of P_s is itself a random variable because it is determined by the sample values x_i and weights w_i. We consider

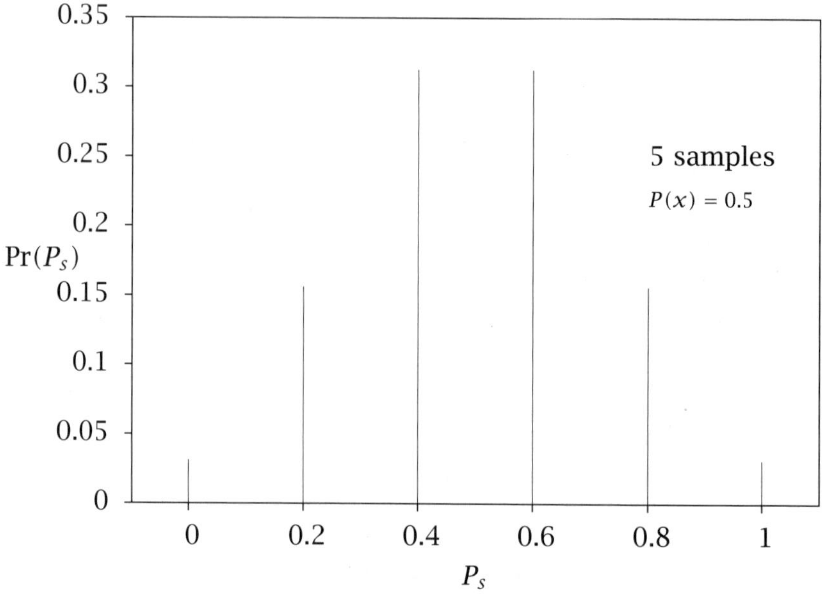

Figure 4.10: Probability density $\Pr(P_s(x;s))$ for x corresponding to $P(x) = 0.5$ and $s = 5$ samples. The distribution is centered at correct value, $P(x)$, but the variance is large.

the case with equal sample weights $w_i = 1/s$ and study the P_s values as a function of s and scalar x. They take values in the range

$$P_s \in \left\{0, \frac{1}{s}, \ldots, \frac{s-1}{s}, 1\right\} \qquad s \geq 1 \quad -\infty < x < \infty$$

Given the sampling process we can readily evaluate the probability of P_s over this set

$$\Pr(P_s(x;s)) = \begin{cases} \binom{i}{s} P(x)^i (1-P(x))^{s-i}, & P_s = \frac{i}{s}, \quad i = 0, \ldots, s \\ 0, & \text{otherwise} \end{cases}$$
$$-\infty < x < \infty \qquad (4.40)$$

These probabilities are calculated as follows. For P_s to take on value zero, for example, all of the samples x_i must be greater than x. The probability that any x_i is greater than x is $1-P(x)$. Because the samples are mutually independent, the probability that all s samples are greater than x is $(1-P(x))^s$, which is the $i = 0$ entry of (4.40). Similarly, for P_s

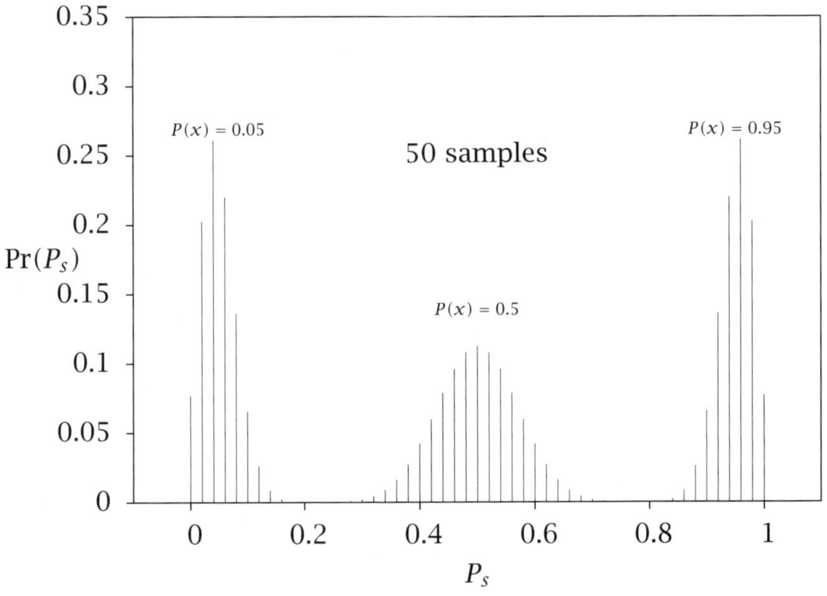

Figure 4.11: Probability density $\Pr(P_s(x;s))$ for three different x corresponding to $P(x) = 0.05, 0.5, 0.95$ and $s = 50$ samples. The three distributions are centered at the correct values, $P(x)$, and the variance is much reduced compared to Figure 4.10.

to have value i/s, i samples must be less than x and $s-i$ samples must be greater than x. This probability is given by $\binom{i}{s}P(x)^i(1-P(x)^{s-i})$, in which $P(x)^i(1-P(x)^{s-i})$ is the probability of having a sample with i values less than x and $s-i$ values greater than x, and $\binom{i}{s}$ accounts for the number of ways such a sample can be drawn from a set of s samples. Figure 4.10 shows the distribution of P_s for a sample size $s = 5$ at the mean, $P(x) = 0.5$. Notice the maximum probability occurs near the value $P_s = P(x)$ but the probability distribution is fairly wide with only 5 samples. The number of samples is increased to 50 in Figure 4.11, and three different x values are shown, at which $P(x) = 0.05, 0.5, 0.95$. The peak for each P_s distribution is near the value $P(x)$, and the distribution is much narrower for 50 samples. The sampled density $P_s(x;s)$ becomes arbitrarily sharply distributed with value $P(x)$

as the sample size s increases.

$$\lim_{s\to\infty} \Pr(P_s(x;s)) = \begin{cases} 1 & P_s = P(x) \\ 0 & \text{otherwise} \end{cases} \qquad -\infty < x < \infty$$

The convergence is often not uniform in x. Achieving a given variance in $P_s(x;s)$ generally requires larger sample sizes for x values in the tails of the density $p(x)$ compared to the sample sizes required to achieve this variance for x values in regions of high density. The nonuniform convergence is perhaps displayed more clearly in Figures 4.12 and 4.13. We have chosen the beta distribution for $P(x)$ and show the spread in the probability of P_s for three x values, corresponding to $P(x) = \{0.1, 0.5, 0.9\}$. Given $s = 25$ samples in Figure 4.12, we see a rather broad probability distribution for the sampled distribution $P_s(x)$. Turning up the number of samples to $s = 250$ gives the tighter probability distribution shown in Figure 4.13.

Finally, we present a classic sampling error distribution result due to Kolmogorov. The measure of sampling error is defined to be

$$D_s = \sup_x |P_s(x;s) - P(x)|$$

and we have the following result on the distribution of D_s for large sample sizes.

Theorem 4.29 (Kolmogoroff (1933)). [6] *Suppose that $P(x)$ is continuous. Then for every fixed $z \geq 0$ as $s \to \infty$*

$$\Pr\left(D_s \leq zs^{-1/2}\right) \to L(z) \tag{4.41}$$

in which $L(z)$ is the cumulative distribution function given for $z > 0$ by

$$L(z) = \sqrt{2\pi}z^{-1} \sum_{\nu=1}^{\infty} e^{-(2\nu-1)^2\pi^2/8z^2} \tag{4.42}$$

and $L(z) = 0$ for $z \leq 0$.

One of the significant features of results such as this one is that the limiting distribution is independent of the details of the sampled distribution $P(x)$ itself. Feller (1948) provides a proof of this theorem and discussion of this and other famous sampling error distribution results due to Smirnov (1939).

---

[6] Kolmogorov's theorem on sampling error was published in an Italian journal with the spelling Kolmogoroff.

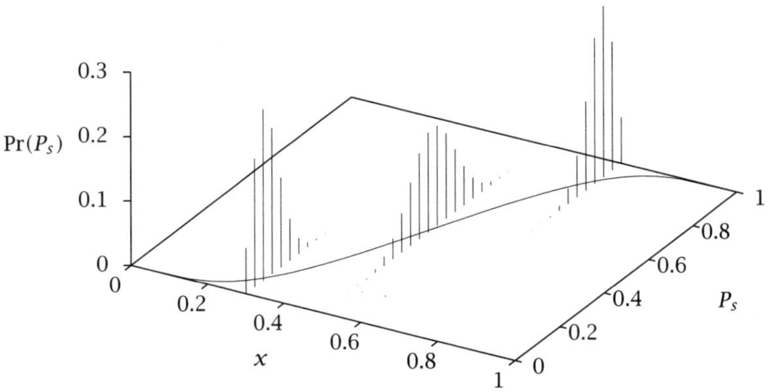

Figure 4.12: Probability density $\Pr(P_s(x;s))$ for $s = 25$ samples at three different x.

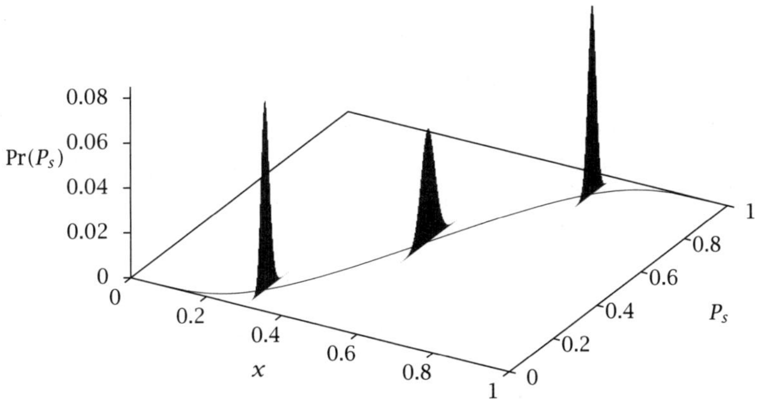

Figure 4.13: Probability density $\Pr(P_s(x;s))$ for $s = 250$ samples. Note the variance is much reduced compared to Figure 4.12.

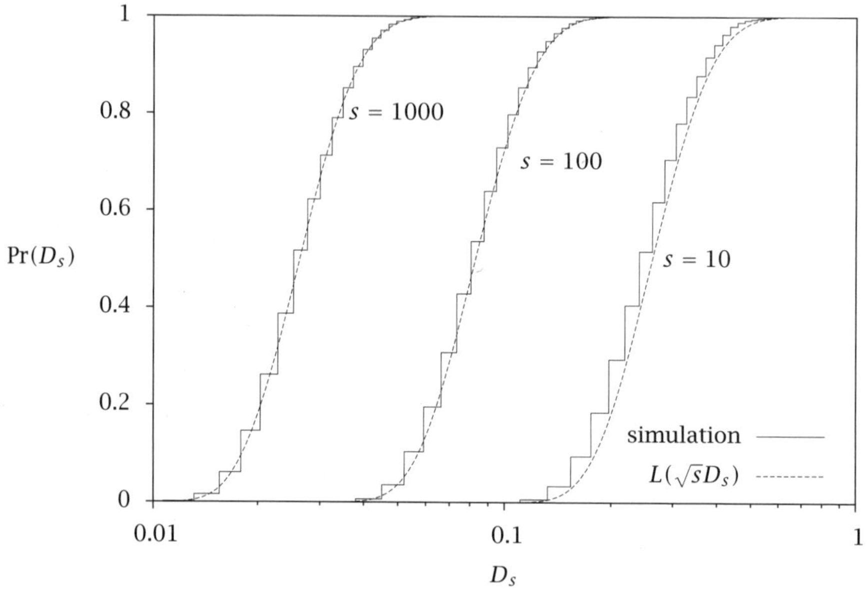

Figure 4.14: Cumulative distribution for the sampling error $\Pr(D_s)$ for three different sample sizes, $s = 10, 100, 1000$. Distribution from simulation using 5000 realizations (solid) and Kolmogorov limiting distribution (dashed).

Example 4.30: Sampling error distribution for many samples

Plot the actual and limiting distributions for D_s for $s = 10, 100, 1000$ when sampling a normal distribution with unit variance. How close is the limiting sampling error distribution to the actual sampling error distribution for these three sample sizes?

Solution

Figure 4.14 displays the result using 5000 realizations of the sampling process to approximate the actual distribution of D_s. Notice that for the small sample size, we can see a slight difference between the Kolmogorov limiting distribution and the one obtained from simulation. This difference is not noticeable for samples sizes greater than $s = 100$. From the argument scaling given in (4.41) we see that the mean of the sampling error decreases by a factor of $\sqrt{10}$ for each factor of 10 increase in sample size (on the log scale, the distribution of D_s is translated to the left by $\sqrt{10}$). Exercise 4.20 discusses this example further. □

Unbiasedness of sampled densities. A sampled density is *unbiased* if it possesses the following property

$$\mathcal{E}_{\text{sa}}(P_s(x;s)) = P(x) \qquad 1 \leq s, \quad -\infty < x < \infty$$

in which the expectation is taken over the probability density of P_s considered as a random variable as discussed previously. As we discuss subsequently, some sampling procedures are unbiased for all s, while others are only asymptotically unbiased as s becomes large. A convenient test for unbiasedness is the following

$$\mathcal{E}_{\text{sa}}\left(\int p_s(x)g(x)dx\right) = \int p(x)g(x)dx \qquad \text{for all } g(\cdot) \tag{4.43}$$

In other words, the *expectation over the sampling process* of integrals of any function g with the sampled density should be equal to the integral of g with the exact density. If the sampling process has the probability given by (4.38), we can verify (4.43) as follows

$$\begin{aligned}
\mathcal{E}_{\text{sa}}\left(\int p_s(x)g(x)dx\right) &= \mathcal{E}_{\text{sa}}\left(\sum_i w_i g(x_i)\right) \\
&= \int p_{\text{sa}}(x_1,\ldots,x_s)\sum_i w_i g(x_i)dx_1\cdots dx_s \\
&= \int p_{\text{sa}}(x_1)\cdots p_{\text{sa}}(x_s)\sum_i w_i g(x_i)dx_1\cdots dx_s \\
&= \int p(x_1)\cdots p(x_s)\sum_i w_i g(x_i)dx_1\cdots dx_s \\
&= \frac{1}{s}\sum_i \int p(x_i)g(x_i)dx_i \prod_{j\neq i}\int p(x_j)dx_j \\
&= \frac{1}{s}\sum_i \int p(x_i)g(x_i)dx_i \\
\mathcal{E}_{\text{sa}}\left(\int p_s(x)g(x)dx\right) &= \int p(x)g(x)dx
\end{aligned}$$

Example 4.31: Sampling independent random variables

Consider two independent random variables ξ, η, whose probability density satisfies

$$p_{\xi,\eta}(x,y) = p_\xi(x)p_\eta(y)$$

and assume we have samples of the two marginals

$$\xi \sim \{x_i, w_{xi}\} \qquad w_{xi} = 1/s_x, \quad i = 1, \ldots, s_x$$
$$\eta \sim \{y_j, w_{yj}\} \qquad w_{yj} = 1/s_y, \quad j = 1, \ldots, s_y$$

We have many valid options for creating samples of the joint density. Here are three useful ones.

(a) Show the following is a valid sample of the joint density

$$\{(x_i, y_j), w_{ij}\} \qquad w_{ij} = 1/(s_x s_y), \quad i = 1, \ldots, s_x, \quad j = 1, \ldots, s_y$$

Notice we have $s_x s_y$ total samples of the joint density.

(b) If $s_x = s_y = s$, show the following is a valid sample of the joint density

$$\{(x_i, y_i), w_i\} \qquad w_i = 1/s, \quad i = 1, \ldots, s$$

Notice we have s total samples of the joint density unlike the previous case in which we would have s^2 samples.

(c) If we have available (or select) only a single sample of ξ's marginal, $s_x = 1$ and $s_y = s$ samples of η's marginal, show the following is a valid sample of the joint density

$$\{(x_1, y_i), w_{yi}\} \qquad w_{yi} = 1/s, \quad i = 1, \ldots, s$$

Here we have generated again s samples of the joint density, but we have allowed unequal numbers of samples of the two marginals.

Solution

Because the two random variables are independent, the probability of drawing a sample with values (x_i, y_j) is given by

$$p_{sa}(x_i, y_j) = p_{sa}(x_i) p_{sa}(y_j) = p_\xi(x_i) p_\eta(y_j) = p_{\xi,\eta}(x_i, y_j)$$

Denote the samples as $z_k = (x_{i(k)}, y_{j(k)})$. We have for all three choices

$$p_{sa}(z_k) = p_{\xi,\eta}(z_k) \qquad k = 1, \ldots, s \tag{4.44}$$

(a) For this case,

$$i(k) = \text{mod}(k-1, s_x) + 1 \qquad j(k) = \text{ceil}(k/s_x)$$
$$w_k = \frac{1}{s_x s_y}, \quad k = 1, \ldots, s_x s_y$$

in which $\text{ceil}(x)$ is the smallest integer not less than x.

(b) For this case

$$i(k) = k \qquad j(k) = k \qquad w_k = 1/s, \quad k = 1, \ldots, s$$

(c) For this case

$$i(k) = 1 \qquad j(k) = k \qquad w_k = 1/s, \quad k = 1, \ldots, s$$

Because all three cases satisfy (4.44) and the weights are equal to each other in each case, these are all valid samples of the joint density. □

If we arrange the s_x ξ samples and s_y η samples in a rectangle, the first option takes all the points in the rectangle, the second option takes the diagonal (for a square), and the third option takes one edge of the rectangle. See Exercise 4.19 for taking a single point in the rectangle. In fact, any set of points in the rectangle is a valid sample of the joint density.

Example 4.32: Sampling a conditional density

The following result proves useful in the later discussion of particle filtering. Consider conditional densities satisfying the following property

$$p(a, b, c|d, e) = p(a|b, d)p(b, c|e) \tag{4.45}$$

We wish to draw samples of $p(a, b, c|d, e)$ and we proceed as follows. We draw samples of $p(b, c|e)$. Call these samples $(b_i, c_i), i = 1, \ldots, s$. Next we draw for each $i = 1, \ldots, s$, one sample of $p(a|b_i, d)$. Call these samples a_i. We assemble the s samples (a_i, b_i, c_i) and claim they are samples of the desired density $p(a, b, c|d, e)$ with uniform weights $w_i = 1/s$. Prove or disprove this claim.

Solution

The claim is true, and to prove it we need to establish that the probability of drawing a sample with value (a_i, b_i, c_i) is equal to the desired density $p(a_i, b_i, c_i|d, e)$. We proceed as follows. From the definition of conditional density, we know

$$p_{\text{sa}}(a_i, b_i, c_i|d, e) = p_{\text{sa}}(a_i|b_i, c_i, d, e)p_{\text{sa}}(b_i, c_i|d, e)$$

For the selection of a_i described previously, we know

$$p_{\text{sa}}(a_i|b_i, c_i, d, e) = p(a_i|b_i, d)$$

The values of c_i and e are irrelevant to the sampling procedure generating the a_i. For the (b_i, c_i) samples, the sampling procedure gives

$$p_{\text{sa}}(b_i, c_i|d, e) = p(b_i, c_i|e)$$

and the value of d is irrelevant to the procedure for generating the (b_i, c_i) samples. Combining these results, we have for the (a_i, b_i, c_i) samples

$$p_{\text{sa}}(a_i, b_i, c_i|d, e) = p(a_i|b_i, d)p(b_i, c_i|e)$$

Equation (4.45) then gives

$$p_{\text{sa}}(a_i, b_i, c_i|d, e) = p(a_i, b_i, c_i|d, e)$$

We conclude the sampling procedure is selecting (a_i, b_i, c_i) samples with the desired probability, and as shown in (4.39), the weights are all equal to $1/s$ under this kind of sampling. □

Importance sampling. Consider next the case in which we have a smooth density $p(x)$ that is easy to *evaluate* but difficult to *sample* with probability given by (4.38). This situation is not unusual. In fact, it arises frequently in applications for the following reason. Many good algorithms are available for generating samples of the uniform density. One simple method to sample an arbitrary density for a scalar random variable is the following. First compute $P(x)$ from $p(x)$ by integration. Let u_i be the samples of the uniform density on the interval $[0, 1]$. Then samples of x_i of density $p(x)$ are given by

$$x_i = P^{-1}(u_i) \qquad u_i = P(x_i)$$

Figures 4.15 and 4.16 give a graphical display of this procedure. We briefly verify that the samples x_i have the claimed density. We show that if μ is a uniform random variable and ξ is defined by the invertible transformation given previously, $\mu = P(\xi)$, then ξ has density $p_\xi(x) = dP(x)/dx$. From (A.30) we have

$$p_\xi(x) = p_\mu(P(x)) \left| \frac{dP(x)}{dx} \right|$$

Since μ is uniformly distributed, $p_\mu = 1$, and $dP(x)/dx \geq 0$, we have

$$p_\xi(x) = \frac{dP(x)}{dx}$$

and the samples have the desired density. But notice this procedure for generating samples of $p(x)$ uses $P(x)$, which requires integration, as well as evaluating $P^{-1}(x)$, which generally requires solving nonlinear equations. Importance sampling is a method for sampling $p(x)$ without performing integration or solving nonlinear equations.

The following idea motivates importance sampling. Consider the random variable ξ to be distributed with density p. Consider a new random variable η to be distributed with density q

$$\xi \sim p(x) \qquad \eta \sim q(x)$$

The density $q(x)$, known as the importance function, is any density that can be readily sampled according to (4.38) and has the same support as p. Examples of such q are uniforms for bounded intervals, lognormals and exponentials for semi-infinite intervals, and normals for infinite intervals. For any function $g(x)$, we have

$$\begin{aligned}
\mathcal{E}_p(g(\xi)) &= \int g(x)p(x)dx \\
&= \int \left[g(x)\frac{p(x)}{q(x)} \right] q(x)dx \\
\mathcal{E}_p(g(\xi)) &= \mathcal{E}_q\left(g(\eta)\frac{p(\eta)}{q(\eta)} \right) \qquad \text{for all } g(\cdot)
\end{aligned}$$

When we can sample p directly, we use for the sampled density

$$p_s = \left\{ x_i, \quad w_i = \frac{1}{s} \right\} \qquad p_{sa}(x_i) = p(x_i)$$

So when we cannot conveniently sample p but can sample q, we use instead

$$\overline{p}_s = \left\{ x_i, \quad w_i = \frac{1}{s}\frac{p(x_i)}{q(x_i)} \right\} \qquad p_{is}(x_i) = q(x_i)$$

Given s samples x_i from $q(x)$, denote the sampled density of q as q_s, and we have defined the importance-sampled density $\overline{p}_s(x)$ as

$$\overline{p}_s(x) = q_s(x)\frac{p(x)}{q(x)}$$

We next show that $\overline{p}_s(x)$ converges to $p(x)$ as sample size increases (Smith and Gelfand, 1992). Using the fact that q_s converges to q gives

$$\lim_{s\to\infty} \overline{p}_s(x) = \lim_{s\to\infty} q_s(x)\frac{p(x)}{q(x)} = p(x)$$

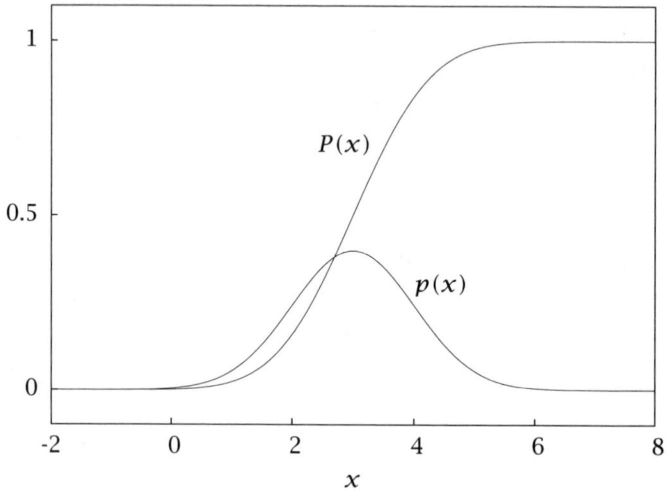

Figure 4.15: Probability density $p(x)$ to be sampled and the corresponding cumulative distribution $P(x)$.

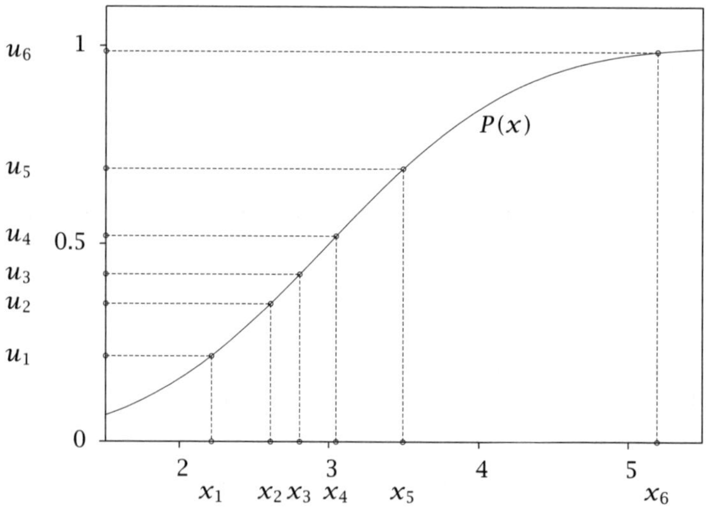

Figure 4.16: Six samples of the uniform density on $[0, 1]$, u_i, and the corresponding samples of $p(x)$, x_i. The samples satisfy $x_i = P^{-1}(u_i)$.

The weighted sample of p is also unbiased for all sample sizes, which we can verify as follows

$$\begin{aligned}
\mathcal{E}_{\text{is}}\left(\overline{p}_s(x)\right) &= \mathcal{E}_{\text{is}}\left(\sum_i w_i \delta(x - x_i)\right) \\
&= \int p_{\text{is}}(x_1, \ldots, x_s) \sum_i w_i \delta(x - x_i) dx_1 \cdots dx_s \\
&= \int q(x_1) \cdots q(x_s) \sum_i w_i \delta(x - x_i) dx_1 \cdots dx_s \\
&= \sum_i \int q(x_i) w_i \delta(x - x_i) dx_i \prod_{j \neq i} \int q(x_j) dx_j \\
&= \sum_i \int q(x_i) \frac{1}{s} \frac{p(x_i)}{q(x_i)} \delta(x - x_i) dx_i \\
&= \frac{1}{s} \sum_i p(x) \\
\mathcal{E}_{\text{is}}\left(\overline{p}_s(x)\right) &= p(x)
\end{aligned}$$

Notice this result holds for all $s \geq 1$.

Using the same development, we can represent any function $h(x)$ (not necessarily a density) having the same support as $q(x)$ as a sampled function

$$\begin{aligned}
\overline{h}_s(x) &= \sum_{i=1}^{s} w_i \delta(x - x_i) \qquad w_i = \frac{1}{s} \frac{h(x_i)}{q(x_i)} \\
\lim_{s \to \infty} \overline{h}_s(x) &= h(x)
\end{aligned} \tag{4.46}$$

The next example demonstrates using importance sampling to generate samples of a multimodal density.

Example 4.33: Importance sampling of a multimodal density

Given the following bimodal distribution

$$p(x) = \frac{1}{2\sqrt{2\pi P_1}} e^{-(x-m_1)^2/P_1} + \frac{1}{2\sqrt{2\pi P_2}} e^{-(x-m_2)^2/P_2}$$
$$m_1 = -4 \quad m_2 = 4 \quad P_1 = P_2 = 1$$

generate samples using the following unimodal importance function

$$q(x) = \frac{1}{\sqrt{2\pi P}} e^{-(x-m)^2/P} \qquad m = 0 \quad P = 4$$

Solution

Figure 4.17 shows the exact and sampled density of the importance function $q(x)$ using 5000 samples. The weighted density for $p(x)$ is shown in Figure 4.18. We obtain a good representation of the bimodal distribution with 5000 samples. Notice also that one should use a broad density for $q(x)$ to obtain sufficient samples in regions where $p(x)$ has significant probability. Using $q(x)$ with variance of $P = 1$ instead of $P = 4$ would require many more samples to obtain an accurate representation of $p(x)$. Of course we cannot choose $q(x)$ too broad or we sample the region of interest too sparsely. Choosing an appropriate importance function for an unknown $p(x)$ is naturally a significant challenge in many applications. □

Importance sampling when the density cannot be evaluated. In many applications we have a density $p(x)$ that is difficult to evaluate directly, but it can be expressed as

$$p(x) = \frac{h(x)}{\int h(x)dx} \qquad p(x) \propto h(x)$$

in which $h(x)$ is readily evaluated. We wish to avoid the task of integration of h to find the normalizing constant. Importance sampling can still be used to sample p in this case, but, as we discuss next, we lose the unbiased property of the sampled density for finite sample sizes. In this case, define the candidate sampled density as

$$\overline{p}_s(x) = \frac{q_s(x)}{d(s)} \frac{h(x)}{q(x)} \qquad d(s) = \frac{1}{s}\sum_j \frac{h(x_j)}{q(x_j)} \tag{4.47}$$

in which the samples are again chosen from the importance function $q(x)$. Summarizing, the candidate sampled density is

$$\overline{p}_s(x) = \sum_i w_i \delta(x - x_i)$$

$$p_{\text{is}}(x_i) = q(x_i) \qquad w_i = \frac{h(x_i)/q(x_i)}{\sum_j h(x_j)/q(x_j)} \qquad i = 1, \ldots, s \tag{4.48}$$

Notice the weights are normalized in the case when we do not know the normalizing constant to convert from $h(x)$ to $p(x)$. We next show that $\overline{p}_s(x)$ converges to $p(x)$ as sample size increases (Smith and Gelfand,

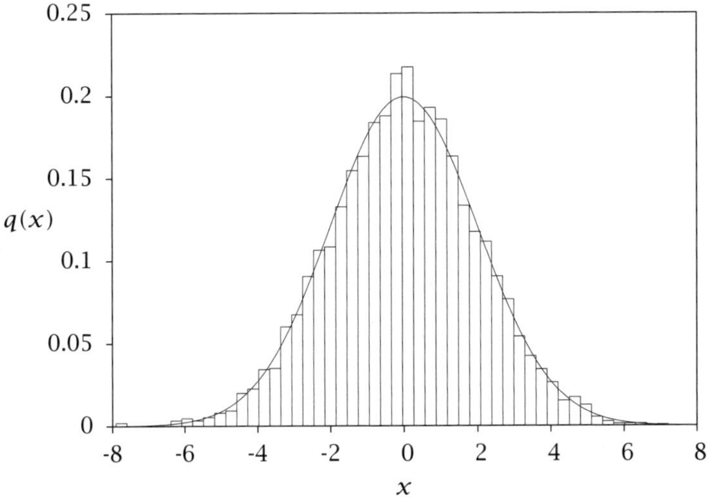

Figure 4.17: Importance function $q(x)$ and its histogram based on 5000 samples.

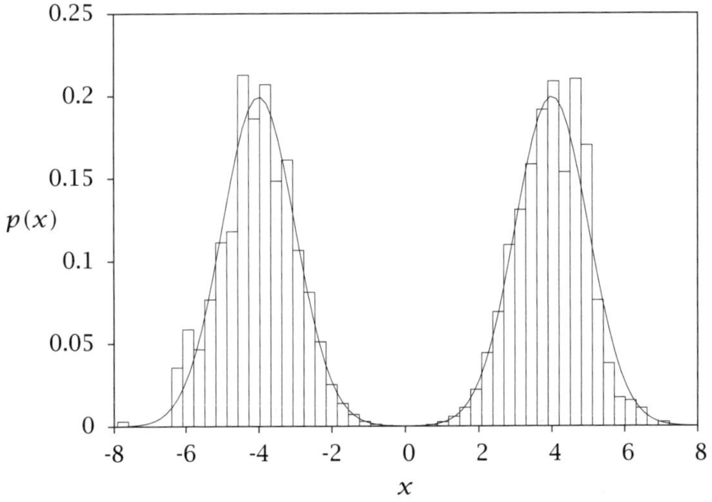

Figure 4.18: Exact density $p(x)$ and its histogram based on 5000 importance samples.

1992). First we express $d(s)$ as

$$
\begin{aligned}
d(s) &= \frac{1}{s}\sum_j \frac{h(x_j)}{q(x_j)} \\
&= \int_{-\infty}^{\infty} \frac{1}{s}\sum_j \frac{h(x_j)}{q(x_j)}\delta(x-x_j)dx \\
&= \int_{-\infty}^{\infty} \frac{1}{s}\sum_j \frac{h(x)}{q(x)}\delta(x-x_j)dx \\
d(s) &= \int_{-\infty}^{\infty} h_s(x)dx
\end{aligned}
$$

Exchanging the order of limit and integral and using (4.46) give

$$
\lim_{s\to\infty} d(s) = \int_{-\infty}^{\infty} \lim_{s\to\infty} h_s(x)dx = \int_{-\infty}^{\infty} h(x)dx
$$

Next we take the limit in (4.47) to obtain

$$
\begin{aligned}
\lim_{s\to\infty} \overline{p}_s(x) &= \lim_{s\to\infty} \frac{q_s(x)}{d(s)}\frac{h(x)}{q(x)} \\
&= \frac{\lim_{s\to\infty} q_s(x)}{\lim_{s\to\infty} d(s)}\frac{h(x)}{q(x)} \\
&= \frac{q(x)}{\int h(x)dx}\frac{h(x)}{q(x)} \\
&= \frac{h(x)}{\int h(x)dx} \\
\lim_{s\to\infty} \overline{p}_s(x) &= p(x)
\end{aligned}
$$

Notice the unbiased property no longer holds for a finite sample size. We can readily show

$$
\mathcal{E}_{\text{is}}\left(\overline{p}_s(x)\right) \neq p(x) \qquad \text{for finite } s \tag{4.49}
$$

For example, take $s = 1$. We have from (4.48) that $w_1 = 1$, and therefore

$$
\begin{aligned}
\mathcal{E}_{\text{is}}\left(\overline{p}_s(x)\right) &= \int p_{\text{is}}(x_1)w_1\delta(x-x_1)dx_1 \\
&= \int q(x_1)\delta(x-x_1)dx_1 \\
\mathcal{E}_{\text{is}}\left(\overline{p}_s(x)\right) &= q(x)
\end{aligned}
$$

and we see that the expectation of the sampling process with a single sample gives back the importance function $q(x)$ rather than the desired $p(x)$. Obviously we should choose many more samples than $s = 1$ for this case to reduce this bias. Consider the next example in which we use a large number of samples.

Example 4.34: Importance sampling of a multimodal function

We revisit Example 4.33 but use the following function $h(x)$

$$h(x) = e^{-(x-m_1)^2/P_1} + e^{-(x-m_2)^2/P_2} \qquad m_1 = -4, m_2 = 4, P_1 = P_2 = 1$$

and we do not have the normalization constant available. We again generate samples using the following importance function

$$q(x) = \frac{1}{\sqrt{2\pi P}} e^{-(x-m)^2/P} \qquad m = 0, P = 4$$

Solution

The exact and sampled density of the importance function $q(x)$ using 5000 samples is the same as Figure 4.17. The weighted density for $p(x)$ is shown in Figure 4.19. Comparing Figure 4.19 to Figure 4.18 shows the representation of the bimodal distribution with 5000 samples using $h(x)$ is of comparable quality to the one using $p(x)$ itself. The bias is not noticeable using 5000 samples. □

Weighted importance sampling. In applications of importance sampling to state estimation, the importance function is often available as a *weighted* sample in which the weights are not all equal. Therefore, as a final topic in importance sampling, we consider the case in which a weighted sample of the importance function is available

$$q_s(x) = \sum_{i=1}^{s} w_i^- \delta(x - x_i) \qquad w_i^- \geq 0$$

We have the two cases of interest covered previously.

(a) We can evaluate $p(x)$. For this case we define the sampled density for $p(x)$ as

$$\overline{p}_s(x) = \sum_{i=1}^{s} w_i \delta(x - x_i) \qquad w_i = w_i^- \frac{p(x_i)}{q(x_i)}$$

For this case, the sampled density is unbiased for all sample sizes and converges to $p(x)$ as the sample size increases.

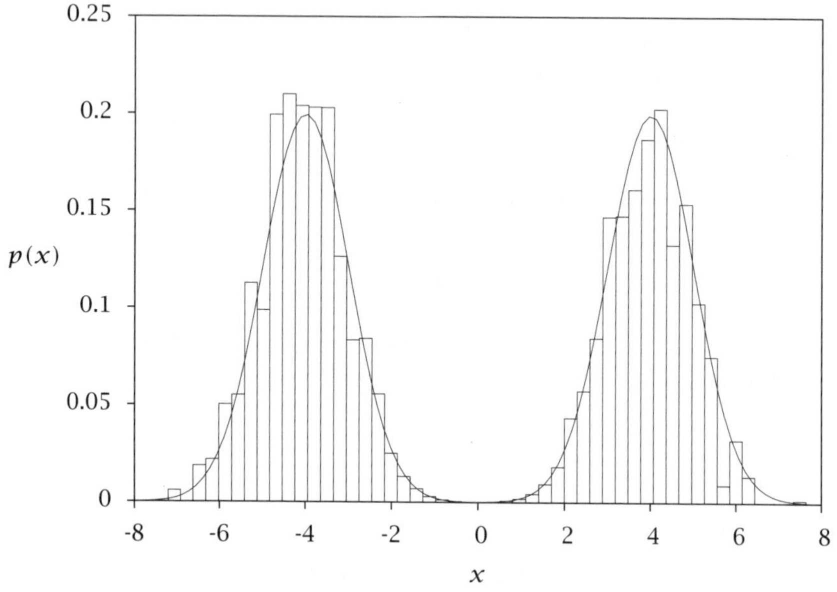

Figure 4.19: Exact density $p(x)$ and its histogram based on 5000 importance samples evaluating $h(x)$ in place of $p(x) = h(x)/\int h(x)dx$.

(b) We cannot evaluate $p(x)$, but can evaluate only $h(x)$ with $p(x) = h(x)/\int h(x)dx$. For this case, we define the sampled density as

$$\overline{p}_s(x) = \sum_{i=1}^{s} \overline{w}_i \delta(x - x_i)$$

$$w_i = w_i^- \frac{h(x_i)}{q(x_i)} \qquad \overline{w}_i = \frac{w_i}{\sum_j w_j} \tag{4.50}$$

For this case, the sampled density is biased for all finite sample sizes, but converges to $p(x)$ as the sample size increases.

The proofs of these properties are covered in Exercises 4.21 and 4.22.

4.7.3 Resampling

Consider a set of samples at $x = x_i$, $i = 1, \dots s$ and associated normalized weights w_i, $w_i \geq 0$, $\sum_{i=1}^{s} w_i = 1$. Define a probability density

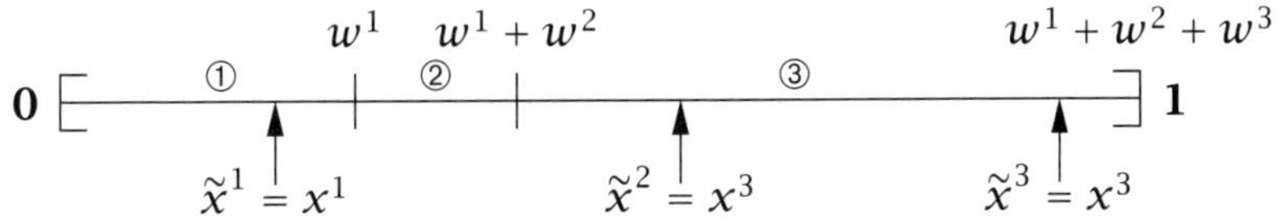

Figure 4.20: Interval $[0, 1]$ partitioned by original sample weights, w_i. The arrows depict the outcome of drawing three uniformly distributed random numbers. For the case depicted here, the new samples are $\tilde{x}_1 = x_1$, $\tilde{x}_2 = x_3$, $\tilde{x}_3 = x_3$ because the first arrow falls into the first interval and the other two arrows both fall into the third interval. Sample x_2 is discarded and sample x_3 is repeated twice in the resample. The new sample's weights are simply $\tilde{w}^1 = \tilde{w}^2 = \tilde{w}^3 = 1/3$.

using these samples and weights by

$$p(x) = \sum_{i=1}^{s} w_i \delta(x - x_i)$$

Consider any function $f(x)$ defined on a set that contains the samples, x_i. Then the integral of f using the defined density is

$$\int f(x)p(x)dx = \sum_{i=1}^{s} w_i f(x_i) = \sum_{i=1}^{s} w_i f_i$$

in which $f_i = f(x_i)$. We now consider a resampling procedure that produces a new set of samples $\tilde{x}_i$ with new weights $\tilde{w}_i$. The resampling procedure is depicted in Figure 4.20 for the case $s = 3$. We partition the interval $[0, 1]$ into s intervals using the original sample weights, w_i, as shown in Figure 4.20, in which the ith interval has width w_i. To choose s resamples, we generate s random numbers from a uniform distribution on $[0, 1]$. Denote these random numbers as u_i, $i = 1, \ldots, s$. For each i, we find the interval in which the drawn random number falls. Denote the interval number as $m(i)$, defined by the relation

$$0 \leq w_1 + w_2 + \cdots + w_{m(i)-1} \leq u_i \leq w_1 + w_2 + \cdots + w_{m(i)} \leq 1$$

We then choose as resamples

$$\tilde{x}_i = x_{m(i)} \qquad i = 1, \ldots s$$

The resampling selects the new sample locations $\widetilde{x}$ in regions of high density. We set all the $\widetilde{w}$ weights equal to $1/s$. The result illustrated in Figure 4.20 is summarized in the following table

Original sample		Resample	
State	Weight	State	Weight
x_1	$w_1 = \frac{3}{10}$	$\widetilde{x}_1 = x_1$	$\widetilde{w}_1 = \frac{1}{3}$
x_2	$w_2 = \frac{1}{10}$	$\widetilde{x}_2 = x_3$	$\widetilde{w}_2 = \frac{1}{3}$
x_3	$w_3 = \frac{6}{10}$	$\widetilde{x}_3 = x_3$	$\widetilde{w}_3 = \frac{1}{3}$

The properties of the resamples are summarized by

$$p_{\mathrm{re}}(\widetilde{x}_i) = \begin{cases} w_j & \widetilde{x}_i = x_j \\ 0 & \widetilde{x}_i \neq x_j \end{cases}$$

$$\widetilde{w}_i = 1/s \quad \text{all } i$$

We can associate with each resampling a sampled probability density

$$\widetilde{p}(x) = \sum_{i=1}^{s} \widetilde{w}_i \delta(x - \widetilde{x}_i)$$

The resampled density is clearly *not the same* as the original sampled density. It is likely that we have moved many of the new samples to places where the original density has large weights. But by resampling in the fashion described here, we have not introduced bias into the estimates.

Consider taking many such resamples. We can calculate for each of these resamples a value of the integral of f as follows

$$\int f(x)\widetilde{p}(x)dx = \sum_{i=1}^{s} \widetilde{w}_i f(\widetilde{x}_i)$$

To show this resampling procedure is valid, we show that the average over these values of the f integrals with $\widetilde{p}(x)$ is equal to the original value of the integral using $p(x)$. We state this result for the resampling procedure described previously as the following theorem.

Theorem 4.35 (Resampling). *Consider a sampled density $p(x)$ with s samples at $x = x_i$ and associated weights w_i*

$$p(x) = \sum_{i=1}^{s} w_i \delta(x - x_i) \qquad w_i \geq 0, \qquad \sum_{i=1}^{s} w_i = 1$$

Consider the resampling procedure that gives a resampled density

$$\widetilde{p}(x) = \sum_{i=1}^{s} \widetilde{w}_i \delta(x - \widetilde{x}_i)$$

in which the $\widetilde{x}_i$ *are chosen according to resample probability* p_{re}

$$p_{\text{re}}(\widetilde{x}_i) = \begin{cases} w_j & \widetilde{x}_i = x_j \\ 0 & \widetilde{x}_i \neq x_j \end{cases}$$

and with uniform weights $\widetilde{w}_i = 1/s$. *Consider a function* $f(\cdot)$ *defined on a set* X *containing the points* x_i.

With this resampling procedure, the expectation over resampling of any integral of the resampled density is equal to that same integral of the original density

$$\mathcal{E}_{\text{re}}\left(\int f(x)\widetilde{p}(x)dx\right) = \int f(x)p(x)dx \qquad \text{all } f$$

The proof of this theorem is discussed in Exercise 4.16. To get a feel for why this resampling procedure works, however, consider the case $s = 2$. There are four possible outcomes of $\widetilde{x}_1, \widetilde{x}_2$ in resampling. Because of the resampling procedure, the random variables $\widetilde{x}_i$ and $\widetilde{x}_j$, $j \neq i$ are independent, and their joint density is

$$p_{\text{re}}(\widetilde{x}_1, \widetilde{x}_2) = \begin{cases} w_1^2 & \widetilde{x}_1 = x_1, \widetilde{x}_2 = x_1 \\ w_1 w_2 & \widetilde{x}_1 = x_1, \widetilde{x}_2 = x_2 \\ w_2 w_1 & \widetilde{x}_1 = x_2, \widetilde{x}_2 = x_1 \\ w_2^2 & \widetilde{x}_1 = x_2, \widetilde{x}_2 = x_2 \end{cases}$$

The values of the integral of f for each of these four outcomes is

$$\sum_{i=1}^{2} \widetilde{w}_i f(\widetilde{x}_i) = \begin{cases} \frac{1}{2}(f_1 + f_1) & \widetilde{x}_1 = x_1, \widetilde{x}_2 = x_1 \\ \frac{1}{2}(f_1 + f_2) & \widetilde{x}_1 = x_1, \widetilde{x}_2 = x_2 \\ \frac{1}{2}(f_2 + f_1) & \widetilde{x}_1 = x_2, \widetilde{x}_2 = x_1 \\ \frac{1}{2}(f_2 + f_2) & \widetilde{x}_1 = x_2, \widetilde{x}_2 = x_2 \end{cases}$$

Notice there are only three different values for the integral of f. Next, calculating the expectation over the resampling process gives

$$\begin{aligned}
\mathcal{E}_{\text{re}}\left(\sum_{i=1}^{2}\widetilde{w}_i f(\widetilde{x}_i)\right) &= w_1^2 f_1 + w_1 w_2 (f_1 + f_2) + w_2^2 f_2^2 \\
&= (w_1^2 + w_1 w_2) f_1 + (w_1 w_2 + w_2^2) f_2 \\
&= w_1 (w_1 + w_2) f_1 + w_2 (w_1 + w_2) f_2 \\
&= w_1 f_1 + w_2 f_2 \\
&= \int f(x) p(x) dx
\end{aligned}$$

and the conclusion of the theorem is established for $s = 2$.

One can also change the total number of samples in resampling without changing the conclusions of Theorem 4.35. Exercise 4.17 explores this issue in detail. In many applications of sampling, we use the resampling process to discard samples with excessively small weights in order to reduce the storage requirements and computational burden associated with a large number of samples.

To make this discussion explicit, consider again the bimodal distribution of Exercise 4.33 shown in Figure 4.18 that was sampled using importance sampling. Many of the samples are located in the interval $[-1, 1]$ because the importance function q has large density in this interval. In fact, 1964 of the 5000 samples fall in this interval given the random sample corresponding to Figure 4.18. But notice the weights in this interval are quite small. If we resample p, we can retain the accuracy with many fewer samples as we show in the next example.

Example 4.36: Resampling a bimodal density

Consider the bimodal sampled density obtained in Example 4.33 using importance sampling. Resample this sampled density with 500 samples. Compare the accuracy to the original density with 5000 samples.

Solution

The histogram of the resampled density with 500 samples is shown in Figure 4.21. The weights in the resampled density are all equal to 1/500. Notice that the accuracy is comparable to Figure 4.18 with one tenth as many samples because most of the samples with small weights have been removed by the resampling process. In fact, none of the 500 resamples fall in the interval $[-1, 1]$. □

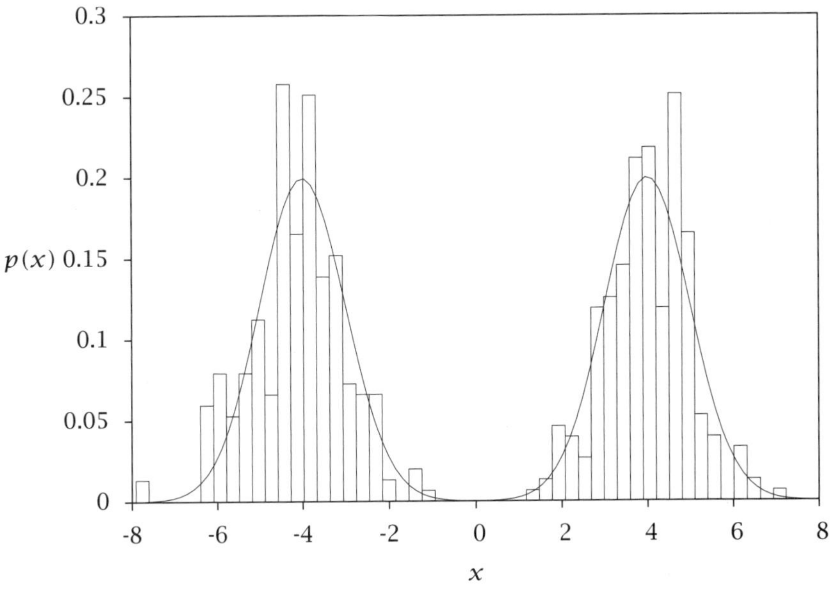

Figure 4.21: Resampled density of Example 4.33 using 500 samples. Compare to Figure 4.18 that uses 5000 samples.

4.7.4 The Simplest Particle Filter

Next we implement these sampling ideas for state estimation. This first version follows the approach given by Gordon, Salmond, and Smith (1993). In state estimation, the density $p(x(k)|\mathbf{y}(k))$ contains the information of most interest. The model is of the form

$$\begin{aligned} x(k+1) &= f(x(k), n(k)) \\ y(k) &= h(x(k), m(k)) \end{aligned}$$

in which f is a possibly nonlinear function of the state and process noise, n, and h is a possibly nonlinear function of the state and measurement noise, m. We assume that the densities of m, n and $x(0)$ are available. To start things off, first assume the conditional density $p(x(k)|\mathbf{y}(k))$ is available as a sampled density

$$p(x(k)|\mathbf{y}(k)) = \sum_{i=1}^{s} w_i(k)\delta(x(k) - x_i(k))$$

and we wish to find samples for $p(x(k+1)|\mathbf{y}(k))$. The state evolution can be considered a noninvertible transformation from $x(k), n(k)$ to $x(k+1)$, in which $n(k)$ is statistically independent of $x(k)$ and $\mathbf{y}(k)$. We generate s samples of $n(k)$, call these $n_i(k)$, and we have s samples of the conditional density $p(x(k), n(k)|\mathbf{y}(k))$ given by $\{x_i(k), n_i(k)\}, i = 1, \ldots, s$. As shown in Section 4.7.1, the sampled density of $p(x(k+1)|\mathbf{y}(k))$ is given by

$$p(x(k+1)|\mathbf{y}(k)) = \{x_i(k+1), w_i^-(k+1)\}$$
$$x_i(k+1) = f(x_i(k), n_i(k)) \qquad w_i^-(k+1) = w_i(k)$$

Next, given the sampled density for the conditional density $p(x(k)|\mathbf{y}(k-1))$

$$p(x(k)|\mathbf{y}(k-1)) = \sum_{i=1}^{s} w_i^-(k)\delta(x(k) - x_i(k))$$

we add the measurement $y(k)$ to obtain the sampled density $p(x(k)|\mathbf{y}(k))$. Notice that $\mathbf{y}(k) = (y(k), \mathbf{y}(k-1))$ and use the relationship (see Exercise 1.47)

$$p_{A|B,C}(a|b,c) = p_{C|A,B}(c|a,b)\frac{p_{A|B}(a|b)}{p_{C|B}(c|b)}$$

to obtain

$$p(x(k)|\mathbf{y}(k)) = \frac{p(y(k)|x(k), \mathbf{y}(k-1))p(x(k)|\mathbf{y}(k-1))}{p(y(k)|\mathbf{y}(k-1))}$$

Because the process is Markov, $p(y(k)|x(k), \mathbf{y}(k-1)) = p(y(k)|x(k))$, and we have

$$p(x(k)|\mathbf{y}(k)) = \frac{p(y(k)|x(k))p(x(k)|\mathbf{y}(k-1))}{p(y(k)|\mathbf{y}(k-1))}$$

The density of interest is in the form

$$p(x(k)|\mathbf{y}(k)) = g(x(k))p(x(k)|\mathbf{y}(k-1))$$
$$g(x(k)) = \frac{p(y(k)|x(k))}{p(y(k)|\mathbf{y}(k-1))}$$

and we have a sampled density for $p(x(k)|\mathbf{y}(k-1))$. If we could conveniently evaluate g, then we could obtain a sampled density using the product rule given in (4.36)

$$p(x(k)|\mathbf{y}(k)) = \{x_i(k), \widetilde{w}_i(k)\}$$

in which

$$\widetilde{w}_i(k) = w_i^-(k) \frac{p(y(k)|x_i(k))}{p(y(k)|\mathbf{y}(k-1))} \tag{4.51}$$

This method would provide an unbiased sampled density, but it is inconvenient to evaluate the term $p(y(k)|\mathbf{y}(k-1))$. So we consider an alternative in which the available sampled density is used as a weighted importance function for the conditional density of interest. If we define the importance function $q(x(k)) = p(x(k)|\mathbf{y}(k-1))$, then the conditional density is of the form

$$p(x(k)|\mathbf{y}(k)) = \frac{h(x(k))}{\int h(x(k))dx(k)}$$
$$h(x(k)) = p(y(k)|x(k))p(x(k)|\mathbf{y}(k-1))$$

We then use weighted importance sampling and (4.50) to obtain

$$p(x(k)|\mathbf{y}(k)) = \{x_i(k), \overline{w}_i(k)\} \qquad w_i(k) = w_i^-(k)p(y(k)|x_i(k))$$
$$\overline{w}_i(k) = \frac{w_i(k)}{\sum_j w_j(k)}$$

By using this form of importance sampling, the sampled density is biased for all finite sample sizes, but converges to $p(x(k)|\mathbf{y}(k))$ as the sample size increases.

Summary. Starting with s samples of $p(n(k))$ and s samples of $p(x(0))$, we assume that we can evaluate $p(y(k)|x(k))$ using the measurement equation. The iteration for the simple particle filter is summarized by the following recursion.

$$\begin{aligned}
p(x(0)) &= \{x_i(0), w_i(0) = 1/s\} \\
p(x(k)|\mathbf{y}(k)) &= \{x_i(k), \overline{w}_i(k)\} \\
w_i(k) &= w_i(k-1)p(y(k)|x_i(k)) \\
\overline{w}_i(k) &= \frac{w_i(k)}{\sum_j w_j(k)} \\
p(x(k+1)|\mathbf{y}(k)) &= \{x_i(k+1), \overline{w}_i(k)\} \\
x_i(k+1) &= f(x_i(k), n_i(k))
\end{aligned}$$

The sampled density of the simplest particle filter converges to the conditional density $p(x(k)|\mathbf{y}(k))$ in the limit of large sample size. The sampled density is biased for all finite sample sizes.

Analysis of the simplest particle filter. The simplest particle filter has well-known weaknesses that limit its use as a practical method for state estimation. The variances in both the particle locations and the filter weights can increase without bound as time increases and more measurements become available. Consider first the particle locations. For even the simple linear model with Gaussian noise, we have

$$x_i(k+1) = Ax_i(k) + Bu(k) + Gw_i(k)$$
$$x_i(0) \sim N(\overline{x}(0), Q_0) \qquad w_i(k) \sim N(0, Q)$$

which gives the following statistical properties for the particle locations

$$\begin{aligned} x_i(k) &\sim N(\overline{x}(k), \overline{P}(k)) \quad i = 1, \ldots, s \\ \overline{x}(k) &= A\overline{x}(k-1) + Bu(k) \\ \overline{P}(k) &= A\overline{P}(k-1)A' + GQG' \end{aligned} \tag{4.52}$$

If A is not strictly stable, the variance of the samples locations, $P(k)$, increases without bound despite the availability of the measurement at every time. In this simplest particle filter, one is expecting the particle weights to carry all the information in the measurements. As we will see in the upcoming example, this idea does not work and after a few time iterations the resulting state estimates are useless.

To analyze the variance of the resulting particle weights, it is helpful to define the following statistical properties and establish the following identities. Consider two random variables A and B. Conditional expectations of A and functions of A and conditional variance of A are defined as

$$\begin{aligned} \mathcal{E}(A|B) &:= \int p_{A|B}(a|b)a\,da \\ \mathcal{E}(A^2|B) &:= \int p_{A|B}(a|b)a^2\,da \\ \mathcal{E}(g(A)|B) &:= \int p_{A|B}(a|b)g(a)\,da \\ \text{var}(A|B) &:= \mathcal{E}(A^2|B) - \mathcal{E}^2(A|B) \end{aligned}$$

in which we assume as usual that B's marginal is nonzero so the conditional density is well defined. We derive a first useful identity

$$E(E(g(A)|B)) = E(g(A)) \tag{4.53}$$

as follows

$$
\begin{aligned}
\mathcal{E}(\mathcal{E}(g(A)|B)) &= \int p_B(b) \int p_{A|B}(a|b) g(a) da db \\
&= \int p_B(b) \int \frac{p_{A,B}(a,b)}{p_B(b)} g(a) da db \\
&= \iint p_{A,B}(a,b) g(a) da db \\
&= \int p_A(a) g(a) da \\
&= \mathcal{E}(g(A))
\end{aligned}
$$

We require a second identity

$$
\text{var}(A) = \mathcal{E}(\text{var}(A|B)) + \text{var}(\mathcal{E}(A|B)) \tag{4.54}
$$

which is known as the conditional variance formula or the law of total variance. We establish this identity as follows. Starting with the definition of variance

$$
\text{var}(A) = \mathcal{E}(A^2) - \mathcal{E}^2(A)
$$

we use (4.53) to obtain

$$
\text{var}(A) = \mathcal{E}(\mathcal{E}(A^2|B)) - \mathcal{E}^2(\mathcal{E}(A|B))
$$

Using the definition of variance on the first term on the right-hand side gives

$$
\begin{aligned}
\text{var}(A) &= \mathcal{E}\left(\text{var}(A|B) + \mathcal{E}^2(A|B)\right) - \mathcal{E}^2(\mathcal{E}(A|B)) \\
&= \mathcal{E}(\text{var}(A|B)) + \mathcal{E}(\mathcal{E}^2(A|B)) - \mathcal{E}^2(\mathcal{E}(A|B))
\end{aligned}
$$

and using the definition of variance again on the last two terms on the right-hand side gives

$$
\text{var}(A) = \mathcal{E}(\text{var}(A|B)) + \text{var}(\mathcal{E}(A|B))
$$

which establishes the result. Notice that since variance is nonnegative, this result also implies the inequality

$$
\text{var}(\mathcal{E}(A|B)) \leq \text{var}(A)
$$

which shows that the conditional expectation of random variable A has less variance than A itself.

We proceed to analyze the simplest particle filter. Actually we analyze the behavior of the weights for the idealized, unbiased case given by (4.51)

$$w_i(k) = w_i(k-1)\frac{p(y(k)|x_i(k))}{p(y(k)|\mathbf{y}(k-1))}$$

in which we consider the random variable $w_i(k)$ to be a function of the random variables $y(k), x_i(k)$. We next consider the conditional density of the random variables $y(k), x_i(k)$ relative to the previous samples $x_i(k-1)$, and the data $\mathbf{y}(k-1)$. We have

$$\begin{aligned}
&p(y(k), x_i(k)|\mathbf{y}(k-1), x_i(k-1)) \\
&\qquad = p(y(k)|\mathbf{y}(k-1), x_i(k-1))p(x_i(k)|\mathbf{y}(k-1), x_i(k-1)) \\
&\qquad = p(y(k)|\mathbf{y}(k-1))p(x_i(k)|x_i(k-1))
\end{aligned}$$

The first equation results from the statistical independence of $y(k)$ and $x_i(k)$, and the second results from the sampling procedure used to generate $x_i(k)$ given $x_i(k-1)$. Note that in the next section, we use a different sampling procedure in which $x_i(k)$ depends on both the new data $y(k)$ as well as the $x_i(k-1)$. Now we take the expectation of the weights at time k conditional on the previous samples and previous measurement trajectory

$$\begin{aligned}
&E(w_i(k)|x_i(k-1), \mathbf{y}(k-1)) \\
&\qquad = \iint w_i(k)p(y(k), x_i(k)|x_i(k-1), \mathbf{y}(k-1))dx_i(k)dy(k) \\
&\qquad = \iint w_i(k)p(y(k)|\mathbf{y}(k-1))p(x_i(k)|x_i(k-1))dx_i(k)dy(k)
\end{aligned}$$

Substituting the weight recursion and simplifying yields

$$\begin{aligned}
&E(w_i(k)|x_i(k-1), \mathbf{y}(k-1)) \\
&\qquad = \iint w_i(k-1)\frac{p(y(k)|x_i(k))}{p(y(k)|\mathbf{y}(k-1))} \\
&\qquad\qquad p(y(k)|\mathbf{y}(k-1))p(x_i(k)|x_i(k-1))dx_i(k)dy(k)
\end{aligned}$$

$$\begin{aligned}
&E(w_i(k)|x_i(k-1), \mathbf{y}(k-1)) \\
&= \iint w_i(k-1)p(y(k)|x_i(k))p(x_i(k)|x_i(k-1))dx_i(k-1)dy(k)
\end{aligned}$$

Taking $w_i(k-1)$ outside the integral and performing the integral over $x_i(k)$ and then $y(k)$ gives

$$E(w_i(k)|x_i(k-1),\mathbf{y}(k-1)) = w_i(k-1)\int p(y(k)|x_i(k-1))dy(k)$$

$$E(w_i(k)|x_i(k-1),\mathbf{y}(k-1)) = w_i(k-1)$$

Taking the variance of both sides and using the conditional variance formula (4.54) gives

$$\begin{aligned} \text{var}(E(w_i(k)|x_i(k-1),\mathbf{y}(k-1))) &= \text{var}(w_i(k-1)) \\ \text{var}(w_i(k)) - E(\text{var}(w_i(k)|x_i(k-1),\mathbf{y}(k-1))) &= \text{var}(w_i(k-1)) \end{aligned}$$

Again, noting that variance is nonnegative gives the inequality

$$\text{var}(w_i(k)) \geq \text{var}(w_i(k-1))$$

and we see that the variance for the unbiased weights of the simplest particle filter increase with time.

Next we present two examples that show the serious practical limitations of the simplest particle filter and the simplest particle filter with resampling.

Example 4.37: What's wrong with the simplest particle filter?

Consider the following linear system with Gaussian noise.

$$A = \begin{bmatrix} \cos\theta & \sin\theta \\ -\sin\theta & \cos\theta \end{bmatrix} \quad \theta = 6 \quad C = \begin{bmatrix} 0.5 & 0.25 \end{bmatrix} \quad G = I \quad B = I$$

$$\overline{x}(0) = \begin{bmatrix} 1 \\ 1 \end{bmatrix} \quad Q_0 = \frac{1}{4}\begin{bmatrix} 7 & 5 \\ 5 & 7 \end{bmatrix} \quad Q = 0.01\, I \quad R = 0.01$$

$$u(0,1,\cdots,5) = \begin{bmatrix} 7 \\ 2 \end{bmatrix}, \begin{bmatrix} 5 \\ 5 \end{bmatrix}, \begin{bmatrix} -1 \\ 2 \end{bmatrix}, \begin{bmatrix} -1 \\ -2 \end{bmatrix}, \begin{bmatrix} 1 \\ -3 \end{bmatrix}$$

(a) Plot the particle locations versus time from $k = 0$ to $k = 5$. Plot also the 95% contour of the true conditional density $p(x(k)|\mathbf{y}(k))$. Discuss the locations of the particles using the simplest particle filter.

(b) Write out the recursions for the conditional density of the particle locations $p(x_i(k)|\mathbf{y}(k))$ as well as the true conditional density $p(x(k)|\mathbf{y}(k))$. Discuss the differences.

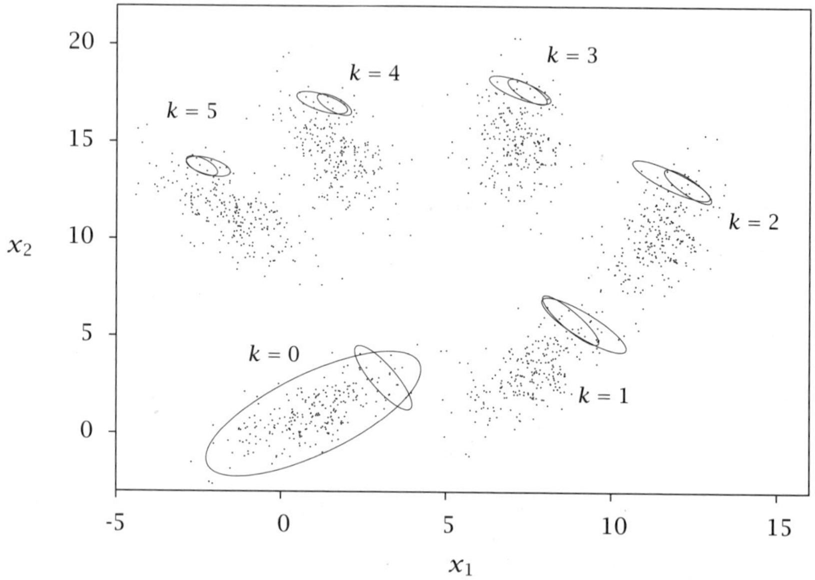

Figure 4.22: Particles' locations versus time for the simplest particle filter; 250 particles. Ellipses show the 95% contour of the true conditional densities before and after measurement.

Solution

(a) The samples and 95% conditional density contour are shown in Figure 4.22. The particles are located properly at $k = 0$ and about 95% of them are inside the state's initial density. But notice that the particles spread out quickly and few particles remain inside the 95% contour of the true conditional density after a few time steps.

(b) The true conditional density is the normal density given by the time-varying Kalman filter recursion. The conditional density of the particle location is given by (4.52) and the samples are identically distributed

$$\begin{aligned}
p(x(k)|\mathbf{y}(k)) &\sim N(\hat{x}(k), P(k)) \\
\hat{x}(k+1) &= A\hat{x}(k) + Bu(k) \underline{+L(k)(y(k) - C(A\hat{x}(k) + Bu(k)))} \\
P(k+1) &= AP(k)A' + GQG' \underline{-L(k+1)C(AP(k)A' + Q)} \\
p(x_i(k)|\mathbf{y}(k)) &\sim N(\overline{x}(k), \overline{P}(k)), \quad i = 1, \ldots, s \\
\overline{x}(k+1) &= A\overline{x}(k) + Bu(k) \\
\overline{P}(k+1) &= A\overline{P}(k)A' + GQG'
\end{aligned}$$

The major differences are underlined. Notice that the mean of the particle samples is independent of $\mathbf{y}(k)$, which causes the samples to drift away from the conditional density's mean with time. Notice that the covariance does not have the reduction term present in the Kalman filter, which causes the variance of the particles to increase with time. Therefore, due to the missing underlined terms, the mean of the samples drifts and the variance increases with time. The particle weights cannot compensate for the inaccurate placement of the particles, and the state estimates from the simplest particle filter are not useful after a few time iterations.

□

Example 4.38: Can resampling fix the simplest particle filter?

Repeat the simulation of Example 4.37, but use resampling after each time step. Discuss the differences.

Solution

Applying the resampling strategy gives the results in Figure 4.23. Notice that resampling prevents the samples from drifting away from the mean of the conditional density. Resampling maintains a high concentration of particles in the 95% probability ellipse. If we repeat this simulation 500 times and compute the fraction of particles within the conditional density's 95% probability contour, we obtain the results shown in Figure 4.24. Notice the dramatic improvement. Without resampling, fewer than 10% of the particles are in the 95% confidence ellipse after only five time steps. With resampling, about 80% of the samples are inside the 95% confidence ellipse. There is one caution against resampling too frequently, however. If the measurement has a small covariance, then the weights computed from

$$w_i(k) = w_i(k-1)p(y(k)|x_i(i))$$

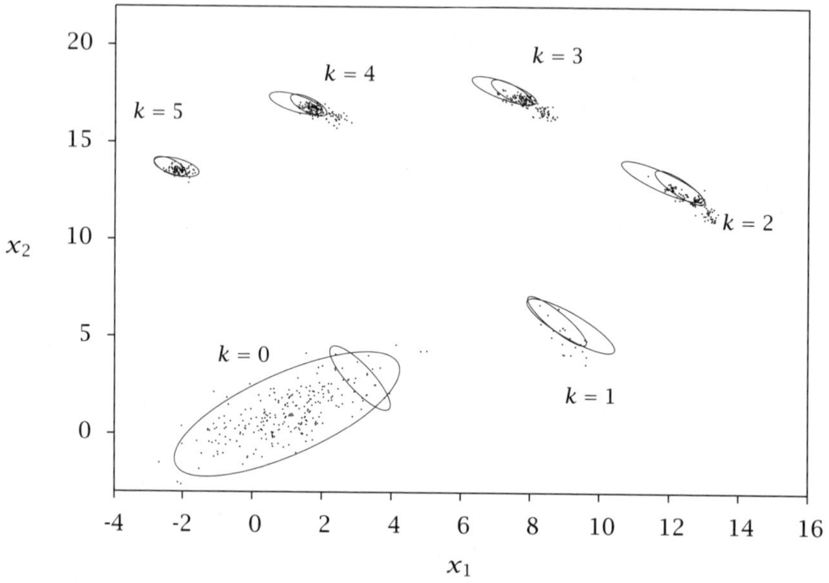

Figure 4.23: Particles' locations versus time for the simplest particle filter with resampling; 250 particles. Ellipses show the 95% contour of the true conditional densities before and after measurement.

will be dominated by only a few particles whose prediction of y is closest to the measurement. Resampling in this situation gives only those few particles repeated many times in the resample. For a sufficiently small covariance, this phenomenon can produce a single x_i value in the resample. This phenomenon is known as sample *impoverishment* (Doucet, Godsill, and Andrieu, 2000; Rawlings and Bakshi, 2006). □

4.7.5 A Particle Filter Based on Importance Sampling

Motivated by the drawbacks of the simplest particle filter of the previous section, researchers have developed alternatives based on a more flexible importance function (Arulampalam, Maskell, Gordon, and Clapp, 2002). We present this approach next. Rather than start with the statistical property of most interest, $p(x(k)|\mathbf{y}(k))$, consider instead the density of the entire *trajectory* of states conditioned on the measurements, $p(\mathbf{x}(k)|\mathbf{y}(k))$, as we did in moving horizon estimation. Our first

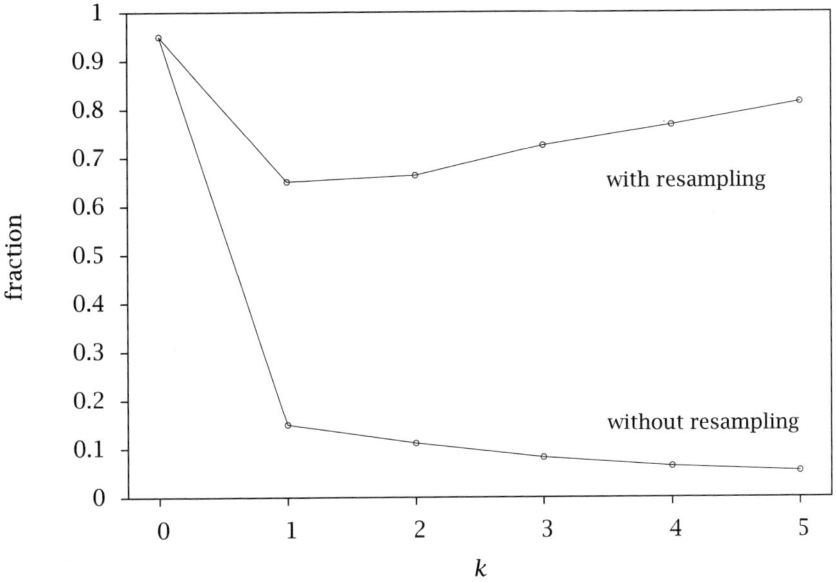

Figure 4.24: Fraction of particles inside the 95% contour of the true conditional density versus time; with and without resampling; average of 500 runs.

objective then is to obtain samples of $p(\mathbf{x}(k+1)|\mathbf{y}(k+1))$ from samples of $p(\mathbf{x}(k)|\mathbf{y}(k))$ and the model. We use importance sampling to accomplish this objective. Assume we have s weighted samples of the trajectory conditioned on measurements up to time k

$$p(\mathbf{x}(k)|\mathbf{y}(k)) = \{\mathbf{x}_i(k), \overline{w}_i(k)\} \quad i = 1, \ldots, s$$

in which the samples have been drawn from an importance function q, whose properties will be chosen as we proceed further. The weights $\overline{w}_i(k)$ are given by

$$\begin{aligned} w_i(k) &= \frac{h(\mathbf{x}_i(k))}{q(\mathbf{x}_i(k)|\mathbf{y}(k))} \\ p(\mathbf{x}_i(k)|\mathbf{y}(k)) &= \frac{h(\mathbf{x}_i(k))}{\int h(\mathbf{x}_i(k))d\mathbf{x}_i(k)} \\ \overline{w}_i(k) &= \frac{w_i(k)}{\sum_j w_j(k)} \end{aligned}$$

Notice $\mathbf{x}_i(k)$ is a set of ks n-vector samples, and, as in full information estimation, the storage requirements grow linearly with time. We remove this drawback subsequently, but for now we wish to obtain samples of $p(\mathbf{x}(k+1)|\mathbf{y}(k+1))$ in which $\mathbf{x}(k+1) = \{x(k+1), \mathbf{x}(k)\}$ and $\mathbf{y}(k+1) = \{y(k+1), \mathbf{y}(k)\}$. We start with

$$p(\mathbf{x}(k+1)|\mathbf{y}(k+1)) = \frac{p(y(k+1)|\mathbf{x}(k+1))p(\mathbf{x}(k+1)|\mathbf{y}(k))}{p(y(k+1)|\mathbf{y}(k))} \tag{4.55}$$

in which we have used the identity (see Exercise 1.47)

$$p_{A|B}(a|b) = \int p_{A|B,C}(a|b,c)p_{B|C}(b|c)dc$$

Because the process is Markov $p(y(k+1)|\mathbf{x}(k+1)) = p(y(k+1)|x(k+1))$. We next use the identity $p_{A,B|C}(a,b|c) = p_{A|B,C}(a|b,c)p_{B|C}(b|c)$ (see Exercise 1.46) and obtain

$$p(\mathbf{x}(k+1)|\mathbf{y}(k)) = p(x(k+1)|\mathbf{x}(k), \mathbf{y}(k))p(\mathbf{x}(k)|\mathbf{y}(k))$$

Again using the Markov property in this equation, we know $p(x(k+1)|\mathbf{x}(k), \mathbf{y}(k)) = p(x(k+1)|x(k))$ and therefore

$$p(\mathbf{x}(k+1)|\mathbf{y}(k)) = p(x(k+1)|x(k))p(\mathbf{x}(k)|\mathbf{y}(k))$$

Substituting these relations into (4.55) gives

$$p(\mathbf{x}(k+1)|\mathbf{y}(k+1)) = \frac{p(y(k+1)|x(k+1))p(x(k+1)|x(k))}{p(y(k+1)|\mathbf{y}(k))}p(\mathbf{x}(k)|\mathbf{y}(k)) \tag{4.56}$$

We use importance sampling to sample this density. Notice the denominator does not depend on $\mathbf{x}(k+1)$ and is therefore not required when using importance sampling. We use instead

$$p(\mathbf{x}(k+1)|\mathbf{y}(k+1)) = \frac{h(\mathbf{x}(k+1))}{\int h(\mathbf{x}(k+1))d\mathbf{x}(k+1)}$$

$$h(\mathbf{x}(k+1)) = p(y(k+1)|x(k+1))p(x(k+1)|x(k))p(\mathbf{x}(k)|\mathbf{y}(k)) \tag{4.57}$$

Note also that using importance sampling here when we do not wish to evaluate the normalizing constant introduces bias for finite sample size as stated in (4.49). We now state the two properties of q that provide a convenient importance function

$$q(x(k+1)|\mathbf{x}(k), \mathbf{y}(k+1)) = q(x(k+1)|x(k), y(k+1))$$

$$q(\mathbf{x}(k+1)|\mathbf{y}(k+1)) = q(x(k+1)|x(k), y(k+1))\, q(\mathbf{x}(k)|\mathbf{y}(k)) \qquad (4.58)$$

The first property of q is satisfied also by the density p, so it is not unusual to pick an importance function to share this behavior. The second property is *not* satisfied by the density, however, and it is chosen strictly for convenience; it allows a recursive evaluation of q at time $k+1$ from the value at time k. See Exercise 4.18 for further discussion of this point.

Next we need to generate the samples of $q(\mathbf{x}(k+1)|\mathbf{y}(k+1))$. Given the second property in (4.58), we have

$$\begin{aligned} q(\mathbf{x}(k+1)|\mathbf{y}(k+1)) &= q(x(k+1), x(k), \mathbf{x}(k-1)|y(k+1), \mathbf{y}(k)) \\ &= q(x(k+1)|x(k), y(k+1))\, q(x(k), \mathbf{x}(k-1)|\mathbf{y}(k)) \end{aligned}$$

which is of the form studied in Example 4.32 with the substitution

$$a = x(k+1) \qquad b = x(k) \qquad c = \mathbf{x}(k-1) \qquad d = y(k+1) \qquad e = \mathbf{y}(k)$$

Using the results of that example, our sampling procedure is as follows. We have available samples of $q(\mathbf{x}(k), \mathbf{y}(k)) = q(x(k), \mathbf{x}(k-1)|\mathbf{y}(k))$. Denote these samples by $(x_i(k), \mathbf{x}_i(k-1))$, $i = 1, \ldots, s$. Then we draw one sample from $q(x(k+1)|x_i(k), y(k+1))$ for each $i = 1, \ldots, s$. Denote these samples as $x_i(k+1)$. Then the samples of $q(\mathbf{x}(k+1)|\mathbf{y}(k+1))$ are given by $(x_i(k+1), x_i(k), \mathbf{x}_i(k-1)) = (x_i(k+1), \mathbf{x}_i(k))$. So we have

$$\mathbf{x}_i(k+1) = (x_i(k+1), \mathbf{x}_i(k)) \qquad i = 1, \ldots, s$$

Next we evaluate the weights for these samples

$$w_i(k+1) = \frac{h(\mathbf{x}_i(k+1)|\mathbf{y}(k+1))}{q(\mathbf{x}_i(k+1)|\mathbf{y}(k+1))}$$

Using (4.57) to evaluate h and the second property of the importance function to evaluate q gives

$$w_i(k+1) = \frac{p(y(k+1)|x_i(k+1))p(x_i(k+1)|x_i(k))}{q(x_i(k+1)|x_i(k), y(k+1))} \frac{h(\mathbf{x}_i(k)|\mathbf{y}(k))}{q(\mathbf{x}_i(k)|\mathbf{y}(k))}$$

$$w_i(k+1) = \frac{p(y(k+1)|x_i(k+1))p(x_i(k+1)|x_i(k))}{q(x_i(k+1)|x_i(k), y(k+1))} w_i(k) \qquad (4.59)$$

$$\overline{w}_i(k+1) = \frac{w_i(k+1)}{\sum_j w_j(k+1)}$$

Notice we obtain a convenient recursion for the weights that depends only on the values of the samples $x_i(k+1)$ and $x_i(k)$ and not the rest of the trajectory contained in the samples $\mathbf{x}_i(k)$. The trajectory's sampled density is given by

$$p(x(k+1), \mathbf{x}(k)|\mathbf{y}(k+1)) = \sum_{i=1}^{s} \overline{w}_i(k+1)\delta(x(k+1) - x_i(k+1))\delta(\mathbf{x}(k) - \mathbf{x}_i(k))$$

Integrating both sides over the $\mathbf{x}(k)$ variables gives the final result

$$p(x(k+1)|\mathbf{y}(k+1)) = \sum_{i=1}^{s} \overline{w}_i(k+1)\delta(x(k+1) - x_i(k+1))$$

Since we generate $x_i(k+1)$ from sampling $q(x(k+1)|x_i(k), y(k+1))$, the trajectory samples, $\mathbf{x}_i(k)$, and measurement trajectory, $\mathbf{y}(k)$, are not required at all, and the particle filter storage requirements do not grow with time. Notice also that if we choose the importance function

$$q(x_i(k+1)|x_i(k), y(k+1)) = p(x_i(k+1)|x_i(k))$$

which ignores the current measurement when sampling, we obtain for the weights

$$w_i(k+1) = w_i(k)\ p(y(k+1)|x_i(k+1))$$

This choice of importance function reduces to the simplest particle filter of the previous section, with its concomitant drawbacks.

Summary. We select an importance function $q(x(k+1)|x(k), y(k+1))$. We start with s samples of $p(x(0))$. We assume that we can evaluate $p(y(k)|x(k))$ using the measurement equation and $p(x(k+

$1)|x(k))$ using the model equation. The importance function particle filter is summarized by the following recursion

$$
\boxed{
\begin{aligned}
p(x(0)|y(0)) &= \{x_i(0), \overline{w}_i(0)\} \\
w_i(0) &= p(y(0)|x_i(0)) \qquad \overline{w}_i(0) = \frac{w_i(0)}{\sum_j w_j(0)} \\
p(x(k)|\mathbf{y}(k)) &= \{x_i(k), \overline{w}_i(k)\} \\
w_i(k+1) &= w_i(k) \, \frac{p(y(k+1)|x_i(k+1))p(x_i(k+1)|x_i(k))}{q(x_i(k+1)|x_i(k), y(k+1))} \\
\overline{w}_i(k+1) &= \frac{w_i(k+1)}{\sum_j w_j(k+1)}
\end{aligned}
}
$$

and $x_i(k+1)$ is a sample of $q(x(k+1)|x_i(k), y(k+1))$, $\quad i = 1, \ldots, s$. The sampled density of the importance-sampled particle filter converges to the conditional density $p(x(k)|\mathbf{y}(k))$ in the limit of infinite samples. Because of the way importance sampling was used, the sampled density is biased for all finite sample sizes.

Exercise 4.23 provides the recursion for the weights in the unbiased particle filter; these weights require the evaluation of $p(y(k)|\mathbf{y}(k-1))$. Exercise 4.24 shows that the variance of the unbiased weights increases with sample size.

4.7.6 Optimal Importance Function

In this section we develop the so-called "optimal" importance function $q(x(k)|x_i(k-1), y(k))$. We start with the weight recursion for the importance function particle filter given in (4.59), repeated here with k replacing $k+1$

$$
w_i(k) = w_i(k-1) \, \frac{p(y(k)|x_i(k))p(x_i(k)|x_i(k-1))}{q(x_i(k)|x_i(k-1), y(k))}
$$

We consider the $w_i(k)$ conditioned on the random variables $x_i(k-1), \mathbf{y}(k)$. The weight $w_i(k)$ is then a function of the random variable $x_i(k)$, which is sampled from the importance function $q(x(k)|x_i(k-$

$1), y(k))$. Taking the expectation gives

$$\begin{aligned}
\mathcal{E}\,(w_i(k)|x_i(k-1), \mathbf{y}(k)) \\
&= \int w_i(k)\; q(x_i(k)|x_i(k-1), y(k))\; dx_i(k) \\
&= \int \frac{p(y(k)|x_i(k)) p(x_i(k)|x_i(k-1))}{q(x_i(k)|x_i(k-1), y(k))} \\
&\qquad w_i(k-1)\; q(x_i(k)|x_i(k-1), y(k))\; dx_i(k) \\
&= \int p(y(k)|x_i(k))\; p(x_i(k)|x_i(k-1))\; w_i(k-1)\; dx_i(k) \\
&= w_i(k-1)\; p(y(k)|x_i(k-1))
\end{aligned}$$

Next we compute the conditional variance of the weights

$$\begin{aligned}
\mathrm{var}(w_i(k)|x_i(k-1), \mathbf{y}(k)) \\
&= \mathcal{E}(w_i^2(k)|x_i(k-1), \mathbf{y}(k)) - \mathcal{E}^2(w_i|x_i(k-1), \mathbf{y}(k))
\end{aligned}$$

Using the recursion in the first term and the expectation just derived in the second term gives

$$\begin{aligned}
\mathrm{var}(w_i(k)|x_i(k-1), \mathbf{y}(k)) = \\
&\int w_i^2(k)\; q(x_i(k)|x_i(k-1), y(k))\; dx_i(k) \\
&\quad - (w_i(k-1)\; p(y(k)|x_i(k-1)))^2
\end{aligned}$$

$$\begin{aligned}
\mathrm{var}(w_i(k)|x_i(k-1), \mathbf{y}(k)) \\
= \int w_i^2(k-1) \frac{(p(y(k)|x_i(k))\; p(x_i(k)|x_i(k-1)))^2}{q^2(x_i(k)|x_i(k-1), y(k))} \\
\quad q(x_i(k)|x_i(k-1), y(k))\; dx_i(k) - (w_i(k-1)\; p(y(k)|x_i(k-1)))^2 \\
= w_i^2(k-1)\left[\int \frac{p^2(y(k)|x_i(k))\; p^2(x_i(k)|x_i(k-1))}{q(x_i(k)|x_i(k-1), y(k))}\, dx_i(k) \right. \\
\left. \quad - p^2(y(k)|x_i(k-1))\right]
\end{aligned}$$

We can now optimize the choice of $q(x_i(k)|x_i(k-1), y(k))$ to minimize this conditional variance. Consider the choice

$$\boxed{q(x_i(k)|x_i(k-1), y(k)) = p(x_i(k)|x_i(k-1), y(k))} \tag{4.60}$$

which makes the samples at k depend on current measurement $y(k)$ as well as the past samples. We know from Bayes's rule and the Markov property

$$\begin{aligned} q(x_i(k)|x_i(k-1), y(k)) &= p(x_i(k)|x_i(k-1), y(k)) \\ &= \frac{p(y(k)|x_i(k), x_i(k-1))p(x_i(k)|x_i(k-1))}{p(y(k)|x_i(k-1))} \\ q(x_i(k)|x_i(k-1), y(k)) &= \frac{p(y(k)|x_i(k))p(x_i(k)|x_i(k-1))}{p(y(k)|x_i(k-1))} \end{aligned}$$

Using this result we have for the integral term

$$\begin{aligned} &\int \frac{p^2(y(k)|x_i(k))\; p^2(x_i(k)|x_i(k-1))}{q(x_i(k)|x_i(k-1), y(k))}\, dx_i(k) \\ &= p(y(k)|x_i(k-1)) \int p(y(k)|x_i(k))\; p(x_i(k)|x_i(k-1))\; dx_i(k) \\ &= p^2(y(k)|x_i(k-1)) \end{aligned}$$

Substituting this result into the previous equation for conditional variance gives

$$\operatorname{var}(w_i(k)|x_i(k-1), \mathbf{y}(k)) = 0$$

Since variance is nonnegative, the choice of importance function given in (4.60) is optimal for reducing the conditional variance of the weights. This choice has the important benefit of making the samples $x_i(k)$ more responsive to the measurement $y(k)$, which we show in the next example is a big improvement over the simplest particle filter.

Example 4.39: Optimal importance function applied to a linear estimation problem

Given the linear system of Example 4.37 and 250 particles, show the particles' locations for times $k = 0, 1, \ldots, 5$ along with the 95% elliptical contour of the true conditional density $p(x(k)|\mathbf{y}(k))$. Perform this calculation with and without resampling after every time step.

Solution

The optimal importance function is given in (4.60)

$$q(x_i(k)|x_i(k-1), y(k)) = p(x_i(k)|x_i(k-1), y(k))$$

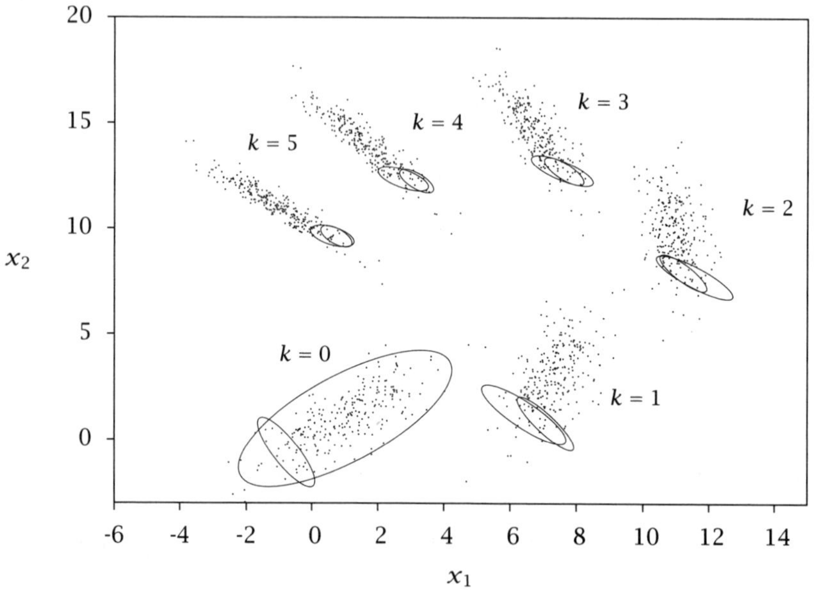

Figure 4.25: Particles' locations versus time using the optimal importance function; 250 particles. Ellipses show the 95% contour of the true conditional densities before and after measurement.

The conditional density on the right-hand side is given by

$$p(x_i(k)|x_i(k-1), y(k)) \sim N(\overline{x}(k), \overline{P})$$

$$\overline{x}(k) = P\left(Q^{-1}(Ax_i(k-1) + Bu(k-1)) + C'R^{-1}y(k)\right)$$
$$\overline{P} = \left(Q^{-1} + C'R^{-1}C\right)^{-1}$$

Exercise 4.25 discusses establishing this result. So the $x_i(k)$ are generated by sampling this normal, and the results are shown in Figure 4.25. We see that the optimal importance function adds a $y(k)$ term to the evolution of the particle mean. This term makes the particles more responsive to the data and the mean particle location better tracks the conditional density's mean. Compare Figure 4.22 for the simplest particle filter with Figure 4.25 to see the improvement. Also the variance no longer increases with time as in the simplest particle filter so the particles do not continue to spread apart.

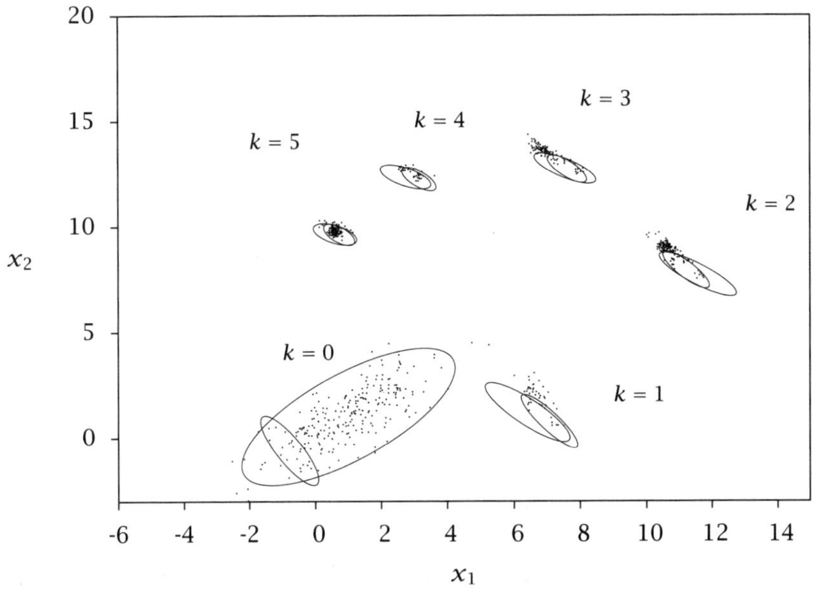

Figure 4.26: Particles' locations versus time using the optimal importance function with resampling; 250 particles. Ellipses show the 95% contour of the true conditional densities before and after measurement.

If we apply resampling at every time step, we obtain the results in Figure 4.26. As we saw in the case of the simplest particle filter, resampling greatly increases the number of samples inside the 95% probability ellipse of the conditional density.

If we rerun the simulation 500 times and plot versus time the fraction of particles that are inside the 95% contour of the true conditional density, we obtain the result shown in Figure 4.27. The optimal importance function is able to maintain about 20% of the particles in the 95% probability ellipse. With the optimal importance function and resampling, more than 90% of the particles are inside the 95% probability ellipse. The earlier warning about sample impoverishment applies here as well. □

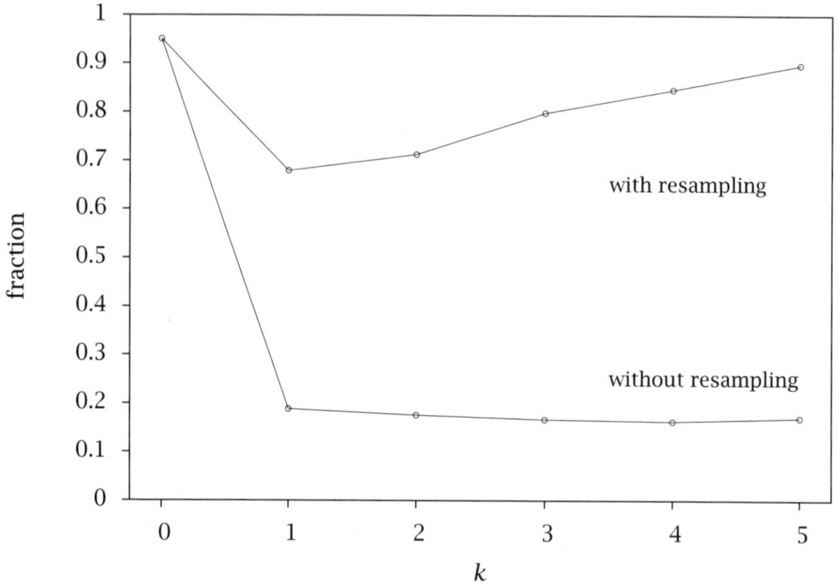

Figure 4.27: Fraction of particles inside the 95% contour of the true conditional density versus time; with and without resampling; average of 500 runs.

4.8 Combined MHE/Particle Filtering

We next propose a new state estimation method that combines some of the best elements of MHE and PF. This type of combination has several design parameters and can take different forms, and we use the general term combined MHE/PF to designate this entire class of state estimators. To motivate the design of MHE/PF, consider the strengths and weaknesses of pure MHE and pure PF. The main *strengths of MHE* are

1. MHE propagates the state using the full nonlinear model.
2. MHE uses optimization to find the most likely estimate. Physical constraints can be included in the optimization.
3. MHE employs a horizon of measurements.

Using the full nonlinear model prevents inaccurate model linearizations from interfering with the fitting of the model to the data. The use of optimization produces the best state or state trajectory to describe the current snapshot of data. Optimization methods generally evaluate a small set of points in the state space to find the best estimate compared to exhaustive enumeration, gridding, and sampling strategies. That becomes a significant strength as the dimension of the state space model increases past $n \approx 2$–3. The use of a moving window of data provides some robustness to unmodeled disturbances entering the system. The goal in most recursive estimation is to consider measurements one at a time. That is often a valid goal, mainly because it allows faster computation of the current estimate given the current measurement. But unmodeled disturbances are often problematic when measurements are considered one at a time. No single measurement is sufficient to conclude that an unmodeled disturbance has shifted the state significantly from its current estimated value. Only when several sequential measurements are considered at once is the evidence sufficient to overturn the current state estimate and move the state a significant distance to better match all of the measurements. MHE has this capability built in.

The main *weaknesses of MHE* are

1. MHE may take significant computation time.

2. MHE uses local instead of global optimization.

Of course attempting global optimization is possible, but that exacerbates weakness 1 significantly and no guarantees of finding a global optimum are available for anything but the simplest nonlinear models. Note that for the special case of linear models, MHE finds the global optimum and weakness 2 is removed.

Particle filtering displays quite different characteristics than those of MHE. The main *strengths of PF* are

1. PF uses the full nonlinear model to propagate the samples.

2. The PF sampled density can represent a general conditional density.

3. PF is simple to program and executes quickly for small sample sizes.

As we have illustrated with simple examples, pure PF also demonstrates significant weaknesses, and these are not remedied by any suggestions in the research literature of which we are aware. The *weaknesses of PF* include

1. PF exhibits significant decrease in performance with increasing state dimension.

2. PF displays poor robustness to unmodeled disturbances.

The lack of robustness is a direct outcome of the sampling strategies. Sampling any of the proposed PF importance functions does not locate the samples close to the true state after a significant and unmodeled disturbance. Once the samples are in the wrong place with respect to the peak in the conditional density, they do not recover. If the samples are in the wrong part of the state space, the weights cannot carry the load and represent the conditional density. Resampling does not successfully reposition the particles if they are already out of place. An appeal to sampled density convergence to the true conditional density with increasing sample number is unrealistic. The number of samples required is simply too large for even reasonably small state dimensions considered in applications; $n > 50$ is not unusual in applications.

In constructing a class of combined methods we propose to

1. Use MHE to locate/relocate the samples.

2. Use PF to obtain fast recursive estimation between MHE optimizations.

We overcome the potentially expensive MHE optimization by using PF to process the measurements and provide rapid online estimates while an MHE computation is underway. We position the samples in regions of high conditional density after every run of the MHE optimization, which allows recovery from unmodeled disturbances as soon as an MHE computation completes. A challenge that is not addressed is the appearance of multiple peaks in the conditional density when using nonlinear models. Handling the multimodal conditional density remains a challenge for any online, and indeed offline, state estimation procedure.

Next we propose a specific state estimator in this general MHE/PF class and examine its performance with some simple computational examples. Because this class of estimators is new, we fully expect significant modifications and improvements to come along. At this early juncture we expect only to be able to illustrate some of the new capabilities of the approach.

Let $\hat{Z}_k(x)$ denote the MHE arrival cost function given in Definition 4.16. We let $\hat{V}_k^0$ denote the optimal cost and $\hat{x}(k)$ the optimal estimate of the last stage at time k. We consider the quadratic approximation of

$\hat{Z}_k(\cdot)$ at the optimum $\hat{x}(k)$

$$V(x) = V_k^0(\hat{x}(k)) + (1/2)(x - \hat{x}(k))'H(x - \hat{x}(k))$$

in which H is the Hessian of $\hat{Z}_k(x)$ evaluated at the optimum $\hat{x}(k)$. We use this function as an importance function for sampling the conditional density. Notice that this procedure is not the same as assuming the conditional density itself is a normal distribution. We are using $N(\hat{x}(k), H^{-1})$ strictly as an importance function for sampling the unknown conditional density. The samples $x_i(k)$ are drawn from $N(\hat{x}(k), H^{-1})$. The weights are given by

$$w_i(k) = V(x_i(k)) \qquad \overline{w}_i(k) = \frac{w_i(k)}{\sum_j w_j(k)} \tag{4.61}$$

and the sampled density is given by

$$p_s(x) = \{x_i(k), \overline{w}_i(k)\}$$

If the conditional density is well represented by the normal approximation, then the normalized weights are all nearly equal to $1/s$. The MHE cost function modifies these ideal weights as shown in (4.61).

Example 4.40: Comparison of MHE, PF, and combined MHE/PF

Consider a well-mixed semibatch chemical reactor in which the following reaction takes place

$$2\text{A} \xrightarrow{k} \text{B} \qquad r = kc_A^2$$

The material balances for the two components are

$$\begin{aligned} \frac{dc_A}{dt} &= -2kc_A^2 + \frac{Q_f}{V} c_{Af} \\ \frac{dc_B}{dt} &= kc_A^2 + \frac{Q_f}{V} c_{Bf} \end{aligned}$$

with constant parameter values

$$\frac{Q_f}{V} = 0.4 \qquad k = 0.16 \qquad c_{Af} = 1 \qquad c_{Bf} = 0$$

The scalar measurement is the total pressure, which is the sum of the two states. The sample time is $\Delta = 0.1$. The initial state is $x(0) = [3\ 1]'$ and the initial prior mean is $\hat{x}(0) = [0.1\ 4.5]'$. Moreover, the input

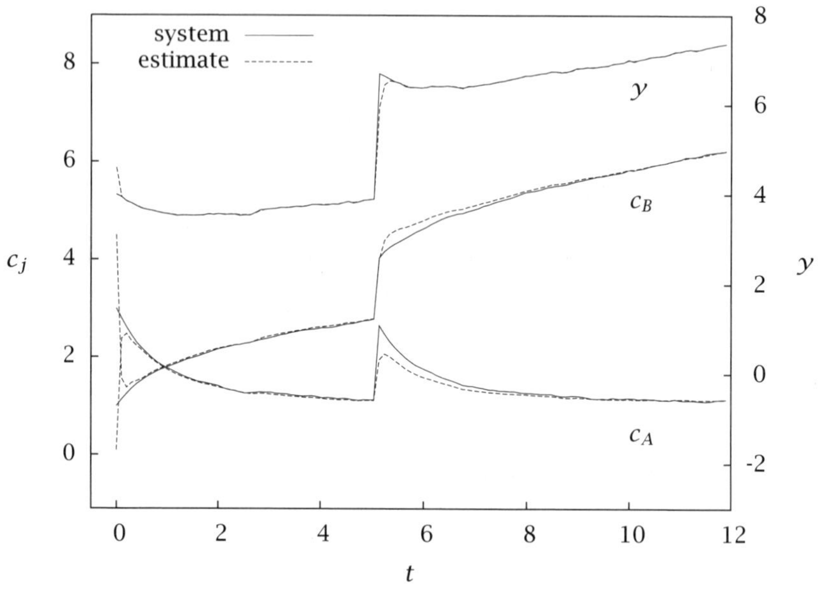

Figure 4.28: Pure MHE.

suffers an unmodeled step disturbance at $t = 5$ for two samples. So this example tests robustness of the estimator to initial state error and unmodeled disturbances.

First we apply MHE to the example and the results are displayed in Figure 4.28. The horizon is chosen as $N = 15$. The initial covariance is chosen to be $P_0 = 10I_2$ to reflect the poor confidence in the initial state. Notice that MHE is able to recover from the poor initial state prior in only 4 or 5 samples.

Next we apply pure particle filtering using 50 particles. We use the optimal importance function because the measurement equation is linear. The particles are initialized using the same initial density as used in the MHE estimator.

$$p_{x(0)}(x) = n(x, \hat{x}(0), P(0))$$

The results are shown in Figure 4.29. The figure shows the state and output mean versus time. We notice two effects. The particle filter is unable to recover from the poor initial samples. The measurement is predicted well but neither state is estimated accurately. The A concentration estimate is also negative, which is physically impossible. The

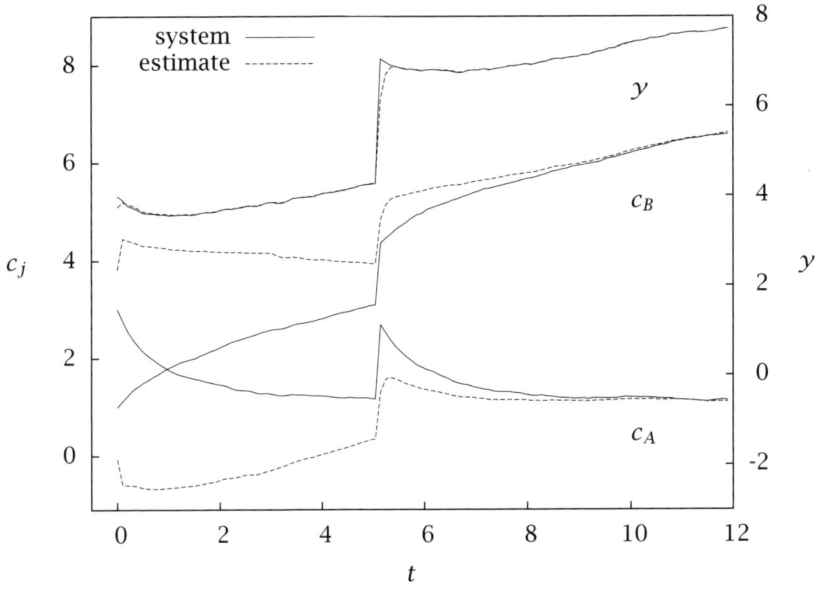

Figure 4.29: Pure PF with optimal importance function.

disturbance at $t = 5$ is fortuitous and helps the PF get back on track.

Next we assume that the MHE optimization cannot finish in one sample, but requires M samples. If we attempt a pure MHE solution in this situation, the estimator falls hopelessly behind; an estimate using data $\mathbf{y}(k - M, k), k \geq M$ is not available until time Mk. Instead we use MHE/PF as follows.

1. At time k run MHE on data $\mathbf{y}(k - M, k)$. This computation is assumed to finish at time $k + M$. For simplicity, assume N large and a noninformative prior.

2. Draw samples from $N(\hat{x}(k), P(k))$. Run the particle filtering update from time k to time $k + M$. For illustrative purposes, we assume this PF step finishes in one sample.

3. Update k to $k + M$ and repeat.

For illustrative purposes, we choose $M = 10$ and apply the combination of MHE and PF with the simple importance function, also using 50 particles as before. The results are shown in Figure 4.30. Notice that again the poor initial samples lead to significant estimate error.

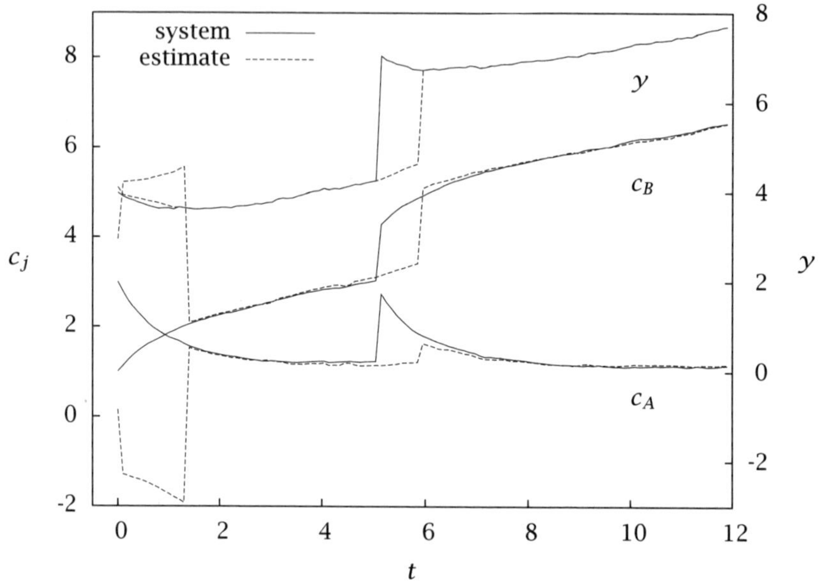

Figure 4.30: Combination MHE/PF with simple importance function.

But the inaccurate sample is repaired after $M = 10$ samples. The MHE calculation completes by about $t = 2$, and the samples are reinitialized from the MHE cost function at $t = 1$, and run forward from $t = 1$. These reinitialized samples converge to the true state shortly after $t = 1$.[7]

The disturbance at $t = 5$ also causes the PF samples with the simple importance function to be in the wrong locations. They do not recover and inaccurate estimates are produced by the PF. Another MHE calculation starts at $t = 5$ and finishes at $t = 6$, and the samples are reinitialized with the MHE cost function at $t = 5$ and run forward. After this resampling, the PF estimates again quickly converge to the true estimates after $t = 6$.

Next we use the combination of MHE and PF with the optimal importance function. These results are shown in Figure 4.31. We see as

[7] Even with only 50 particles, we find that particle filtering is not so much faster than MHE, that its computation time can be neglected as we have done here. The two computations take about the *same* time with 50 particles. The computational expense in PF arises from calling an ODE solver 50 times at each sample time. No attempt was made to tailor the ODE solver for efficiency by exploiting the fact that the sample time is small. Note, however, that tailoring the ODE solver would speed up MHE as well as PF.

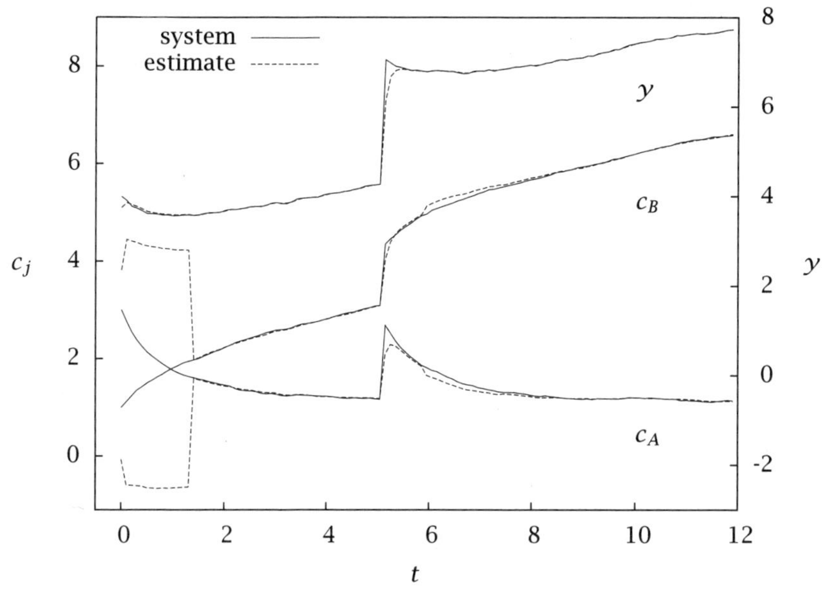

Figure 4.31: Combination MHE/PF with optimal importance function.

in the early part of Figure 4.29 that the samples cannot recover from the poor initial state prior and resampling from the MHE cost function takes place at $t = 1$ after the first MHE calculation finishes at $t = 2$. But as in the case of pure PF with the optimal importance function, the disturbance does not move the state so far from the samples that they are unable to recover and continue to provide accurate estimates. The MHE resampling that takes place at $t = 5$ after MHE finishes at $t = 6$ does not modify significantly the PF samples that are already well placed. □

Of course, the simulations shown in Figures 4.28–4.31 display the outcome of only single random realizations. A full characterization of the behavior of the four estimators is determined by running many such random simulations and computing the statistics of interest. We have not compiled these statistics because the single simulations are rather time consuming. After running several random simulations for each estimator, these single simulations were selected manually as representative behavior of the different estimators.

4.9 Notes

State estimation is a fundamental topic appearing in many branches of science and engineering, and has a large literature. A nice and brief annotated bibliography describing the early contributions to optimal state estimation of the *linear Gaussian* system is provided by Åström (1970, pp. 252-255). Kailath (1974) provides a comprehensive and historical review of *linear* filtering theory including the historical development of Wiener-Kolmogorov theory for filtering and prediction that preceded Kalman filtering (Wiener, 1949; Kolmogorov, 1941).

Jazwinski (1970) provides an early and comprehensive treatment of the optimal stochastic state estimation problem for linear and *nonlinear* systems. As mentioned in Section 4.2.1, Jazwinski (1970) proves stability of the optimal time-varying state estimator for the linear Gaussian case using $V(k,x) = x'P(k)^{-1}x$ as the Lyapunov function for the linear time-varying system governing estimate error. Note that this dynamic system is time-varying even if the model is time invariant because the optimal estimator gains are time varying. This choice of Lyapunov function has been used to establish estimator stability in many subsequent textbooks (Stengel, 1994, pp.474-475). Kailath (1974, p.152) remarks that the known proofs that the optimal filter is stable "are somewhat difficult, and it is significant that only a small fraction of the vast literature on the Kalman filter deals with this problem." Perhaps the stability analysis developed in Section 4.2 can alleviate the difficulties associated with developing Lyapunov function arguments in optimal estimation.

For establishing stability of the *steady-state* optimal linear estimator, simpler arguments suffice because the estimate error equation is time invariant. Establishing duality with the optimal regulator is a favorite technique for establishing estimator stability in this case. See, for example, Kwakernaak and Sivan (1972, Theorem 4.11) for a general steady-state stability theorem for the linear Gaussian case. This result is proved by establishing properties of the steady-state solution to the Riccati equation for regulation and, by duality, estimation.

Many of the full information and MHE results in this chapter are given by Rao (2000) and Rao, Rawlings, and Mayne (2003). The full information analysis given here is more general because (i) we assume nonlinear detectability rather than nonlinear observability, and (ii) we establish asymptotic stability under process and measurement disturbances, which were neglected in previous analysis.

Muske, Rawlings, and Lee (1993) and Meadows, Muske, and Rawlings (1993) apparently were the first to use the increasing property of the optimal cost to establish asymptotic stability for full information estimation for linear models with constraints. Robertson and Lee (2002) present the interesting statistical interpretation of MHE for the constrained linear system. Michalska and Mayne (1995) establish stability of moving horizon estimation with zero prior weighting for the continuous time nonlinear system.

4.10 Exercises

Exercise 4.1: Input to state stability and convergence

Assume the nonlinear system

$$x^+ = f(x, u)$$

is input to state stable (ISS) so that for all $x_0 \in \mathbb{R}^n$, input sequences $\mathbf{u}$, and $k \geq 0$

$$|x(k; x_0, \mathbf{u})| \leq \beta(|x_0|, k) + \gamma(\|\mathbf{u}\|)$$

in which $x(k; x_0, \mathbf{u})$ is the solution to the system equation at time k starting at state x_0 using input sequence $\mathbf{u}$, and $\gamma \in \mathcal{K}$ and $\beta \in \mathcal{KL}$.

(a) Show that the ISS property also implies

$$|x(k; x_0, \mathbf{u})| \leq \beta(|x_0|, k) + \gamma(\|\mathbf{u}\|_{0:k})$$

in which $\|\mathbf{u}\|_{0:k} = \max_{0 \leq j \leq k} |u(j)|$.

(b) Show that the ISS property implies the "converging-input converging-state" property (Jiang and Wang, 2001), (Sontag, 1998a, p. 330), i.e., show that if the system is ISS, then $u(k) \to 0$ implies $x(k) \to 0$.

Exercise 4.2: Output to state stability and convergence

Assume the nonlinear system

$$x^+ = f(x) \qquad y = h(x)$$

is output to state stable (OSS) so that for all $x_0 \in \mathbb{R}^n$ and $k \geq 0$

$$|x(k; x_0)| \leq \beta(|x_0|, k) + \gamma(\|\mathbf{y}\|_{0:k})$$

in which $x(k; x_0)$ is the solution to the system equation at time k starting at state x_0, and $\gamma \in \mathcal{K}$ and $\beta \in \mathcal{KL}$.

Show that the OSS property implies the "converging-output converging-state" property (Sontag and Wang, 1997, p. 281) i.e., show that if the system is OSS, then $y(k) \to 0$ implies $x(k) \to 0$.

Exercise 4.3: i-IOSS and convergence

Prove Proposition 4.2, which states that if system

$$x^+ = f(x, w) \qquad y = g(x)$$

is i-IOSS, and $w_1(k) \to w_2(k)$ and $y_1(k) \to y_2(k)$ as $k \to \infty$, then

$$x(k; z_1, \mathbf{w}_1) \to x(k; z_2, \mathbf{w}_2) \qquad \text{for all } z_1, z_2$$

Exercise 4.4: Observability and detectability of linear time-invariant systems and OSS

Consider the linear time-invariant system

$$x^+ = Ax \qquad y = Cx$$

(a) Show that if the system is observable, then the system is OSS.

(b) Show that the system is detectable if and only if the system is OSS.

Exercise 4.5: Observability and detectability of linear time-invariant system and IOSS

Consider the linear time-invariant system with input

$$x^+ = Ax + Gw \qquad y = Cx$$

(a) Show that if the system is observable, then the system is IOSS.

(b) Show that the system is detectable if and only if the system is IOSS.

Exercise 4.6: Max or sum?

Given $\gamma_1, \gamma_2 \in \mathcal{K}$, show there exists $a, b > 0$ such that

$$a(\gamma_1(x) + \gamma_2(y)) \leq \max(\gamma_1(x), \gamma_2(y)) \leq b(\gamma_1(x) + \gamma_2(y))$$

for all $x \in \mathbb{R}^n$, $y \in \mathbb{R}^m$. Therefore it is equivalent if ISS or OSS is defined in terms of inequalities using the max or the sum.

Exercise 4.7: Linear systems and incremental stability

Show that for a linear time-invariant system, i-ISS (i-OSS, i-IOSS) is equivalent to ISS (OSS, IOSS).

Exercise 4.8: Nonlinear observability and Lipschitz continuity implies i-OSS

Consider the following definition of observability for nonlinear systems in which f and h are Lipschitz continuous. A system

$$x^+ = f(x) \qquad y = h(x)$$

is observable if there exists $N_o \in \mathbb{I}_{\geq 1}$ and $\mathcal{K}$-function γ such that

$$\sum_{k=0}^{N_o - 1} \left| y(k; x_1) - y(k; x_2) \right| \geq \gamma(|x_1 - x_2|) \tag{4.62}$$

holds for all $x_1, x_2 \in \mathbb{R}^n$. This definition was used by Rao et al. (2003) in showing stability of nonlinear MHE to initial condition error under zero state and measurement disturbances.

(a) Show that this form of nonlinear observability implies i-OSS.

(b) Show that i-OSS does not imply this form of nonlinear observability and, therefore, i-OSS is a weaker assumption.

The i-OSS concept generalizes the linear system concept of detectability to nonlinear systems.

Exercise 4.9: Robust GAS implies GAS in estimation

Show that robust GAS of an estimator implies GAS for the estimator.

Exercise 4.10: Relationships between observability, FSO, MHE detectability and i-IOSS

Show that for the nonlinear system $x^+ = f(x, w)$, $y = h(x)$ with Lipschitz continuous f and h, the following relationships hold between observability, FSO, MHE detectability, and i-IOSS (detectability).

$$\begin{aligned}
&\text{observable} \implies \text{FSO} \implies \text{MHE detectable} \implies \text{i-IOSS} \\
&\text{observable} \not\Leftarrow \text{FSO} \not\Leftarrow \text{MHE detectable} \not\Leftarrow \text{i-IOSS}
\end{aligned}$$

Exercise 4.11: Observability, FSO, and detectability of linear systems

Consider the linear time-invariant system

$$x^+ = Ax \qquad y = Cx$$

and its observability canonical form. What conditions must the system satisfy to be

(a) observable?

(b) final-state observable (FSO)?

(c) detectable?

Exercise 4.12: Dynamic programming recursion for Kalman predictor

In the Kalman predictor, we use forward DP to solve at stage k

$$\min_{x,w} \ell(x, w) + V_k^-(x) \qquad \text{s.t. } z = Ax + w$$

in which x is the state at the current stage and z is the state at the next stage. The stage cost and arrival cost are given by

$$\ell(x, w) = (1/2)(\,|y(k) - Cx|^2_{R^{-1}} + w'Q^{-1}w) \qquad V_k^-(x) = (1/2)\,|x - \hat{x}^-(k)|^2_{(P^-(k))^{-1}}$$

and we wish to find the value function $V^0(z)$, which we denote $V_{k+1}^-(z)$ in the Kalman predictor estimation problem.

(a) Combine the two x terms to obtain

$$\min_{x,w} \frac{1}{2}\left(w'Q^{-1}w + (x - \hat{x}(k))'P(k)^{-1}(x - \hat{x}(k))\right) \qquad \text{s.t. } z = Ax + w$$

and, using the third part of Example 1.1, show

$$\begin{aligned}
P(k) &= P^-(k) - P^-(k)C'(CP^-(k)C' + R)^{-1}CP^-(k) \\
L(k) &= P^-(k)C'(CP^-(k)C' + R)^{-1}C'R^{-1} \\
\hat{x}(k) &= \hat{x}^-(k) + L(k)(y(k) - C\hat{x}^-(k))
\end{aligned}$$

(b) Add the w term and use the inverse form in Exercise 1.18 to show the optimal cost is given by

$$\begin{aligned}
V^0(z) &= (1/2)(z - A\hat{x}^-(k+1))'(P^-(k+1))^{-1}(z - A\hat{x}^-(k+1)) \\
\hat{x}^-(k+1) &= A\hat{x}(k) \\
P^-(k+1) &= AP(k)A' + Q
\end{aligned}$$

Substitute the results for $\hat{x}(k)$ and $P(k)$ above and show

$$\begin{aligned} V_{k+1}^-(z) &= (1/2)(z - \hat{x}^-(k+1))'(P^-(k+1))^{-1}(z - \hat{x}(k+1)) \\ P^-(k+1) &= Q + AP^-(k)A' - AP^-(k)C'(CP^-(k)C' + R)^{-1}CP^-(k)A \\ \hat{x}^-(k+1) &= A\hat{x}^-(k) + \widetilde{L}(k)(y(k) - C\hat{x}^-(k)) \\ \widetilde{L}(k) &= AP^-(k)C'(CP^-(k)C' + R)^{-1} \end{aligned}$$

(c) Compare and contrast this form of the estimation problem to the one given in Exercise 1.29 that describes the Kalman filter.

Exercise 4.13: Duality, cost to go, and covariance

Using the duality variables of Table 4.2, translate Theorem 4.10 into the version that is relevant to the state estimation problem.

Exercise 4.14: Estimator convergence for (A, G) not stabilizable

What happens to the stability of the optimal estimator if we violate the condition

$$(A, G) \text{ stabilizable}$$

(a) Is the steady-state Kalman filter a stable estimator? Is the full information estimator a stable estimator? Are these two answers contradictory? Work out the results for the case $A = 1, G = 0, C = 1, P^-(0) = 1, Q = 1, R = 1$.
Hint: you may want to consult de Souza, Gevers, and Goodwin (1986).

(b) Can this phenomenon happen in the LQ regulator? Provide the interpretation of the time-varying regulator that corresponds to the time-varying filter given above. Does this make sense as a regulation problem?

Exercise 4.15: Exponential stability of the Kalman predictor

Establish that the Kalman predictor defined in Section 4.2.1 is a globally exponentially stable estimator. What is the corresponding linear quadratic regulator?

Exercise 4.16: The resampling theorem

Generalize the proof of Theorem 4.35 to cover any number of samples.

Hint: you may find the multinomial expansion formula useful

$$(x_1 + x_2 + \cdots + x_s)^k = \sum_{r_1=0}^{k} \sum_{r_2=0}^{k} \cdots \sum_{r_s=0}^{k} a(r_1, r_2, \ldots, r_s) x_1^{r_1} x_2^{r_2} \cdots x_s^{r_s}$$

in which the coefficients in the expansion formula are given by Feller (1968, p.37)

$$a(r_1, r_2, \ldots, r_s) = \begin{cases} \dfrac{k!}{r_1! r_2! \cdots r_s!} & r_1 + r_2 + \cdots + r_s = k \\ 0 & r_1 + r_2 + \cdots + r_s \neq k \end{cases} \tag{4.63}$$

Exercise 4.17: Pruning while resampling

Sometimes it is convenient in a simulation to reduce the number of samples when resampling a density. In many discrete processes, for example, the number of possible states that may be reached in the simulation increases with time. To keep the number of samples constant, we may wish to remove samples at each time through the resampling process. Consider a modification of Theorem 4.35 in which the number of resamples is $\tilde{s}$, which does not have to be equal to s.

Theorem 4.41 (Resampling and pruning). *Consider a sampled density $p(x)$ with s samples at $x = x_i$ and associated weights w_i*

$$p(x) = \sum_{i=1}^{s} w_i \delta(x - x_i) \qquad w_i \geq 0 \qquad \sum_{i=1}^{s} w_i = 1$$

Consider the resampling procedure that gives a resampled density with $\tilde{s} > 0$ samples

$$\tilde{p}(x) = \sum_{i=1}^{\tilde{s}} \tilde{w}_i \delta(x - \tilde{x}_i)$$

in which the $\tilde{x}_i$ are chosen according to resample probability p_r

$$p_r(\tilde{x}_i) = \begin{cases} w_j, & \tilde{x}_i = x_j \\ 0, & \tilde{x}_i \neq x_j \end{cases}$$

and with uniform weights $\tilde{w}_i = 1/\tilde{s}$. Consider a function $f(\cdot)$ defined on a set X containing the points x_i.

Under this resampling procedure, the expectation over resampling of any integral of the resampled density is equal to that same integral of the original density

$$\mathcal{E}_r\left(\int f(x)\tilde{p}(x)dx\right) = \int f(x)p(x)dx \qquad \text{all } f$$

(a) Is the proposed theorem correct? If so, prove it. If not, provide a counterexample.

(b) What do you suppose happens in a simulation if we perform aggressive pruning by always choosing $\tilde{s} = 1$?

Exercise 4.18: Properties of the importance function

It is stated in the chapter that $p(\mathbf{x}(k+1)|\mathbf{y}(k+1))$ does not satisfy the second importance function property listed in (4.58)

$$q(\mathbf{x}(k+1)|\mathbf{y}(k+1)) = q(x(k+1)|x(k), y(k+1))\; q(\mathbf{x}(k)|\mathbf{y}(k)) \tag{4.64}$$

Derive a similar property that $p(\mathbf{x}(k+1)|\mathbf{y}(k+1))$ *does* satisfy. What has been altered in (4.64)? Why do you think this change has been made?

Exercise 4.19: A single sample of joint density

Consider again Example 4.31 in which we have s_x and s_y samples of the marginals of independent random variables ξ and η, respectively

$$\xi \sim \{x_i, w_{xi}\} \qquad w_{xi} = 1/s_x, \quad i = 1, \ldots, s_x$$
$$\eta \sim \{y_j, w_{yj}\} \qquad w_{yj} = 1/s_y, \quad j = 1, \ldots, s_y$$

and wish to sample the joint density $p_{\xi,\eta}(x,y) = p_\xi(x)p_\eta(y)$. Show that selecting any single sample is a valid sample of the joint density

$$\{(x_1, y_1), w\}, \quad w = 1$$

Exercise 4.20: Kolmogorov-Smirnov limit theorem for sampling error

Consider again s mutually independent samples taken from cumulative distribution $P(x)$ to produce the sampled cumulative distribution $P_s(x; s)$ as discussed in Section 4.7.2. Define sampling error as in the chapter

$$D_s = \sup_x |P_s(x; s) - P(x)|$$

(a) Reproduce the results of Example 4.30. Plot the actual and limiting distributions for D_s for $s = 10, 100, 1000$ when sampling a normal distribution with unit variance. Your result should resemble Figure 4.14

(b) Now compute the actual and limiting probability *densities* of the sampling error $p(D_s)$ rather than the distribution $\Pr(D_s)$. Give a formula for $l(z) = dL(z)/dz$. Plot $p(D_s)$ for $s = 10, 100, 1000$ samples for sampling the normal distribution with unit variance.

Exercise 4.21: Sampled density from a weighted importance function

Given a weighted sample of an importance function $q(x)$

$$q_s(x) = \sum_{i=1}^{s} w_i^- \delta(x - x_i) \qquad \sum_i w_i^- = 1$$

(a) Show that the sampled density

$$\overline{p}_s(x) = \sum_{i=1}^{s} w_i \delta(x - x_i) \qquad w_i = w_i^- \frac{p(x_i)}{q(x_i)}$$

converges to $p(x)$ as sample size increases.

(b) Show that the sampled density is unbiased for all samples sizes.

Exercise 4.22: Sampled density from a weighted importance function when unable to evaluate the density

Given a weighted sample of an importance function $q(x)$

$$q_s(x) = \sum_{i=1}^{s} w_i^- \delta(x - x_i) \qquad \sum_i w_i^- = 1$$

and a density of the following form

$$p(x) = \frac{h(x)}{\int h(x)dx}$$

in which $p(x)$ cannot be conveniently evaluated but $h(x)$ can be evaluated.

(a) Show that the sampled density

$$\overline{p}_s(x) = \sum_{i=1}^{s} \overline{w}_i \delta(x - x_i) \qquad w_i = w_i^- \frac{h(x_i)}{q(x_i)} \qquad \overline{w}_i = \frac{w_i}{\sum_j w_j}$$

converges to $p(x)$ as sample size increases.

(b) Show that the sampled density is biased for all finite sample sizes.

Exercise 4.23: Unbiased particle filter with importance sampling

Show that an unbiased particle filter using importance sampling is given by

$$\overline{p}_s(x(k)|\mathbf{y}(k)) = \{x_i(k), \widetilde{w}_i(k)\}$$

$$\widetilde{w}_i(k+1) = \widetilde{w}_i(k) \frac{p(y(k+1)|x_i(k+1))\, p(x_i(k+1)|x_i(k))}{p(y(k+1)|\mathbf{y}(k))\, q(x_i(k+1)|x_i(k), y(k+1))}$$

in which $x_i(k)$ are samples of the importance function $q(x(k)|x_i(k-1), y(k))$. Note that normalization of $\widetilde{w}_i$ is not required in this form of a particle filter, but evaluation of $p(y(k+1)|\mathbf{y}(k))$ is required.

Exercise 4.24: Variance of the unbiased particle filter with importance sampling

Show that the variance of the weights of the unbiased particle filter given in Exercise 4.23 increases with time.

Exercise 4.25: Optimal importance function for a linear system

The optimal importance function is given in (4.60), repeated here

$$q(x_i(k)|x_i(k-1), y(k)) = p(x_i(k)|x_i(k-1), y(k))$$

For the linear time-invariant model, this conditional density is the following normal density (Doucet et al., 2000)

$$p(x_i(k)|x_i(k-1), y(k)) \sim N(\overline{x}(k), \overline{P})$$

$$\overline{x}(k) = PQ^{-1}(Ax_i(k-1) + Bu(k-1)) + PC'R^{-1}y(k)$$

$$\overline{P} = \left(Q^{-1} + C'R^{-1}C\right)^{-1}$$

Establish this result by first considering the linear transformation between $(x_i(k), y(k))$ and $x_i(k-1), w(k), v(k)$, and then using the formulas for taking conditional densities of normals.

Bibliography

M. S. Arulampalam, S. Maskell, N. J. Gordon, and T. Clapp. A tutorial on particle filters for online nonlinear/non-Gaussian Bayesian tracking. *IEEE Trans. Signal Process.*, 50(2):174–188, February 2002.

K. J. Åström. *Introduction to Stochastic Control Theory*. Academic Press, San Diego, California, 1970.

D. P. Bertsekas. *Dynamic Programming*. Prentice-Hall, Inc., Englewood Cliffs, New Jersey, 1987.

A. E. Bryson and Y. Ho. *Applied Optimal Control*. Hemisphere Publishing, New York, 1975.

F. M. Callier and C. A. Desoer. *Linear System Theory*. Springer-Verlag, New York, 1991.

M. Chaves and E. D. Sontag. State-estimators for chemical reaction networks of Feinberg-Horn-Jackson zero deficiency type. *Eur. J. Control*, 8(4):343–359, 2002.

C. E. de Souza, M. R. Gevers, and G. C. Goodwin. Riccati equation in optimal filtering of nonstabilizable systems having singular state transition matrices. *IEEE Trans. Auto. Cont.*, 31(9):831–838, September 1986.

A. Doucet, S. Godsill, and C. Andrieu. On sequential Monte Carlo sampling methods for Bayesian filtering. *Stat. and Comput.*, 10:197–208, 2000.

W. Feller. On the Kolmogorov-Smirnov limit theorems for empirical distributions. *Ann. Math. Stat.*, 19(2):177–189, 1948.

W. Feller. *An Introduction to Probability Theory and Its Applications: Volume I*. John Wiley & Sons, New York, third edition, 1968.

A. Gelb, editor. *Applied Optimal Estimation*. The M.I.T. Press, Cambridge, Massachusetts, 1974.

N. J. Gordon, D. J. Salmond, and A. F. M. Smith. Novel approach to nonlinear/non-Gaussian Bayesian state estimation. *IEE Proc. F-Radar and Signal Processing*, 140(2):107–113, April 1993.

R. Gudi, S. Shah, and M. Gray. Multirate state and parameter estimation in an antibiotic fermentation with delayed measurements. *Biotech. Bioeng.*, 44: 1271–1278, 1994.

J. E. Handschin and D. Q. Mayne. Monte Carlo techniques to estimate the conditional expectation in multistage nonlinear filtering. *Int. J. Control*, 9 (5):547–559, 1969.

A. H. Jazwinski. *Stochastic Processes and Filtering Theory.* Academic Press, New York, 1970.

Z.-P. Jiang and Y. Wang. Input-to-state stability for discrete-time nonlinear systems. *Automatica*, 37:857–869, 2001.

S. J. Julier and J. K. Uhlmann. A new extension of the Kalman filter to nonlinear systems. In *International Symposium Aerospace/Defense Sensing, Simulation and Controls*, pages 182–193, 1997.

S. J. Julier and J. K. Uhlmann. Author's reply. *IEEE Trans. Auto. Cont.*, 47(8): 1408–1409, August 2002.

S. J. Julier and J. K. Uhlmann. Unscented filtering and nonlinear estimation. *Proc. IEEE*, 92(3):401–422, March 2004a.

S. J. Julier and J. K. Uhlmann. Corrections to unscented filtering and nonlinear estimation. *Proc. IEEE*, 92(12):1958, December 2004b.

S. J. Julier, J. K. Uhlmann, and H. F. Durrant-Whyte. A new method for the nonlinear transformation of means and covariances in filters and estimators. *IEEE Trans. Auto. Cont.*, 45(3):477–482, March 2000.

T. Kailath. A view of three decades of linear filtering theory. *IEEE Trans. Inform. Theory*, IT-20(2):146–181, March 1974.

R. Kandepu, L. Imsland, and B. A. Foss. Constrained state estimation using the unscented kalman filter. In *Procedings of the 16th Mediterranean Conference on Control and Automation*, pages 1453–1458, Ajaccio, France, June 2008.

A. Kolmogoroff. Sulla determinazione empirica di una legge di distribuzione. *Giorn. Ist. Ital. Attuari*, 4:1–11, 1933.

A. N. Kolmogorov. Interpolation and extrapolation of stationary random sequences. *Bull. Moscow Univ., USSR, Ser. Math. 5*, 1941.

H. Kwakernaak and R. Sivan. *Linear Optimal Control Systems.* John Wiley and Sons, New York, 1972.

T. Lefebvre, H. Bruyninckx, and J. De Schutter. Comment on "A new method for the nonlinear transformation of means and covariances in filters and estimators". *IEEE Trans. Auto. Cont.*, 47(8):1406–1408, August 2002.

E. S. Meadows, K. R. Muske, and J. B. Rawlings. Constrained state estimation and discontinuous feedback in model predictive control. In *Proceedings of the 1993 European Control Conference*, pages 2308–2312, 1993.

H. Michalska and D. Q. Mayne. Moving horizon observers and observer-based control. *IEEE Trans. Auto. Cont.*, 40(6):995–1006, 1995.

S. A. Middlebrooks and J. B. Rawlings. State estimation approach for determining composition and growth rate of $Si_{1-x}Ge_x$ chemical vapor deposition utilizing real-time ellipsometric measurements. *Applied Opt.*, 45:7043–7055, 2006.

K. R. Muske, J. B. Rawlings, and J. H. Lee. Receding horizon recursive state estimation. In *Proceedings of the 1993 American Control Conference*, pages 900–904, June 1993.

M. Nørgaard, N. K. Poulsen, and O. Ravn. New developments in state estimation for nonlinear systems. *Automatica*, 36:1627–1638, 2000.

V. Prasad, M. Schley, L. P. Russo, and B. W. Bequette. Product property and production rate control of styrene polymerization. *J. Proc. Cont.*, 12(3):353–372, 2002.

C. C. Qu and J. Hahn. Computation of arrival cost for moving horizon estimation via unscent Kalman filtering. *J. Proc. Cont.*, 19(2):358–363, 2009.

C. V. Rao. *Moving Horizon Strategies for the Constrained Monitoring and Control of Nonlinear Discrete-Time Systems*. PhD thesis, University of Wisconsin–Madison, 2000.

C. V. Rao, J. B. Rawlings, and J. H. Lee. Constrained linear state estimation – a moving horizon approach. *Automatica*, 37(10):1619–1628, 2001.

C. V. Rao, J. B. Rawlings, and D. Q. Mayne. Constrained state estimation for nonlinear discrete-time systems: stability and moving horizon approximations. *IEEE Trans. Auto. Cont.*, 48(2):246–258, February 2003.

H. E. Rauch, F. Tung, and C. T. Striebel. Maximum likelihood estimates of linear dynamic systems. *AIAA J.*, 3(8):1445–1450, 1965.

J. B. Rawlings and B. R. Bakshi. Particle filtering and moving horizon estimation. *Comput. Chem. Eng.*, 30:1529–1541, 2006.

K. Reif and R. Unbehauen. The extended Kalman filter as an exponential observer for nonlinear systems. *IEEE Trans. Signal Process.*, 47(8):2324–2328, August 1999.

K. Reif, S. Günther, E. Yaz, and R. Unbehauen. Stochastic stability of the discrete-time extended Kalman filter. *IEEE Trans. Auto. Cont.*, 44(4):714–728, April 1999.

K. Reif, S. Günther, E. Yaz, and R. Unbehauen. Stochastic stability of the continuous-time extended Kalman filter. *IEE Proceedings-Control Theory and Applications*, 147(1):45–52, January 2000.

D. G. Robertson and J. H. Lee. On the use of constraints in least squares estimation and control. *Automatica*, 38(7):1113–1124, 2002.

A. Romanenko and J. A. A. M. Castro. The unscented filter as an alternative to the EKF for nonlinear state estimation: a simulation case study. *Comput. Chem. Eng.*, 28(3):347–355, March 15 2004.

A. Romanenko, L. O. Santos, and P. A. F. N. A. Afonso. Unscented Kalman filtering of a simulated pH system. *Ind. Eng. Chem. Res.*, 43:7531–7538, 2004.

N. Smirnov. On the estimation of the discrepancy between empirical curves of distribution for two independent samples. *Bulletin Mathématique de l'Univesité de Moscou*, 2, 1939. fasc. 2.

A. F. M. Smith and A. E. Gelfand. Bayesian statistics without tears: A sampling-resampling perspective. *Amer. Statist.*, 46(2):84–88, 1992.

E. D. Sontag. *Mathematical Control Theory*. Springer-Verlag, New York, second edition, 1998a.

E. D. Sontag. Comments on integral variants of ISS. *Sys. Cont. Let.*, 34:93–100, 1998b.

E. D. Sontag and Y. Wang. Output-to-state stability and detectability of nonlinear systems. *Sys. Cont. Let.*, 29:279–290, 1997.

R. F. Stengel. *Optimal Control and Estimation*. Dover Publications, Inc., 1994.

B. O. S. Teixeira, L. A. B. Tôrres, L. A. Aguirre, and D. S. Bernstein. Unscented filtering for interval-constrained nonlinear systems. In *Proceedings of the 47th IEEE Conference on Decision and Control*, pages 5116–5121, Cancun, Mexico, December 9-11 2008.

P. Vachhani, S. Narasimhan, and R. Rengaswamy. Robust and reliable estimation via unscented recursive nonlinear dynamic data reconciliation. *J. Proc. Cont.*, 16(10):1075–1086, December 2006.

R. van der Merwe, A. Doucet, N. de Freitas, and E. Wan. The unscented particle filter. Technical Report CUED/F-INFENG/TR 380, Cambridge University Engineering Department, August 2000.

N. Wiener. *The Extrapolation, Interpolation, and Smoothing of Stationary Time Series with Engineering Applications.* Wiley, New York, 1949. Originally issued as a classified MIT Rad. Lab. Report in February 1942.

D. I. Wilson, M. Agarwal, and D. W. T. Rippin. Experiences implementing the extended Kalman filter on an industrial batch reactor. *Comput. Chem. Eng.*, 22(11):1653–1672, 1998.

5

Output Model Predictive Control

5.1 Introduction

In Chapter 2 we show how model predictive control (MPC) may be employed to control a *deterministic* system, that is, a system in which there are no uncertainties and the state is known. In Chapter 3 we show how to control an *uncertain* system in which uncertainties are present but the state is known. Here we address the problem of MPC of an uncertain system in which the state is *not* fully known. We assume that there are outputs available that may be used to estimate the state as shown in Chapter 4. These outputs are used by the model predictive controller to generate control actions; hence the name *output MPC*.

Output feedback control is, in general, more complex than state feedback control since knowledge of the state provides considerable information. If the state is known, optimal control is, in general, a time-varying function of the current state *even if* the system is uncertain as, for example, when it is subject to an additive disturbance. In this case, the state must include the state of the disturbance.

Generally, however, the state is not known; instead, a noisy measurement $y(t)$ of the state is available at each time t. Since the state x is not known, it is replaced by a hyperstate p that summarizes all prior information (previous inputs and outputs and the prior distribution of the initial state) and that has the "state" property: future values of p can be determined from the current value of p and current and future inputs and outputs. Usually $p(t)$ is the conditional density of $x(t)$ given the prior density $p(0)$ of $x(0)$ and the current available "information" $I(t) := \{y(0), y(1), \ldots, y(t-1), u(0), u(1), \ldots, u(t-1)\}$. If the current hyperstate is known, future hyperstates have to be predicted since future noisy measurements of the state are not known. So the hyperstate satisfies an uncertain difference equation of the form

$$p^+ = \phi(p, u, \psi) \tag{5.1}$$

where $\{\psi(t)\}$ is a sequence of random variables; the problem of controlling a system with unknown state x is transformed into the problem of controlling an uncertain system with known state p. For example, if the underlying system is described by

$$\begin{aligned} x^+ &= Ax + Bu + w \\ y &= Cx + v \end{aligned}$$

where $\{w(t)\}$ and $\{v(t)\}$ are sequences of zero mean normal independent random variables with variances Σ_w and Σ_v, respectively, and if the prior density $p(0)$ of $x(0)$ is normal with density $n(\bar{x}_0, \Sigma_0)$ then, as is well known, $p(t)$ is the normal density $n(\hat{x}(t), \Sigma(t))$ so that the hyperstate $p(t)$ is finitely parameterized by $(\hat{x}(t), \Sigma(t))$; hence the evolution equation for $p(t)$ is defined by the evolution equation for $(\hat{x}, \Sigma)$, that is by:

$$\hat{x}(t+1) = A\hat{x}(t) + Bu + L(t)\psi(t) \tag{5.2}$$

$$\Sigma(t+1) = \Phi(\Sigma(t)) \tag{5.3}$$

in which

$$\begin{aligned} \Phi(\Sigma) &:= A\Sigma A' - A\Sigma C'(C'\Sigma C + \Sigma_v)^{-1}C\Sigma A' + \Sigma_w \\ \psi(t) &:= y(t) - C\hat{x}(t) = C\tilde{x}(t) + v(t) \\ \tilde{x}(t) &:= x(t) - \hat{x}(t) \end{aligned}$$

The initial conditions for (5.2) and (5.3) are

$$\hat{x}(0) = \bar{x}_0 \qquad \Sigma(0) = \Sigma_0$$

These are, of course, the celebrated Kalman filter equations derived in Chapter 1. The random variables $\tilde{x}$ and ψ have the following densities: $\tilde{x}(t) \sim n(0, \Sigma(t))$ and $\psi(t) \sim n(0, \Sigma_v + C'\Sigma(t)C)$. The finite dimensional equations (5.2) and (5.3) replace the difference equation (5.1) for the hyperstate p that is a conditional density and, therefore, infinite dimensional in general. The sequence $\{\psi(t)\}$ is known as the *innovation* sequence; $\psi(t)$ is the "new" information contained in $y(t)$.

Output control, in general, requires control of the hyperstate p which may be computed and is, therefore, known, but which satisfies a complex evolution equation $p^+ = \phi(p, u, \psi)$ where ψ is a random disturbance. Controlling p is a problem of the same type as that considered in Chapter 3, but considerably more complex since the function $p(\cdot)$ is infinite dimensional. Because of the complexity of the evolution equation for p, the separation principle is often invoked; assuming that the state x is known, a stabilizing controller $u = \kappa(x)$ and an observer or filter yielding an estimate $\hat{x}$ of the state are separately designed; the control $u = \kappa(\hat{x})$ is then applied to the plant. Indeed, this form of control is actually optimal for the linear quadratic Gaussian (LQG) optimal control problem considered briefly above but is not necessarily stabilizing when the system is nonlinear and constrained. We propose a variant of this procedure, modified to cope with state and control constraints. The state estimate $\hat{x}$ satisfies an uncertain difference equation with an additive disturbance of the same type as that considered in Chapter 3. Hence we employ tube MPC, similar to that employed in Chapter 3, but modified to ensure that state estimation error, not considered in Chapter 3, does not result in transgression of the control and state constraints. An advantage of the method presented here is that its online complexity is comparable to that of conventional MPC.

As in Chapter 3, a caveat is necessary. Because of the inherent complexity of output MPC, different compromises between simplicity and efficiency are possible; for this reason, output MPC remains an active research area and alternative methods, available or yet to be developed, may be preferred.

5.2 A Method for Output MPC

Suppose the system to be controlled is described by

$$x^+ = f(x, u, w) \tag{5.4}$$

$$y = h(x, v) \tag{5.5}$$

where $x \in \mathbb{R}^n$, $u \in \mathbb{R}^m$ and $y \in \mathbb{R}^p$; the disturbance w lies in $\mathbb{R}^n$, and the measurement noise v lies in $\mathbb{R}^p$ A prime requirement for simplification is replacement of the infinite dimensional hyperstate p, which is $n(\hat{x}, \Sigma)$ in the linear Gaussian case, by something considerably simpler. The hyperstate p, being a conditional density, may be regarded as a continuum of nested confidence regions, each of which is a subset of $\mathbb{R}^n$. Our initial simplification is the replacement, if this is possible, of this continuum of confidence regions by a single region of the form $\{\bar{x}\} \oplus \mathbb{\Sigma} \subseteq \mathbb{R}^n$, where $\bar{x}$ is the "center" of the confidence region and $\mathbb{\Sigma}$

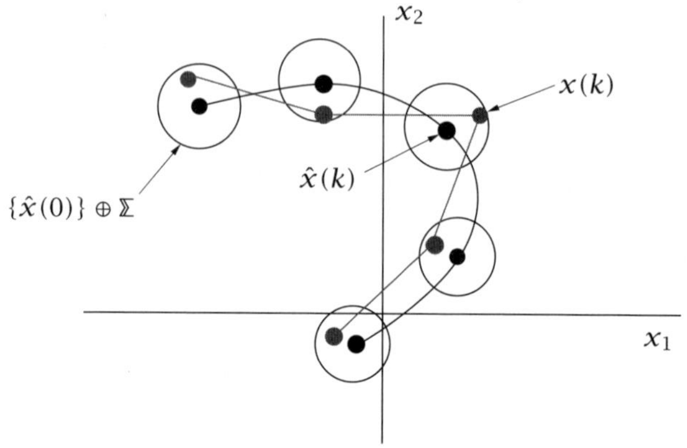

Figure 5.1: State estimator tube.

now denotes a subset of $\mathbb{R}^n$ rather than a variance. If the problem is stochastic, $\{\hat{x}\} \oplus \Sigma$ may be a β confidence region for p, i.e., a region satisfying $\Pr\{x \in \{\hat{x}\} \oplus \Sigma | I\} = \beta$. When all disturbances are bounded, the usual assumption in robust MPC, Σ is chosen to ensure that all possible values of x lie in the set $\{\hat{x}\} \oplus \Sigma$. The finite dimensional variable $(\hat{x}, \Sigma)$ replaces the infinite dimensional object p.[1] In the linear time-invariant case, the state estimator $(\hat{x}, \Sigma)$ evolves, as shown in the sequel, according to

$$\hat{x}^+ = \phi(\hat{x}, u, \psi) \tag{5.6}$$

$$\Sigma^+ = \Phi(\Sigma) \tag{5.7}$$

in which ψ is a random variable in the stochastic case and a bounded disturbance taking values in Ψ when w and v are bounded. In the latter case, $x \in \{\hat{x}\} \oplus \Sigma$ implies $x^+ \in \{\hat{x}^+\} \oplus \Sigma^+$ for all $\psi \in \Psi$.

More generally, let $X \subseteq \mathbb{R}^n$ denote the set of states consistent with the current information I; although $X = \{\hat{x}\} \oplus \Sigma$ in the linear case, the evolution of X, in the nonlinear case, is more complex than (5.6) and (5.7). The hope remains that X has an outer approximation of the form $\{\hat{x}\} \oplus \Sigma$ where $\hat{x}$ may be obtained by one of the methods described in Chapter 4; however Σ may no longer be independent of the observation sequence.

As illustrated in Figure 5.1, the evolution equations generate a *tube*, which is the set sequence $\{\{\hat{x}(t)\} \oplus \Sigma(t)\}$; at time t the center of the

[1]The object $\{\hat{x}\} \oplus \Sigma$ may be regarded (Moitié, Quincampoix, and Veliov, 2002) as the "state" for the output MPC problem.

tube is $\hat{x}(t)$ and the "cross section" is $\mathbb{\Sigma}(t)$. When the disturbances are bounded, which is the only case we consider in the sequel, all possible realizations of the state trajectory $\{x(t)\}$ lie in the set $\{\{\hat{x}(t)\} \oplus \mathbb{\Sigma}(t)\}$ for all t; the dashed line is a sample trajectory of $x(t)$.

From (5.6), the estimator trajectory $\{\hat{x}(t)\}$ is influenced both by the control that is applied and by the disturbance sequence $\{\psi(t)\}$. If the trajectory were influenced only by the control, we could choose the control to satisfy both the control constraints and to cause the estimator tube to lie in a region such that the state constraints are satisfied by all possible realizations of the state trajectory. Hence the output MPC problem would reduce to a conventional MPC problem with modified constraints in which the state is $\hat{x}$, rather than x. The new state constraint is $\hat{x} \in \hat{\mathbb{X}}$ where $\hat{\mathbb{X}}$ is chosen to ensure that $\hat{x} \in \hat{\mathbb{X}}$ implies $x \in \mathbb{X}$ and, therefore, satisfies $\hat{\mathbb{X}} \subseteq \mathbb{X} \ominus \mathbb{\Sigma}$ if $\mathbb{\Sigma}$ does not vary with time t.

But the estimator state $\{\hat{x}(t)\}$ is influenced by the disturbance ψ (see (5.6)), so it cannot be precisely controlled. The problem of controlling the system described by (5.6) is the same type of problem studied in Chapter 3, where the system was described by $x^+ = f(x, u, w)$ with the estimator state $\hat{x}$, which is accessible, replacing the actual state x. Hence we may use the techniques presented in Chapter 3 to choose a control that forces $\hat{x}$ to lie in another tube $\{\{z(t)\} \oplus \mathbb{S}(t)\}$ where the set sequence $\{\mathbb{S}(t)\}$ that defines the cross section of the tube is precomputed, and $\{z(t)\}$ that defines the center of the tube is the state trajectory of the nominal (deterministic) system defined by

$$z^+ = \phi(z, u, 0) \tag{5.8}$$

which is the nominal version of (5.6). Equations (5.8) is obtained by replacing ψ by 0 in the original equations. Thus we get two tubes, one embedded in the other. At time t the estimator state $\hat{x}(t)$ lies in the set $\{z(t)\} \oplus \mathbb{S}(t)$, and $x(t)$ lies in the set $\{\hat{x}(t)\} \oplus \mathbb{\Sigma}(t)$, so that for all t

$$x(t) \in \{z(t)\} \oplus \mathbb{\Gamma}(t) \qquad \mathbb{\Gamma}(t) := \mathbb{\Sigma}(t) \oplus \mathbb{S}(t)$$

The tubes $\{\{z(t)\} \oplus \mathbb{S}(t)\}$, in which the trajectory $\{\hat{x}(t)\}$ lies, and $\{\{z(t)\} \oplus \mathbb{\Gamma}(t)\}$, in which the state trajectory $\{x(t)\}$ lies, are shown in Figure 5.2. The tube $\{\{z(t)\} \oplus \mathbb{S}(t)\}$ is embedded in the larger tube $\{\{z(t)\} \oplus \mathbb{\Gamma}(t)\}$.

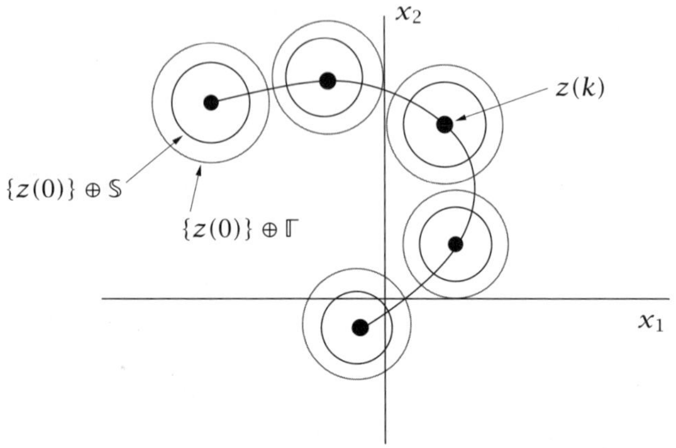

Figure 5.2: State tube.

5.3 Linear Constrained Systems: Time-Invariant Case

5.3.1 Introduction

We consider the following uncertain linear time-invariant system

$$\begin{aligned} x^+ &= Ax + Bu + w \\ y &= Cx + v \end{aligned} \tag{5.9}$$

in which $x \in \mathbb{R}^n$ is the current state, $u \in \mathbb{R}^m$ is the current control action, x^+ is the successor state, $w \in \mathbb{R}^n$ is an unknown state disturbance, $y \in \mathbb{R}^p$ is the current measured output, $v \in \mathbb{R}^p$ is an unknown output disturbance, the pair (A, B) is assumed to be controllable, and the pair (A, C) observable. The state and additive disturbances w and v are known only to the extent that they lie, respectively, in the C-sets[2] $\mathbb{W} \subseteq \mathbb{R}^n$ and $\mathbb{N} \subseteq \mathbb{R}^p$. Let $\phi(i; x(0), \mathbf{u}, \mathbf{w})$ denote the solution of (5.9) at time i if the initial state at time 0 is $x(0)$, and the control and disturbance sequences are, respectively, $\mathbf{u} := \{u(0), u(1), \ldots\}$ and $\mathbf{w} := \{w(0), w(1), \ldots\}$. The system (5.9) is subject to the following set of hard state and control constraints

$$x \in \mathbb{X} \qquad u \in \mathbb{U} \tag{5.10}$$

where $\mathbb{X} \subseteq \mathbb{R}^n$ and $\mathbb{U} \subseteq \mathbb{R}^m$ are polyhedral and polytopic sets respectively; both sets contain the origin as an interior point.

[2]Recall, a C-set is a convex, compact set containing the origin.

5.3.2 State Estimator

To estimate the state a simple Luenberger observer is employed

$$\hat{x}^+ = A\hat{x} + Bu + L(y - \hat{y})$$
$$\hat{y} = C\hat{x} \tag{5.11}$$

where $\hat{x} \in \mathbb{R}^n$ is the current observer state (state estimate), $u \in \mathbb{R}^m$ is the current control action, $\hat{x}^+$ is the successor state of the observer system, $\hat{y} \in \mathbb{R}^p$ is the current observer output, and $L \in \mathbb{R}^{n \times p}$. The output injection matrix L is chosen to satisfy $\rho(A_L) < 1$ where $A_L := A - LC$.

The estimated state $\hat{x}$ therefore satisfies the following uncertain difference equation

$$\hat{x}^+ = A\hat{x} + Bu + L(C\widetilde{x} + v)$$

where the state estimation error $\widetilde{x}$ is defined by $\widetilde{x} := x - \hat{x}$ so that $x = \hat{x} + \widetilde{x}$. Since $x^+ = Ax + Bu + w$, the state estimation error $\widetilde{x}$ satisfies

$$\widetilde{x}^+ = A_L\widetilde{x} + \widetilde{w} \qquad \widetilde{w} := w - Lv \tag{5.12}$$

Because w and v are bounded, so is $\widetilde{w}$; in fact, $\widetilde{w}$ takes values in the C-set $\widetilde{\mathbb{W}}$ defined by

$$\widetilde{\mathbb{W}} := \mathbb{W} \oplus (-L\mathbb{N})$$

We recall the following standard definitions (Blanchini, 1999):

Definition 5.1 (Positive invariance; robust positive invariance). A set $\Omega \subseteq \mathbb{R}^n$ is *positive invariant* for the system $x^+ = f(x)$ and the constraint set $\mathbb{X}$ if $\Omega \subseteq \mathbb{X}$ and $f(x) \in \Omega, \ \forall x \in \Omega$.
A set $\Omega \subseteq \mathbb{R}^n$ is *robust positive invariant* for the system $x^+ = f(x, w)$ and the constraint set $(\mathbb{X}, \mathbb{W})$ if $\Omega \subseteq \mathbb{X}$ and $f(x, w) \in \Omega, \ \forall w \in \mathbb{W}$, $\forall x \in \Omega$.

Since $\rho(A_L) < 1$ and $\widetilde{\mathbb{W}}$ is compact, there exists, as shown in Kolmanovsky and Gilbert (1998), Theorem 4.1, a robust positive invariant set $\Sigma \subseteq \mathbb{R}^n$, satisfying

$$A_L\Sigma \oplus \widetilde{\mathbb{W}} = \Sigma \tag{5.13}$$

Hence, for all $\widetilde{x} \in \Sigma$, $\widetilde{x}^+ = A_L\widetilde{x} + \widetilde{w} \in \Sigma$ for all $\widetilde{w} \in \widetilde{\mathbb{W}}$; the term *robust* in the description of Σ refers to this property. In fact, Σ is the *minimal* robust, positive invariant set for $\widetilde{x}^+ = A_L\widetilde{x} + \widetilde{w}$, $\widetilde{w} \in \widetilde{\mathbb{W}}$, i.e., a set that is a subset of all robust positive invariant sets. There exist techniques

(Raković, Kerrigan, Kouramas, and Mayne, 2005) for obtaining, for every $\epsilon > 0$, a polytopic, nonminimal, robust, positive invariant set Σ^0 that satisfies $d_H(\Sigma, \Sigma^0) \leq \epsilon$ where $d_H(\cdot, \cdot)$ is the Hausdorff metric. An immediate consequence of (5.13) is:

Proposition 5.2 (Proximity of state and state estimate). *If the initial system and observer states, $x(0)$ and $\hat{x}(0)$ respectively, satisfy $\{x(0)\} \in \{\hat{x}(0)\} \oplus \Sigma$, then $x(i) \in \{\hat{x}(i)\} \oplus \Sigma$ for all $i \in \mathbb{I}_{\geq 0}$, and all admissible disturbance sequences* $\mathbf{w}$ *and* $\mathbf{v}$.

The assumption that $\tilde{x}(i) \in \Sigma$ for all i is a *steady-state* assumption; if $\tilde{x}(0) \in \Sigma$, then $\tilde{x}(i) \in \Sigma$ for all i. If, on the other hand, $\tilde{x}(0) \in \Sigma(0)$ where $\Sigma(0) \supseteq \Sigma$, then it is possible to show that $\tilde{x}(i) \in \Sigma(i)$ for all $i \in \mathbb{I}_{\geq 0}$ where $\Sigma(i) \to \Sigma$ in the Hausdorff metric as $i \to \infty$; the sequence $\{\Sigma(i)\}$ satisfies $\Sigma(0) \supseteq \Sigma(1) \supseteq \Sigma(2) \supseteq \cdots \supseteq \Sigma$. Hence, it is reasonable to assume that if the estimator has been running for a "long" time, it is in steady state.

Hence we have obtained a state estimator, with "state" $(\hat{x}, \Sigma)$ satisfying

$$\begin{aligned} \hat{x}^+ &= A\hat{x} + Bu + L(y - \hat{y}) \\ \Sigma^+ &= \Sigma \end{aligned} \tag{5.14}$$

and $x(i) \in \hat{x}(i) \oplus \Sigma$ for all $i \in \mathbb{I}_{\geq 0}$, thus meeting the requirements specified in Section 5.2. Knowing this, our remaining task is to control $\hat{x}(i)$ so that the resultant closed-loop system is stable and satisfies all constraints.

5.3.3 Controlling $\hat{x}$

Since $\tilde{x}(i) \in \Sigma$ for all i, we seek a method for controlling the observer state $\hat{x}(i)$ in such a way that $x(i) = \hat{x}(i) + \tilde{x}(i)$ satisfies the state constraint $x(i) \in \mathbb{X}$ for all i. The state constraint $x(i) \in \mathbb{X}$ will be satisfied if we control the estimator state to satisfy $\hat{x}(i) \in \mathbb{X} \ominus \Sigma$ for all i. The estimator state satisfies (5.14) which can be written in the form

$$\hat{x}^+ = A\hat{x} + Bu + \delta \tag{5.15}$$

where the disturbance δ is defined by

$$\delta := L(y - \hat{y}) = L(C\tilde{x} + v)$$

and, therefore, always lies in the C-set Δ defined by

$$\Delta := L(C\Sigma \oplus \mathbb{N})$$

The problem of controlling $\hat{x}$ is, therefore, the same as that of controlling an uncertain system with known state. This problem was extensively discussed in Chapter 3. We can therefore use the approach of Chapter 3 here with $\hat{x}$ replacing x, δ replacing w, $\mathbb{X} \ominus \Sigma$ replacing $\mathbb{X}$ and $\mathbb{A}$ replacing $\mathbb{W}$.

To control (5.15) we use, as in Chapter 3, a combination of open-loop and feedback control, i.e., we choose the control u as follows

$$u = v + Ke \qquad e := \hat{x} - z \tag{5.16}$$

where z is the state of a nominal (deterministic) system that we shall shortly specify; v is the feedforward component of the control u, and Ke is the feedback component. The matrix K is chosen to satisfy $\rho(A_K) < 1$ where $A_K := A + BK$. The feedforward component v of the control u generates, as we show subsequently, a trajectory $\{z(i)\}$, which is the center of the tube in which the state estimator trajectory $\{\hat{x}(i)\}$ lies. The feedback component Ke attempts to steer the trajectory $\{\hat{x}(i)\}$ of the state estimate toward the center of the tube and thereby controls the cross section of the tube. The controller is *dynamic* since it incorporates the nominal dynamic system.

With this control, $\hat{x}$ satisfies the following difference equation

$$\hat{x}^+ = A\hat{x} + Bv + BKe + \delta \qquad \delta \in \mathbb{A} \tag{5.17}$$

The nominal (deterministic) system describing the evolution of z is obtained by neglecting the disturbances BKe and δ in (5.17) yielding

$$z^+ = Az + Bv$$

The deviation $e = \hat{x} - z$ between the state $\hat{x}$ of the estimator and the state z of the nominal system satisfies

$$e^+ = A_K e + \delta \qquad A_K := A + BK \tag{5.18}$$

The feedforward component v of the control u generates the trajectory $\{z(i)\}$, which is the center of the tube in which the state estimator trajectory $\{\hat{x}(i)\}$ lies. Because $\mathbb{A}$ is a C-set and $\rho(A_K) < 1$, there exists a robust positive invariant C-set $\mathbb{S}$ satisfying

$$A_K \mathbb{S} \oplus \mathbb{A} = \mathbb{S}$$

An immediate consequence is the following.

Proposition 5.3 (Proximity of state estimate and nominal state). *If the initial states of the estimator and nominal system, $\hat{x}(0)$ and $z(0)$ respectively, satisfy $\hat{x}(0) \in \{z(0)\} \oplus \mathbb{S}$, then $\hat{x}(i) \in \{z(i)\} \oplus \mathbb{S}$ and $u(i) \in \{v(i)\} \oplus K\mathbb{S}$ for all $i \in \mathbb{I}_{\geq 0}$, and all admissible disturbance sequences $\mathbf{w}$ and $\mathbf{v}$.*

It follows from Proposition 5.3 that the state estimator trajectory $\hat{\mathbf{x}}$ remains in the tube $\hat{\mathbf{X}}(z(0), \mathbf{v}) := \{\{z(i)\} \oplus \mathbb{S} \mid i \in \mathbb{I}_{\geq 0}\}$ and the control trajectory $\mathbf{v}$ remains in the tube $\hat{\mathbf{V}}(\mathbf{v}) := \{\{v(i)\} \oplus K\mathbb{S} \mid i \in \mathbb{I}_{\geq 0}\}$ provided that $e(0) \in \mathbb{S}$. Hence, from Propositions 5.2 and 5.3, the state trajectory $\mathbf{x}$ lies in the tube $\mathbf{X}(z(0), \mathbf{v}) := \{\{z(i)\} \oplus \mathbb{\Gamma} \mid i \in \mathbb{I}_{\geq 0}\}$ where $\mathbb{\Gamma} := \mathbb{S} \oplus \mathbb{\Sigma}$ provided that $\widetilde{x}(0) = \hat{x}(0) - z(0) \in \mathbb{\Sigma}$ *and* $e(0) \in \mathbb{S}$. This information may be used to construct a robust output feedback model predictive controller using the procedures outlined in Chapter 3 for robust state feedback MPC of systems; the major difference is that we now control the estimator state $\hat{x}$ and use the fact that the actual state x lies in $\{\hat{x}\} \oplus \mathbb{\Sigma}$.

5.3.4 Output MPC

Model predictive controllers can now be constructed as described in Chapter 3, which dealt with robust control when the state was known. There is an obvious difference in that we are now concerned with controlling $\hat{x}$ whereas, in Chapter 3, our concern was control of x. We describe here the appropriate modification of the simple model predictive controller presented in Section 3.4.2. We adopt the same procedure of defining a nominal optimal control problem with tighter constraints than in the original problem. The solution to this problem defines the center of a tube in which solutions to the original system lie, and the tighter constraints in the nominal problem ensure that the original constraints are satisfied by the actual system.

The nominal system is described by

$$z^+ = Az + Bv \tag{5.19}$$

The nominal optimal control problem is the minimization of the cost function $\bar{V}_N(z, \mathbf{v})$ where

$$\bar{V}_N(z, \mathbf{v}) := \sum_{k=0}^{N-1} \ell(z(k), v(k)) + V_f(z(N)) \tag{5.20}$$

subject to satisfaction by the state and control sequences of (5.19) and the *tighter* constraints

$$z(i) \in \mathbb{Z} \subseteq \mathbb{X} \ominus \mathbb{\Gamma} \qquad \mathbb{\Gamma} := \mathbb{S} \oplus \mathbb{\Sigma} \tag{5.21}$$

$$v(i) \in \mathbb{V} \subseteq \mathbb{U} \ominus K\mathbb{S} \tag{5.22}$$

as well as a terminal constraint $z(N) \in \mathbb{Z}_f \subseteq \mathbb{Z}$. Notice that $\mathbb{\Gamma}$ appears in (5.21) whereas $\mathbb{S}$, the set in which $e = \hat{x} - z$ lies, appears in (5.22); this differs from the case studied in Chapter 3 where the same set appears in both equations. The sets $\mathbb{W}$ and $\mathbb{N}$ are assumed to be sufficiently small to ensure the following condition.

Assumption 5.4 (Constraint bounds). $\mathbb{\Gamma} = \mathbb{S} \oplus \mathbb{\Sigma} \subseteq \mathbb{X}$ and $K\mathbb{S} \subseteq \mathbb{U}$.

If Assumption 5.4 holds, the sets on the right-hand side of (5.21) and (5.22) are not empty; it can be seen from their definitions that the sets $\mathbb{\Sigma}$ and $\mathbb{S}$ tend to the set $\{0\}$ as $\mathbb{W}$ and $\mathbb{N}$ tend to the set $\{0\}$ in the sense that $d_H(\mathbb{W}, \{0\}) \to 0$ and $d_H(\mathbb{N}, \{0\}) \to 0$.

It follows from Propositions 5.2 and 5.3, if Assumption 5.4 holds, that satisfaction of the constraints (5.21) and (5.22) by the nominal system ensures satisfaction of the constraints (5.10) by the original system. The nominal optimal control problem is, therefore,

$$\mathbb{P}_N(z): \quad \bar{V}_N^0(z) = \min_{\mathbf{v}}\{\bar{V}_N(z, \mathbf{v}) \mid \mathbf{v} \in \mathcal{V}_N(z)\}$$

where the constraint set $\mathcal{V}_N(z)$ is defined by

$$\mathcal{V}_N(z) := \{\mathbf{v} \mid v(k) \in \mathbb{V} \text{ and } \bar{\phi}(k; z, \mathbf{v}) \in \mathbb{Z} \ \forall k \in \{0, 1, \ldots, N-1\}, \\ \bar{\phi}(N; z, \mathbf{v}) \in \mathbb{Z}_f\} \tag{5.23}$$

In (5.23), $\mathbb{Z}_f \subseteq \mathbb{Z}$ is the terminal constraint set, and $\bar{\phi}(k; z, \mathbf{v})$ denotes the solution of $z^+ = Az + Bv$ at time k if the initial state at time 0 is z and the control sequence is $\mathbf{v} = \{v(0), v(1), \ldots, v(N-1)\}$. Let $\mathbf{v}^0(z)$ denote the minimizing control sequence; the stage cost $\ell(\cdot)$ is chosen to ensure uniqueness of $\mathbf{v}^0(z)$. The implicit MPC control law for the nominal system is $\bar{\kappa}_N(\cdot)$ defined by

$$\bar{\kappa}_N(z) := v^0(0; z)$$

where $v^0(0; z)$ is the first element in the sequence $\mathbf{v}^0(z)$. The domain of $V_N^0(\cdot)$ and $\mathbf{v}^0(\cdot)$, and, hence, of $\kappa_N(\cdot)$, is $\mathcal{Z}_N$ defined by

$$\mathcal{Z}_N := \{z \in \mathbb{Z} \mid \mathcal{V}_N^0(z) \neq \emptyset\} \tag{5.24}$$

$\mathcal{Z}_N$ is the set of initial states z that can be steered to $\mathbb{Z}_f$ by an admissible control $\mathbf{v}$ that satisfies the state and control constraints, (5.21) and (5.22), and the terminal constraint. From (5.16), the implicit control law for the state estimator $\hat{x}^+ = A\hat{x} + Bu + \delta$, $y = Cx$ is $\kappa_N(\cdot)$ defined by

$$\kappa_N(\hat{x}, z) := \bar{\kappa}_N(z) + K(\hat{x} - z)$$

The controlled composite system with state $(\hat{x}, z)$ satisfies

$$\hat{x}^+ = A\hat{x} + B\kappa_N(\hat{x}, z) + \delta \tag{5.25}$$

$$z^+ = Az + B\bar{\kappa}_N(z) \tag{5.26}$$

with initial state $(\hat{x}(0), z(0))$ satisfying $\hat{x}(0) \in \{z(0)\} \oplus \mathbb{S}$, $z(0) \in \mathcal{Z}_N$. These constraints are satisfied if $z(0) = \hat{x}(0) \in \mathcal{Z}_N$. The control algorithm may be formally stated as follows:

Robust control algorithm (linear constrained systems).

Initialization: At time 0, set $i = 0$, set $\hat{x} = \hat{x}(0)$ and set $z = \hat{x}$.

Step 1: At time i, solve the nominal optimal control problem $\bar{\mathbb{P}}_N(z)$ to obtain the current nominal control action $v = \bar{\kappa}_N(z)$ and the control $u = v + K(x - z)$.

Step 2: If $\hat{x} \notin \{z\} \oplus \mathbb{S}$ or $u \notin \{v\} \oplus K\mathbb{S}$, set $z = \hat{x}$ and re-solve $\bar{\mathbb{P}}_N(z)$ to obtain $v = \bar{\kappa}_N(z)$ and $u = v$.

Step 3: Apply the control u to the system being controlled.

Step 4: (a) Compute the successor state estimate $\hat{x}^+ = A\hat{x} + Bu + L(y - C\hat{x})$. (b) Compute the successor state $z^+ = f(z, v)$ of the nominal system.

Step 5: Set $(\hat{x}, z) = (\hat{x}^+, z^+)$, set $i = i + 1$ and go to Step 1.

In normal operation, Step 2 is not activated; Propositions 5.2 and 5.3 ensure that the constraints are satisfied. In the event of an unanticipated event, Step 2 is activated, the controller is reinitialized and normal operation resumed. If Step 2 is activated, $v = \bar{\kappa}_N(\hat{x})$ and $u = v$. Hence *nominal* MPC would ensue if Step 2 were activated at each sample.

If the terminal cost $V_f(\cdot)$ and terminal constraint set $\mathbb{Z}_f$ satisfy the stability Assumptions 2.12 and 2.13 of Chapter 2, and if Assumption 5.4 is satisfied, the value function $\bar{V}_N^0(\cdot)$ satisfies

$$\begin{aligned} \bar{V}_N^0(z) &\geq \ell(z, \bar{\kappa}_N(z)) && \forall z \in \mathcal{Z}_N \\ \Delta\bar{V}_N^0(z) &\leq -\ell(z, \bar{\kappa}_N(z)) && \forall z \in \mathcal{Z}_N \\ \bar{V}_N^0(z) &\leq V_f(z) && \forall z \in \mathbb{Z}_f \end{aligned}$$

in which $\Delta\bar{V}_N^0(z) := \bar{V}_N^0(f(z, \bar{\kappa}_N(z))) - \bar{V}_N^0(z)$.

As shown in Section 3.4.3, if, in addition to Assumption 5.4, (i) the stability Assumptions 2.12 and 2.13 are satisfied, (ii) $\ell(z, v) = (1/2)(|z|_Q^2 + |v|_R^2)$ where Q and R are positive definite, (iii) $V_f(z) = (1/2)\,|Z|_{P_f}^2$ where P_f is positive definite, and (iv) $\mathcal{Z}_N$ is a C-set, then there exist positive constants c_1 and c_2 such that

$$\begin{aligned} \bar{V}_N^0(z) &\geq c_1 |z|^2 && \forall z \in \mathcal{Z}_N \\ \Delta\bar{V}_N^0(z) &\leq -c_1 |z|^2 && \forall z \in \mathcal{Z}_N \\ \bar{V}_N^0(z) &\leq c_2 |z|^2 && \forall z \in \mathcal{Z}_N \end{aligned}$$

It follows from Chapter 2 that the origin is exponentially stable for the nominal system $z^+ = Az + B\bar{\kappa}_N(z)$ with a region of attraction $\mathcal{Z}_N$ so that there exists a $c > 0$ and a $\gamma \in (0, 1)$ such that

$$|z(i)| \leq c|z(0)|\gamma^i$$

for all $z(0) \in \mathcal{Z}_N$, all $i \in \mathbb{I}_{\geq 0}$. Also $z(i) \in \mathcal{Z}_N$ for all $i \in \mathbb{I}_{\geq 0}$ if $z(0) \in \mathcal{Z}_N$ so that problem $\mathbb{P}_N(z(i))$ is always feasible. Because the state $\hat{x}(i)$ of the state estimator always lies in $\{z(i)\} \oplus \mathbb{S}$, and the state $x(i)$ of the system being controlled always lies in $\{z(i)\} \oplus \Gamma$, it follows that $\hat{x}(i)$ converges robustly and exponentially fast to $\mathbb{S}$, and $x(i)$ converges robustly and exponentially fast to Γ. We are now in a position to establish exponential stability of $\mathcal{A} := \mathbb{S} \times \{0\}$ with a region of attraction $(\mathcal{Z}_N \oplus \mathbb{S}) \times \mathcal{Z}_N$ for the composite system (5.25) and (5.26).

Proposition 5.5 (Exponential stability of output MPC). *The set* $\mathcal{A} := \mathbb{S} \times \{0\}$ *is exponentially stable with a region of attraction* $(\mathcal{Z}_N \oplus \mathbb{S}) \times \mathcal{Z}_N$ *for the composite system* (5.25) *and* (5.26).

Proof. Let $\phi := (\hat{x}, z)$ denote the state of the composite system. Then $|\phi|_{\mathcal{A}}$ is defined by

$$|\phi|_{\mathcal{A}} = |\hat{x}|_{\mathbb{S}} + |z|$$

where $|\hat{x}|_{\mathbb{S}} := d(\hat{x}, \mathbb{S})$. But $\hat{x} \in \{z\} \oplus \mathbb{S}$ implies

$$|\hat{x}|_{\mathbb{S}} = d(\hat{x}, \mathbb{S}) = d(z + e, \mathbb{S}) \leq d(z + e, e) = |z|$$

since $e \in \mathbb{S}$. Hence $|\phi|_{\mathcal{A}} \le 2|z|$ so that

$$|\phi(i)|_{\mathcal{A}} \le 2|z(i)| \le 2c|z(0)|\gamma^i \le 2c|\phi(0)|\gamma^i$$

for all $\phi(0) \in (\mathcal{Z}_N \oplus \mathbb{S}) \times \mathcal{Z}_N$. Since, for all $z(0) \in \mathcal{Z}_N$, $z(i) \in \mathbb{Z}$ and $v(i) \in \mathbb{V}$, it follows that $\hat{x}(i) \in \{z(i)\} \oplus \mathbb{S}$, $x(i) \in \mathbb{X}$, and $u(i) \in \mathbb{U}$ for all $i \in \mathbb{I}_{\ge 0}$. Thus $\mathcal{A} := \mathbb{S} \times \{0\}$ is exponentially stable with a region of attraction $(\mathcal{Z}_N \oplus \mathbb{S}) \times \mathcal{Z}_N$ for the composite system (5.25) and (5.26). ■

It follows from Proposition 5.5 that $x(i)$, which lies in the set $\{z(i)\} \oplus \Gamma$, $\Gamma := \mathbb{S} \oplus \Sigma$, converges to the set Γ. In fact $x(i)$ converges to a set that is, in general, smaller than Γ since Γ is a conservative bound on $\tilde{x}(i) + e(i)$. We determine this smaller set as follows. Let $\phi := (\tilde{x}, e)$ and let $\psi := (w, v)$; ϕ is the state of the two error systems and ψ is a bounded disturbance lying in a C-set $\Psi := \mathbb{W} \times \mathbb{N}$. Then, from (5.12) and (5.18) the state ϕ evolves according to

$$\phi^+ = \tilde{A}\phi + \tilde{B}\psi \tag{5.27}$$

where

$$\tilde{A} := \begin{bmatrix} A_L & 0 \\ LC & A_K \end{bmatrix} \qquad \tilde{B} := \begin{bmatrix} I & -L \\ 0 & L \end{bmatrix}$$

Because $\rho(A_L) < 1$ and $\rho(A_K) < 1$, it follows that $\rho(\tilde{A}) < 1$. Since $\rho(\tilde{A}) < 1$ and Ψ is compact, there exists a robust positive invariant set $\Phi \subseteq \mathbb{R}^n \times \mathbb{R}^n$ for (5.27) satisfying

$$\tilde{A}\Phi \oplus \tilde{B}\Psi = \Phi$$

Hence $\phi(i) \in \Phi$ for all $i \in \mathbb{I}_{\ge 0}$ if $\phi(0) \in \Gamma$. Since $x(i) = z(i) + e(i) + \tilde{x}(i)$, it follows that $x(i) \in \{z(i)\} \oplus H\Phi$, $H := \begin{bmatrix} I_n & I_n \end{bmatrix}$, for all $i \in \mathbb{I}_{\ge 0}$ provided that $x(0)$, $\hat{x}(0)$ and $z(0)$ satisfy $(\tilde{x}(0), e(0)) \in \Phi$ where $\tilde{x}(0) = x(0) - \hat{x}(0)$ and $e(0) = \hat{x}(0) - z(0)$. If these initial conditions are satisfied, $x(i)$ converges robustly and exponentially fast to the set $H\Phi$.

The remaining robust controllers presented in Section 3.4 of Chapter 3 may be similarly modified to obtain a robust output model predictive controller.

5.4 Linear Constrained Systems: Time-Varying Case

5.4.1 Introduction

In the previous section we considered the case when the state estimator was time-invariant in the sense that the state estimation error $\tilde{x}(i)$ lies in a constant set $\mathbb{\Sigma}$ for all i. The state estimator, in this case, is analogous to the steady-state Kalman filter for which the state estimation error has constant variance. In this section we consider the case where the initial state estimation error $\widetilde{x}(0)$ lies in a set $\mathbb{\Sigma}(0)$, which is larger than the time-invariant set $\mathbb{\Sigma}$ considered in Section 5.3. We show subsequently that in the time-varying case, the estimation error lies in a set $\mathbb{\Sigma}(i)$ at time i where $\mathbb{\Sigma}(i)$ converges to $\mathbb{\Sigma}$ as i tends to infinity. Because the set $\mathbb{\Sigma}(i)$ in which the state estimation error lies is now time varying, the nominal optimal control problem has time-varying constraints and thus requires a different approach. Although this section shows that extension of the tube-based controller to the time-varying case is theoretically relatively simple, implementation is considerably more complex; readers whose main interest is in controllers that may be implemented simply should omit this section. To deal with the time-varying case, we extend slightly standard definitions of positive invariance and robust positive invariance.

Definition 5.6 (Positive invariance; time-varying case)**.** A sequence $\{\mathbb{\Sigma}(i)\}$ of sets is positive invariant for the time-varying system $x^+ = f(x,i)$, $i^+ = i+1$ if, for all $i \in \mathbb{I}_{\geq 0}$, all $x \in \mathbb{\Sigma}(i)$, $f(x,i) \in \mathbb{\Sigma}(i+1)$.

Definition 5.7 (Robust positive invariance; time-varying case)**.** A sequence $\{\mathbb{\Sigma}(i)\}$ of sets is robust positive invariant for the time-varying system $x^+ = f(x,w,i)$, $i^+ = i+1$ where the disturbance w lies in the set $\mathbb{W}$ if, for all $i \in \mathbb{I}_{\geq 0}$, all $x \in \mathbb{\Sigma}(i)$, $f(x,w,i) \in \mathbb{\Sigma}(i+1)$ for all $w \in \mathbb{W}$.

We assume, as before, that (A,B,C) is stabilizable and detectable.

5.4.2 State Estimator

The state estimator is defined as in (5.11) in Section 5.3.2. The state estimate $\hat{x}$ satisfies the difference equation

$$\hat{x}^+ = A\hat{x} + Bu + \delta \qquad \delta := L(y - C\hat{x})$$

and the state estimation error $\widetilde{x}$ satisfies

$$\widetilde{x}^+ = A_L\widetilde{x} + \widetilde{w} \qquad \widetilde{w} := w - Lv$$

5.4.3 Controlling x and $\hat{x}$

As before, we use MPC to control the state estimator and the system $x^+ = Ax + Bu$ by controlling the nominal system

$$z^+ = Az + Bv$$

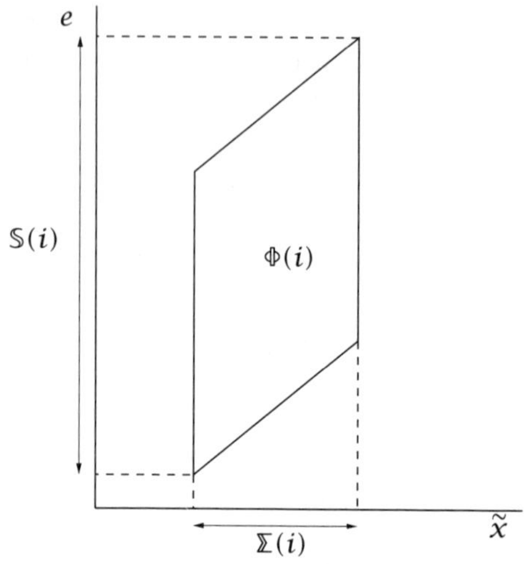

Figure 5.3: The sets $\Phi(i)$, $\Sigma(i)$ and $\mathbb{S}(i)$.

and setting $u = v + Ke$, $e := \hat{x} - z$. With this control, the composite system whose state is $\phi := (\widetilde{x}, e)$ satisfies

$$\begin{aligned} \widetilde{x}^+ &= A_L\widetilde{x} + w - Lv & \quad w \in \mathbb{W} \\ e^+ &= A_K e + LC\widetilde{x} + Lv & \quad v \in \mathbb{N} \end{aligned}$$

where $A_K := A - BK$. The difference equations for the composite system may be written in the more compact form

$$\phi^+ = \widetilde{A}\phi + \widetilde{B}\psi$$

where the composite state $\phi := (\widetilde{x}, e)$ lies in $\mathbb{R}^n \times \mathbb{R}^n$ and the bounded disturbance $\psi := (w, v)$ lies in the constant compact set $\Psi := \mathbb{W} \times \mathbb{N}$; the state matrix $\widetilde{A}$ and the disturbance matrix $\widetilde{B}$ are defined by

$$\widetilde{A} = \begin{bmatrix} A_L & 0 \\ LC & A_K \end{bmatrix} \qquad \widetilde{B} = \begin{bmatrix} I & -L \\ 0 & L \end{bmatrix}$$

We assume that K and L are such that $\rho(A_K) < 1$ and $\rho(A_L) < 1$; hence $\rho(\widetilde{A}) < 1$. Consider the set sequence $\{\Phi(i)\}$ defined by

$$\Phi(i+1) = \widetilde{A}\Phi(i) \oplus \widetilde{B}\Psi$$

with initial condition $\Phi(0) = \Sigma(0) \times \mathbb{S}(0)$ where $\widetilde{e}(0)$ lies $\Sigma(0)$, the initial state uncertainty set, and $e(0)$ lies in $\mathbb{S}(0)$. A comprehensive analysis of this coupled set of equations is provided in Raković (2007). It follows that the sequence $\{\Phi(i)\}$ is robust positive invariant for $\phi^+ = \widetilde{A}\phi + \widetilde{B}\psi$, $\psi \in \Psi$; if $\phi(0) = (\widetilde{x}(0), e(0)) \in \Phi(0)$; then $\phi(i) = (\widetilde{x}(0), e(0)) \in \Phi(i)$, $\widetilde{x}(i) \in \Sigma(i) := \begin{bmatrix} I_n & 0 \end{bmatrix} \Phi(i)$ and $e(i) \in \mathbb{S}(i) := \begin{bmatrix} 0 & I_n \end{bmatrix} \Phi(i)$ for all $i \in \mathbb{I}_{\geq 0}$. Since $x(i) = z(i) + e(i) + \widetilde{x}(i)$ and $u(i) = v(i) + Ke(i)$, it follows that $x(i) \in \{z(i)\} \oplus \Gamma(i)$, $\Gamma_i := \begin{bmatrix} I_n & I_n \end{bmatrix} \Phi(i)$, and $u(i) \in \{v(i)\} \oplus K\mathbb{S}(i)$ for all $i \in \mathbb{I}_{\geq 0}$. See Figure 5.3. The following result provides further properties of the sequence $\{\Phi(i)\}$ that we will require in the sequel.

Proposition 5.8 (Properties of composite system). *If $\Phi(0)$ is compact and $0 \in \Phi(0)$, then $0 \in \Phi(i)$ for all $i \in \mathbb{I}_{\geq 0}$ and the sequence $\{\Phi(i)\}$ converges, in the Hausdorff metric to Φ, the compact, minimal robust positive invariant set for $\phi^+ = \widetilde{A}\phi + \widetilde{B}\psi$, $\psi \in \Psi$. Moreover, $0 \in \Phi$, Φ satisfies $\Phi = \widetilde{A}\Phi \oplus \widetilde{B}\Psi$, and there exist $c > 0$ and $\lambda \in (0,1)$ such that $d_H(\Phi(i), \Phi) \leq c d_H(\Phi(0), \Phi)\lambda^i$ for all $i \in \mathbb{I}_{\geq 0}$. If, in addition, $\Phi(0)$ is a robust positive invariant set for the system $\phi^+ = \widetilde{A}\phi + \widetilde{B}\psi$, $\psi \in \Psi$, then, for each $i \in \mathbb{I}_{\geq 0}$, $\Phi(i)$ is robust positive invariant for $\phi^+ = \widetilde{A}\phi + \widetilde{B}\psi$, $\psi \in \Psi$, and $\{\Phi(i)\}$ is a monotonically nonincreasing sequence satisfying $0 \in \Phi(i)$ and $\Phi(i+1) \subseteq \Phi(i)$ for all $i \in \mathbb{I}_{\geq 0}$.*

Proof. It follows from the definition of the sequence $\{\Phi(i)\}$ that

$$\Phi(i) = \widetilde{A}^i \Phi(0) \oplus \mathcal{A}(i) \qquad \mathcal{A}(i) := \sum_{j=0}^{i-1} \widetilde{A}^j \widetilde{B}\Psi$$

where $\rho(\widetilde{A}) < 1$. The family of compact sets in $\mathbb{R}^n$ endowed with the Hausdorff metric is a complete space so any Cauchy sequence has a limit in this space. As shown in Kolmanovsky and Gilbert (1998), $\{\mathcal{A}(i)\}$ is a Cauchy sequence which, therefore, converges in the Hausdorff metric to the compact set Φ that satisfies $\Phi = \widetilde{A}\Phi \oplus \widetilde{B}\Psi$ and is the minimal robust positive invariant set for $\phi^+ = \widetilde{A}\phi + \widetilde{B}\psi$, $\psi \in \Psi$. Because $\Phi(0)$ is compact, and contains the origin, the set $\widetilde{A}^i \Phi(0)$ converges to $\{0\}$. Hence $\Phi(i)$ converges in the Hausdorff metric to Φ. Clearly $0 \in \mathcal{A}(i)$ for all $i \in \mathbb{I}_{\geq 0}$. Because $0 \in \Phi(0)$, it follows that $0 \in \Phi(i)$ for all $i \in \mathbb{I}_{\geq 0}$; because Φ is closed, $0 \in \Phi$. By hypothesis $\Phi(0)$ is robust positive invariant for $\phi^+ = \widetilde{A}\phi + \widetilde{B}\psi$, $\psi \in \Psi(0)$ so that

$$\Phi(1) = \widetilde{A}\Phi(0) + \widetilde{B}\Psi \subseteq \Phi(0)$$

Let $i \in \mathbb{I}_{\geq 0}$ be arbitrary and assume that $\Phi(i)$ is robust positive invariant for $\phi^+ = \widetilde{A}\phi + \widetilde{B}\psi$, $\psi \in \Psi$. Then

$$\Phi(i+1) = \widetilde{A}\Phi(i) \oplus \widetilde{B}\Psi \subseteq \Phi(i)$$

so that $\Phi(i)$ is robust positive invariant for $\phi^+ = \widetilde{A}\phi + \widetilde{B}\psi$, $\psi \in \Psi$ and $\Phi(i+1) \subseteq \Phi(i)$. By induction, $\Phi(i)$ is robust positive invariant for $\phi^+ = \widetilde{A}\phi + \widetilde{B}\psi$, $\psi \in \Psi$ and $\Phi(i+1) \subseteq \Phi(i)$ for all $i \in \mathbb{I}_{\geq 0}$. Hence $\{\Phi(i)\}$ is a monotonically nonincreasing sequence. The proof that $d_H(\Phi(i), \Phi) \leq c d_H(\Phi(0), \Phi)\gamma^i$ is left as Exercise 5.8. ■

It follows that $\Sigma(i) \to \Sigma$ and $\mathbb{S}(i) \to \mathbb{S}$ as $i \to \infty$ where Σ and $\mathbb{S}$ satisfy

$$\Sigma = A_L \Sigma \oplus (\mathbb{W} \oplus (-L\mathbb{N}))$$
$$\mathbb{S} = A_K \mathbb{S} \oplus L(C\Sigma \oplus \mathbb{N})$$

and are the minimal robust positive invariant sets for, respectively, $\widetilde{x}^+ = A_L \widetilde{x} + \widetilde{w}$, $\widetilde{w} \in (\mathbb{W} \oplus (-L\mathbb{N}))$ and $e^+ = A_K e + \delta$, $\delta \in L(C\Sigma \oplus \mathbb{N})$; the sequences $\{\Sigma(i)\}$ and $\{\mathbb{S}(i)\}$ are nonincreasing and converge in the Hausdorff metric to $\Sigma = \begin{bmatrix} I_n & 0 \end{bmatrix} \Phi$ and $\mathbb{S} = \begin{bmatrix} 0 & I_n \end{bmatrix} \Phi$, respectively.

5.4.4 Control of the Nominal System

Since $x(i) \in \{z(i)\} \oplus \Gamma(i)$ and $u(i) \in \{v(i)\} \oplus K\mathbb{S}(i)$ for all i, we can use MPC to control the sequences $\{z(i)\}$ and $\{v(i)\}$ so that $x(i) \in \mathbb{X}$ and $u(i) \in \mathbb{U}$ for all i. The constraints on x and u are satisfied if z and v are required to satisfy the tighter time-varying constraints

$$z(i) \in \mathbb{Z}_i := \mathbb{X} \ominus \Gamma(i) \qquad v(i) \in \mathbb{V}_i := \mathbb{U} \ominus K\mathbb{S}(i)$$

for all i; $\mathbb{Z}_i$ and $\mathbb{V}_i$ may be replaced by outer approximating sets. For this to be possible, we assume

Assumption 5.9 (Constraint bounds; time-varying case). $\Gamma(0) \subset \mathbb{X}$ and $K\mathbb{S}(0) \subset \mathbb{U}$.

Since both $\{\Gamma(i)\}$ and $\{\mathbb{S}(i)\}$ are nonincreasing sequences, $\{\mathbb{Z}_i\}$ and $\{\mathbb{S}_i\}$ are nondecreasing sequences so that satisfaction of Assumption 5.9 ensures that $\mathbb{Z}_i$ and $\mathbb{V}_i$ are not empty for all $i \in \mathbb{I}_{\geq 0}$. The constraints are time varying, so the nominal MPC problem at time k, state z is $\mathbb{P}_N(z, k)$ defined by

$$\mathbb{P}_N(z,k): \quad \bar{V}_N^0(z,k) = \min_{\mathbf{v}} \{\bar{V}_N(z, \mathbf{v}) \mid \mathbf{v} \in \mathcal{V}_N(z,k)\}$$

where the cost function $\bar{V}_N(\cdot)$ is defined by

$$\bar{V}_N(z, \mathbf{v}) := \sum_{k=0}^{N-1} \ell(z(k), v(k)) + V_f(z(N))$$

and the constraint set $\mathcal{V}_N(z,k)$ by

$$\mathcal{V}_N(z,k) := \{\mathbf{v} \mid v(i) \in \mathbb{V}_{k+i},\ z(i) \in \mathbb{Z}_{k+i},\ \forall i \in \{0,1,\ldots,N\},\ z(N) \in \mathbb{Z}_f\} \quad (5.28)$$

where, for all i, $z(i) := \bar{\phi}(i;z,\mathbf{v})$, the solution of $z^+ = Ax + Bv$ at time i if the initial state at time 0 is z and the nominal control sequence is $\mathbf{v} = \{v(0), v(1), \ldots, v(N-1)\}$. In (5.28), $\mathbb{Z}_f \subseteq \mathbb{Z}_N$ is the terminal constraint set and $z(i)$ is the predicted state at time $k+i$ which is why $z(i)$ is required to lie in the set $\mathbb{Z}_{k+i}$ and $v(i)$ to lie in $\mathbb{V}_{k+i}$; clearly $\mathbb{Z}_f \subseteq \mathbb{Z}_i$ for all $i \geq N$ so there is no need to make the terminal constraint set time varying. Let $\mathbf{v}^0(z,k) = \{v^0(0;z,k), v^0(1;z,k), \ldots, v^0(N;z,k)\}$ denote the minimizing control sequence; the stage cost $\ell(\cdot)$ is chosen to ensure uniqueness of $\mathbf{v}^0(z,k)$. The implicit MPC control law for the nominal system is $\bar{\kappa}_N(\cdot)$ defined by

$$\bar{\kappa}_N(z,k) := v^0(0;z,k)$$

where $v^0(0;z,k)$ is the first element in the sequence $\mathbf{v}^0(z,k)$. The domain of $\bar{V}_N^0(\cdot,k)$ and $\mathbf{v}(\cdot,k)$ and, hence, of $\kappa_N(\cdot,k)$, is $\mathcal{Z}_N(k)$ defined by

$$\mathcal{Z}_N(k) := \{z \in \mathbb{Z}_k \mid \mathcal{V}_N^0(z,k) \neq \emptyset\}$$

$\mathcal{Z}_N(k)$ is the set of states z at time k that can be robustly steered to $\mathbb{Z}_f$ in N steps by an admissible control $\mathbf{v}$. Because the constraints become weaker with time, the domain $\mathcal{Z}_N(k+1)$ of $\bar{V}_N^0(\cdot,k+1)$ is larger than the domain $\mathcal{Z}_N(k)$ of $\bar{V}_N^0(\cdot,k)$ for all $k > 0$; the sequence $\{\mathcal{Z}_N(k)\}$ is monotonically nondecreasing.

If the terminal cost $V_f(\cdot)$ and terminal constraint set $\mathbb{Z}_f$ satisfy the stability Assumptions 2.12 and 2.13 of Chapter 2, and if Assumption 5.9 is satisfied, the value function $\bar{V}_N^0(\cdot)$ satisfies, for all $k \in \mathbb{I}_{\geq 0}$

$$\begin{aligned} \bar{V}_N^0(z,k) &\geq \ell(z,\bar{\kappa}_N(z,k)) && \forall z \in \mathcal{Z}_N(k) \\ \Delta V_N^0(z,k) &\leq -\ell(z,\bar{\kappa}_N(z,k)) && \forall z \in \mathcal{Z}_N(k) \\ V_N^0(z,k) &\leq V_f(z) && \forall z \in \mathbb{Z}_f \end{aligned}$$

where $\Delta V_N^0(z,k) := V_N^0(f(z,\kappa_N(z)),k+1) - V_N^0(z,k)$.

If, in addition, we assume that $\ell(z,v) = (1/2)(|z|_Q^2 + |v|_R^2)$ where Q and R are positive definite and $V_f(z) = (1/2)|z|_{P_f}^2$ where P_f is positive definite, and if $\mathcal{Z}_N(0)$ is a C-set, then, as shown in Section 3.4.2, there exist positive constants c_1 and c_2 such that

$$\begin{aligned}\bar{V}_N^0(z,k) &\geq c_1|z|^2\\ \Delta\bar{V}_N^0(z,k) &\leq -c_1|z|^2\\ \bar{V}_N^0(z) &\leq c_2|z|^2\end{aligned}$$

for all $z \in \mathcal{Z}_N(k)$, all $k \in \mathbb{I}_{\geq 0}$. It follows from Chapter 2 that the origin is uniformly (in time k) exponentially stable for the nominal system $z^+ = Az + B\bar{\kappa}_N(z,k)$ with a region of attraction $\mathcal{Z}_N(0)$, and that $z(k) \in \mathcal{Z}_N(0) \subseteq \mathcal{Z}_N(k)$ for all $k \in \mathbb{I}_{\geq 0}$ if $z(0) \in \mathcal{Z}_N(0)$ so that problem $\mathbb{P}_N(z(k),k)$ is always feasible; here $z(k)$ is the solution of $z^+ = Az + B\kappa_N(z,k)$ at time k if the initial state is $z(0)$. There exists a $c > 0$ and a $\lambda \in (0,1)$ such that $|z(k)| \leq c|z(0)|\lambda^k$ for all $k \in \mathbb{I}_{\geq 0}$, all $z(0) \in \mathcal{Z}_N(0)$.

5.4.5 Control of the State Estimator

The implicit control law for the state estimator is $\kappa_N(\cdot)$ defined by

$$\kappa_N(\hat{x},z,k) := \bar{\kappa}_N(z,k) + K(\hat{x} - z)$$

Hence, the composite system with state $(\hat{x}, z)$ satisfies

$$\hat{x}^+ = A\hat{x} + B\kappa_N(\hat{x},z,k) + \delta(k) \tag{5.29}$$

$$z^+ = Az + B\bar{\kappa}_N(z,k) \tag{5.30}$$

$$k^+ = k + 1 \tag{5.31}$$

with initial state $(\hat{x}(0), z(0))$ satisfying $\hat{x}(0) \in \{z(0)\} \oplus \mathbb{S}(0)$, $z(0) \in \mathcal{Z}_N(0)$; these constraints are satisfied if $z(0) = \hat{x}(0) \in \mathcal{Z}_N(0)$.

Also, from Proposition 5.8, the sequences $\{\Phi(k)\}$, $\{\Gamma(k)\}$ and $\{\mathbb{S}(k)\}$ converge exponentially fast to Φ, Γ and $\mathbb{S}$, respectively. We have the following result for robust time-varying output MPC:

Proposition 5.10 (Exponential convergence of output MPC: time-varying case). *There exists a $c > 0$ and a $\gamma \in (0,1)$ such that $|z(k)| \leq c|z(0)|\gamma^k$ and $d(x(k),\Gamma) \leq c(|z(0)| + 1)\gamma^k$ for all $k \in \mathbb{I}_{\geq 0}$, all $x(0)$, $\hat{x}(0)$, $z(0)$ such that $(x(0) - \hat{x}(0), \hat{x}(0) - z(0)) \in \Phi(0)$, $z(0) \in \mathcal{Z}_N(0)$.*

Proof. If $z(0) \in \mathcal{Z}_N(0)$, we have $x(k) \in \{z(k)\} \oplus \Gamma(k)$ for all $k \in \mathbb{I}_{\geq 0}$. From Proposition 5.8, there exists a $c > 0$ and a $\gamma \in (0,1)$ such that

$d_H(\mathbb{\Gamma}(k), \mathbb{\Gamma}) \leq c\gamma^k$ and $|z(k)| \leq c\,|z(0)|\,\gamma^k$, $z(0) \in \mathcal{Z}_N(0)$, for all $k \in \mathbb{I}_{\geq 0}$. Hence

$$\begin{aligned} d(x(k), \mathbb{\Gamma}) &\leq d_H(\{z(k)\} \oplus \mathbb{\Gamma}(k), \mathbb{\Gamma}) \\ &\leq |z(k)| + d_H(\mathbb{\Gamma}(k), \mathbb{\Gamma}) \\ &\leq c(|z(0)| + 1)\gamma^k \end{aligned} \tag{5.32}$$

for all $k \in \mathbb{I}_{\geq 0}$, all $(x(0), \hat{x}(0))$ such that $\phi(0) = (x(0) - \hat{x}(0), \hat{x}(0) - z(0)) \in \Phi(0)$ ■

Similarly it can be shown that there exist a possibly different $c > 0$ and $\gamma \in (0, 1)$ such that

$$d(\hat{x}(k), \mathbb{S}) \leq c(|z(0)| + 1)\gamma^k$$

for all $k \in \mathbb{I}_{\geq 0}$. This result is not as strong as the corresponding result in Proposition 5.5 where exponential stability of $\mathbb{S} \times \{0\}$ with a region of attraction $(\mathcal{Z}_N \oplus \mathbb{S}) \times \mathcal{Z}_N$ is established for the composite system (5.25) and (5.26); the time-varying nature of the problem appears to preclude a stronger result.

5.5 Offset-Free MPC

We are now in a position to give a more realistic solution to the problem of offset-free MPC, briefly introduced in Chapter 2 in a deterministic context. Suppose the system to be controlled is described by

$$\begin{aligned} x^+ &= Ax + B_d d + Bu + w_x \\ y &= Cx + C_d d + \nu \\ r &= Hy \qquad \tilde{r} = r - \bar{r} \end{aligned}$$

where w_x and ν are unknown bounded disturbances taking values, respectively, in the compact sets $\mathbb{W}_x$ and $\mathbb{N}$ containing the origin in their interiors. We assume d is constant, or almost constant, but unknown, and models an additive disturbance; $y = Cx + C_d d$ is the output of the system being controlled, r is the controlled variable and $\bar{r}$ is its setpoint. The variable $\tilde{r}$ is the tracking error that we wish to minimize. We assume, for purposes of determining a control, that d satisfies

$$d^+ = d + w_d$$

where w_d is a bounded disturbance taking values in the compact set $\mathbb{W}_d$; in practice d is bounded although this is not implied by our model.

Set	Definition	Membership
$\mathbb{X}$	state constraint set	$x \in \mathbb{X}$
$\mathbb{U}$	input constraint set	$u \in \mathbb{U}$
$\mathbb{W}_x$	state disturbance set	$w_x \in \mathbb{W}_x$
$\mathbb{W}_d$	integrating disturbance set	$w_d \in \mathbb{W}_d$
$\mathbb{W}$	total state disturbance set, $\mathbb{W}_x \times \mathbb{W}_d$	$w \in \mathbb{W}$
$\mathbb{N}$	measurement error set	$\nu \in \mathbb{N}$
$\widetilde{\mathbb{W}}$	estimate error disturbance set, $\mathbb{W} \oplus (-L\mathbb{N})$	$\widetilde{w} \in \widetilde{\mathbb{W}}$
Φ	total estimate error disturbance set, $\Phi = \widetilde{A}_L \Phi \oplus \widetilde{\mathbb{W}}$	$\phi \in \Phi$
Σ_x	state estimate error disturbance set, $\begin{bmatrix} I_n & 0 \end{bmatrix} \Phi$	$\widetilde{x} \in \Sigma_x$
Σ_d	integrating disturbance estimate error set, $\begin{bmatrix} 0 & I_p \end{bmatrix} \Phi$	$\widetilde{d} \in \Sigma_d$
Δ	innovation set, $L(\widetilde{C}\Phi \oplus \mathbb{N})$	$L\widetilde{y} \in \Delta$
Δ_x	set containing state component of innovation, $L_x(\widetilde{C}\Phi \oplus \mathbb{N})$	$L_x\widetilde{y} \in \Delta_x$
Δ_d	set containing integrating disturbance component of innovation, $L_d(\widetilde{C}\Phi \oplus \mathbb{N})$	$L_d\widetilde{y} \in \Delta_d$
$\mathbb{S}$	nominal state tracking error invariance set, $A_K\mathbb{S} \oplus \Delta_x = \mathbb{S}$	$e \in \mathbb{S}$ $\hat{x} \in \{z\} + \mathbb{S}$
Γ	state tracking error invariance set, $\mathbb{S} + \Sigma_x$	$x \in \{z\} + \Gamma$
$\mathbb{V}$	nominal input constraint set, $\mathbb{V} = \mathbb{U} \ominus K\mathbb{S}$	$v \in \mathbb{V}$
$\mathbb{Z}$	nominal state constraint set, $\mathbb{Z} = \mathbb{X} \ominus \Gamma$	$z \in \mathbb{Z}$

Table 5.1: Summary of the sets and variables used in output MPC.

We assume that $x \in \mathbb{R}^n$, $d \in \mathbb{R}^p$,$u \in \mathbb{R}^m$, $y \in \mathbb{R}^r$, and $e \in \mathbb{R}^q$, $q \leq r$ and that the system to be controlled is subject to the usual state and control constraints

$$x \in \mathbb{X} \qquad u \in \mathbb{U}$$

where $\mathbb{X}$ is polyhedral and $\mathbb{U}$ is polytopic.

5.5.1 Estimation

Given the numerous sets that are required to specify the output feedback case we are about to develop, Table 5.1 may serve as a reference for the sets defined in the chapter and the variables that are members of these sets.

Since both x and d are unknown, it is necessary to estimate them. For estimation purposes, it is convenient to work with the composite system whose state is $\phi := (x, d)$. This system may be described more compactly by

$$\phi^+ = \widetilde{A}\phi + \widetilde{B}u + w$$
$$y = \widetilde{C}\phi + \nu$$

in which

$$\widetilde{A} := \begin{bmatrix} A & B_d \\ 0 & I \end{bmatrix} \qquad \widetilde{B} := \begin{bmatrix} B \\ 0 \end{bmatrix} \qquad \widetilde{C} := \begin{bmatrix} C & C_d \end{bmatrix}$$

and $w := (w_x, w_d)$ takes values in $\mathbb{W} = \mathbb{W}_x \times \mathbb{W}_d$. A necessary and sufficient condition for the detectability of $(\widetilde{A}, \widetilde{C})$ is given in Lemma 1.8 in Chapter 1; a sufficient condition is detectability of (A, C) coupled with invertibility of C_d. If $(\widetilde{A}, \widetilde{C})$ is detectable, the state may be estimated using the time-invariant observer or filter described by

$$\hat{\phi}^+ = \widetilde{A}\hat{\phi} + \widetilde{B}u + \delta \qquad \delta := L(y - \widetilde{C}\hat{\phi})$$

in which L is such that $\rho(\widetilde{A}_L) < 1$ where $\widetilde{A}_L := \widetilde{A} - L\widetilde{C}$. Clearly $\delta = L\widetilde{y}$ where $\widetilde{y} = \widetilde{C}\phi + \nu$. The estimation error $\widetilde{\phi} := \phi - \hat{\phi}$ satisfies

$$\widetilde{\phi}^+ = \widetilde{A}\widetilde{\phi} + w - L(\widetilde{C}\widetilde{\phi} + \nu)$$

satisfies or, in simpler form

$$\widetilde{\phi}^+ = \widetilde{A}_L\widetilde{\phi} + \widetilde{w} \qquad \widetilde{w} := w - L\nu$$

Clearly $\widetilde{w} = L(\widetilde{C}\widetilde{\phi} + \nu)$ takes values in the compact set $\widetilde{\mathbb{W}}$ defined by

$$\widetilde{\mathbb{W}} := \mathbb{W} \oplus (-L\mathbb{N})$$

If w and ν are zero, $\widetilde{\phi}$ decays to zero exponentially fast. Since $\rho(\widetilde{A}_L) < 1$ and $\mathbb{A}$ is compact, there exists a robust positive invariant set Φ for $\widetilde{\phi}^+ = \widetilde{A}_L\widetilde{\phi} + \delta$, $\delta \in \mathbb{A}$ satisfying

$$\Phi = \widetilde{A}_L\Phi \oplus \widetilde{\mathbb{W}}$$

Hence $\widetilde{\phi}(i) \in \Phi$ for all $i \in \mathbb{I}_{\geq 0}$ if $\widetilde{\phi}(0) \in \Phi$. Since $\widetilde{\phi} = (\widetilde{x}, \widetilde{d}) \in \mathbb{R}^n \times \mathbb{R}^p$ where $\widetilde{x} := x - \hat{x}$ and $\widetilde{d} := d - \hat{d}$, we define the sets Σ_x and Σ_d as follows

$$\Sigma_x := \begin{bmatrix} I_n & 0 \end{bmatrix}\Phi \qquad \Sigma_d := \begin{bmatrix} 0 & I_p \end{bmatrix}\Phi$$

It follows that $\widetilde{x}(i) \in \Sigma_x$ and $\widetilde{d}(i) \in \Sigma_d$ for all $i \in \mathbb{I}_{\geq 0}$ if $\widetilde{\phi}(0) = (\widetilde{x}(0), \widetilde{d}(0)) \in \Phi$. That $\widetilde{\phi}(0) \in \Phi$ is a steady-state assumption.

5.5.2 Control

The estimation problem has a solution similar to previous solutions. The control problem is more difficult. As before, we control the estimator state, making allowance for state estimation error. The estimator state $\hat{\phi}$ satisfies the difference equation

$$\hat{\phi}^+ = \widetilde{A}\hat{\phi} + \widetilde{B}u + \delta$$

where the disturbance δ is defined by

$$\delta := L\widetilde{y} = L(\widetilde{C}\widetilde{\phi} + \nu)$$

The disturbance $\delta = (\delta_x, \delta_d)$ lies in the C-set $\mathbb{\Delta}$ defined by

$$\mathbb{\Delta} := L(\widetilde{C}\Phi \oplus \mathbb{N})$$

where the set Φ is defined in Section 5.5.1. The system $\hat{\phi}^+ = \widetilde{A}\hat{\phi} + \widetilde{B}u + \delta$ is not stabilizable, however, so we examine the subsystems with states $\hat{x}$ and $\hat{d}$

$$\hat{x}^+ = A\hat{x} + Bu + \delta_x$$
$$\hat{d}^+ = \hat{d} + \delta_d$$

where the disturbances δ_x and δ_d are components of δ ($\delta = (\delta_x, \delta_d)$) and are defined by

$$\delta_x := L_x\widetilde{y} = L_x(\widetilde{C}\widetilde{\phi} + \nu) \qquad \delta_d := L_d\widetilde{y} = L_d(\widetilde{C}\widetilde{\phi} + \nu)$$

The matrices L_x and L_d are the corresponding components of L. The disturbance δ_x and δ_d lie in the C-sets $\mathbb{\Delta}_x$ and $\mathbb{\Delta}_d$ defined by

$$\mathbb{\Delta}_x := \begin{bmatrix} I_n & 0 \end{bmatrix} \mathbb{\Delta} = L_x[\widetilde{C}\Phi \oplus \mathbb{N}] \qquad \mathbb{\Delta}_d := \begin{bmatrix} 0 & I_p \end{bmatrix} \mathbb{\Delta} = L_d[\widetilde{C}\Phi \oplus \mathbb{N}]$$

We assume that (A, B) is a stabilizable pair so the tube methodology may be employed to control $\hat{x}$. The system $\hat{d}^+ = \hat{d} + \delta_d$ is uncontrollable. The central trajectory is therefore described by

$$z^+ = Az + Bv$$
$$\hat{d}^+ = \hat{d}$$

We obtain $v = \bar{\kappa}_N(z, \hat{d}, \bar{r})$ by solving a nominal optimal control problem defined later and set $u = v + Ke$, $e := \hat{x} - z$ where K is chosen so that

$\rho(A_K) < 1$, $A_K := A - BK$; this is possible since (A, B) is assumed to be stabilizable. It follows that $e := \hat{x} - z$ satisfies the difference equation

$$e^+ = A_K e + \delta_x \qquad \delta_x \in \mathbb{\Delta}_x$$

Because $\mathbb{\Delta}_x$ is compact and $\rho(A_K) < 1$, there exists a robust positive invariant set $\mathbb{S}$ for $e^+ = A_K e + \delta_x$, $\delta_x \in \mathbb{\Delta}_x$ satisfying

$$A_K \mathbb{S} \oplus \mathbb{\Delta}_x = \mathbb{S}$$

Hence $e(i) \in \mathbb{S}$ for all $i \in \mathbb{I}_{\geq 0}$ if $e(0) \in \mathbb{S}$. So, as in Proposition 5.3, the states and controls of the estimator and nominal system satisfy $\hat{x}(i) \in \{z(i)\} \oplus \mathbb{S}$ and $u(i) \in \{v(i)\} \oplus K\mathbb{S}$ for all $i \in \mathbb{I}_{\geq 0}$ if the initial states $\hat{x}(0)$ and $z(0)$ satisfy $\hat{x}(0) \in \{z(0)\} \oplus \mathbb{S}$. Using the fact established previously that $\tilde{x}(i) \in \Sigma_x$ for all i, we can also conclude that $x(i) = z(i) + e(i) + \tilde{x}(i) \in \{z(i)\} \oplus \mathbb{\Gamma}$ and that $u(i) = v(i) + Ke(i) \in \{v(i)\} + K\mathbb{S}$ for all i where $\mathbb{\Gamma} := \mathbb{S} \oplus \Sigma_x$ provided, of course, that $\phi(0) \in \{\hat{\phi}(0)\} \oplus \Phi$ and $x(0) \in \{\hat{x}(0)\} \oplus \mathbb{S}$. These conditions are equivalent to $\tilde{\phi}(0) \in \Phi$ and $e(0) \in \mathbb{S}$ where, for all i, $e(i) := \hat{x}(i) - z(i)$. Hence $x(i)$ lies in $\mathbb{X}$ and $u(i)$ lies in $\mathbb{U}$ if $z(i) \in \mathbb{Z} := \mathbb{X} \ominus \mathbb{\Gamma}$ and $v(i) \in \mathbb{V} := \mathbb{U} \ominus K\mathbb{S}$.

Thus $\hat{x}(i)$ and $x(i)$ evolve in known neighborhoods of the central state $z(i)$ that we can control. Although we know that the uncontrollable state $d(i)$ lies, for all i, in the set $\{\hat{d}(i)\} \oplus \Sigma_d$, the evolution of $\hat{d}(i)$ is an uncontrollable random walk and is, therefore, unbounded; if the initial value of $\hat{d}$ at time 0 is $\hat{d}_0$, then $\hat{d}(i)$ lies in the set $\{\hat{d}_0\} \oplus \mathbb{W}_d$ that increases without bound as i increases. This behavior is a defect in our model for the disturbance d; the model is useful for estimation purposes, but is unrealistic in permitting unbounded values for d. Hence we assume in the sequel that d evolves in a compact C−set X_d. We can modify the observer to ensure that $\hat{d}$ lies in X_d but find it simpler to observe that, if d lies in X_d, $\hat{d}$ must lie in $X_d \oplus \Sigma_d$.

Target Calculation. We are now in a position to specify the optimal control problem whose solution yields $v = \bar{\kappa}_N(\hat{x}, z)$ and, hence, $u = v + K(\hat{x} - z)$. Our first task is to determine the target state $\bar{z}$ and associated control $\bar{v}$; we require our estimate of the tracking error $\tilde{r} = r - \bar{r}$ to be zero. Since our estimate of the measurement noise ν is 0 and since our best estimate of d when the target state is reached is $\hat{d}$, we require

$$\hat{r} - \bar{r} = H(C\bar{z} + C_d\hat{d}) - \bar{r} = 0$$

We also require the target state to be an equilibrium state satisfying, therefore, $\bar{z} = A\bar{z} + B\bar{v}$ for some control $\bar{v}$. Given $(\hat{d}, \bar{r})$, the target equilibrium pair $(\bar{z}, \bar{v})(\hat{d}, \bar{r})$ is computed as follows.

$$(\bar{z}, \bar{v})(\hat{d}, \bar{r}) = \arg\min_{z,v}\{L(z, v) \mid z = Az + Bv,\ H(Cz + C_d\hat{d}) = \bar{r},$$
$$z \in \mathbb{Z},\ v \in \mathbb{V}\}$$

where $L(\cdot)$ is an appropriate cost function; e.g. $L(v) = (1/2)|v|^2_{\bar{R}}$. The equality constraints in this optimization problem can be satisfied if the matrix $\begin{bmatrix} I-A & -B \\ HC & 0 \end{bmatrix}$ has full rank. As the notation indicates, the target equilibrium pair $(\bar{z}, \bar{v})(\hat{d}, \bar{r})$ is not constant but varies with the estimate of the disturbance state d.

MPC algorithm. The control objective is to steer the central state z to the target state $\bar{z}(\hat{d}, \bar{r})$ while satisfying the state and control constraints $x \in \mathbb{X}$ and $u \in \mathbb{U}$. It is desirable that $z(i)$ converges to $\bar{z}(\hat{d}, \bar{r})$ if $\hat{d}$ remains constant in which case $x(i)$ converges to the set $\{\bar{z}(\hat{d}, \bar{r})\} \oplus \Gamma$. To achieve this objective, we define the deterministic optimal control problem

$$\bar{\mathbb{P}}_N(z, \hat{d}, \bar{r}): \quad V_N^0(z, \hat{d}, \bar{r}) := \min_{\mathbf{v}}\{V_N(z, \hat{d}, \mathbf{v}) \mid \mathbf{v} \in \mathcal{V}_N(z, \hat{d}, \bar{r})\}$$

in which the cost $V_N(\cdot)$ and the constraint set $\mathcal{V}_N(z, \hat{d})$ are defined by

$$V_N(z, \hat{d}, \bar{r}, \mathbf{v}) := \sum_{i=0}^{N-1} \ell(z(i) - \bar{z}(\hat{d}, \bar{r}), v(i) - \bar{v}(\hat{d}, \bar{r})) + V_f(z(N), \bar{z}(\hat{d}, \bar{r}))$$
$$\mathcal{V}_N(z, \hat{d}, \bar{r}) := \{\mathbf{v} \mid z(i) \in \mathbb{Z}, v(i) \in \mathbb{V}\ \forall i \in \mathbb{I}_{0:N-1}, z(N) \in \mathbb{Z}_f(\bar{z}(\hat{d}, \bar{r}))\}$$

where, for each i, $z(i) = \bar{\phi}(i; z, \mathbf{v})$, the solution of $z^+ = Az + Bv$ when the initial state is z and the control sequence is $\mathbf{v}$. The terminal cost is zero when the terminal state is equal to the target state and the target state lies in the center of the terminal constraint set. The solution to $\mathbb{P}_N(z, \hat{d}, \bar{r})$ is

$$\mathbf{v}^0(z, \hat{d}, \bar{r}) = \{v^0(0; z, \hat{d}, \bar{r}), v^0(1; z, \hat{d}, \bar{r}), \ldots, v^0(N-1; z, \hat{d}, \bar{r})\}$$

and the implicit model control law $\bar{\kappa}_N(\cdot)$ is defined by

$$\bar{\kappa}_N(z, \hat{d}, \bar{r}) := v^0(0; z, \hat{d}, \bar{r})$$

where $v^0(0; z, \hat{d}, \bar{r})$ is the first element in the sequence $\mathbf{v}^0(z, \hat{d}, \bar{r})$. The control u applied to the plant and the observer is $u = \kappa_N(\hat{x}, z, \hat{d})$ where $\kappa_N(\cdot)$ is defined by

$$\kappa_N(\hat{x}, z, \hat{d}, \bar{r}) := \bar{\kappa}_N(z, \hat{d}, \bar{r}) + K(\hat{x} - z)$$

Although the optimal control problem $\mathbb{P}_N(z, \hat{d}, \bar{r})$ is deterministic, $\hat{d}$ is random, so that the sequence $\{z(i)\}$ is random. The control algorithm may now be formally stated:

Robust control algorithm (offset-free MPC).

Initialization: At time 0, set $i = 0$, set $\hat{\phi} = \hat{\phi}(0)$ in which $\hat{\phi} = (\hat{x}, \hat{d})$ and set $z = \hat{x}$.

Step 1 (Compute control): At time i, solve the "nominal" optimal control problem $\bar{\mathbb{P}}_N(z, \hat{d}, \bar{r})$ to obtain the current "nominal" control action $v = \bar{\kappa}_N(z, \hat{d}, \bar{r})$ and the control action $u = v + K(\hat{x} - z)$.

Step 2 (Check): If $\bar{\mathbb{P}}_N(z, \hat{d}, \bar{r})$ is infeasible, adopt safety/recovery procedure.

Step 3 (Apply control): Apply the control u to the system being controlled.

Step 4 (Update): (a) Compute the successor state estimate $\hat{\phi}^+ = \widetilde{A}\hat{x} + \widetilde{B}u + L(y - \widetilde{C}\hat{\phi})$. (b) Compute the successor state $z^+ = f(z, v)$ of the nominal system.

Step 5: Set $(\hat{\phi}, z) = (\hat{\phi}^+, z^+)$, set $i = i + 1$, and go to Step 1.

In normal operation, Step 2 is not activated; Propositions 5.2 and 5.3 ensure that the constraints $\hat{x} \in \{z\} \oplus \mathbb{S}$ and $u \in \{v\} \oplus K\mathbb{S}$ are satisfied. If an unanticipated event occurs and Step 2 is activated, the controller can be reinitialized by setting $v = \bar{\kappa}_N(\hat{x}, \hat{d}, \bar{r})$, setting $u = v$ and relaxing constraints if necessary.

5.5.3 Stability Analysis

We give here an informal discussion of the stability properties of the controller because offset-free MPC of constrained uncertain systems remains an area of current research. The controller described above is motivated by the following consideration: nominal MPC is able to handle "slow" uncertainties such as the drift of a target point if the value function $V_N^0(\cdot)$ is Lipschitz continuous."Fast" uncertainties, however, are better handled by the tube controller that generates, using MPC, a suitable central trajectory that uses a "fast" ancillary controller to steer trajectories of the uncertain system toward the central trajectory. As shown above, the controller ensures that $x(i) \in \{z(i)\} \oplus \mathbb{\Gamma}$ for all i; its success therefore depends on the ability of the controlled nominal system $z^+ = Az + B\bar{\kappa}_N(z,\hat{d})$, $v = \bar{\kappa}_N(z,\hat{d})$, to track the target $\bar{z}(\hat{d},\bar{r})$, which varies as $\hat{d}$ evolves.

Assuming that the standard stability assumptions are satisfied for the nominal optimal control problem $\bar{\mathbb{P}}_N(z,\hat{d})$ defined above, we have

$$\begin{aligned} V_N^0(z,\hat{d},\bar{r}) &\geq c_1|z-\bar{z}(\hat{d},\bar{r})|^2 \\ \Delta V_N^0(z,\hat{d},\bar{r}) &\leq -c_1|z-\bar{z}(\hat{d},\bar{r})|^2 \\ V_N^0(z,\hat{d},\bar{r}) &\leq c_2|z-\bar{z}(\hat{d},\bar{r})|^2 \end{aligned}$$

for all $z \in \mathcal{Z}_N(\hat{d},\bar{r})$ where, since $(\hat{d},\bar{r})$ is constant,

$$\Delta V_N^0(z,\hat{d},\bar{r}) := V_N^0(Az + B\bar{\kappa}_N(z,\hat{d},\bar{r}),\hat{d},\bar{r}) - V_N^0(z,\hat{d},\bar{r})$$

and, for each $(\hat{d},\bar{r})$, $\mathcal{Z}_N(\hat{d},\bar{r}) = \{z \mid \mathcal{V}_N(z,\hat{d},\bar{r}) \neq \emptyset\}$ is the domain of $V_N^0(\cdot,\hat{d},\bar{r})$.

Constant $\hat{d}$. If $\hat{d}$ remains constant, $\bar{z}(\hat{d},\bar{r})$ is exponentially stable for $z^+ = Az + B\bar{\kappa}_N(z,\hat{d},\bar{r})$ with a region of attraction $\mathcal{Z}_N(\hat{d},\bar{r})$. It can be shown, as in the proof of Proposition 5.5, that the set $\mathcal{A}(\hat{d},\bar{r}) := (\{\bar{z}(\hat{d},\bar{r})\} \oplus \mathbb{S}) \times \{\bar{z}(\hat{x},\bar{r})\}$ is exponentially stable for the composite system $\hat{x}^+ = A\hat{x} + B\kappa_N(\hat{x},z,\hat{d},\bar{r}) + \delta_x$, $z^+ = Az + B\bar{\kappa}_N(z,\hat{d})$, $\delta_x \in \mathbb{\Delta}_x$, with a region of attraction $(\mathcal{Z}_N(\hat{d},\bar{r}) \oplus \mathbb{S}) \times \mathcal{Z}_N(\hat{d},\bar{r})$. Hence $x(i) \in \{z(i)\} \oplus \mathbb{\Gamma}$ tends to the set $\{\bar{z}(\hat{d},\bar{r})\} \oplus \mathbb{\Gamma}$ exponentially fast. If the external disturbance w is zero, $\mathbb{W} = \{0\}$. If, in addition, $\mathbb{N} = \{0\}$, then $\mathbb{\Delta} = \{0\}$ and $\mathbb{S} = \{0\}$ and $x(i) \to \bar{z}(\hat{d},\bar{r})$ exponentially fast so that the tracking error $\tilde{r}(i) \to 0$ as $i \to \infty$.

Slowly varying $\hat{d}$. If $\hat{d}$ is varying, the inequality for $\Delta V_N^0(\cdot)$ must be replaced by:

$$\begin{aligned} \Delta V_N^0(z,\hat{d},\bar{r},w_d) &= V_N^0(Az + B\bar{\kappa}_N(z,\hat{d},\bar{r}),\hat{d}+w_d) - V_N^0(z,\hat{d},\bar{r}) \\ &\leq -c_1\left|z-\bar{z}(\hat{d},\bar{r})\right|^2 + \bar{k}\,|w_d| \end{aligned}$$

where $\bar{k}$ is a Lipschitz constant for $V_N^0(\cdot,\hat{d},\bar{r})$ for all $(\hat{d},\bar{r}) \in X_d \times X_r$, and X_r is the set of permissible values for $\bar{r}$. Employing the approach adopted in Section 3.2.4, it can be shown, if $\mathbb{W}_d$ is sufficiently small, that there exist two sublevel sets $S_b(\hat{d},\bar{r}) := \{z \mid V_N^0(z,\hat{d},\bar{r}) \leq b\}$ and $S_c(\hat{d},\bar{r}) := \{z \mid V_N^0(z,\hat{d},\bar{r}) \leq c\}$, $c > b$ such that $z(0) \in S_c(\hat{d}(0),\bar{r})$ implies the existence of a finite time i_0 such that $z(i) \in S_b(\hat{d}(i),\bar{r})$ for all $i \geq i_0$. The center of each set is the target state $\bar{z}(\hat{d}(i),\bar{r})$ so that, if $\mathbb{W}_d$ is small and recursive feasibility is maintained, $z(i)$ remains close to the target state; $x(i) \in \{z(i)\} \oplus \mathbb{\Gamma}$ also remains close if, in addition, $\mathbb{W}_x$ and $\mathbb{N}$ are small. Recursive feasibility is ensured if there are no state or terminal constraints in the nominal optimal control problem.

5.6 Nonlinear Constrained Systems

5.6.1 Introduction

For simplicity, we consider here the following uncertain, discrete time, nonlinear system

$$x^+ = f(x, u) + w \qquad y = h(x) + v \tag{5.33}$$

where $x \in \mathbb{R}^n$ is the current state, $u \in \mathbb{R}^m$ is the current control action, x^+ is the successor state, $w \in \mathbb{R}^n$ is an unknown state disturbance, $y \in \mathbb{R}^p$ is the current measured output, and $v \in \mathbb{R}^p$ is an unknown output disturbance. The state and additive disturbances w and v are known only to the extent that they lie, respectively, in the C sets $\mathbb{W} \subseteq \mathbb{R}^n$ and $\mathbb{N} \subseteq \mathbb{R}^p$. Let $\phi(i; x(0), \mathbf{u}, \mathbf{w})$ denote the solution of (5.9) at time i if the initial state at time 0 is $x(0)$, and the control and disturbance sequences are, respectively, $\mathbf{u} := \{u(0), u(1), \ldots\}$ and $\mathbf{w} := \{w(0), w(1), \ldots\}$. The system (5.33) is subject to the following set of hard state and control constraints

$$(x, u) \in \mathbb{X} \times \mathbb{U}$$

in which $\mathbb{X} \subseteq \mathbb{R}^n$ and $\mathbb{U} \subseteq \mathbb{R}^m$ are polyhedral and polytopic sets, respectively, with each set containing the origin in its interior. Output MPC of nonlinear systems remains an active area of research; the proposals to follow are speculative.

5.6.2 State Estimator

Several state estimators for nonlinear systems are described in Chapter 4. For each t, let $\mathcal{I}(t)$ denote the information available to the state estimator at time t: for a full information estimator

$$\mathcal{I}(t) := \{(y(j), u(j)) \mid j \in \mathbb{I}_{-\infty:t}\}$$

whereas for a moving horizon estimator

$$\mathcal{I}(t) := \{(y(j), u(j)) \mid j \in \mathbb{I}_{t-T:t}\}$$

where T is the horizon. For each t, j let $\hat{x}(t|j)$ denote the estimate of $x(t)$ give data $\mathcal{I}(j)$; for simplicity, we use $\hat{x}(t)$ to denote $\hat{x}(t|t-1)$.

We make the strong assumption that we have available an estimator satisfying the following difference equation

$$\hat{x}(t+1) = f(\hat{x}(t), u(t)) + \delta$$

where $\delta \in \mathbb{\Delta}$ and $\mathbb{\Delta}$ is a compact subset of $\mathbb{R}^n$. Since

$$\hat{x}(t+1) = f(\hat{x}(t|t), u(t)) + w(t) = f(\hat{x}(t), u(t)) + \delta(t)$$

where

$$\delta(t) := [f(\hat{x}(t|t), u(t)) - f(\hat{x}(t), u(t))] + w(t)$$

the form of the evolution equation for $\hat{x}(t)$ is acceptable; the assumption that $\mathbb{\Delta}$ is constant is conservative. However controlling a random system with a time-varying bound on the disturbance would be considerably more complicated.

Our second assumption is the the state estimation error $\tilde{x}(t) := x(t) - \hat{x}(t)$ lies in a compact set $\mathbb{\Sigma}_x$. This is also a conservative assumption, made for simplicity.

Before proceeding to propose a tube-based controller, we examine briefly nominal MPC.

5.6.3 Nominal MPC

In nominal output MPC, the control u is determined by solving an optimal control problem $\bar{\mathbb{P}}_N(\hat{x})$ for the nominal deterministic system defined by

$$z^+ = f(z, u) \qquad z(0) = \hat{x}$$

where $\hat{x}$ is the current estimate of the state x. This yields the implicit control law $\bar{\kappa}_N(\cdot)$ so the control u applied to the system $x^+ = f(x, u) + w$ when the current state estimate is $\hat{x}$ is

$$u = \bar{\kappa}_N(\hat{x})$$

Because the evolution of the state x differs from the evolution of the state estimate $\hat{x}$, the control $u = \bar{\kappa}_N(\hat{x})$ is not necessarily stabilizing. If the ingredients $V_f(\cdot)$ and $\mathbb{Z}_f$ of the optimal control problem $\bar{\mathbb{P}}_N(\hat{x})$ are chosen appropriately, and $\ell(\cdot)$ is quadratic and positive definite, the value function $\bar{V}_N^0(\cdot)$ satisfies the usual inequalities:

$$c_1|z|^2 \leq \bar{V}_N^0(z) \leq c_2|z|^2$$
$$\bar{V}_N^0(z^+) \leq \bar{V}_N^0(z) - c_1|z|^2$$

where $z^+ = f(z, \bar{\kappa}_N(z))$. These inequalities are sufficient to establish the exponential stability of the origin for the nominal system $z^+ = f(z, \bar{\kappa}_N(z))$ with a region of attraction $\mathcal{Z}_N$ which is the domain of the value function if bounded, or an appropriate level set of the value function otherwise.

5.6.4 Tube-Based Output MPC

We apply the methodology of Chapter 3, Section 3.6 to the control of the uncertain system $\hat{x}^+ = f(\hat{x}, u) + \delta$, $\delta \in \mathbb{A}$, making allowance for the fact that $x(i)$ lies in $\{\hat{x}(i)\} \oplus \Sigma_x$ for all i. This method of control permits, in principle, larger disturbances.

We assume, therefore, that we have an implicit, stabilizing, control law $v = \bar{\kappa}_N(z)$ for the nominal system $z^+ = f(z, v)$. This control law is chosen to satisfy the tightened constraints

$$z \in \mathbb{Z} \qquad v \in \mathbb{V}$$

We discuss the choice of $\mathbb{Z}$ and $\mathbb{V}$ later. The control law is obtained by solving the nominal control problem $\bar{\mathbb{P}}_N(z)$ whose solution also yields the "central" state and control trajectories $\{\mathbf{z}^*(i; z)\}$ and $\{\mathbf{u}^*(i; z)\}$; these trajectories are the solutions of

$$z^+ = f(z, \bar{\kappa}_N(z)), \qquad v = \bar{\kappa}_N(z)$$

with initial state $z(0) = z$.

The second ingredient of the tube-based controller is the ancillary controller that attempts to steer the trajectories of the uncertain system $\hat{x}^+ = f(\hat{x}, u) + \delta$ toward the central path defined above. This determines u by solving the ancillary problem $\mathbb{P}_N(\hat{x}, z)$ defined by

$$\bar{V}_N^0(\hat{x}, z) = \min_{\mathbf{u}} \{V_N(\hat{x}, z, \mathbf{u}) \mid \mathbf{u} \in \mathcal{U}_N(\hat{x}, z)\}$$

in which the cost function $\bar{V}_N^0(\cdot)$ is defined, as in Chapter 3, by

$$V_N(\hat{x}, z, \mathbf{u}) := \sum_{i=0}^{N-1} \ell(\hat{x}(i) - z^*(i; z), u(i) - v^*(i; z))$$

where $\hat{x}(i) := \bar{\phi}(i; \hat{x}, \mathbf{u})$, the solution at time i of the *nominal* system $z^+ = f(z, u)$ with initial state $\hat{x}$ and control sequence $\mathbf{u}$; $z^*(i; z) := \bar{\phi}(i; z, \bar{\kappa}_N(\cdot))$, the solution at time i of the controlled nominal system

$z^+ = f(z, \bar{\kappa}_N(z))$ MPC with initial state z, and $v^*(i; z) = \bar{\kappa}_N(z^*(i; z))$. The constraint set $\mathcal{U}_N(\hat{x}, z)$ is defined by

$$\mathcal{U}_N(\hat{x}, z) := \{\mathbf{u} \in \mathbb{R}^{Nm} \mid \bar{\phi}(N; \hat{x}, \mathbf{u}) = z^*(N; z)\}$$

The terminal equality constraint is chosen for simplicity. Because of the terminal equality constraint, there is no terminal cost. The terminal constraint $\hat{x}(N) = z^*(N; z)$ induces the implicit constraint $\mathbf{u} \in \mathcal{U}_N(\hat{x}, z)$ on the control sequence $\mathbf{u}$. For each $z \in \mathcal{Z}_N$, the domain of the value function $\bar{V}_N^0(\cdot, z)$, and of the minimizer $\mathbf{u}^0(\cdot, z)$, is the set $\hat{X}_N(z)$ defined by

$$\hat{X}_N(z) := \{\hat{x} \mid \mathcal{U}_N(\hat{x}, z) \neq \emptyset\}$$

The minimizing control sequence is

$$\mathbf{u}^0(\hat{x}, z) = \{u^0(0; \hat{x}, z), u^0(1; \hat{x}, z), \ldots, u^0(N-1; \hat{x}, z)\}$$

and the control applied to the estimator system (when the estimator state is $\hat{x}$ and the state of the nominal system is z) is $u^0(0; \hat{x}, z)$, the first element in this sequence. The corresponding optimal state sequence is

$$\hat{\mathbf{x}}^0(\hat{x}, z) = \{\hat{x}^0(0; \hat{x}, z), \hat{x}^0(1; \hat{x}, z), \ldots, \hat{x}^0(N; \hat{x}, z)\}$$

The implicit ancillary control law is, therefore, $\kappa_N(\cdot)$ defined by

$$\kappa_N(\hat{x}, z) := u^0(0; \hat{x}, z)$$

The controlled composite system satisfies

$$\begin{aligned} \hat{x}^+ &= f(\hat{x}, \kappa_N(\hat{x}, z)) + \delta \\ z^+ &= f(z, \bar{\kappa}_N(z)) \end{aligned}$$

For each $c > 0$, each $z \in \mathcal{Z}_N$, let $S_c(z) := \{\hat{x} \mid \bar{V}_N^0(\hat{x}, z) \leq c\}$. With appropriate assumptions, there exists a $c \in (0, \infty)$ such that if $\hat{x}(0) \in S_c(z(0))$, then $\hat{x}(i) \in S_c(z^0(i; z(0)))$ for all $i \in \mathbb{I}_{\geq 0}$ and all admissible disturbance and measurement noise sequences, $\mathbf{w}$ and $\boldsymbol{\nu}$. In other words, c is such that $S_c(\cdot)$ is $\hat{x}$-robust positive invariant for the controlled composite system. It follows from the discussion previously that the solutions $\hat{x}(i)$ and $z(i)$ of the controlled composite system satisfy

$$\begin{aligned} & z(i) \to 0 \text{ as } i \to \infty \\ & \hat{x}(i) \in S_c(z(i)) \quad \forall i \in \mathbb{I}_{\geq 0} \\ & x(i) \in S_c(z(i)) \oplus \Sigma_x \quad \forall i \in \mathbb{I}_{\geq 0} \end{aligned}$$

provided that $z(0) \in \mathcal{Z}_N$ and $\hat{x}(0) \in S_c(z(0))$. Thus, the constraint sets $\mathbb{Z}$ and $\mathbb{V}$ required for the nominal optimal control problem should satisfy

$$\begin{aligned}
&\mathbb{Z}_f \subseteq \mathbb{Z} \\
&S_c(z) \oplus \Sigma_x \subseteq \mathbb{X} \quad \forall z \in \mathcal{Z}_N \\
&\kappa_N(\hat{x}, z) \in \mathbb{U} \ \forall \hat{x} \in S_c(z), \ \forall z \in \mathcal{Z}_N
\end{aligned}$$

If these conditions are satisfied, the solutions $\hat{x}(i)$ and $z(i)$ of the controlled composite system and the associated control $u(i) = \kappa_N(\hat{x}(i), z(i))$ satisfy

$$\begin{aligned}
&z(i) \to 0 \text{ as } i \to \infty \\
&x(i) \in S_c(z(i)) \oplus \Sigma_x \subseteq \mathbb{X} \quad \forall i \in \mathbb{I}_{\geq 0} \\
&u(i) \in \mathbb{U} \quad \forall i \in \mathbb{I}_{\geq 0}
\end{aligned}$$

Compared with the corresponding conditions in Chapter 3, we see that the state constraint set $\mathbb{Z}$ now has to satisfy a stronger requirement than the condition $S_c(z) \subseteq \mathbb{X}$ for all $z \in \mathbb{Z}$ precisely because of the state estimation error, which is bounded by Σ_x. Hence the state constraint set $\mathbb{Z}$ required here is smaller than that required in Chapter 3.

5.6.5 Choosing $\mathbb{Z}$ and $\mathbb{V}$

Because the sets Σ_x and $S_d(z)$ cannot be easily computed, a pragmatic approach is required for choosing $\mathbb{Z}$ and $\mathbb{V}$. One simple, if conservative, possibility is to set $\mathbb{Z} = \alpha\mathbb{X}$ and $\mathbb{V} = \beta\mathbb{U}$ where α and β lie in $(0, 1)$. The tuning parameters α and β may be adjusted using data obtained by Monte Carlo simulation or from operation. If constraints are violated in the simulation, or in operation, α and β may be reduced; if the constraints are too conservative, α and β may be increased.

5.7 Notes

The problem of output feedback control has been extensively discussed in the general control literature. It is well known that, for linear systems, a stabilizing state feedback controller and an observer may be separately designed and combined to give a stabilizing output feedback controller (the separation principle). For nonlinear systems, Teel and Praly (1994) show that global stabilizability and complete uniform observability are sufficient to guarantee semiglobal stabilizability using a dynamic observer and provide useful references to related work on this topic.

Although output MPC in which nominal MPC is combined with a separately designed observer is widely used in industry since the state is seldom available, it has received relatively little attention in the literature because of the inherent difficulty in establishing asymptotic stability. An extra complexity in MPC is the presence of hard constraints. A useful survey, more comprehensive than these notes, is provided in Findeisen, Imsland, Allgöwer, and Foss (2003). Thus Michalska and Mayne (1995) show for deterministic systems that, for any subset of the region of attraction of the full state feedback system, there exists a sampling time and convergence rate for the observer such that the subset also lies in the region of attraction of the output feedback system. A more sophisticated analysis in Imsland, Findeisen, Allgöwer, and Foss (2003) using continuous time MPC shows that the region of attraction and rate of convergence of the output feedback system can approach that of the state feedback system as observer gain increases.

We consider systems with input disturbances and noisy state measurement; we employ the "tube" methodology that has its roots in the work of Bertsekas and Rhodes (1971), and Glover and Schweppe (1971) on constrained discrete time systems subject to bounded disturbances. Reachability of a "target set" and a "target tube" are considered in these papers. These concepts were substantially developed in the context of continuous time systems in Khurzhanski and Valyi (1997); Aubin (1991); Kurzhanski and Filippova (1993). The theory for discrete time systems is considerably simpler; a modern tube-based theory for optimal control of discrete time uncertain systems with imperfect state measurement appears in Moitié et al. (2002). As in this chapter, they regard a set X of states x that are consistent with past measurements as the "state" of the optimal control problem. The set X satisfies an uncertain "full information" difference equation of the form $X^+ = f^*(X, u, \mathbb{W}, v)$ so the output feedback optimal control problem reduces to robust control of an uncertain system with known state X. The optimal control problem remains difficult because the state X, a subset of $\mathbb{R}^n$, is difficult to obtain numerically and determination of a control law as a function of (X, t) prohibitive. In Mayne, Raković, Findeisen, and Allgöwer (2006, 2009) the output feedback problem is simplified considerably by replacing $X(t)$ by a simple outer approximation $\{\hat{x}(t)\} \oplus \Sigma_x$ in the time-invariant case and by $\{\hat{x}(t)\} \oplus \Sigma_x(t)$ in the time-varying case. The set $\mathbb{X}_X$, or the sequence $\{\mathbb{X}_X(t)\}$, may be precomputed so the difficult evolution equation for X is replaced by a simple evolution equation for $\hat{x}$; in the linear case, the Luenberger observer or Kalman filter describes the evolution of $\hat{x}$. The output feedback control problem reduces to control of an uncertain system with known state $\hat{x}$.

5.8 Exercises

Exercise 5.1: Hausdorff distance between a set and a subset

Show that $d_H(\mathbb{A}, \mathbb{B}) = \max_{a \in \mathbb{A}} d(a, \mathbb{B})$ if $\mathbb{A}$ and $\mathbb{B}$ are two compact subsets of $\mathbb{R}^n$ satisfying $\mathbb{B} \subseteq \mathbb{A}$.

Exercise 5.2: Hausdorff distance between sets $\mathbb{A} \oplus \mathbb{B}$ and $\mathbb{B}$

Show that $d_H(\mathbb{A} \oplus \mathbb{B}, \mathbb{A}) \leq |\mathbb{B}|$ if $\mathbb{A}$ and $\mathbb{B}$ are two compact subsets of $\mathbb{R}^n$ satisfying $0 \in \mathbb{B}$ in which $|\mathbb{B}| := \max_b \{|b| \mid b \in \mathbb{B}\}$.

Exercise 5.3: Hausdorff distance between sets $\{z\} \oplus \mathbb{B}$ and $\mathbb{A}$

Show that $d_H(\{z\} \oplus \mathbb{B}, \mathbb{A}) \leq |z| + d_H(\mathbb{A}, \mathbb{B})$ if $\mathbb{A}$ and $\mathbb{B}$ are two compact sets in $\mathbb{R}^n$.

Exercise 5.4: Hausdorff distance between sets $\{z\} \oplus \mathbb{A}$ and $\mathbb{A}$

Show that $d_H(\{z\} \oplus \mathbb{A}, \mathbb{A}) = |z|$ if z is a point and $\mathbb{A}$ is a compact set in $\mathbb{R}^n$.

Exercise 5.5: Hausdorff distance between sets $\mathbb{A} \oplus \mathbb{C}$ and $\mathbb{B} \oplus \mathbb{C}$

Show that $d_H(\mathbb{A} \oplus \mathbb{C}, \mathbb{B} \oplus \mathbb{C}) = d_H(\mathbb{A}, \mathbb{B})$ if $\mathbb{A}$, $\mathbb{B}$ and $\mathbb{C}$ are compact subsets of $\mathbb{R}^n$ satisfying $\mathbb{B} \subseteq \mathbb{A}$.

Exercise 5.6: Hausdorff distance between sets $F\mathbb{A}$ and $F\mathbb{B}$

Let $\mathbb{A}$ and $\mathbb{B}$ be two compact sets in $\mathbb{R}^n$ satisfying $\mathbb{A} \subseteq \mathbb{B}$, and let $F \in \mathbb{R}^{n \times n}$. Show that $d_H(F\mathbb{A}, F\mathbb{B}) \leq |F| d_H(\mathbb{A}, \mathbb{B})$ in which $|F|$ is the induced norm of F satisfying $|Fx| \leq |F|\ |x|$ and $|x| := d(x, 0)$.

Exercise 5.7: Linear combination of sets; $\lambda_1 \mathbb{W} \oplus \lambda_2 \mathbb{W} = (\lambda_1 + \lambda_2)\mathbb{W}$

If $\mathbb{W}$ is a convex set, show that $\lambda_1 \mathbb{W} \oplus \lambda_2 \mathbb{W} = (\lambda_1 + \lambda_2)\mathbb{W}$ for any $\lambda_1, \lambda_2 \in \mathbb{R}_{\geq 0}$. Hence show $\mathbb{W} \oplus \lambda \mathbb{W} \oplus \lambda^2 \mathbb{W} \oplus \cdots = (1 - \lambda)^{-1} \mathbb{W}$ if $\lambda \in [0, 1)$.

Exercise 5.8: Hausdorff distance between the sets $\Phi(i)$ and Φ

Show that there exist $c > 0$ and $\gamma \in (0, 1)$ such that

$$d_H(\Phi(i), \Phi) \leq c\, d_H(\Phi(0), \Phi)\gamma^i$$

in which

$$\Phi(i) = \widetilde{A}\Phi(i-1) + \widetilde{B}\Psi$$
$$\Phi = \widetilde{A}\Phi + \widetilde{B}\Psi$$

and $\widetilde{A}$ is a stable matrix ($\rho(\widetilde{A}) < 1$).

Bibliography

J. P. Aubin. *Viability Theory*. Systems & Control: Foundations & Applications. Birkhauser, Boston, Basel, Berlin, 1991.

D. P. Bertsekas and I. B. Rhodes. Recursive state estimation for a set-membership description of uncertainty. *IEEE Trans. Auto. Cont.*, 16:117–128, 1971.

F. Blanchini. Set invariance in control. *Automatica*, 35:1747–1767, 1999.

R. Findeisen, L. Imsland, F. Allgöwer, and B. A. Foss. State and output feedback nonlinear model predictive control: An overview. *Eur. J. Control*, 9(2–3):190–206, 2003. Survey paper.

J. D. Glover and F. C. Schweppe. Control of linear dynamic systems with set constrained disturbances. *IEEE Trans. Auto. Cont.*, 16:411–423, 1971.

L. Imsland, R. Findeisen, F. Allgöwer, and B. A. Foss. A note on stability, robustness and performance of output feedback nonlinear model predictive control. *J. Proc. Cont.*, 13:633–644, 2003.

A. B. Khurzhanski and I. Valyi. *Ellipsoidal-valued dynamics for estimation and control.* Systems & Control: Foundations & Applications. Birkhauser, Boston, Basel, Berlin, 1997.

I. Kolmanovsky and E. G. Gilbert. Theory and computation of disturbance invariant sets for discrete-time linear systems. *Math. Probl. Eng.*, 4(4):317–367, 1998.

A. B. Kurzhanski and T. F. Filippova. On the theory of trajectory tubes: A mathematical formalism for uncertain dynamics, viability and control. In A. B. Kurzhanski, editor, *Advances in Nonlinear Dynamics and Control: A Report from Russia*, volume 17 of *PSCT*, pages 122–188. Birkhauser, Boston, Basel, Berlin, 1993.

D. Q. Mayne, S. V. Raković, R. Findeisen, and F. Allgöwer. Robust output feedback model predictive control of constrained linear systems. *Automatica*, 42(7):1217–1222, July 2006.

D. Q. Mayne, S. V. Raković, R. Findeisen, and F. Allgöwer. Robust output feedback model predictive control of constrained linear systems: time varying case. *Automatica*, 2009. Accepted.

H. Michalska and D. Q. Mayne. Moving horizon observers and observer-based control. *IEEE Trans. Auto. Cont.*, 40(6):995–1006, 1995.

R. Moitié, M. Quincampoix, and V. M. Veliov. Optimal control of discrete-time uncertain systems with imperfect measurement. *IEEE Trans. Auto. Cont.*, 47 (11):1909–1914, November 2002.

S. V. Raković. Output feedback robust positive invariance: claims and proofs. Technical note; unpublished, May 2007.

S. V. Raković, E. C. Kerrigan, K. I. Kouramas, and D. Q. Mayne. Invariant approximations of the minimal robustly positively invariant sets. *IEEE Trans. Auto. Cont.*, 50(3):406–410, 2005.

A. R. Teel and L. Praly. Global stabilizability and observability implies semiglobal stabilizability by output feedback. *Sys. Cont. Let.*, 22:313–325, 1994.

6

Distributed Model Predictive Control

6.1 Introduction and Preliminary Results

In many large-scale control applications, it becomes convenient to break the large plantwide problem into a set of smaller and simpler subproblems in which the local inputs are used to regulate the local outputs. The overall plantwide control is then accomplished by the composite behavior of the interacting, local controllers. There are many ways to design the local controllers, some of which produce guaranteed properties of the overall plantwide system. We consider four control approaches in this chapter: decentralized, noncooperative, cooperative, and centralized control. The first three methods require the local controllers to optimize over only their local inputs. Their computational requirements are identical. The communication overhead is different, however. Decentralized control requires no communication between subsystems. Noncooperative and cooperative control require the input sequences and the current states or state estimates for all the other local subsystems. Centralized control solves the large, complex plantwide optimization over all the inputs. Communication is not a relevant property for centralized control because all information is available in the single plantwide controller. We use centralized control in this chapter to provide a benchmark of comparison for the distributed controllers.

We have established the basic properties of centralized MPC, both with and without state estimation, in Chapters 2, 3, and 5. In this chapter, we analyze some basic properties of the three distributed approaches: decentralized, noncooperative, and cooperative MPC. We show that the conditions required for closed-loop stability of decentralized control and noncooperative control are often violated for models of chemical plants under reasonable decompositions into subsystems. For ensuring closed-loop stability of a wide class of plantwide models and decomposition choices, cooperative control emerges as the most attractive option for distributed MPC. We then establish the closed-loop properties of cooperative MPC for unconstrained and constrained linear systems with and without state estimation. We also discuss current challenges facing this method, such as input constraints that are coupled between subsystems.

In our development of distributed MPC, we require some basic results on two topics: how to organize and solve the linear algebra of linear MPC, and how to ensure stability when using suboptimal MPC. We cover these two topics in the next sections, and then turn to the distributed MPC approaches.

6.1.1 Least Squares Solution

In comparing various forms of linear distributed MPC it proves convenient to see the MPC quadratic program for the sequence of states and inputs as a single large linear algebra problem. To develop this linear algebra problem, we consider first the *unconstrained* LQ problem of Chapter 1, which we solved efficiently with dynamic programming (DP) in Section 1.3.3

$$V(x(0),\mathbf{u}) = \frac{1}{2}\sum_{k=0}^{N-1} (x(k)'Qx(k) + u(k)'Ru(k)) + (1/2)x(N)'P_f x(N)$$

subject to

$$x^+ = Ax + Bu$$

In this section, we first take the direct but brute-force approach to finding the optimal control law. We write the model solution as

$$\begin{bmatrix} x(1) \\ x(2) \\ \vdots \\ x(N) \end{bmatrix} = \underbrace{\begin{bmatrix} A \\ A^2 \\ \vdots \\ A^N \end{bmatrix}}_{\mathcal{A}} x(0) + \underbrace{\begin{bmatrix} B & 0 & \cdots & 0 \\ AB & B & \cdots & 0 \\ \vdots & \vdots & \ddots & \vdots \\ A^{N-1}B & A^{N-2}B & \cdots & B \end{bmatrix}}_{\mathcal{B}} \begin{bmatrix} u(0) \\ u(1) \\ \vdots \\ u(N-1) \end{bmatrix} \tag{6.1}$$

or using the input and state sequences

$$\mathbf{x} = \mathcal{A}x(0) + \mathcal{B}\mathbf{u}$$

The objective function can be expressed as

$$V(x(0),\mathbf{u}) = (1/2)\left(x'(0)Qx(0) + \mathbf{x}'\mathcal{Q}\mathbf{x} + \mathbf{u}'\mathcal{R}\mathbf{u}\right)$$

in which

$$\mathcal{Q} = \text{diag}\left(\begin{bmatrix} Q & Q & \dots & P_f \end{bmatrix}\right) \in \mathbb{R}^{Nn \times Nn}$$
$$\mathcal{R} = \text{diag}\left(\begin{bmatrix} R & R & \dots & R \end{bmatrix}\right) \in \mathbb{R}^{Nm \times Nm} \tag{6.2}$$

Eliminating the state sequence. Substituting the model into the objective function and *eliminating* the state sequence gives a quadratic function of $\mathbf{u}$

$$V(x(0), \mathbf{u}) = (1/2)x'(0)(Q + \mathcal{A}'\mathcal{Q}\mathcal{A})x(0) + \mathbf{u}'(\mathcal{B}'\mathcal{Q}\mathcal{A})x(0) + (1/2)\mathbf{u}'(\mathcal{B}'\mathcal{Q}\mathcal{B} + \mathcal{R})\mathbf{u} \tag{6.3}$$

and the optimal solution for the entire set of inputs is obtained in one shot

$$\mathbf{u}^0(x(0)) = -(\mathcal{B}'\mathcal{Q}\mathcal{B} + \mathcal{R})^{-1}\mathcal{B}'\mathcal{Q}\mathcal{A}\, x(0)$$

and the optimal cost is

$$V^0(x(0)) = \left(\frac{1}{2}\right)x'(0)\left(Q + \mathcal{A}'\mathcal{Q}\mathcal{A} - \mathcal{A}'\mathcal{Q}\mathcal{B}(\mathcal{B}'\mathcal{Q}\mathcal{B} + \mathcal{R})^{-1}\mathcal{B}'\mathcal{Q}\mathcal{A}\right)x(0)$$

If used explicitly, this procedure for computing $\mathbf{u}^0$ would be inefficient because $\mathcal{B}'\mathcal{Q}\mathcal{B} + \mathcal{R}$ is an $(mN \times mN)$ matrix. Notice that in the DP formulation one has to invert instead an $(m \times m)$ matrix N times, which is computationally less expensive.[1] Notice also that unlike DP, the least squares approach provides *all* input moves as a function of the *initial* state, $x(0)$. The gain for the control law comes from the first input move in the sequence

$$K(0) = -\begin{bmatrix} I_m & 0 & \cdots & 0 \end{bmatrix}(\mathcal{B}'\mathcal{Q}\mathcal{B} + \mathcal{R})^{-1}\mathcal{B}'\mathcal{Q}\mathcal{A}$$

It is not immediately clear that the $K(0)$ and V^0 given above from the least squares approach are equivalent to the result from the Riccati iteration, (1.11)–(1.15) of Chapter 1, but since we have solved the same optimization problem, the two results are the same.[2]

Retaining the state sequence. In this section we set up the least squares problem again, but with an eye toward improving its efficiency. Retaining the state sequence and adjoining the model equations as

---

[1] Would you prefer to invert by hand 100 (1×1) matrices or a single (100×100) dense matrix?

[2] Establishing this result directly is an exercise in using the partitioned matrix inversion formula. The next section provides another way to show they are equivalent.

equality constraints is a central idea in optimal control and is described in standard texts (Bryson and Ho, 1975, p. 44). We apply this standard approach here. Wright (1997) provides a discussion of this problem in the linear model MPC context and the extensions required for the quadratic programming problem when there are inequality constraints on the states and inputs. Including the state with the input in the sequence of unknowns, we define the enlarged vector $\mathbf{z}$ to be

$$\mathbf{z} = \begin{bmatrix} u(0) \\ x(1) \\ u(1) \\ x(2) \\ \vdots \\ u(N-1) \\ x(N) \end{bmatrix}$$

The objective function is

$$\min_{\mathbf{u}} (1/2)(x'(0)Qx(0) + \mathbf{z}'H\mathbf{z})$$

in which

$$H = \operatorname{diag}\left(\begin{bmatrix} R & Q & R & Q & \cdots & R & P_f \end{bmatrix}\right)$$

The constraints are

$$D\mathbf{z} = d$$

in which

$$D = -\begin{bmatrix} B & -I & & & & \\ & A & B & -I & & \\ & & & \ddots & & \\ & & & A & B & -I \end{bmatrix} \qquad d = \begin{bmatrix} A \\ 0 \\ \vdots \\ 0 \end{bmatrix} x(0)$$

We now substitute these results into (1.58) and obtain the linear algebra problem

$$\begin{bmatrix}
R & & & & & & & B' & & & \\
& Q & & & & & & -I & A' & & \\
& & R & & & & & & B' & & \\
& & & Q & & & & & -I & & \\
& & & & \ddots & & & & & \ddots & \\
& & & & & R & & & & & B' \\
& & & & & & P_f & & & & -I \\
B & -I & & & & & & & & & \\
& A & B & -I & & & & & & & \\
& & & & \ddots & & & & & & \\
& & & & & B & -I & & & &
\end{bmatrix}
\begin{bmatrix} u(0) \\ x(1) \\ u(1) \\ x(2) \\ \vdots \\ u(N-1) \\ x(N) \\ \lambda(1) \\ \lambda(2) \\ \vdots \\ \lambda(N) \end{bmatrix}
=
\begin{bmatrix} 0 \\ 0 \\ 0 \\ 0 \\ \vdots \\ 0 \\ 0 \\ -A \\ 0 \\ \vdots \\ 0 \end{bmatrix} x(0)$$

Method	FLOPs
dynamic programming (DP)	Nm^3
dense least squares	N^3m^3
banded least squares	$N(2n+m)(3n+m)^2$

Table 6.1: Computational cost of solving finite horizon LQR problem.

This equation is rather cumbersome, but if we reorder the unknown vector to put the Lagrange multiplier together with the state and input from the same time index, and reorder the equations, we obtain the following banded matrix problem

$$\begin{bmatrix} R & B' & & & & & & & & & \\ B & & -I & & & & & & & & \\ & -I & Q & & & & & & & & \\ & & & R & & & & & & & \\ & & & \ddots & \ddots & & & & & & \\ & & & & & R & B' & & & & \\ & & & & A & B & & -I & & & \\ & & & & & & -I & Q & & A' & \\ & & & & & & & & R & B' & \\ & & & & & & & A & B & & -I \\ & & & & & & & & & -I & P_f \end{bmatrix} \begin{bmatrix} u(0) \\ \lambda(1) \\ x(1) \\ u(1) \\ \vdots \\ u(N-2) \\ \lambda(N-1) \\ x(N-1) \\ u(N-1) \\ \lambda(N) \\ x(N) \end{bmatrix} = \begin{bmatrix} 0 \\ -A \\ 0 \\ 0 \\ \vdots \\ 0 \\ 0 \\ 0 \\ 0 \\ 0 \\ 0 \end{bmatrix} x(0) \quad (6.4)$$

The banded structure allows a more efficient solution procedure. The floating operation (FLOP) count for the factorization of a banded matrix is $O(LM^2)$ in which L is the dimension of the matrix and M is the bandwidth. This compares to the regular FLOP count of $O(L^3)$ for the factorization of a regular dense matrix. The bandwidth of the matrix in (6.4) is $3n+m$ and the dimension of the matrix is $N(2n+m)$. Therefore the FLOP count for solving this equation is $O(N(2n+m)(3n+m)^2)$. Notice that this approach reduces the N^3 dependence of the previous MPC solution method. That is the computational advantage provided by these adjoint methods for treating the model constraints. Table 6.1 summarizes the computational cost of the three approaches. As shown in the table, DP is highly efficient. When we add input and state inequality constraints to the control problem and the state dimension is large, however, we cannot conveniently apply DP. The dense least squares computational cost is high if we wish to compute a large number of moves in the horizon. Note the cost of dense least squares scales with the third power of horizon length N. As we have discussed in Chapter 2, considerations of control theory favor large N. Another factor increasing the computational cost is the trend in industrial MPC implementations to larger multivariable control problems with more states and inputs, i.e., larger m and n. Therefore, the adjoint approach using banded least squares method becomes important for industrial applications in which the problems are large and a solid theoretical foundation for the control method is desirable.

We might obtain more efficiency than the banded structure if we view (6.4) as a block tridiagonal matrix and use the method provided by Golub and Van Loan (1996, p. 174). The final fine tuning of the solution method for this class of problems is a topic of current research, but the important point is that, whatever final procedure is selected, the computational cost will be linear in N as in DP instead of cubic in N as in dense least squares.

Furthermore, if we wish to see the connection to the DP solution, we can proceed as follows. Substitute $\Pi(N) = P_f$ as in (1.12) of Chapter 1 and consider the last three-equation block of the matrix appearing in (6.4)

$$\begin{bmatrix} & R & B' & & \\ A & B & & -I & \\ & & -I & \Pi(N) & \end{bmatrix} \begin{bmatrix} x(N-1) \\ u(N-1) \\ \lambda(N) \\ x(N) \end{bmatrix} = \begin{bmatrix} 0 \\ 0 \\ 0 \end{bmatrix}$$

We can eliminate this small set of equations and solve for $u(N-1)$, $\lambda(N)$, $x(N)$ in terms of $x(N-1)$, resulting in

$$\begin{bmatrix} u(N-1) \\ \lambda(N) \\ x(N) \end{bmatrix} = \begin{bmatrix} -(B'\Pi(N)B + R)^{-1}B'\Pi(N)A \\ \Pi(N)(I - B(B'\Pi(N)B + R)^{-1}B'\Pi(N))A \\ (I - B(B'\Pi(N)B + R)^{-1}B'\Pi(N))A \end{bmatrix} x(N-1)$$

Notice that in terms of the Riccati matrix, we also have the relationship

$$A'\lambda(N) = \Pi(N-1)x(N-1) - Qx(N-1)$$

We then proceed to the next to last block of three equations

$$\begin{bmatrix} & R & B' & & & \\ A & B & & -I & & \\ & & -I & Q & & A' \end{bmatrix} \begin{bmatrix} x(N-2) \\ u(N-2) \\ \lambda(N-1) \\ x(N-1) \\ u(N-1) \\ \lambda(N) \end{bmatrix} = \begin{bmatrix} 0 \\ 0 \\ 0 \end{bmatrix}$$

Note that the last equation gives

$$\lambda(N-1) = Qx(N-1) + A'\lambda(N) = \Pi(N-1)x(N-1)$$

Using this relationship and continuing on to solve for $x(N-1)$, $\lambda(N-1)$, $u(N-2)$ in terms of $x(N-2)$ gives

$$\begin{bmatrix} u(N-2) \\ \lambda(N-1) \\ x(N-1) \end{bmatrix} = \begin{bmatrix} -(B'\Pi(N-1)B+R)^{-1}B'\Pi(N-1)A \\ \Pi(N-1)(I - B(B'\Pi(N-1)B+R)^{-1}B'\Pi(N-1))A \\ (I - B(B'\Pi(N-1)B+R)^{-1}B'\Pi(N-1))A \end{bmatrix} x(N-2)$$

Continuing on through each previous block of three equations produces the Riccati iteration and feedback gains of (1.11)-(1.14). The other unknowns, the multipliers, are simply

$$\lambda(k) = \Pi(k)x(k) \qquad k = 1, 2, \ldots, N$$

so the cost to go at each stage is simply $x(k)'\lambda(k)$, and we see the nice connection between the Lagrange multipliers and the cost of the LQR control problem.

6.1.2 Stability of Suboptimal MPC

When using distributed MPC, it may be necessary or convenient to implement the control without solving the complete optimization. We then have a form of suboptimal MPC, which was first considered in Chapter 2, Section 2.8. Before adding the complexity of the distributed version, we wish to further develop a few features of suboptimal MPC in the centralized, single-player setting. These same features arise in the distributed, many-player setting as we discuss subsequently.

We consider a specific variation of suboptimal MPC in which a starting guess is available from the control trajectory at the previous time and we take a fixed number of steps of an optimization algorithm. The exact nature of the optimization method is not essential, but we do restrict the method so that each iteration is feasible and decreases the value of the cost function. To initialize the suboptimal controller, we are given an initial state $x(0) = x_0$, and we generate an initial control sequence $\mathbf{u}(0) = \mathbf{h}(x_0)$. We consider input constraints $u(i) \in \mathbb{U} \subseteq \mathbb{R}^m$, $i \in \mathbb{I}_{0:N-1}$, which we also write as $\mathbf{u} \in \mathbb{U}^N \subseteq \mathbb{R}^N$. As in Chapter 2 we denote the set of feasible states as $\mathcal{X}_N$. These are the states for which the initial control sequence $\mathbf{h}(x_0)$ is well defined. The suboptimal MPC algorithm is as follows.

Suboptimal MPC algorithm.

Data: Integer N_{iter}.

Initialize: Set current state $x = x_0$, current control sequence, $\mathbf{u} = \mathbf{h}(x_0)$.

Step 1 (State evolution): Apply control $u = u(0)$ to the system. Obtain state at next sample, x^+. For the nominal system

$$x^+ = f(x, u(0))$$

Step 2 (Warm start): Denote the warm start for the next sample time as $\widetilde{\mathbf{u}}^+$. We use

$$\widetilde{\mathbf{u}}^+ = \{u(1), u(2), \ldots, u(N-1), 0\}$$

in which $x(N) = \phi(N; x, \mathbf{u})$. The warm start $\widetilde{\mathbf{u}}^+$ therefore is a function of $(x, \mathbf{u})$. This warm start is a simplified version of the one considered in Chapter 2, in which the final control move in the warm start was determined by the control law $\kappa_f(x)$. In distributed MPC it is simpler to use zero for the final control move in the warm start.

Step 3 (Iteration of an optimization method): The controller performs N_{iter} iterations of a feasible path optimization algorithm to obtain an improved control sequence using initial state x^+. The final input sequence $\mathbf{u}^+$ is a function of the state initial condition and the warm start $(x^+, \widetilde{\mathbf{u}})$. Noting that x^+ and $\widetilde{\mathbf{u}}$ are both functions of $(x, \mathbf{u})$, the input sequence $\mathbf{u}^+$ can also be expressed as function of only $(x, \mathbf{u})$

$$\mathbf{u}^+ = g(x, \mathbf{u})$$

Step 4 (Next time step): Update state and input sequence: $x \leftarrow x^+$, $\mathbf{u} \leftarrow \mathbf{u}^+$. Go to Step 1.

We establish later in the chapter that the system cost function $V(x, \mathbf{u})$ satisfies the following properties for the form of suboptimal MPC generated by distributed MPC. There exist constants $a, b, c > 0$ such that

$$a\,|(x, \mathbf{u})|^2 \leq V(x, \mathbf{u}) \leq b\,|(x, \mathbf{u})|^2$$
$$V(x^+, \mathbf{u}^+) - V(x, \mathbf{u}) \leq -c\,|(x, u(0))|^2$$

These properties are similar to those required for a valid Lyapunov function. The difference is that the cost decrease here does not depend on the size of $\mathbf{u}$, but only x and the first element of $\mathbf{u}$, $u(0)$. This cost decrease is sufficient to establish that $x(k)$ and $u(k)$ converge to zero, but allows the possibility that $\mathbf{u}(k)$ is large even though $x(k)$ is small. That fact prevents us from establishing the solution $x(k) = 0$ for all k is Lyapunov stable. We can establish that the solution $x(k) = 0$ for all k is Lyapunov stable at $k = 0$ only. We cannot establish uniform Lyapunov stability nor Lyapunov stability for any $k > 0$. The problem is not that our proof technique is deficient. There is no reason to *expect* that the solution $x(k) = 0$ for all k is Lyapunov stable for suboptimal MPC. The lack of Lyapunov stability of $x(k) = 0$ for all k is a subtle issue and warrants some discussion. To make these matters more precise, consider the following standard definitions of Lyapunov stability at time k and uniform Lyapunov stability (Vidyasagar, 1993, p. 136).

Definition 6.1 (Lyapunov stability)**.** The zero solution $x(k) = 0$ for all k is stable (in the sense of Lyapunov) at $k = k_0$ if for any $\varepsilon > 0$ there exists a $\delta(k_0, \varepsilon) > 0$ such that

$$|x(k_0)| < \delta \implies |x(k)| < \varepsilon \quad \forall k \geq k_0 \tag{6.5}$$

Lyapunov stability is defined at a time k_0. Uniform stability is the concept that guarantees that the zero solution is not losing stability with time. For a uniformly stable zero solution, δ in Definition 6.1 is *not* a function of k_0, so that (6.5) holds for all k_0.

Definition 6.2 (Uniform Lyapunov stability)**.** The zero solution $x(k) = 0$ for all k is uniformly stable (in the sense of Lyapunov) if for any $\varepsilon > 0$ there exists a $\delta(\varepsilon) > 0$ such that

$$|x(k_0)| < \delta \implies |x(k)| < \varepsilon \quad \forall k \geq k_0 \quad \forall k_0$$

Exercise 6.6 gives an example of a linear system for which $x(k)$ converges exponentially to zero with increasing k for all $x(0)$, but the zero solution $x(k) = 0$ for all k is Lyapunov stable only at $k = 0$. It is not uniformly Lyapunov stable nor Lyapunov stable for any $k > 0$. Without further restrictions, suboptimal MPC admits this same type of behavior.

To ensure uniform Lyapunov stability, we add requirements to suboptimal MPC beyond obtaining only a cost decrease. Here we impose the constraint

$$|\mathbf{u}| \leq d\,|x| \qquad x \in r\mathcal{B}$$

in which $d, r > 0$. This type of constraint was first introduced in (2.43) of Chapter 2. In that arrangement of suboptimal MPC it was simplest to switch to local controller $u = \kappa_f(x)$ when the state entered $\mathbb{X}_f$ to automatically enforce this constraint. In this chapter we instead include the constraint explicitly in the distributed MPC optimization problem and do not switch to a local controller. Both alternatives provide (uniform) Lyapunov stability of the solution $x(k) = 0$ for all k. The following lemma summarizes the conditions we use later in the chapter for establishing exponential stability of distributed MPC. A similar lemma establishing asymptotic stability of suboptimal MPC was given by Scokaert, Mayne, and Rawlings (1999) (Theorem 1).

Definition 6.3 (Exponential stability). Let $\mathbb{X}$ be positive invariant for $x^+ = f(x)$. Then the origin is exponentially stable for $x^+ = f(x)$ with a region of attraction $\mathbb{X}$ if there exists $c > 0$ and $\gamma < 1$ such that

$$\left|\phi(i;x)\right| \leq c\,|x|\,\gamma^i$$

for all $i \geq 0, x \in \mathbb{X}$.

Consider next the suboptimal MPC controller. Let the system satisfy $(x^+, \mathbf{u}^+) = (f(x,\mathbf{u}), g(x,\mathbf{u}))$ with initial sequence $\mathbf{u}(0) = \mathbf{h}(x(0))$. The controller constraints are $x(i) \in \mathbb{X} \subseteq \mathbb{R}^n$ for all $i \in \mathbb{I}_{0:N}$ and $u(i) \in \mathbb{U} \subseteq \mathbb{R}^m$ for all $i \in \mathbb{I}_{0:N-1}$. Let $\mathcal{X}_N$ denote the set of states for which the MPC controller is feasible. The suboptimal MPC system satisfies the following. Given $r > 0$, there exist $a, b, c > 0$ such that

$$\begin{aligned}
a\,|(x,\mathbf{u})|^2 \leq V(x,\mathbf{u}) &\leq b\,|(x,\mathbf{u})|^2 & x \in \mathcal{X}_N \quad \mathbf{u} \in \mathbb{U}^N \\
V(x^+,\mathbf{u}^+) - V(x,\mathbf{u}) &\leq -c\,|(x,u(0))|^2 & x \in \mathcal{X}_N \quad \mathbf{u} \in \mathbb{U}^N \\
|\mathbf{u}| &\leq d\,|x| & x \in r\mathcal{B}
\end{aligned}$$

Lemma 6.4 (Exponential stability of suboptimal MPC). *The origin is exponentially stable for the closed-loop system under suboptimal MPC with region of attraction $\mathcal{X}_N$ if either of the following assumptions holds*

(a) $\mathbb{U}$ is compact. In this case, $\mathcal{X}_N$ may be unbounded.

(b) $\mathcal{X}_N$ is compact. In this case $\mathbb{U}$ may be unbounded.

Exercises 6.7 and 6.8 explore what to conclude about exponential stability when both $\mathbb{U}$ and $\mathcal{X}_N$ are unbounded.

Proof. First we show that the origin of the extended state $(x, \mathbf{u})$ is exponentially stable for $x(0) \in \mathcal{X}_N$.

(a) For the case $\mathbb{U}$ compact, we use the same argument used to prove Proposition 2.18 of Chapter 2. We have $|\mathbf{u}| \leq d\,|x|\,, x \in r\mathcal{B}$. Consider the optimization

$$\max_{\mathbf{u}\in\mathbb{U}^N} |\mathbf{u}| = s > 0$$

The solution exists by the Weierstrass theorem since $\mathbb{U}$ is compact, which implies $\mathbb{U}^N$ is compact. Then we have $|\mathbf{u}| \leq (s/r)\,|x|$ for $x \in \mathcal{X}_N \setminus r\mathcal{B}$, so we have $|\mathbf{u}| \leq d'\,|x|$ for $x \in \mathcal{X}_N$ in which $d' = \max(d, s/r)$.

(b) For the case $\mathcal{X}_N$ compact, consider the optimization

$$\max_{x\in\mathcal{X}_N} V(x, \mathbf{h}(x)) = \bar{V} > 0$$

The solution exists because $\mathcal{X}_N$ is compact and $\mathbf{h}(\cdot)$ and $V(\cdot)$ are continuous. Define the compact set $\bar{\mathbb{U}}$ by

$$\bar{\mathbb{U}} = \{\mathbf{u} \mid V(x,\mathbf{u}) \leq \bar{V}, \quad x \in \mathcal{X}_N\}$$

The set is bounded because $V(x,\mathbf{u}) \geq a\,|(x,\mathbf{u})|^2 \geq a\,|\mathbf{u}|^2$. The set is closed because V is continuous. The significance of this set is that for all $k \geq 0$ and all $x \in \mathcal{X}_N$, $\mathbf{u}(k) \in \bar{\mathbb{U}}$. Therefore we have established that $\mathcal{X}_N$ compact implies $\mathbf{u}(k)$ evolves in a compact set as in the previous case when $\mathbb{U}$ is assumed compact. Using the same argument as in that case, we have established that there exists $d' > 0$ such that $|\mathbf{u}| \leq d'\,|x|$ for all $x \in \mathcal{X}_N$.

For the two cases, we therefore have established for all $x \in \mathcal{X}_N$, $\mathbf{u} \in \mathbb{U}^N$ (case (a)) or $\mathbf{u} \in \bar{\mathbb{U}}$ (case (b))

$$|(x,\mathbf{u})| \leq |x| + |\mathbf{u}| \leq |x| + d'\,|x| \leq (1 + d')\,|x|$$

which gives $|x| \geq c'\,|(x,\mathbf{u})|$ with $c' = 1/(1 + d') > 0$. Hence, there exists $a_3 = c(c')^2$ such that $V(x^+,\mathbf{u}^+) - V(x,\mathbf{u}) \leq -a_3\,|(x,\mathbf{u})|^2$ for all $x \in \mathcal{X}_N$. Therefore the extended state $(x,\mathbf{u})$ satisfies the standard conditions of an exponential stability Lyapunov function (see Theorem B.14 in Appendix B) with $a_1 = a, a_2 = b, a_3 = c(c')^2, \sigma = 2$ for $(x,\mathbf{u}) \in \mathcal{X}_N \times \mathbb{U}^N$ (case (a)) or $\mathcal{X}_N \times \bar{\mathbb{U}}$ (case (b)). Therefore for all $x(0) \in \mathcal{X}_N$, $k \geq 0$,

$$|(x(k),\mathbf{u}(k))| \leq \alpha\,|(x(0),\mathbf{u}(0))|\,\gamma^k$$

in which $\alpha > 0$ and $0 < \gamma < 1$.

Finally we remove the input sequence and establish that the origin for the state (rather than the extended state) is exponentially stable for the closed-loop system. We have for all $x(0) \in \mathcal{X}_N$ and $k \geq 0$

$$\begin{aligned} |x(k)| &\leq |(x(k), \mathbf{u}(k))| \leq \alpha \, |(x(0), \mathbf{u}(0))| \, \gamma^k \\ &\leq \alpha(|x(0)| + |\mathbf{u}(0)|)\gamma^k \leq \alpha(1 + d') \, |x(0)| \, \gamma^k \\ &\leq \alpha' \, |x(0)| \, \gamma^k \end{aligned}$$

in which $\alpha' = \alpha(1 + d') > 0$, and we have established exponential stability of the origin on the feasible set $\mathcal{X}_N$. ■

We also consider later in the chapter the effects of state estimation error on the closed-loop properties of distributed MPC. For analyzing stability under perturbations, the following lemma is useful. Here e plays the role of estimation error.

Lemma 6.5 (Exponential stability with mixed powers of norm). *Consider a dynamic system*

$$(x^+, e^+) = f(x, e)$$

with a zero steady-state solution, $f(0,0) = (0,0)$. Assume there exists a function $V : \mathbb{R}^{n+m} \to \mathbb{R}_{\geq 0}$ that satisfies the following for all $(x, e) \in \mathbb{R}^n \times \mathbb{R}^m$

$$a(|x|^\sigma + |e|^\gamma) \leq V((x,e)) \leq b(|x|^\sigma + |e|^\gamma) \tag{6.6}$$

$$V(f(x,e)) - V((x,e)) \leq -c(|x|^\sigma + |e|^\gamma) \tag{6.7}$$

with constants $a, b, c, \sigma, \gamma > 0$. Then the zero steady-state solution is globally exponentially stable for $(x^+, e^+) = f(x, e)$.

The proof of this lemma is discussed in Exercise 6.9. We also require a converse theorem for exponential stability.

Lemma 6.6 (Converse theorem for exponential stability). *If the zero steady-state solution of $x^+ = f(x)$ is globally exponentially stable, then there exists Lipschitz continuous $V : \mathbb{R}^n \to \mathbb{R}_{\geq 0}$ that satisfies the following: there exist constants $a, b, c, \sigma > 0$, such that for all $x \in \mathbb{R}^n$*

$$\begin{aligned} a \, |x|^\sigma &\leq V(x) \leq b \, |x|^\sigma \\ V(f(x)) - V(x) &\leq -c \, |x|^\sigma \end{aligned}$$

Moreover, any $\sigma > 0$ is valid, and the constant c can be chosen as large as one wishes.

The proof of this lemma is discussed in Exercise B.3.

6.2 Unconstrained Two-Player Game

To introduce clearly the concepts and notation required to analyze distributed MPC, we start with a two-player game. We then generalize to an M-player game in the next section.

Let (A_{11}, B_{11}, C_{11}) be a minimal state space realization of the (u_1, y_1) input-output pair. Similarly, let (A_{12}, B_{12}, C_{12}) be a minimal state space realization of the (u_2, y_1) input-output pair. The dimensions are $u_1 \in \mathbb{R}^{m_1}$, $y_1 \in \mathbb{R}^{p_1}$, $x_{11} \in \mathbb{R}^{n_{11}}$, $x_{12} \in \mathbb{R}^{n_{12}}$ with $n_1 = n_{11} + n_{12}$. Output y_1 can then be represented as the following, possibly nonminimal, state space model

$$\begin{bmatrix} x_{11} \\ x_{12} \end{bmatrix}^+ = \begin{bmatrix} A_{11} & 0 \\ 0 & A_{12} \end{bmatrix} \begin{bmatrix} x_{11} \\ x_{12} \end{bmatrix} + \begin{bmatrix} B_{11} \\ 0 \end{bmatrix} u_1 + \begin{bmatrix} 0 \\ B_{12} \end{bmatrix} u_2$$
$$y_1 = \begin{bmatrix} C_{11} & C_{12} \end{bmatrix} \begin{bmatrix} x_{11} \\ x_{12} \end{bmatrix}$$

Proceeding in an analogous fashion with output y_2 and inputs u_1 and u_2, we model y_2 with the following state space model

$$\begin{bmatrix} x_{22} \\ x_{21} \end{bmatrix}^+ = \begin{bmatrix} A_{22} & 0 \\ 0 & A_{21} \end{bmatrix} \begin{bmatrix} x_{22} \\ x_{21} \end{bmatrix} + \begin{bmatrix} B_{22} \\ 0 \end{bmatrix} u_2 + \begin{bmatrix} 0 \\ B_{21} \end{bmatrix} u_1$$
$$y_2 = \begin{bmatrix} C_{22} & C_{21} \end{bmatrix} \begin{bmatrix} x_{22} \\ x_{21} \end{bmatrix}$$

We next define player one's local cost functions

$$V_1(x_1(0), \mathbf{u}_1, \mathbf{u}_2) = \sum_{k=0}^{N-1} \ell_1(x_1(k), u_1(k)) + V_{1f}(x_1(N))$$

in which

$$x_1 = \begin{bmatrix} x_{11} \\ x_{12} \end{bmatrix}$$

Note that the first local objective is affected by the second player's inputs through the model evolution of x_1, i.e., through the x_{12} states. We choose the stage cost to account for the first player's inputs and outputs

$$\ell_1(x_1, u_1) = (1/2)(y_1' \overline{Q}_1 y_1 + u_1' R_1 u_1)$$
$$\ell_1(x_1, u_1) = (1/2)(x_1' Q_1 x_1 + u_1' R_1 u_1)$$

in which

$$Q_1 = C_1' \overline{Q}_1 C_1 \qquad C_1 = \begin{bmatrix} C_{11} & C_{12} \end{bmatrix}$$

Motivated by the warm start to be described later, for stable systems, we choose the terminal penalty to be the infinite horizon cost to go under zero control

$$V_{1f}(x_1(N)) = (1/2) x_1'(N) P_{1f} x_1(N)$$

We choose P_{1f} as the solution to the following Lyapunov equation assuming A_1 is stable

$$A_1' P_{1f} A_1 - P_{1f} = -Q_1 \tag{6.8}$$

We proceed analogously to define player two's local objective function and penalties

$$V_2(x_2(0), \mathbf{u}_1, \mathbf{u}_2) = \sum_{k=0}^{N-1} \ell_2(x_2(k), u_2(k)) + V_{2f}(x_2(N))$$

In centralized control and the cooperative game, the two players share a common objective, which can be considered to be the overall plant objective

$$V(x_1(0), x_2(0), \mathbf{u}_1, \mathbf{u}_2) = \rho_1 V_1(x_1(0), \mathbf{u}_1, \mathbf{u}_2) + \rho_2 V_2(x_2(0), \mathbf{u}_2, \mathbf{u}_1)$$

in which the parameters ρ_1, ρ_2 are used to specify the relative weights of the two subsystems in the overall plant objective. Their values are restricted so $\rho_1, \rho_2 > 0$, $\rho_1 + \rho_2 = 1$ so that both local objectives must have some nonzero effect on the overall plant objective.

6.2.1 Centralized Control

Centralized control requires the solution of the systemwide control problem. It can be stated as

$$\begin{aligned} \min_{\mathbf{u}_1, \mathbf{u}_2} \; & V(x_1(0), x_2(0), \mathbf{u}_1, \mathbf{u}_2) \\ \text{s.t. } x_1^+ &= A_1 x_1 + \overline{B}_{11} u_1 + \overline{B}_{12} u_2 \\ x_2^+ &= A_2 x_2 + \overline{B}_{22} u_2 + \overline{B}_{21} u_1 \end{aligned}$$

in which

$$A_1 = \begin{bmatrix} A_{11} & 0 \\ 0 & A_{12} \end{bmatrix} \qquad A_2 = \begin{bmatrix} A_{22} & 0 \\ 0 & A_{21} \end{bmatrix}$$

$$\overline{B}_{11} = \begin{bmatrix} B_{11} \\ 0 \end{bmatrix} \qquad \overline{B}_{12} = \begin{bmatrix} 0 \\ B_{12} \end{bmatrix} \qquad \overline{B}_{21} = \begin{bmatrix} 0 \\ B_{21} \end{bmatrix} \qquad \overline{B}_{22} = \begin{bmatrix} B_{22} \\ 0 \end{bmatrix}$$

This optimal control problem is more complex than all of the distributed cases to follow because the decision variables include both $\mathbf{u}_1$ and $\mathbf{u}_2$. Because the performance is optimal, centralized control is a natural benchmark against which to compare the distributed cases: cooperative, noncooperative, and decentralized MPC. The plantwide stage cost and terminal cost can be expressed as quadratic functions of the subsystem states and inputs

$$\ell(x,u) = (1/2)(x'Qx + u'Ru)$$
$$V_f(x) = (1/2)x'P_f x$$

in which

$$x = \begin{bmatrix} x_1 \\ x_2 \end{bmatrix} \quad u = \begin{bmatrix} u_1 \\ u_2 \end{bmatrix} \quad Q = \begin{bmatrix} \rho_1 Q_1 & 0 \\ 0 & \rho_2 Q_2 \end{bmatrix}$$
$$R = \begin{bmatrix} \rho_1 R_1 & 0 \\ 0 & \rho_2 R_2 \end{bmatrix} \quad P_f = \begin{bmatrix} \rho_1 P_{1f} & 0 \\ 0 & \rho_2 P_{2f} \end{bmatrix} \tag{6.9}$$

and we have the standard MPC problem considered in Chapters 1 and 2

$$\min_{\mathbf{u}} V(x(0), \mathbf{u})$$
$$\text{s.t. } x^+ = Ax + Bu \tag{6.10}$$

in which

$$A = \begin{bmatrix} A_1 & 0 \\ 0 & A_2 \end{bmatrix} \quad B = \begin{bmatrix} \overline{B}_{11} & \overline{B}_{12} \\ \overline{B}_{21} & \overline{B}_{22} \end{bmatrix} \tag{6.11}$$

Given the terminal penalty in (6.8), stability of the closed-loop centralized system is guaranteed for all choices of system models and tuning parameters subject to the usual stabilizability assumption on the system model.

6.2.2 Decentralized Control

Centralized and decentralized control define the two extremes in distributing the decision making in a large-scale system. Centralized control has full information and optimizes the full control problem over all decision variables. Decentralized control, on the other hand, optimizes only the local objectives and has no information about the actions of the other subsystems. Player one's objective function is

$$V_1(x_1(0), \mathbf{u}_1) = \sum_{k=0}^{N-1} \ell_1(x_1(k), u_1(k)) + V_{1f}(x_1(N))$$

We then have player one's decentralized control problem

$$\min_{\mathbf{u}_1} V_1(x_1(0), \mathbf{u}_1)$$
$$\text{s.t. } x_1^+ = A_1 x_1 + \overline{B}_{11} u_1$$

We know the optimal solution for this kind of LQ problem is a linear feedback law

$$u_1^0 = K_1 x_1(0)$$

Notice that in decentralized control, player one's model does not account for the inputs of player two, and already contains model error. In the decentralized problem, player one requires no information about player two. The communication overhead for decentralized control is therefore minimal, which is an implementation advantage, but the resulting performance may be quite poor for systems with reasonably strong coupling. We compute an optimal K_1 for system one $(A_1, \overline{B}_{11}, Q_1, R_1)$ and optimal K_2 for system 2. The closed-loop system evolution is then

$$\begin{bmatrix} x_1 \\ x_2 \end{bmatrix}^+ = \begin{bmatrix} A_1 + \overline{B}_{11} K_1 & \overline{B}_{12} K_2 \\ \overline{B}_{21} K_1 & A_2 + \overline{B}_{22} K_2 \end{bmatrix} \begin{bmatrix} x_1 \\ x_2 \end{bmatrix}$$

and we know only that $A_{11} + \overline{B}_{11} K_1$ and $A_{22} + \overline{B}_{22} K_2$ are stable matrices. Obviously the stability of the closed-loop, decentralized system is fragile and depends in a sensitive way on the sizes of the interaction terms $\overline{B}_{12}$ and $\overline{B}_{21}$ and feedback gains K_1, K_2.

6.2.3 Noncooperative Game

In the noncooperative game, player one optimizes $V_1(x_1(0), \mathbf{u}_1, \mathbf{u}_2)$ over $\mathbf{u}_1$ and player two optimizes $V_2(x_2(0), \mathbf{u}_1, \mathbf{u}_2)$ over $\mathbf{u}_2$. From player one's perspective, player two's planned inputs $\mathbf{u}_2$ are known disturbances affecting player one's output through the dynamic model. Part of player one's optimal control problem is therefore to compensate for player two's inputs with his optimal $\mathbf{u}_1$ sequence in order to optimize his local objective V_1. Similarly, player two considers player one's inputs as a known disturbance and solves an optimal control problem that removes their effect in his local objective V_2. Because this game is noncooperative ($V_1 \neq V_2$), the struggle between players one and two can produce an outcome that is bad for both of them as we show subsequently. Notice that unlike decentralized control, there is no model error in the noncooperative game. Player one knows exactly the effect of the actions of player two and vice versa. Any poor nominal performance is caused by the noncooperative game, not model error.

Summarizing the noncooperative control problem statement, player one's model is

$$x_1^+ = A_1 x_1 + \overline{B}_{11} u_1 + \overline{B}_{12} u_2$$

and player one's objective function is

$$V_1(x_1(0), \mathbf{u}_1, \mathbf{u}_2) = \sum_{k=0}^{N-1} \ell_1(x_1(k), u_1(k)) + V_{1f}(x_1(N))$$

Note that V_1 here depends on $\mathbf{u}_2$ because the state trajectory $x_1(k), k \geq 1$ depends on $\mathbf{u}_2$ as shown in player one's dynamic model. We then have player one's noncooperative control problem

$$\begin{aligned} &\min_{\mathbf{u}_1} V_1(x_1(0), \mathbf{u}_1, \mathbf{u}_2) \\ &\text{s.t. } x_1^+ = A_1 x_1 + \overline{B}_{11} u_1 + \overline{B}_{12} u_2 \end{aligned}$$

Solution to player one's optimal control problem. We now solve player one's optimal control problem. Proceeding as in Section 6.1.1 we define

$$\mathbf{z} = \begin{bmatrix} u_1(0) \\ x_1(1) \\ \vdots \\ u_1(N-1) \\ x_1(N) \end{bmatrix} \qquad H = \text{diag}\left(\begin{bmatrix} R_1 & Q_1 & \cdots & R_1 & P_{1f} \end{bmatrix}\right)$$

and can express player one's optimal control problem as

$$\begin{aligned} &\min_{\mathbf{z}} (1/2)(\mathbf{z}' H \mathbf{z} + x_1(0)' Q_1 x_1(0)) \\ &\text{s.t. } D\mathbf{z} = d \end{aligned}$$

in which

$$D = -\begin{bmatrix} \overline{B}_{11} & -I & & & & & \\ & A_1 & \overline{B}_{11} & -I & & & \\ & & & \ddots & & & \\ & & & & A_1 & \overline{B}_{11} & -I \end{bmatrix}$$

$$d = \begin{bmatrix} A_1 x_1(0) + \overline{B}_{12} u_2(0) \\ \overline{B}_{12} u_2(1) \\ \vdots \\ \overline{B}_{12} u_2(N-1) \end{bmatrix}$$

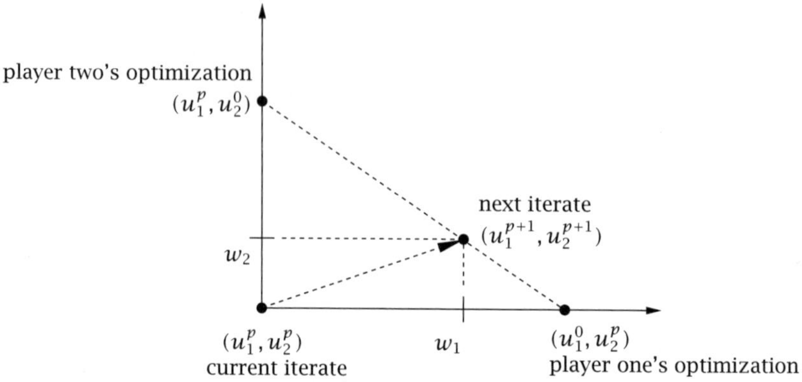

Figure 6.1: Convex step from (u_1^p, u_2^p) to (u_1^{p+1}, u_2^{p+1}); the parameters w_1, w_2 with $w_1 + w_2 = 1$ determine location of next iterate on line joining the two players' optimizations: (u_1^0, u_2^p) and (u_1^p, u_2^0).

We then apply (1.58) to obtain

$$\begin{bmatrix} H & -D' \\ -D & 0 \end{bmatrix} \begin{bmatrix} \mathbf{z} \\ \boldsymbol{\lambda} \end{bmatrix} = \begin{bmatrix} 0 \\ -\widetilde{A}_1 \end{bmatrix} x_1(0) + \begin{bmatrix} 0 \\ -\widetilde{B}_{12} \end{bmatrix} \mathbf{u}_2 \tag{6.12}$$

in which we have defined

$$\boldsymbol{\lambda} = \begin{bmatrix} \lambda(1) \\ \lambda(2) \\ \vdots \\ \lambda(N) \end{bmatrix} \qquad \widetilde{A}_1 = \begin{bmatrix} A_1 \\ 0 \\ \vdots \\ 0 \end{bmatrix} \qquad \widetilde{B}_{12} = \begin{bmatrix} \overline{B}_{12} & & & \\ & \overline{B}_{12} & & \\ & & \ddots & \\ & & & \overline{B}_{12} \end{bmatrix}$$

Solving this equation and picking out the rows of $\mathbf{z}$ corresponding to the elements of $\mathbf{u}_1$ gives

$$\mathbf{u}_1^0 = K_1 x_1(0) + L_1 \mathbf{u}_2$$

and we see player one's optimal decision depends linearly on his initial state, but also on *player two's decision.* This is the key difference between decentralized control and noncooperative control. In noncooperative control, player two's decisions are communicated to player one and player one accounts for them in optimizing the local objective.

Convex step. Let $p \in \mathbb{I}_{\geq 0}$ denote the integer-valued iteration in the optimization problem. Looking ahead to the M-player game, we do not take the full step, but a convex combination of the the current optimal solution, $\mathbf{u}_1^0$, and the current iterate, $\mathbf{u}_1^p$

$$\mathbf{u}_1^{p+1} = w_1 \mathbf{u}_1^0 + (1 - w_1)\mathbf{u}_1^p \qquad 0 < w_1 < 1$$

This iteration is displayed in Figure 6.1. Notice we have chosen a distributed optimization of the Gauss-Jacobi type (see Bertsekas and Tsitsiklis, 1997, pp.219–223).

We place restrictions on the systems under consideration before analyzing stability of the controller.

Assumption 6.7 (Unconstrained two-player game).

(a) All subsystems, $A_{ij}, i = 1,2, j = 1,2$, are stable.

(b) The controller penalties Q_1, Q_2, R_1, R_2 are positive definite.

The assumption of stable models is purely for convenience of exposition. We treat unstable, stabilizable systems in Section 6.3.

Convergence of the players' iteration. To understand the convergence of the two players' iterations, we express both players' moves as follows

$$\begin{aligned}
\mathbf{u}_1^{p+1} &= w_1 \mathbf{u}_1^0 + (1 - w_1)\mathbf{u}_1^p \\
\mathbf{u}_2^{p+1} &= w_2 \mathbf{u}_2^0 + (1 - w_2)\mathbf{u}_2^p \\
1 &= w_1 + w_2 \qquad 0 < w_1, w_2 < 1
\end{aligned}$$

or

$$\begin{bmatrix} \mathbf{u}_1 \\ \mathbf{u}_2 \end{bmatrix}^{p+1} = \begin{bmatrix} w_1 I & 0 \\ 0 & w_2 I \end{bmatrix} \begin{bmatrix} \mathbf{u}_1^0 \\ \mathbf{u}_2^0 \end{bmatrix} + \begin{bmatrix} (1-w_1)I & 0 \\ 0 & (1-w_2)I \end{bmatrix} \begin{bmatrix} \mathbf{u}_1 \\ \mathbf{u}_2 \end{bmatrix}^p$$

The optimal control for each player is

$$\begin{bmatrix} \mathbf{u}_1^0 \\ \mathbf{u}_2^0 \end{bmatrix} = \begin{bmatrix} K_1 & 0 \\ 0 & K_2 \end{bmatrix} \begin{bmatrix} x_1(0) \\ x_2(0) \end{bmatrix} + \begin{bmatrix} 0 & L_1 \\ L_2 & 0 \end{bmatrix} \begin{bmatrix} \mathbf{u}_1 \\ \mathbf{u}_2 \end{bmatrix}^p$$

Substituting the optimal control into the iteration gives

$$\begin{bmatrix} \mathbf{u}_1 \\ \mathbf{u}_2 \end{bmatrix}^{p+1} = \underbrace{\begin{bmatrix} w_1 K_1 & 0 \\ 0 & w_2 K_2 \end{bmatrix}}_{\overline{K}} \begin{bmatrix} x_1(0) \\ x_2(0) \end{bmatrix} + \underbrace{\begin{bmatrix} (1-w_1)I & w_1 L_1 \\ w_2 L_2 & (1-w_2)I \end{bmatrix}}_{L} \begin{bmatrix} \mathbf{u}_1 \\ \mathbf{u}_2 \end{bmatrix}^p$$

Finally writing this equation in the plantwide notation, we express the iteration as

$$\mathbf{u}^{p+1} = \overline{K}x(0) + L\mathbf{u}^p$$

The convergence of the two players' control iteration is governed by the eigenvalues of L. If L is stable, the control sequence converges to

$$\mathbf{u}^{\infty} = (I - L)^{-1}\overline{K}x(0) \qquad |\lambda| < 1 \text{ for } \lambda \in \text{eig}(L)$$

in which

$$(I - L)^{-1}\overline{K} = \begin{bmatrix} w_1 I & -w_1 L_1 \\ -w_2 L_2 & w_2 I \end{bmatrix}^{-1} \begin{bmatrix} w_1 K_1 & 0 \\ 0 & w_2 K_2 \end{bmatrix}$$

$$(I - L)^{-1}\overline{K} = \begin{bmatrix} I & -L_1 \\ -L_2 & I \end{bmatrix}^{-1} \begin{bmatrix} K_1 & 0 \\ 0 & K_2 \end{bmatrix}$$

Note that the weights w_1, w_2 do not appear in the converged input sequence. The $\mathbf{u}_1^{\infty}$, $\mathbf{u}_2^{\infty}$ pair have the equilibrium property that neither player can improve his position given the other player's current decision. This point is called a Nash equilibrium (Başar and Olsder, 1999, p. 4). Notice that the distributed MPC game does not have a Nash equilibrium if the eigenvalues of L are on or outside the unit circle. If the controllers have sufficient time during the control system's sample time to iterate to convergence, then the effect of the initial control sequence is removed by using the converged control sequence. If the iteration has to be stopped before convergence, the solution is

$$\mathbf{u}^{p+1} = L^p \mathbf{u}^{[0]} + \sum_{j=0}^{p-1} L^j \overline{K}x(0) \qquad 0 \leq p$$

in which $\mathbf{u}^{[0]}$ is the $p = 0$ (initial) input sequence. We use the brackets with $p = 0$ to distinguish this initial input sequence from an optimal input sequence.

Stability of the closed-loop system. We assume the Nash equilibrium is stable and there is sufficient computation time to iterate to convergence.

We require a matrix of zeros and ones to select the first move from the input sequence for injection into the plant. For the first player, the required matrix is

$$u_1(0) = E_1 \mathbf{u}_1$$
$$E_1 = \begin{bmatrix} I_{m_1} & 0_{m_1} & \dots & 0_{m_1} \end{bmatrix} \qquad m_1 \times m_1 N \text{ matrix}$$

The closed-loop system is then

$$\begin{bmatrix} x_1 \\ x_2 \end{bmatrix}^+ = \underbrace{\begin{bmatrix} A_1 & 0 \\ 0 & A_2 \end{bmatrix}}_{A} \begin{bmatrix} x_1 \\ x_2 \end{bmatrix} +$$
$$\underbrace{\begin{bmatrix} \overline{B}_{11} & \overline{B}_{12} \\ \overline{B}_{21} & \overline{B}_{22} \end{bmatrix}}_{B} \underbrace{\begin{bmatrix} E_1 & 0 \\ 0 & E_2 \end{bmatrix} \begin{bmatrix} I & -L_1 \\ -L_2 & I \end{bmatrix}^{-1} \begin{bmatrix} K_1 & 0 \\ 0 & K_2 \end{bmatrix}}_{K} \begin{bmatrix} x_1 \\ x_2 \end{bmatrix}$$

Using the plantwide notation for this equation and defining the feedback gain K gives

$$x^+ = (A + BK)x$$

The stability of the closed loop with converged, noncooperative control is therefore determined by the eigenvalues of $(A + BK)$.

We next present three simple examples to show that (i) the Nash equilibrium may not be stable (L is unstable), (ii) the Nash equilibrium may be stable but the closed loop is unstable (L is stable, $A + BK$ is unstable), and (iii) the Nash equilibrium may be stable and the closed loop is stable (L is stable, $A + BK$ is stable). Which situation arises depends in a nonobvious way on all of the problem data: A_1, A_2, $\overline{B}_{11}$, $\overline{B}_{12}$, $\overline{B}_{21}$, $\overline{B}_{22}$, Q_1, Q_2, P_{1f}, P_{2f}, R_1, R_2, w_1, w_2, N. One has to examine the eigenvalues of L and $A + BK$ for each application of interest to know how the noncooperative distributed MPC is going to perform. Even for a fixed dynamic model, when changing tuning parameters such as Q, P_f, R, w, one has to examine eigenvalues of L and $A + BK$ to know the effect on the closed-loop system. This is the main drawback of the noncooperative game. In many control system design methods, such as all forms of MPC presented in Chapter 2, closed-loop properties such as exponential stability are guaranteed for the *nominal* system for all choices of performance tuning parameters. Noncooperative distributed MPC does not have this feature and a stability analysis is required. We show in the next section that cooperative MPC does not suffer from this drawback, at the cost of slightly more information exchange.

Example 6.8: Nash equilibrium is unstable

Consider the following transfer function matrix for a simple two-input two-output system

$$\begin{bmatrix} y_1(s) \\ y_2(s) \end{bmatrix} = \begin{bmatrix} G_{11}(s) & G_{12}(s) \\ G_{21}(s) & G_{22}(s) \end{bmatrix} \begin{bmatrix} u_1(s) \\ u_2(s) \end{bmatrix}$$

in which

$$G(s) = \begin{bmatrix} \dfrac{1}{s^2 + 2(0.2)s + 1} & \dfrac{0.5}{0.225s + 1} \\ \dfrac{-0.5}{(0.5s + 1)(0.25s + 1)} & \dfrac{1.5}{0.75s^2 + 2(0.8)(0.75)s + 1} \end{bmatrix}$$

Obtain discrete time models (A_{ij}, B_{ij}, C_{ij}) for each of the four transfer functions $G_{ij}(s)$ using a sample time of $T = 0.2$ and zero-order holds on the inputs. Set the control cost function parameters to be

$$\overline{Q}_1 = \overline{Q}_2 = 1 \qquad \overline{P}_{1f} = \overline{P}_{2f} = 0 \qquad R_1 = R_2 = 0.01$$
$$N = 30 \qquad w_1 = w_2 = 0.5$$

Compute the eigenvalues of the L matrix for this system using noncooperative MPC. Show the Nash equilibrium is unstable and the closed-loop system is therefore unstable. Discuss why this system is problematic for noncooperative control.

Solution

For this problem L is a 60×60 matrix $(N(m_1 + m_2))$. The magnitudes of the largest eigenvalues are

$$\left|\text{eig}(L)\right| = \begin{bmatrix} 1.11 & 1.11 & 1.03 & 1.03 & 0.914 & 0.914 & \cdots \end{bmatrix}$$

The noncooperative iteration does not converge. The steady-state gains for this system are

$$G(0) = \begin{bmatrix} 1 & 0.5 \\ -0.5 & 1.5 \end{bmatrix}$$

and we see that the diagonal elements are reasonably large compared to the nondiagonal elements. So the *steady-state* coupling between the two systems is relatively weak. The dynamic coupling is unfavorable, however. The response of y_1 to u_2 is more than four times faster than the response of y_1 to u_1. The faster input is the disturbance and the slower input is used for control. Likewise the response of y_2 to u_1 is three times faster than the response of y_2 to u_2. Also in the second loop, the faster input is the disturbance and the slower input is used for control. These pairings are unfavorable dynamically, and that fact is revealed in the instability of L and lack of a Nash equilibrium for the noncooperative dynamic regulation problem. □

Example 6.9: Nash equilibrium is stable but closed loop is unstable

Switch the outputs for the previous example and compute the eigenvalues of L and $(A+BK)$ for the noncooperative distributed MPC regulator for the system

$$G(s) = \begin{bmatrix} \dfrac{-0.5}{(0.5s+1)(0.25s+1)} & \dfrac{1.5}{0.75s^2+2(0.8)(0.75)s+1} \\ \dfrac{1}{s^2+2(0.2)s+1} & \dfrac{0.5}{0.225s+1} \end{bmatrix}$$

Show in this case that the Nash equilibrium is stable, but the noncooperative regulator destabilizes the system. Discuss why this system is problematic for noncooperative control.

Solution

For this case the largest magnitude eigenvalues of L are

$$|\operatorname{eig}(L)| = \begin{bmatrix} 0.63 & 0.63 & 0.62 & 0.62 & 0.59 & 0.59 & \cdots \end{bmatrix}$$

and we see the Nash equilibrium for the noncooperative game is stable. So we have removed the first source of closed-loop instability by switching the input-output pairings of the two subsystems. There are seven states in the complete system model, and the magnitudes of the eigenvalues of the closed-loop regulator $(A+BK)$ are

$$|\operatorname{eig}(A+BK)| = \begin{bmatrix} 1.03 & 1.03 & 0.37 & 0.37 & 0.77 & 0.77 & 0.04 \end{bmatrix}$$

which also gives an unstable closed-loop system. We see the distributed noncooperative regulator has destabilized a stable open-loop system. The problem with this pairing is the steady-state gains are now

$$G(0) = \begin{bmatrix} -0.5 & 1.5 \\ 1 & 0.5 \end{bmatrix}$$

If one computes any steady-state interaction measure, such as the relative gain array (RGA), we see the new pairings are poor from a steady-state interaction perspective

$$\text{RGA} = \begin{bmatrix} 0.14 & 0.86 \\ 0.86 & 0.14 \end{bmatrix}$$

Neither pairing of the inputs and outputs is closed-loop stable with noncooperative distributed MPC.

Decentralized control with this pairing is discussed in Exercise 6.10. □

Example 6.10: Nash equilibrium is stable and the closed loop is stable

Next consider the system

$$G(s) = \begin{bmatrix} \dfrac{1}{s^2 + 2(0.2)s + 1} & \dfrac{0.5}{0.9s + 1} \\ \dfrac{-0.5}{(2s+1)(s+1)} & \dfrac{1.5}{0.75s^2 + 2(0.8)(0.75)s + 1} \end{bmatrix}$$

Compute the eigenvalues of L and $A + BK$ for this system. What do you conclude about noncooperative distributed MPC for this system?

Solution

This system is not difficult to handle with distributed control. The gains are the same as in the original pairing in Example 6.8, and the steady-state coupling between the two subsystems is reasonably weak. Unlike Example 6.8, however, the responses of y_1 to u_2 and y_2 to u_1 have been slowed so they are not faster than the responses of y_1 to u_1 and y_2 to u_2, respectively. Computing the eigenvalues of L and $A + BK$ for noncooperative control gives

$$\left|\text{eig}(L)\right| = \begin{bmatrix} 0.61 & 0.61 & 0.59 & 0.59 & 0.56 & 0.56 & 0.53 & 0.53 \cdots \end{bmatrix}$$

$$\left|\text{eig}(A + BK)\right| = \begin{bmatrix} 0.88 & 0.88 & 0.74 & 0.67 & 0.67 & 0.53 & 0.53 \end{bmatrix}$$

The Nash equilibrium is stable since L is stable, and the closed loop is stable since both L and $A + BK$ are stable. □

These examples reveal the simple fact that communicating the actions of the other controllers does not guarantee acceptable closed-loop behavior. If the coupling of the subsystems is weak enough, both dynamically and in steady state, then the closed loop is stable. In this sense, noncooperative MPC has few advantages over completely decentralized control, which has this same basic property.

We next show how to obtain much better closed-loop properties while maintaining the small size of the distributed control problems.

6.2.4 Cooperative Game

In the cooperative game, the two players share a common objective, which can be considered to be the overall plant objective

$$V(x_1(0), x_2(0), \mathbf{u}_1, \mathbf{u}_2) = \rho_1 V_1(x_1(0), \mathbf{u}_1, \mathbf{u}_2) + \rho_2 V_2(x_2(0), \mathbf{u}_2, \mathbf{u}_1)$$

in which the parameters ρ_1, ρ_2 are used to specify the relative weights of the two subsystems in the overall plant objective. In the cooperative problem, each player keeps track of *how his input affects the other player's output* as well as his own output. We can implement this cooperative game in several ways. The implementation leading to the simplest notation is to combine x_1 and x_2 into a single model

$$\begin{bmatrix} x_1 \\ x_2 \end{bmatrix}^+ = \begin{bmatrix} A_1 & 0 \\ 0 & A_2 \end{bmatrix} \begin{bmatrix} x_1 \\ x_2 \end{bmatrix} + \begin{bmatrix} \overline{B}_{11} \\ \overline{B}_{21} \end{bmatrix} u_1 + \begin{bmatrix} \overline{B}_{12} \\ \overline{B}_{22} \end{bmatrix} u_2$$

and then express player one's stage cost as

$$\ell_1(x_1, x_2, u_1) = \frac{1}{2} \begin{bmatrix} x_1 \\ x_2 \end{bmatrix}' \begin{bmatrix} \rho_1 Q_1 & 0 \\ 0 & \rho_2 Q_2 \end{bmatrix} \begin{bmatrix} x_1 \\ x_2 \end{bmatrix} + \frac{1}{2} u_1'(\rho_1 R_1) u_1 + \text{const.}$$

$$V_{1f}(x_1, x_2) = \frac{1}{2} \begin{bmatrix} x_1 \\ x_2 \end{bmatrix}' \begin{bmatrix} \rho_1 P_{1f} & 0 \\ 0 & \rho_2 P_{2f} \end{bmatrix} \begin{bmatrix} x_1 \\ x_2 \end{bmatrix}$$

Notice that u_2 does not appear because the contribution of u_2 to the stage cost cannot be affected by player one, and can therefore be neglected. The cost function is then expressed as

$$V(x_1(0), x_2(0), \mathbf{u}_1, \mathbf{u}_2) = \sum_{k=0}^{N-1} \ell_1(x_1(k), x_2(k), u_1(k)) + V_{1f}(x_1(N), x_2(N))$$

Player one's optimal control problem is

$$\min_{\mathbf{u}_1} V(x_1(0), x_2(0), \mathbf{u}_1, \mathbf{u}_2)$$

$$\text{s.t.} \quad \begin{bmatrix} x_1 \\ x_2 \end{bmatrix}^+ = \begin{bmatrix} A_1 & 0 \\ 0 & A_2 \end{bmatrix} \begin{bmatrix} x_1 \\ x_2 \end{bmatrix} + \begin{bmatrix} \overline{B}_{11} \\ \overline{B}_{21} \end{bmatrix} u_1 + \begin{bmatrix} \overline{B}_{12} \\ \overline{B}_{22} \end{bmatrix} u_2$$

Note that this form is identical to the noncooperative form presented previously if we redefine the terms (noncooperative $\longrightarrow$ cooperative)

$$x_1 \to \begin{bmatrix} x_1 \\ x_2 \end{bmatrix} \quad A_1 \to \begin{bmatrix} A_1 & 0 \\ 0 & A_2 \end{bmatrix} \quad \overline{B}_{11} \to \begin{bmatrix} \overline{B}_{11} \\ \overline{B}_{21} \end{bmatrix} \quad \overline{B}_{12} \to \begin{bmatrix} \overline{B}_{12} \\ \overline{B}_{22} \end{bmatrix}$$

$$Q_1 \to \begin{bmatrix} \rho_1 Q_1 & 0 \\ 0 & \rho_2 Q_2 \end{bmatrix} \qquad R_1 \to \rho_1 R_1 \quad P_{1f} \to \begin{bmatrix} \rho_1 P_{1f} & 0 \\ 0 & \rho_2 P_{2f} \end{bmatrix}$$

Any computational program written to solve either the cooperative or noncooperative optimal control problem can be used to solve the other.

Eliminating states $\mathbf{x}_2$. An alternative implementation is to remove states $x_2(k), k \geq 1$ from player one's optimal control problem by substituting the dynamic model of system two. This implementation reduces the size of the dynamic model because only states x_1 are retained. This reduction in model size may be important in applications with many players. The removal of states $x_2(k), k \geq 1$ also introduces linear terms into player one's objective function. We start by using the dynamic model for x_2 to obtain

$$\begin{bmatrix} x_2(1) \\ x_2(2) \\ \vdots \\ x_2(N) \end{bmatrix} = \begin{bmatrix} A_2 \\ A_2^2 \\ \vdots \\ A_2^N \end{bmatrix} x_2(0) + \begin{bmatrix} \overline{B}_{21} & & & \\ A_2\overline{B}_{21} & \overline{B}_{21} & & \\ \vdots & \vdots & \ddots & \\ A_2^{N-1}\overline{B}_{21} & A_2^{N-2}\overline{B}_{21} & \dots & \overline{B}_{21} \end{bmatrix} \begin{bmatrix} u_1(0) \\ u_1(1) \\ \vdots \\ u_1(N-1) \end{bmatrix} + \begin{bmatrix} \overline{B}_{22} & & & \\ A_2\overline{B}_{22} & \overline{B}_{22} & & \\ \vdots & \vdots & \ddots & \\ A_2^{N-1}\overline{B}_{22} & A_2^{N-2}\overline{B}_{22} & \dots & \overline{B}_{22} \end{bmatrix} \begin{bmatrix} u_2(0) \\ u_2(1) \\ \vdots \\ u_2(N-1) \end{bmatrix}$$

Using more compact notation, we have

$$\mathbf{x}_2 = \mathcal{A}_2 x_2(0) + \mathcal{B}_{21}\mathbf{u}_1 + \mathcal{B}_{22}\mathbf{u}_2$$

We can use this relation to replace the cost contribution of $\mathbf{x}_2$ with linear and quadratic terms in $\mathbf{u}_1$ as follows

$$\sum_{k=0}^{N-1} x_2(k)' Q_2 x_2(k) + x_2(N)' P_{2f} x_2(N) =$$

$$\mathbf{u}_1' \left[\mathcal{B}_{21}' \mathcal{Q}_2 \mathcal{B}_{21}\right] \mathbf{u}_1 + 2\left[x_2(0)'\mathcal{A}_2' + \mathbf{u}_2'\mathcal{B}_{22}'\right] \mathcal{Q}_2 \mathcal{B}_{21}\, \mathbf{u}_1 + \text{constant}$$

in which

$$\mathcal{Q}_2 = \text{diag}\left(\begin{bmatrix} Q_2 & Q_2 & \dots & P_{2f} \end{bmatrix}\right) \qquad Nn_2 \times Nn_2 \text{ matrix}$$

and the constant term contains products of $x_2(0)$ and $\mathbf{u}_2$, which are constant with respect to player one's decision variables and can therefore be neglected.

Next we insert the new terms created by eliminating $\mathbf{x}_2$ into the cost function. Assembling the cost function gives

$$\min_{\mathbf{z}} (1/2)\mathbf{z}'\widetilde{H}\mathbf{z} + h'\mathbf{z}$$
$$\text{s.t. } D\mathbf{z} = d$$

and (1.58) again gives the necessary and sufficient conditions for the optimal solution

$$\begin{bmatrix} \widetilde{H} & -D' \\ -D & 0 \end{bmatrix} \begin{bmatrix} \mathbf{z} \\ \boldsymbol{\lambda} \end{bmatrix} = \begin{bmatrix} 0 \\ -\widetilde{A}_1 \end{bmatrix} x_1(0) + \begin{bmatrix} -\widetilde{A}_2 \\ 0 \end{bmatrix} x_2(0) + \begin{bmatrix} -\widetilde{B}_{22} \\ -\widetilde{B}_{12} \end{bmatrix} \mathbf{u}_2 \quad (6.13)$$

in which

$$\widetilde{H} = H + E'\mathcal{B}_{21}'\mathcal{Q}_2\mathcal{B}_{21}E \qquad \widetilde{B}_{22} = E'\mathcal{B}_{21}'\mathcal{Q}_2\mathcal{B}_{22} \qquad \widetilde{A}_2 = E'\mathcal{B}_{21}'\mathcal{Q}_2\mathcal{A}_2$$
$$E = I_N \otimes \begin{bmatrix} I_{m_1} & 0_{m_1,n_1} \end{bmatrix}$$

See also Exercise 6.13 for details on constructing the padding matrix E. Comparing the cooperative and noncooperative dynamic games, (6.13) and (6.12), we see the cooperative game has made three changes: (i) the quadratic penalty H has been modified, (ii) the effect of $x_2(0)$ has been included with the term $\widetilde{A}_2$, and (iii) the influence of $\mathbf{u}_2$ has been modified with the term $\widetilde{B}_{22}$. Notice that the size of the vector $\mathbf{z}$ has not changed, and we have accomplished the goal of keeping player one's dynamic model in the cooperative game the same size as his dynamic model in the noncooperative game.

Regardless of the implementation choice, the cooperative optimal control problem is no more complex than the noncooperative game considered previously. The extra information required by player one in the cooperative game is $x_2(0)$. Player one requires $\mathbf{u}_2$ in both the cooperative and noncooperative games. Only in decentralized control does player one not require player two's input sequence $\mathbf{u}_2$. The other extra required information, $A_2, B_{21}, Q_2, R_2, P_{2f}$, are fixed parameters and making their values available to player one is a minor communication overhead.

Proceeding as before, we solve this equation for $\mathbf{z}^0$ and pick out the rows corresponding to the elements of $\mathbf{u}_1^0$ giving

$$\mathbf{u}_1^0(x(0), \mathbf{u}_2) = \begin{bmatrix} K_{11} & K_{12} \end{bmatrix} \begin{bmatrix} x_1(0) \\ x_2(0) \end{bmatrix} + L_1\mathbf{u}_2$$

Combining the optimal control laws for each player gives

$$\begin{bmatrix} \mathbf{u}_1^0 \\ \mathbf{u}_2^0 \end{bmatrix} = \begin{bmatrix} K_{11} & K_{12} \\ K_{21} & K_{22} \end{bmatrix} \begin{bmatrix} x_1(0) \\ x_2(0) \end{bmatrix} + \begin{bmatrix} 0 & L_1 \\ L_2 & 0 \end{bmatrix} \begin{bmatrix} \mathbf{u}_1 \\ \mathbf{u}_2 \end{bmatrix}^p$$

in which the gain matrix multiplying the state is a full matrix for the cooperative game. Substituting the optimal control into the iteration gives

$$\begin{bmatrix} \mathbf{u}_1 \\ \mathbf{u}_2 \end{bmatrix}^{p+1} = \underbrace{\begin{bmatrix} w_1 K_{11} & w_1 K_{12} \\ w_2 K_{21} & w_2 K_{22} \end{bmatrix}}_{\overline{K}} \begin{bmatrix} x_1(0) \\ x_2(0) \end{bmatrix} + \underbrace{\begin{bmatrix} (1-w_1)I & w_1 L_1 \\ w_2 L_2 & (1-w_2)I \end{bmatrix}}_{L} \begin{bmatrix} \mathbf{u}_1 \\ \mathbf{u}_2 \end{bmatrix}^p$$

Finally writing this equation in the plantwide notation, we express the iteration as

$$\mathbf{u}^{p+1} = \overline{K}x(0) + L\mathbf{u}^p$$

Exponential stability of the closed-loop system. In the case of cooperative control, we consider the closed-loop system with a finite number of iterations, p. With finite iterations, distributed MPC becomes a form of *suboptimal* MPC as discussed in Sections 6.1.2 and 2.8. To analyze the behavior of the cooperative controller with a finite number of iterations, we require the cost decrease achieved by a single iteration, which we derive next. First we write the complete system evolution as in (6.10)

$$x^+ = Ax + Bu$$

in which A and B are defined in (6.11). We can then use (6.3) to express the overall cost function

$$V(x(0), \mathbf{u}) = (1/2)x'(0)(Q + \mathcal{A}'\mathcal{Q}\mathcal{A})x(0) + \mathbf{u}'(\mathcal{B}'\mathcal{Q}\mathcal{A})x(0) + (1/2)\mathbf{u}'H_{\mathbf{u}}\mathbf{u}$$

in which $\mathcal{A}$ and $\mathcal{B}$ are given in (6.1), the cost penalties $\mathcal{Q}$ and $\mathcal{R}$ are given in (6.2) and (6.9), and

$$H_{\mathbf{u}} = \mathcal{B}'\mathcal{Q}\mathcal{B} + \mathcal{R}$$

The overall cost is a positive definite quadratic function in $\mathbf{u}$ because R_1 and R_2 are positive definite, and therefore so are $\mathcal{R}_1$, $\mathcal{R}_2$, and $\mathcal{R}$.

The iteration in the two players' moves satisfies

$$\begin{aligned} (\mathbf{u}_1^{p+1}, \mathbf{u}_2^{p+1}) &= \Big((w_1\mathbf{u}_1^0 + (1-w_1)\mathbf{u}_1^p), (w_2\mathbf{u}_2^0 + (1-w_2)\mathbf{u}_2^p)\Big) \\ &= (w_1\mathbf{u}_1^0, (1-w_2)\mathbf{u}_2^p) + ((1-w_1)\mathbf{u}_1^p, w_2\mathbf{u}_2^0) \\ (\mathbf{u}_1^{p+1}, \mathbf{u}_2^{p+1}) &= w_1(\mathbf{u}_1^0, \mathbf{u}_2^p) + w_2(\mathbf{u}_1^p, \mathbf{u}_2^0) \end{aligned} \tag{6.14}$$

Exercise 6.18 analyzes the cost decrease for a convex step with a positive definite quadratic function and shows

$$V(x(0), \mathbf{u}_1^{p+1}, \mathbf{u}_2^{p+1}) = V(x(0), \mathbf{u}_1^p, \mathbf{u}_2^p) - \frac{1}{2}\left[\mathbf{u}^p - \mathbf{u}^0(x(0))\right]' P \left[\mathbf{u}^p - \mathbf{u}^0(x(0))\right] \quad (6.15)$$

in which $P > 0$ is given by

$$P = H_{\mathbf{u}} D^{-1} \widetilde{H} D^{-1} H_{\mathbf{u}} \qquad \widetilde{H} = D - N$$

$$D = \begin{bmatrix} w_1^{-1} H_{\mathbf{u},11} & 0 \\ 0 & w_2^{-1} H_{\mathbf{u},22} \end{bmatrix} \qquad N = \begin{bmatrix} -w_1^{-1} w_2 H_{\mathbf{u},11} & H_{\mathbf{u},12} \\ H_{\mathbf{u},21} & -w_1 w_2^{-1} H_{\mathbf{u},22} \end{bmatrix}$$

and $H_{\mathbf{u}}$ is partitioned for the two players' input sequences. Notice that the cost decrease achieved in a single iteration is quadratic in the distance from the optimum. An important conclusion is that *each iteration in the cooperative game reduces the systemwide cost.* This cost reduction is the key property that gives cooperative MPC its excellent convergence properties, as we show next.

The two players' warm starts at the next sample are given by

$$\widetilde{\mathbf{u}}_1^+ = \{u_1(1), u_1(2), \ldots, u_1(N-1), 0\}$$
$$\widetilde{\mathbf{u}}_2^+ = \{u_2(1), u_2(2), \ldots, u_2(N-1), 0\}$$

We define the following linear time-invariant functions g_1^p and g_2^p as the outcome of applying the control iteration procedure p times

$$\mathbf{u}_1^p = g_1^p(x_1, x_2, \mathbf{u}_1, \mathbf{u}_2)$$
$$\mathbf{u}_2^p = g_2^p(x_1, x_2, \mathbf{u}_1, \mathbf{u}_2)$$

in which $p \geq 0$ is an integer, x_1 and x_2 are the states, and $\mathbf{u}_1, \mathbf{u}_2$ are the input sequences from the previous sample, used to generate the warm start for the iteration. Here we consider p to be constant with time, but Exercise 6.20 considers the case in which the controller iterations may vary with sample time. The system evolution is then given by

$$x_1^+ = A_1 x_1 + \overline{B}_{11} u_1 + \overline{B}_{12} u_2 \qquad x_2^+ = A_2 x_2 + \overline{B}_{21} u_1 + \overline{B}_{22} u_2$$
$$\mathbf{u}_1^+ = g_1^p(x_1, x_2, \mathbf{u}_1, \mathbf{u}_2) \qquad \mathbf{u}_2^+ = g_2^p(x_1, x_2, \mathbf{u}_1, \mathbf{u}_2) \quad (6.16)$$

By the construction of the warm start, $\widetilde{\mathbf{u}}_1^+, \widetilde{\mathbf{u}}_2^+$, we have

$$V(x_1^+, x_2^+, \widetilde{\mathbf{u}}_1^+, \widetilde{\mathbf{u}}_2^+) = V(x_1, x_2, \mathbf{u}_1, \mathbf{u}_2) - \rho_1 \ell_1(x_1, u_1) - \rho_2 \ell_2(x_2, u_2)$$
$$+ x_1(N)' \left[A_1' P_{1f} A_1 - P_{1f} + Q_1\right] x_1(N)$$
$$+ x_2(N)' \left[A_2' P_{2f} A_2 - P_{2f} + Q_2\right] x_2(N)$$

From our choice of terminal penalty satisfying (6.8), the last two terms are zero giving

$$V(x_1^+, x_2^+, \tilde{\mathbf{u}}_1^+, \tilde{\mathbf{u}}_2^+) = V(x_1, x_2, \mathbf{u}_1, \mathbf{u}_2) - \rho_1 \ell_1(x_1, u_1) - \rho_2 \ell_2(x_2, u_2) \quad (6.17)$$

No optimization, $p = 0$. If we do no further optimization, then we have $\mathbf{u}_1^+ = \tilde{\mathbf{u}}_1^+$, $\mathbf{u}_2^+ = \tilde{\mathbf{u}}_2^+$, and the equality

$$V(x_1^+, x_2^+, \mathbf{u}_1^+, \mathbf{u}_2^+) = V(x_1, x_2, \mathbf{u}_1, \mathbf{u}_2) - \rho_1 \ell_1(x_1, u_1) - \rho_2 \ell_2(x_2, u_2)$$

The input sequences add a zero at each sample until $\mathbf{u}_1 = \mathbf{u}_2 = 0$ at time $k = N$. The system decays exponentially under zero control and the closed loop is exponentially stable.

Further optimization, $p \geq 1$. We next consider the case in which optimization is performed. Equation 6.15 then gives

$$V(x_1^+, x_2^+, \mathbf{u}_1^+, \mathbf{u}_2^+) \leq V(x_1^+, x_2^+, \tilde{\mathbf{u}}_1^+, \tilde{\mathbf{u}}_2^+) - \left[\tilde{\mathbf{u}}^+ - \mathbf{u}^0(x^+)\right]' P \left[\tilde{\mathbf{u}}^+ - \mathbf{u}^0(x^+)\right] \qquad p \geq 1$$

with equality holding for $p = 1$. Using this result in (6.17) gives

$$V(x_1^+, x_2^+, \mathbf{u}_1^+, \mathbf{u}_2^+) \leq V(x_1, x_2, \mathbf{u}_1, \mathbf{u}_2) - \rho_1 \ell_1(x_1, u_1) - \rho_2 \ell_2(x_2, u_2) - \left[\tilde{\mathbf{u}}^+ - \mathbf{u}^0(x^+)\right]' P \left[\tilde{\mathbf{u}}^+ - \mathbf{u}^0(x^+)\right]$$

Since V is bounded below by zero and ℓ_1 and ℓ_2 are positive functions, we conclude the time sequence $V(x_1(k), x_2(k), \mathbf{u}_1(k), \mathbf{u}_2(k))$ converges. and therefore $x_1(k)$, $x_2(k)$, $u_1(k)$, and $u_2(k)$ converge to zero. Moreover, since $P > 0$, the last term implies that $\tilde{\mathbf{u}}^+$ converges to $\mathbf{u}^0(x^+)$, which converges to zero because x^+ converges to zero. Therefore, the entire input sequence $\mathbf{u}$ converges to zero. Because the total system evolution is a linear time-invariant system, the convergence is exponential. Even though we are considering here a form of *suboptimal* MPC, we do not require an additional inequality constraint on $\mathbf{u}$ because the problem considered here is *unconstrained* and the iterations satisfy (6.15).

6.2.5 Tracking Nonzero Setpoints

For tracking nonzero setpoints, we compute steady-state targets as discussed in Section 1.5. The steady-state input-output model is given by

$$y_s = G u_s \qquad G = C(I - A)^{-1} B$$

in which G is the steady-state gain of the system. The two subsystems are denoted

$$\begin{bmatrix} y_{1s} \\ y_{2s} \end{bmatrix} = \begin{bmatrix} G_{11} & G_{12} \\ G_{21} & G_{22} \end{bmatrix} \begin{bmatrix} u_{1s} \\ u_{2s} \end{bmatrix}$$

For simplicity, we assume that the targets are chosen to be the measurements ($H = I$). Further, we assume that both local systems are square, and that the local targets can be reached exactly with the local inputs. This assumption means that G_{11} and G_{22} are square matrices of full rank. We remove all of these assumptions when we treat the constrained two-player game in the next section. If there is model error, integrating disturbance models are required as discussed in Chapter 1. We discuss these later.

The target problem also can be solved with any of the four approaches discussed so far. We consider each.

Centralized case. The centralized problem gives in one shot both inputs required to meet both output setpoints

$$u_s = G^{-1} y_{\text{sp}}$$
$$y_s = y_{\text{sp}}$$

Decentralized case. The decentralized problem considers only the diagonal terms and computes the following steady inputs

$$u_s = \begin{bmatrix} G_{11}^{-1} & \\ & G_{22}^{-1} \end{bmatrix} y_{\text{sp}}$$

Notice these inputs produce offset in both output setpoints

$$y_s = \begin{bmatrix} I & G_{12}G_{22}^{-1} \\ G_{21}G_{11}^{-1} & I \end{bmatrix} y_{\text{sp}}$$

Noncooperative case. In the noncooperative game, each player attempts to remove offset in only its outputs. Player one solves the following problem

$$\min_{u_1} (y_1 - y_{1sp})' \overline{Q}_1 (y_1 - y_{1sp})$$
$$\text{s.t. } y_1 = G_{11}u_1 + G_{12}u_2$$

Because the target can be reached exactly, the optimal solution is to find u_1 such that $y_1 = y_{1sp}$, which gives

$$u_{1s}^0 = G_{11}^{-1} \left(y_{1sp} - G_{12}u_2^p \right)$$

Player two solves the analogous problem. If we iterate on the two players' solutions, we obtain

$$\begin{bmatrix} u_{1s} \\ u_{2s} \end{bmatrix}^{p+1} = \underbrace{\begin{bmatrix} w_1 G_{11}^{-1} & \\ & w_2 G_{22}^{-1} \end{bmatrix}}_{\overline{K}_s} \begin{bmatrix} y_{1sp} \\ y_{2sp} \end{bmatrix} +$$
$$\underbrace{\begin{bmatrix} w_2 I & -w_1 G_{11}^{-1} G_{12} \\ -w_2 G_{22}^{-1} G_{21} & w_1 I \end{bmatrix}}_{L_s} \begin{bmatrix} u_{1s} \\ u_{2s} \end{bmatrix}^{p}$$

This iteration can be summarized by

$$u_s^{p+1} = \overline{K}_s y_{\text{sp}} + L_s u_s^p$$

If L_s is stable, this iteration converges to

$$\begin{aligned} u_s^{\infty} &= (I - L_s)^{-1} \overline{K}_s y_{\text{sp}} \\ u_s^{\infty} &= G^{-1} y_{\text{sp}} \end{aligned}$$

and we have no offset. We already have seen that we cannot expect the dynamic noncooperative iteration to converge. The next several examples explore the issue of whether we can expect at least the steady-state iteration to be stable.

Cooperative case. In the cooperative case, both players work on minimizing the offset in both outputs. Player one solves

$$\begin{aligned} \min_{u_1} (1/2) & \begin{bmatrix} y_1 - y_{1sp} \\ y_2 - y_{2sp} \end{bmatrix}' \begin{bmatrix} \rho_1 \overline{Q}_1 & \\ & \rho_2 \overline{Q}_2 \end{bmatrix} \begin{bmatrix} y_1 - y_{1sp} \\ y_2 - y_{2sp} \end{bmatrix} \\ \text{s.t.} & \begin{bmatrix} y_1 \\ y_2 \end{bmatrix} = \begin{bmatrix} G_{11} \\ G_{21} \end{bmatrix} u_1 + \begin{bmatrix} G_{12} \\ G_{22} \end{bmatrix} u_2 \end{aligned}$$

We can write this in the general form

$$\begin{aligned} &\min_{r_s} (1/2) r_s' H r_s + h' r_s \\ &\text{s.t. } D r_s = d \end{aligned}$$

in which

$$r_s = \begin{bmatrix} y_{1s} \\ y_{2s} \\ u_{1s} \end{bmatrix} \qquad H = \begin{bmatrix} \rho_1 \overline{Q}_1 & & \\ & \rho_2 \overline{Q}_2 & \\ & & 0 \end{bmatrix} \qquad h = \begin{bmatrix} -Q y_{\text{sp}} \\ 0 \end{bmatrix}$$

$$D = \begin{bmatrix} I & -G_1 \end{bmatrix} \qquad d = G_2 u_2 \qquad G_1 = \begin{bmatrix} G_{11} \\ G_{12} \end{bmatrix} \qquad G_2 = \begin{bmatrix} G_{12} \\ G_{22} \end{bmatrix}$$

We can then solve the linear algebra problem

$$\begin{bmatrix} H & -D' \\ -D & 0 \end{bmatrix} \begin{bmatrix} r_s \\ \lambda_s \end{bmatrix} = -\begin{bmatrix} h \\ d \end{bmatrix}$$

and identify the linear gains between the optimal u_{1s} and the setpoint y_{sp} and player two's input u_{2s}

$$u_{1s}^0 = K_{1s} y_{\text{sp}} + L_{1s} u_{2s}^p$$

Combining the optimal control laws for each player gives

$$\begin{bmatrix} u_{1s}^0 \\ u_{2s}^0 \end{bmatrix} = \begin{bmatrix} K_{1s} \\ K_{2s} \end{bmatrix} y_{\text{sp}} + \begin{bmatrix} 0 & L_{1s} \\ L_{2s} & 0 \end{bmatrix} \begin{bmatrix} u_{1s} \\ u_{2s} \end{bmatrix}^p$$

Substituting the optimal control into the iteration gives

$$\begin{bmatrix} u_{1s} \\ u_{2s} \end{bmatrix}^{p+1} = \underbrace{\begin{bmatrix} w_1 K_{1s} \\ w_2 K_{2s} \end{bmatrix}}_{\overline{K}_s} y_{\text{sp}} + \underbrace{\begin{bmatrix} (1-w_1)I & w_1 L_{1s} \\ w_2 L_{2s} & (1-w_2)I \end{bmatrix}}_{L_s} \begin{bmatrix} u_{1s} \\ u_{2s} \end{bmatrix}^p$$

Finally writing this equation in the plantwide notation, we express the iteration as

$$u_s^{p+1} = \overline{K}_s y_{\text{sp}} + L_s u_s^p$$

As we did with the cooperative regulation problem, we can analyze the optimization problem to show that this iteration is always stable and converges to the centralized target. Next we explore the use of these approaches in some illustrative examples.

Example 6.11: Stability and offset in the distributed target calculation

Consider the following two-input, two-output system with steady-state gain matrix and setpoint

$$\begin{bmatrix} y_{1s} \\ y_{2s} \end{bmatrix} = \begin{bmatrix} -0.5 & 1.0 \\ 2.0 & 1.0 \end{bmatrix} \begin{bmatrix} u_{1s} \\ u_{2s} \end{bmatrix} \qquad \begin{bmatrix} y_{1sp} \\ y_{2sp} \end{bmatrix} = \begin{bmatrix} 1 \\ 1 \end{bmatrix}$$

(a) Show the first 10 iterations of the noncooperative and cooperative steady-state cases starting with the decentralized solution as the initial guess.

Describe the differences. Compute the eigenvalues of L for the cooperative and noncooperative cases. Discuss the relationship between these eigenvalues and the result of the iteration calculations.

Mark also the solution to the centralized and decentralized cases on your plots.

(b) Switch the pairings and repeat the previous part. Explain your results.

Solution

(a) The first 10 iterations of the noncooperative steady-state calculation are shown in Figure 6.2. Notice the iteration is unstable and the steady-state target does not converge. The cooperative case is shown in Figure 6.3. This case is stable and the iterations converge to the centralized target and achieve zero offset. The magnitudes of the eigenvalues of L_s for the noncooperative (nc) and cooperative (co) cases are given by

$$|\text{eig}(L_{snc})| = \{1.12, 1.12\} \qquad |\text{eig}(L_{sco})| = \{0.757, 0.243\}$$

Stability of the iteration is determined by the magnitudes of the eigenvalues of L_s.

(b) Reversing the pairings leads to the following gain matrix in which we have reversed the labels of the outputs for the two systems

$$\begin{bmatrix} y_{1s} \\ y_{2s} \end{bmatrix} = \begin{bmatrix} 2.0 & 1.0 \\ -0.5 & 1.0 \end{bmatrix} \begin{bmatrix} u_{1s} \\ u_{2s} \end{bmatrix}$$

The first 10 iterations of the noncooperative and cooperative games are shown in Figures 6.4 and 6.5. For this pairing, the noncooperative case also converges to the centralized target. The eigenvalues are given by

$$|\text{eig}(L_{snc})| = \{0.559, 0.559\} \qquad |\text{eig}(L_{sco})| = \{0.757, 0.243\}$$

The eigenvalues of the cooperative case are unaffected by the reversal of pairings. □

Given the stability analysis of the simple unconstrained two-player game, we remove from further consideration two options we have been

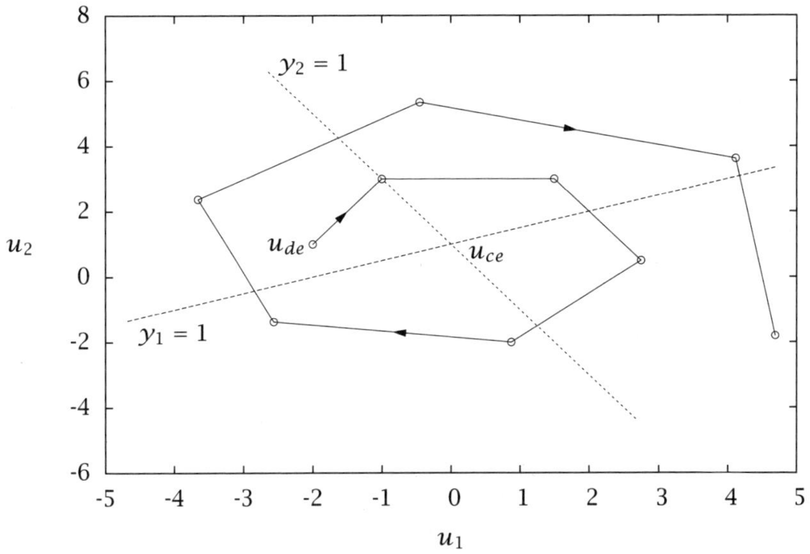

Figure 6.2: Ten iterations of the noncooperative steady-state calculation, $u^{[0]} = u_{de}$; iterations are unstable, $u^p \to \infty$.

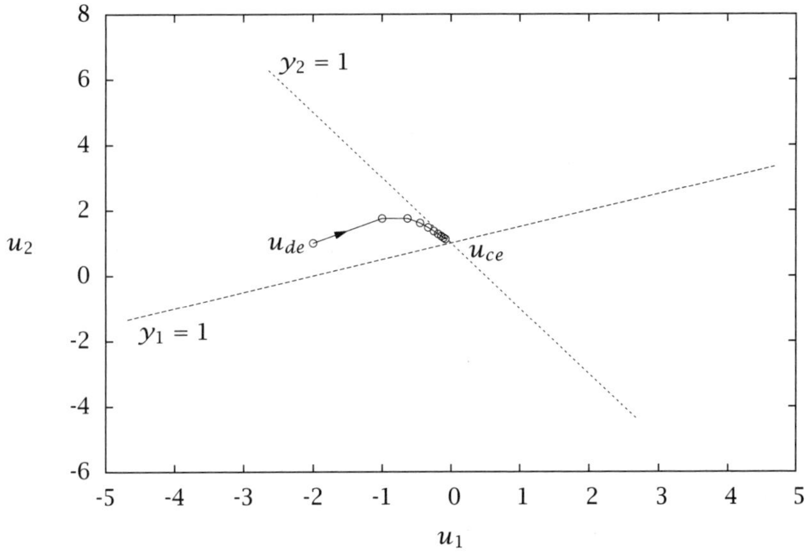

Figure 6.3: Ten iterations of the cooperative steady-state calculation, $u^{[0]} = u_{de}$; iterations are stable, $u^p \to u_{ce}$.

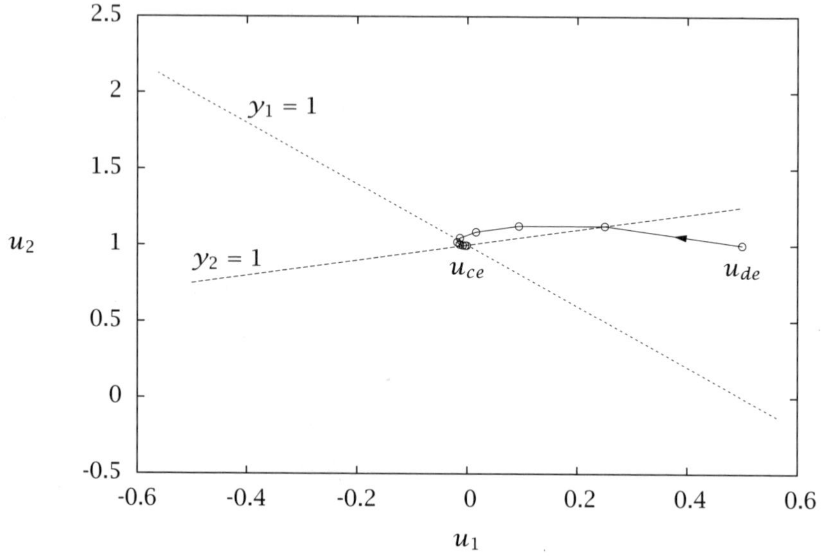

Figure 6.4: Ten iterations of the noncooperative steady-state calculation, $u^{[0]} = u_{de}$; iterations are now stable with reversed pairing.

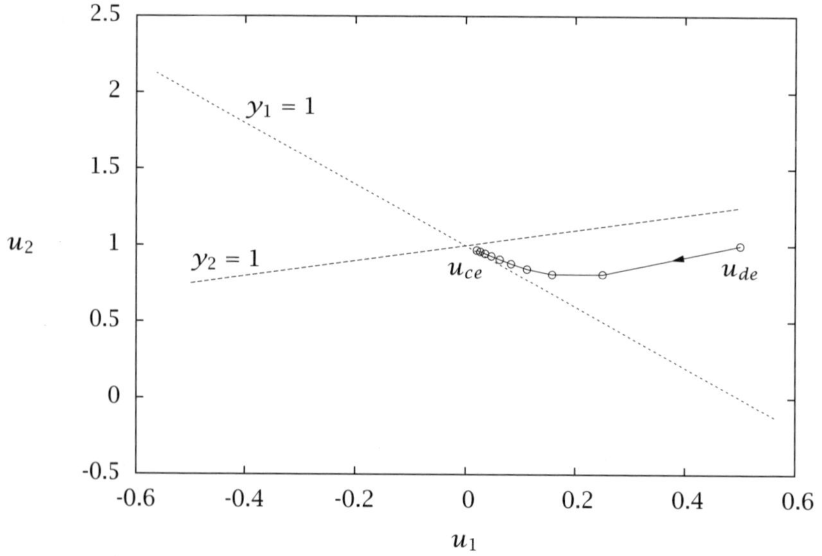

Figure 6.5: Ten iterations of the cooperative steady-state calculation, $u^{[0]} = u_{de}$; iterations remain stable with reversed pairing.

discussing to this point: noncooperative control and decentralized control. We next further develop the theory of cooperative MPC and compare its performance to centralized MPC in more general and challenging situations.

6.2.6 State Estimation

Given output measurements, we can express the state estimation problem also in distributed form. Player one uses local measurements of y_1 and knowledge of both inputs u_1 and u_2 to estimate state x_1

$$\hat{x}_1^+ = A_1\hat{x}_1 + \overline{B}_{11}u_1 + \overline{B}_{12}u_2 + L_1(y_1 - C_1\hat{x}_1)$$

Defining estimate error to be $e_1 = x_1 - \hat{x}_1$ gives

$$e_1^+ = (A_1 - L_1C_1)e_1$$

Because all the subsystems are stable, we know L_1 exists so that $A_1 - L_1C_1$ is stable and player one's local estimator is stable. The estimate error for the two subsystems is then given by

$$\begin{bmatrix} e_1 \\ e_2 \end{bmatrix}^+ = \begin{bmatrix} A_{L1} & \\ & A_{L2} \end{bmatrix} \begin{bmatrix} e_1 \\ e_2 \end{bmatrix} \tag{6.18}$$

in which $A_{Li} = A_i - L_iC_i$.

Closed-Loop Stability. The dynamics of the estimator are given by

$$\begin{bmatrix} \hat{x}_1 \\ \hat{x}_2 \end{bmatrix}^+ = \begin{bmatrix} A_1 & \\ & A_2 \end{bmatrix} \begin{bmatrix} \hat{x}_1 \\ \hat{x}_2 \end{bmatrix} + \begin{bmatrix} \overline{B}_{11} & \overline{B}_{12} \\ \overline{B}_{21} & \overline{B}_{22} \end{bmatrix} \begin{bmatrix} u_1 \\ u_2 \end{bmatrix} + \begin{bmatrix} L_1C_1 & \\ & L_2C_2 \end{bmatrix} \begin{bmatrix} e_1 \\ e_2 \end{bmatrix}$$

In the control law we use the state estimate in place of the state, which is unmeasured and unknown. We consider two cases.

Converged controller. In this case the distributed control law converges to the centralized controller, and we have

$$\begin{bmatrix} u_1 \\ u_2 \end{bmatrix} = \begin{bmatrix} K_{11} & K_{12} \\ K_{21} & K_{22} \end{bmatrix} \begin{bmatrix} \hat{x}_1 \\ \hat{x}_2 \end{bmatrix}$$

The closed-loop system evolves according to

$$\begin{bmatrix} \hat{x}_1 \\ \hat{x}_2 \end{bmatrix}^+ = \left\{ \begin{bmatrix} A_1 & \\ & A_2 \end{bmatrix} + \begin{bmatrix} \overline{B}_{11} & \overline{B}_{12} \\ \overline{B}_{21} & \overline{B}_{22} \end{bmatrix} \begin{bmatrix} K_{11} & K_{12} \\ K_{21} & K_{22} \end{bmatrix} \right\} \begin{bmatrix} \hat{x}_1 \\ \hat{x}_2 \end{bmatrix} + \begin{bmatrix} L_1C_1 & \\ & L_2C_2 \end{bmatrix} \begin{bmatrix} e_1 \\ e_2 \end{bmatrix}$$

The $A+BK$ term is stable because this term is the same as in the stabilizing centralized controller. The perturbation is exponentially decaying because the distributed estimators are stable. Therefore $\hat{x}$ goes to zero exponentially, which, along with e going to zero exponentially, implies x goes to zero exponentially.

Finite iterations. Here we use the state plus input sequence description given in (6.16), which, as we have already noted, is a linear time-invariant system. With estimate error, the system equation is

$$\begin{bmatrix} \hat{x}_1^+ \\ \hat{x}_2^+ \\ \mathbf{u}_1^+ \\ \mathbf{u}_2^+ \end{bmatrix} = \begin{bmatrix} A_1\hat{x}_1 + \overline{B}_{11}u_1 + \overline{B}_{12}u_2 \\ A_2\hat{x}_2 + \overline{B}_{21}u_1 + \overline{B}_{22}u_2 \\ g_1^p(\hat{x}_1, \hat{x}_2, \mathbf{u}_1, \mathbf{u}_2) \\ g_2^p(\hat{x}_1, \hat{x}_2, \mathbf{u}_1, \mathbf{u}_2) \end{bmatrix} + \begin{bmatrix} L_1C_1e_1 \\ L_2C_2e_2 \\ 0 \\ 0 \end{bmatrix}$$

Because there is again only one-way coupling between the estimate error evolution, (6.18), and the system evolution given above, the composite system is exponentially stable.

6.3 Constrained Two-Player Game

Now that we have introduced most of the notation and the fundamental ideas, we consider more general cases. Because we are interested in establishing stability properties of the controlled systems, we focus exclusively on *cooperative distributed MPC* from this point forward. In this section we consider convex input constraints on the two players. We assume output constraints have been softened with exact soft constraints and added to the objective function, so do not consider output constraints explicitly. The input constraints break into two significant categories: coupled and uncoupled constraints. We treat each of these in turn.

We also allow unstable systems and replace Assumption 6.7 with the following more general restrictions on the systems and controller parameters.

Assumption 6.12 (Constrained two-player game).

(a) The systems (A_i, B_i), $i = 1, 2$ are stabilizable.

(b) The systems (A_i, C_i), $i = 1, 2$ are detectable.

(c) The input penalties R_1, R_2 are positive definite, and the state penalties Q_1, Q_2 are semidefinite.

(d) The systems (A_1, Q_1) and (A_2, Q_2) are detectable.

(e) The horizon is chosen sufficiently long to zero the unstable modes, $N \geq \max_{j \in \mathbb{I}_{1:2}}(n_{1j}^u + n_{2j}^u)$, in which n_{ij}^u is the number of unstable modes of A_{ij}, i.e., number of $\lambda \in \text{eig}(A_{ij})$ such that $|\lambda| \geq 1$.

Assumption (b) implies that we have L_i such that $(A_i - L_i C_i), i = 1, 2$ is stable. Note that the stabilizable and detectable conditions of Assumption 6.12 are automatically satisfied if we obtain the state space models from a minimal realization of the input/output models for (u_i, y_j), $i, j = 1, 2$.

Unstable modes. To handle unstable systems, we add constraints to zero the unstable modes at the end of the horizon. To set up this constraint, consider the real Schur decomposition of A_{ij} for $i, j \in \mathbb{I}_{1:2}$

$$A_{ij} = \begin{bmatrix} S_{ij}^s & S_{ij}^u \end{bmatrix} \begin{bmatrix} A_{ij}^s & - \\ & A_{ij}^u \end{bmatrix} \begin{bmatrix} {S_{ij}^s}' \\ {S_{ij}^u}' \end{bmatrix} \tag{6.19}$$

in which A_{ij}^s is upper triangular and stable, and A_{ij}^u is upper triangular with all unstable eigenvalues.[3] Let Σ_{ij} denote the solution of the Lyapunov equation

$$A_{ij}^{s'} \Sigma_{ij} A_{ij}^s - \Sigma_{ij} = -{S_{ij}^s}' Q_{ij} S_{ij}^s$$

Given the Schur decomposition (6.19), we define the matrices

$$\begin{aligned} S_i^s &= \text{diag}(S_{i1}^s, S_{i2}^s) & A_i^s &= \text{diag}(A_{i1}^s, A_{i2}^s) \quad i \in \mathbb{I}_{1:2} \\ S_i^u &= \text{diag}(S_{i1}^u, S_{i2}^u) & A_i^u &= \text{diag}(A_{i1}^u, A_{i2}^u) \quad i \in \mathbb{I}_{1:2} \end{aligned}$$

These matrices satisfy the Schur decompositions

$$A_i = \begin{bmatrix} S_i^s & S_i^u \end{bmatrix} \begin{bmatrix} A_i^s & - \\ & A_i^u \end{bmatrix} \begin{bmatrix} S_i^{s'} \\ S_i^{u'} \end{bmatrix} \quad i \in \mathbb{I}_{1:2}$$

[3] If A_{ij} is stable, then there is no A_{ij}^u and S_{ij}^u.

We further define the matrices

$$\Sigma_i = \text{diag}(\Sigma_{i1}, \Sigma_{i2}) \quad i \in \mathbb{I}_{1:2}$$

These matrices satisfy the Lyapunov equations

$$A_1^{s\prime}\Sigma_1 A_1^s - \Sigma_1 = -S_1^{s\prime} Q_1 S_1^s \qquad A_2^{s\prime}\Sigma_2 A_2^s - \Sigma_2 = -S_2^{s\prime} Q_2 S_2^s \tag{6.20}$$

We then choose the terminal penalty for each subsystem to be the cost to go under zero control

$$P_{1f} = S_1^s \Sigma_1 S_1^{s\prime} \qquad\qquad P_{2f} = S_2^s \Sigma_2 S_2^{s\prime}$$

6.3.1 Uncoupled Input Constraints

We consider convex input constraints of the following form

$$Hu(k) \leq h \qquad k = 0, 1, \ldots, N$$

Defining convex set $\mathbb{U}$

$$\mathbb{U} = \{u | Hu \leq h\}$$

we express the input constraints as

$$u(k) \in \mathbb{U} \qquad k = 0, 1, \ldots, N$$

We drop the time index and indicate the constraints are applied over the entire input sequence using the notation $\mathbf{u} \in \mathbb{U}$. In the uncoupled constraint case, the two players' inputs must satisfy

$$u_1 \in \mathbb{U}_1 \qquad u_2 \in \mathbb{U}_2$$

in which $\mathbb{U}_1$ and $\mathbb{U}_2$ are convex subsets of $\mathbb{R}^{m_1}$ and $\mathbb{R}^{m_2}$, respectively. The constraints are termed *uncoupled* because there is no interaction or coupling of the inputs in the constraint relation. Player one then solves the following constrained optimization

$$\min_{\mathbf{u}_1} V(x_1(0), x_2(0), \mathbf{u}_1, \mathbf{u}_2)$$

$$\text{s.t.} \quad \begin{bmatrix} x_1 \\ x_2 \end{bmatrix}^+ = \begin{bmatrix} A_1 & 0 \\ 0 & A_2 \end{bmatrix} \begin{bmatrix} x_1 \\ x_2 \end{bmatrix} + \begin{bmatrix} \overline{B}_{11} \\ \overline{B}_{21} \end{bmatrix} u_1 + \begin{bmatrix} \overline{B}_{12} \\ \overline{B}_{22} \end{bmatrix} u_2$$

$$\mathbf{u}_1 \in \mathbb{U}_1$$

$$S_{j1}^{u\prime} x_{j1}(N) = 0 \quad j \in \mathbb{I}_{1:2}$$

$$|\mathbf{u}_1| \leq d_1(|x_{11}(0)| + |x_{21}(0)|) \quad x_{11}(0), x_{21}(0) \in r\mathcal{B}$$

in which we include the system's hard input constraints, the stability constraint on the unstable modes, and the Lyapunov stability constraints. Exercise 6.22 discusses how to write the constraint $|\mathbf{u}_1| \leq d_1 |x_1(0)|$ as a set of linear inequalities on $\mathbf{u}_1$. Similarly, player two solves

$$\min_{\mathbf{u}_2} V(x_1(0), x_2(0), \mathbf{u}_1, \mathbf{u}_2)$$

$$\text{s.t.} \quad \begin{bmatrix} x_1 \\ x_2 \end{bmatrix}^+ = \begin{bmatrix} A_1 & 0 \\ 0 & A_2 \end{bmatrix} \begin{bmatrix} x_1 \\ x_2 \end{bmatrix} + \begin{bmatrix} \overline{B}_{11} \\ \overline{B}_{21} \end{bmatrix} u_1 + \begin{bmatrix} \overline{B}_{12} \\ \overline{B}_{22} \end{bmatrix} u_2$$

$$\mathbf{u}_2 \in \mathbb{U}_2$$

$$S_{j2}^{u\,\prime} x_{j2}(N) = 0 \quad j \in \mathbb{I}_{1:2}$$

$$|\mathbf{u}_2| \leq d_2(|x_{21}(0)| + |x_{22}(0)|) \quad x_{12}(0), x_{22}(0) \in r\mathcal{B}$$

We denote the solutions to these problems as

$$\mathbf{u}_1^0(x_1(0), x_2(0), \mathbf{u}_2) \qquad \mathbf{u}_2^0(x_1(0), x_2(0), \mathbf{u}_1)$$

The feasible set $\mathcal{X}_N$ for the unstable system is the set of states for which the unstable modes can be brought to zero in N moves while satisfying the input constraints.

Given an initial iterate, $(\mathbf{u}_1^p, \mathbf{u}_2^p)$, the next iterate is defined to be

$$(\mathbf{u}_1, \mathbf{u}_2)^{p+1} = w_1(\mathbf{u}_1^0(x_1(0), x_2(0), \mathbf{u}_2^p), \mathbf{u}_2^p) + \\ w_2(\mathbf{u}_1^p, \mathbf{u}_2^0(x_1(0), x_2(0), \mathbf{u}_1^p))$$

To reduce the notational burden we denote this as

$$(\mathbf{u}_1, \mathbf{u}_2)^{p+1} = w_1(\mathbf{u}_1^0, \mathbf{u}_2^p) + w_2(\mathbf{u}_1^p, \mathbf{u}_2^0)$$

and the functional dependencies of $\mathbf{u}_1^0$ and $\mathbf{u}_2^0$ should be kept in mind.

This procedure provides three important properties, which we establish next.

1. The iterates are feasible: $(\mathbf{u}_1, \mathbf{u}_2)^p \in (\mathbb{U}_1, \mathbb{U}_2)$ implies $(\mathbf{u}_1, \mathbf{u}_2)^{p+1} \in (\mathbb{U}_1, \mathbb{U}_2)$. This follows from convexity of $\mathbb{U}_1$, $\mathbb{U}_2$ and the convex combination of the feasible points $(\mathbf{u}_1^p, \mathbf{u}_2^p)$ and $(\mathbf{u}_1^0, \mathbf{u}_2^0)$ to make $(\mathbf{u}_1, \mathbf{u}_2)^{p+1}$.

2. The cost decreases on iteration: $V(x_1(0), x_2(0), (\mathbf{u}_1, \mathbf{u}_2)^{p+1}) \leq V(x_1(0), x_2(0), (\mathbf{u}_1, \mathbf{u}_2)^p)$ for all $x_1(0)$, $x_2(0)$, and for all feasible

$(\mathbf{u}_1, \mathbf{u}_2)^p \in (\mathbb{U}_1, \mathbb{U}_2)$. The systemwide cost satisfies the following inequalities

$$\begin{aligned}
V(x(0), \mathbf{u}_1^{p+1}, \mathbf{u}_2^{p+1}) &= V\left(x(0), \left(w_1(\mathbf{u}_1^0, \mathbf{u}_2^p) + w_2(\mathbf{u}_1^p, \mathbf{u}_2^0)\right)\right) \\
&\leq w_1 V(x(0), (\mathbf{u}_1^0, \mathbf{u}_2^p)) + w_2 V(x(0), (\mathbf{u}_1^p, \mathbf{u}_2^0)) \\
&\leq w_1 V(x(0), (\mathbf{u}_1^p, \mathbf{u}_2^p)) + w_2 V(x(0), (\mathbf{u}_1^p, \mathbf{u}_2^p)) \\
&= V(x(0), \mathbf{u}_1^p, \mathbf{u}_2^p)
\end{aligned}$$

The first equality follows from (6.14). The next inequality follows from convexity of V. The next follows from optimality of $\mathbf{u}_1^0$ and $\mathbf{u}_2^0$, and the last follows from $w_1 + w_2 = 1$. Because the cost is bounded below, the cost iteration converges.

3. The converged solution of the cooperative problem is equal to the optimal solution of the centralized problem. Establishing this property is discussed in Exercise 6.26.

Exponential stability of the closed-loop system. We next consider the closed-loop system. The two players' warm starts at the next sample are as defined previously

$$\begin{aligned}
\widetilde{\mathbf{u}}_1^+ &= \{u_1(1), u_1(2), \ldots, u_1(N-1), 0\} \\
\widetilde{\mathbf{u}}_2^+ &= \{u_2(1), u_2(2), \ldots, u_2(N-1), 0\}
\end{aligned}$$

We define again the functions g_1^p, g_2^p as the outcome of applying the control iteration procedure p times

$$\begin{aligned}
\mathbf{u}_1^p &= g_1^p(x_1, x_2, \mathbf{u}_1, \mathbf{u}_2) \\
\mathbf{u}_2^p &= g_2^p(x_1, x_2, \mathbf{u}_1, \mathbf{u}_2)
\end{aligned}$$

The important difference between the previous unconstrained and this constrained case is that the functions g_1^p, g_2^p are nonlinear due to the input constraints. The system evolution is then given by

$$\begin{aligned}
x_1^+ &= A_1 x_1 + \overline{B}_{11} u_1 + \overline{B}_{12} u_2 & x_2^+ &= A_2 x_2 + \overline{B}_{21} u_1 + \overline{B}_{22} u_2 \\
\mathbf{u}_1^+ &= g_1^p(x_1, x_2, \mathbf{u}_1, \mathbf{u}_2) & \mathbf{u}_2^+ &= g_2^p(x_1, x_2, \mathbf{u}_1, \mathbf{u}_2)
\end{aligned}$$

We have the following cost using the warm start at the next sample

$$\begin{aligned}
V(x_1^+, x_2^+, \widetilde{\mathbf{u}}_1^+, \widetilde{\mathbf{u}}_2^+) &= V(x_1, x_2, \mathbf{u}_1, \mathbf{u}_2) - \rho_1 \ell_1(x_1, u_1) - \rho_2 \ell_2(x_2, u_2) \\
&\quad + (1/2) x_1(N)' \left[A_1' P_{1f} A_1 - P_{1f} + Q_1\right] x_1(N) \\
&\quad + (1/2) x_2(N)' \left[A_2' P_{2f} A_2 - P_{2f} + Q_2\right] x_2(N)
\end{aligned}$$

Using the Schur decomposition (6.19) and the constraints $S_{ji}^{u\prime} x_{ji}(N) = 0$ for $i, j \in \mathbb{I}_{1:2}$, the last two terms can be written as

$$\begin{aligned} &(1/2)x_1(N)' S_1^s \left[A_1^{s\prime} \Sigma_1 A_1^s - \Sigma_1 + S_1^{s\prime} Q_1 S_1^s \right] S_1^{s\prime} x_1(N) \\ &+(1/2)x_2(N)' S_2^s \left[A_2^{s\prime} \Sigma_2 A_2^s - \Sigma_2 + S_2^{s\prime} Q_2 S_2^s \right] S_2^{s\prime} x_2(N) \end{aligned}$$

These terms are zero because of (6.20). Using this result and applying the iteration for the controllers gives

$$V(x_1^+, x_2^+, \mathbf{u}_1^+, \mathbf{u}_2^+) \leq V(x_1, x_2, \mathbf{u}_1, \mathbf{u}_2) - \rho_1 \ell_1(x_1, u_1) - \rho_2 \ell_2(x_2, u_2)$$

The Lyapunov stability constraints give (see also Exercise 6.28)

$$|(\mathbf{u}_1, \mathbf{u}_2)| \leq 2 \max(d_1, d_2) \, |x_1, x_2| \qquad (x_1, x_2) \in r\mathcal{B}$$

Given the cost decrease and this constraint on the size of the input sequence, we satisfy the conditions of Lemma 6.4, and conclude the solution $x(k) = 0$ for all k is exponentially stable on all of $\mathcal{X}_N$ if either $\mathcal{X}_N$ is compact or $\mathbb{U}$ is compact.

6.3.2 Coupled Input Constraints

By contrast, in the coupled constraint case, the constraints are of the form

$$H_1 \mathbf{u}_1 + H_2 \mathbf{u}_2 \leq h \quad \text{or} \quad (\mathbf{u}_1, \mathbf{u}_2) \in \mathbb{U} \tag{6.21}$$

These constraints represent the players sharing some common resource. An example would be different subsystems in a chemical plant drawing steam or some other utility from a single plantwide generation plant. The total utility used by the different subsystems to meet their control objectives is constrained by the generation capacity.

The players solve the same optimization problems as in the uncoupled constraint case, with the exception that both players' input constraints are given by (6.21). This modified game provides only two of the three properties established for the uncoupled constraint case. These are

1. The iterates are feasible: $(\mathbf{u}_1, \mathbf{u}_2)^p \in \mathbb{U}$ implies $(\mathbf{u}_1, \mathbf{u}_2)^{p+1} \in \mathbb{U}$. This follows from convexity of $\mathbb{U}$ and the convex combination of the feasible points $(\mathbf{u}_1^p, \mathbf{u}_2^p)$ and $(\mathbf{u}_1^0, \mathbf{u}_2^0)$ to make $(\mathbf{u}_1, \mathbf{u}_2)^{p+1}$.

2. The cost decreases on iteration: $V(x_1(0), x_2(0), (\mathbf{u}_1, \mathbf{u}_2)^{p+1}) \leq V(x_1(0), x_2(0), (\mathbf{u}_1, \mathbf{u}_2)^p)$ for all $x_1(0)$, $x_2(0)$, and for all feasible

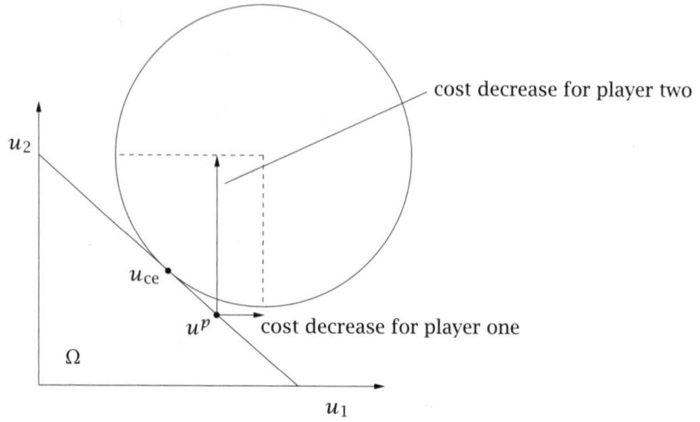

Figure 6.6: Cooperative control stuck on the boundary of $\mathbb{U}$ under coupled constraints; $u^{p+1} = u^p \neq u_{ce}$.

$(\mathbf{u}_1, \mathbf{u}_2)^p \in \mathbb{U}$. The systemwide cost satisfies the same inequalities established for the uncoupled constraint case giving

$$V(x(0), \mathbf{u}_1^{p+1}, \mathbf{u}_2^{p+1}) \leq V(x(0), \mathbf{u}_1^p, \mathbf{u}_2^p)$$

Because the cost is bounded below, the cost iteration converges.

The converged solution of the cooperative problem is *not* equal to the optimal solution of the centralized problem, however. We have lost property 3 of the uncoupled case. To see how the convergence property is lost, consider Figure 6.6. Region $\mathbb{U}$ is indicated by the triangle and its interior. Consider point u^p on the boundary of $\mathbb{U}$. Neither player one nor player two can improve upon the current point u^p so the iteration has converged. But the converged point is clearly not the optimal point, u_{ce}.

Because of property 2, the nominal stability properties for the coupled and uncoupled cases are identical. The differences arise when the performance of cooperative control is compared to the benchmark of centralized control. Improving the performance of cooperative control in the case of coupled constraints is therefore a topic of current research. Current approaches include adding another player to the game, whose sole objective is to parcel out the coupled resource to the other players in a way that achieves optimality on iteration. This approach also makes sense from an engineering perspective because it is commonplace to design a dedicated control system for managing a shared resource such as steam or power among many plant units. The design of this single unit's control system is a reasonably narrow and well defined task compared to the design of a centralized controller for the entire plant.

6.3.3 Exponential Stability with Estimate Error

Consider next the constrained system evolution with estimate error

$$\begin{bmatrix} \hat{x}^+ \\ \mathbf{u}^+ \\ e^+ \end{bmatrix} = \begin{bmatrix} A\hat{x} + \overline{B}_1 u_1 + \overline{B}_2 u_2 + Le \\ g^p(\hat{x}, \mathbf{u}) \\ A_L e \end{bmatrix} \tag{6.22}$$

The estimate error is globally exponentially stable so we know from Lemma 6.6 that there exists a Lipschitz continuous Lyapunov function $J(\cdot)$ such that for all $e \in \mathbb{R}^n$

$$\overline{a}\,|e| \leq J(e) \leq \overline{b}\,|e|$$
$$J(e^+) - J(e) \leq -\overline{c}\,|e|$$

in which $\overline{b} > 0$, $\overline{a} > 0$, and we can choose constant $\overline{c} > 0$ as large as desired. In the subsequent development, we require this Lyapunov function to be based on the first power of the norm rather than the usual square of the norm to align with Lipschitz continuity of the Lyapunov function. From the stability of the solution $x(k) = 0$ for all k for the *nominal* system, the cost function $V(\hat{x}, \mathbf{u})$ satisfies for all $\hat{x} \in \mathcal{X}_N$, $\mathbf{u} \in \mathbb{U}^N$

$$\widetilde{a}\,|(\hat{x}, \mathbf{u})|^2 \leq V(\hat{x}, \mathbf{u}) \leq \widetilde{b}\,|(\hat{x}, \mathbf{u})|^2$$
$$V(A\hat{x} + \overline{B}_1 u_1 + \overline{B}_2 u_2, \mathbf{u}^+) - V(\hat{x}, \mathbf{u}) \leq -\widetilde{c}\,|\hat{x}|^2$$
$$|\mathbf{u}| \leq d\,|\hat{x}| \qquad \hat{x} \in \widetilde{r}\mathcal{B}$$

in which $\widetilde{a}, \widetilde{b}, \widetilde{c}, \widetilde{r} > 0$. We propose $W(\hat{x}, \mathbf{u}, e) = V(\hat{x}, \mathbf{u}) + J(e)$ as a Lyapunov function candidate for the perturbed system. We next derive the required properties of $W(\cdot)$ to establish exponential stability of the solution $(x(k), e(k)) = 0$. From the definition of $W(\cdot)$ we have for all $(\hat{x}, \mathbf{u}, e) \in \mathcal{X}_N \times \mathbb{U}^N \times \mathbb{R}^n$

$$\widetilde{a}\,|(\hat{x}, \mathbf{u})|^2 + \overline{a}\,|e| \leq W(\hat{x}, \mathbf{u}, e) \leq \widetilde{b}\,|(\hat{x}, \mathbf{u})|^2 + \overline{b}\,|e|$$
$$a(|(\hat{x}, \mathbf{u})|^2 + |e|) \leq W(\hat{x}, \mathbf{u}, e) \leq b(|(\hat{x}, \mathbf{u})|^2 + |e|) \tag{6.23}$$

in which $a = \min(\widetilde{a}, \overline{a}) > 0$, $b = \max(\widetilde{b}, \overline{b})$. Next we compute the cost change

$$W(\hat{x}^+, \mathbf{u}^+, e^+) - W(\hat{x}, \mathbf{u}, e) = V(\hat{x}^+, \mathbf{u}^+) - V(\hat{x}, \mathbf{u}) + J(e^+) - J(e)$$

The Lyapunov function V is quadratic in $(x, \mathbf{u})$ and therefore Lipschitz continuous on bounded sets. Therefore, for all $\hat{x}, u_1, u_2, \mathbf{u}^+, e$ in some bounded set,

$$\left| V(A\hat{x} + \overline{B}_1 u_1 + \overline{B}_2 u_2 + Le, \mathbf{u}^+) - V(A\hat{x} + \overline{B}_1 u_1 + \overline{B}_2 u_2, \mathbf{u}^+) \right| \leq L_V \left| Le \right|$$

in which L_V is the Lipschitz constant for V with respect to its first argument. Using the system evolution we have

$$V(\hat{x}^+, \mathbf{u}^+) \leq V(A\hat{x} + \overline{B}_1 u_1 + \overline{B}_2 u_2, \mathbf{u}^+) + L_V' \left| e \right|$$

in which $L_V' = L_V \left| L \right|$. Subtracting $V(\hat{x}, \mathbf{u})$ from both sides gives

$$V(\hat{x}^+, \mathbf{u}^+) - V(\hat{x}, \mathbf{u}) \leq -\widetilde{c} \left| \hat{x} \right|^2 + L_V' \left| e \right|$$

Substituting this result into the equation for the change in W gives

$$\begin{aligned} W(\hat{x}^+, \mathbf{u}^+, e^+) - W(\hat{x}, \mathbf{u}, e) &\leq -\widetilde{c} \left| \hat{x} \right|^2 + L_V' \left| e \right| - \overline{c} \left| e \right| \\ &\leq -\widetilde{c} \left| \hat{x} \right|^2 - (\overline{c} - L_V') \left| e \right| \\ W(\hat{x}^+, \mathbf{u}^+, e^+) - W(\hat{x}, \mathbf{u}, e) &\leq -c(\left| \hat{x} \right|^2 + \left| e \right|) \end{aligned} \tag{6.24}$$

in which we choose $\overline{c} > L_V'$, which is possible because we may choose $\overline{c}$ as large as we wish, and $c = \min(\widetilde{c}, \overline{c} - L_V') > 0$. Notice this step is what motivated using the first power of the norm in $J(\cdot)$. Lastly, we require the constraint

$$\left| \mathbf{u} \right| \leq d \left| \hat{x} \right| \qquad \hat{x} \in \widetilde{r}\mathcal{B} \tag{6.25}$$

Lemma 6.13 (Exponential stability of perturbed system). *If either $\mathcal{X}_N$ or $\mathbb{U}$ is compact, the origin for the state plus estimate error (x, e) is exponentially stable for system* (6.22) *under cooperative distributed MPC.*

The proof is based on the properties (6.23), (6.24) and (6.25) of function $W(\hat{x}, \mathbf{u}, e)$ and is similar to the proof of Lemma 6.4. The region of attraction is the set of states and initial estimate errors for which the unstable modes of the two subsystems can be brought to zero in N moves while satisfying the respective input constraints. If both subsystems are stable, for example, the region of attraction is $(x, e) \in \mathcal{X}_N \times \mathbb{R}^n$.

6.3.4 Disturbance Models and Zero Offset

Integrating disturbance model. As discussed in Chapter 1, we model the disturbance with an integrator to remove steady offset. The augmented models for the local systems are

$$\begin{bmatrix} x_i \\ d_i \end{bmatrix}^+ = \begin{bmatrix} A_i & B_{di} \\ 0 & I \end{bmatrix} \begin{bmatrix} x_i \\ d_i \end{bmatrix} + \begin{bmatrix} \overline{B}_{i1} \\ 0 \end{bmatrix} u_1 + \begin{bmatrix} \overline{B}_{i2} \\ 0 \end{bmatrix} u_2$$

$$y_i = \begin{bmatrix} C_i & C_{di} \end{bmatrix} \begin{bmatrix} x_i \\ d_i \end{bmatrix} \qquad i = 1, 2$$

We wish to estimate both x_i and d_i from measurements y_i. To ensure this goal is possible, we make the following restriction on the disturbance models

Assumption 6.14 (Disturbance models).

$$\text{rank} \begin{bmatrix} I - A_i & -B_{di} \\ C_i & C_{di} \end{bmatrix} = n_i + p_i \qquad i = 1, 2$$

It is always possible to satisfy this assumption by proper choice of B_{di}, C_{di}. From Assumption 6.12 (b), (A_i, C_i) is detectable, which implies that the first n_i columns of the square $(n_i + p_i) \times (n_i + p_i)$ matrix in Assumption 6.14 are linearly independent. Therefore the columns of $\begin{bmatrix} -B_{di} \\ C_{di} \end{bmatrix}$ can be chosen so that the entire matrix has rank $n_i + p_i$. Assumption 6.14 is equivalent to detectability of the following augmented system.

Lemma 6.15 (Detectability of distributed disturbance model). *Consider the augmented systems*

$$\widetilde{A}_i = \begin{bmatrix} A_i & B_{di} \\ 0 & I \end{bmatrix} \qquad \widetilde{C}_i = \begin{bmatrix} C_i & C_{di} \end{bmatrix} \qquad i = 1, 2$$

The augmented systems $(\widetilde{A}_i, \widetilde{C}_i), i = 1, 2$ *are detectable if and only if Assumption 6.14 is satisfied.*

Proving this lemma is discussed in Exercise 6.29. The detectability assumption then establishes the existence of $\widetilde{L}_i$ such that $(\widetilde{A}_i - \widetilde{L}_i \widetilde{C}_i), i = 1, 2$ are stable and the local integrating disturbances can be estimated from the local measurements.

Centralized target problem. We can solve the target problem at the plantwide level or as a distributed target problem at the subunit controller level. Consider first the centralized target problem with the disturbance model discussed in Chapter 1, (1.46)

$$\min_{x_s,u_s} \frac{1}{2} \left| u_s - u_{\text{sp}} \right|^2_{R_s} + \frac{1}{2} \left| Cx_s + C_d\hat{d}(k) - y_{\text{sp}} \right|^2_{Q_s}$$

subject to:

$$\begin{bmatrix} I-A & -B \\ HC & 0 \end{bmatrix} \begin{bmatrix} x_s \\ u_s \end{bmatrix} = \begin{bmatrix} B_d\hat{d}(k) \\ r_{\text{sp}} - HC_d\hat{d}(k) \end{bmatrix}$$
$$Eu_s \leq e$$

in which we have removed the state inequality constraints to be consistent with the regulator problem. We denote the solution to this problem $(x_s(k), u_s(k))$. Notice first that the solution of the target problem depends only on the disturbance estimate, $\hat{d}(k)$, and not the solution of the control problem. So we can analyze the behavior of the target by considering only the exponential convergence of the estimator. We restrict the plant disturbance d so that the target problem is feasible, and denote the solution to the target problem for the plant disturbance, $\hat{d}(k) = d$, as (x_s^*, u_s^*). Because the estimator is exponentially stable, we know that $\hat{d}(k) \to d$ as $k \to \infty$. Because the target problem is a positive definite QP, we know the solution is Lipschitz continuous on bounded sets in the term $\hat{d}(k)$, which appears linearly in the objective function and the right-hand side of the equality constraint. Therefore, if we also restrict the initial disturbance estimate error so that the target problem remains feasible for all time, we know $(x_s(k), u_s(k)) \to (x_s^*, u_s^*)$ and the rate of convergence is exponential.

Distributed target problem. Consider next the cooperative approach, in which we assume the input inequality constraints are uncoupled. In the constrained case, we try to set things up so each player solves a local target problem

$$\min_{x_{1s},u_{1s}} \frac{1}{2} \begin{bmatrix} y_{1s} - y_{1sp} \\ y_{2s} - y_{2sp} \end{bmatrix}' \begin{bmatrix} Q_{1s} & \\ & Q_{2s} \end{bmatrix} \begin{bmatrix} y_{1s} - y_{1sp} \\ y_{2s} - y_{2sp} \end{bmatrix} +$$
$$\frac{1}{2} \begin{bmatrix} u_{1s} - u_{1sp} \\ u_{2s} - u_{2sp} \end{bmatrix}' \begin{bmatrix} R_{1s} & \\ & R_{2s} \end{bmatrix} \begin{bmatrix} u_{1s} - u_{1sp} \\ u_{2s} - u_{2sp} \end{bmatrix}$$

subject to

$$\begin{bmatrix} I - A_1 & & -\overline{B}_{11} & -\overline{B}_{12} \\ & I - A_2 & -\overline{B}_{21} & -\overline{B}_{22} \\ H_1 C_1 & & & \\ & H_2 C_2 & & \end{bmatrix} \begin{bmatrix} x_{1s} \\ x_{2s} \\ u_{1s} \\ u_{2s} \end{bmatrix} = \begin{bmatrix} B_{d1}\hat{d}_1(k) \\ B_{d2}\hat{d}_2(k) \\ r_{1sp} - H_1 C_{d1}\hat{d}_1(k) \\ r_{2sp} - H_2 C_{d2}\hat{d}_2(k) \end{bmatrix}$$

$$E_1 u_{1s} \leq e_1$$

in which

$$y_{1s} = C_1 x_{1s} + C_{d1}\hat{d}_1(k) \qquad y_{2s} = C_2 x_{2s} + C_{d2}\hat{d}_2(k) \tag{6.27}$$

But here we run into several problems. First, the constraints to ensure zero offset in both players' controlled variables are not feasible with only the u_{1s} decision variables. We require also u_{2s}, which is not available to player one. We can consider deleting the zero offset condition for player two's controlled variables, the last equality constraint. But if we do that for both players, then the two players have *different and coupled* equality constraints. That is a path to instability as we have seen in the noncooperative target problem. To resolve this issue, we move the controlled variables to the objective function, and player one solves instead the following

$$\min_{x_{1s}, u_{1s}} \frac{1}{2} \begin{bmatrix} H_1 y_{1s} - r_{1sp} \\ H_2 y_{2s} - r_{2sp} \end{bmatrix}' \begin{bmatrix} T_{1s} & \\ & T_{2s} \end{bmatrix} \begin{bmatrix} H_1 y_{1s} - r_{1sp} \\ H_2 y_{2s} - r_{2sp} \end{bmatrix}$$

subject to (6.27) and

$$\begin{bmatrix} I - A_1 & & -\overline{B}_{11} & -\overline{B}_{12} \\ & I - A_2 & -\overline{B}_{21} & -\overline{B}_{22} \end{bmatrix} \begin{bmatrix} x_{1s} \\ x_{2s} \\ u_{1s} \\ u_{2s} \end{bmatrix} = \begin{bmatrix} B_{d1}\hat{d}_1(k) \\ B_{d2}\hat{d}_2(k) \end{bmatrix}$$

$$E_1 u_{1s} \leq e_1 \tag{6.28}$$

The equality constraints for the two players appear coupled when written in this form. Coupled constraints admit the potential for the optimization to become stuck on the boundary of the feasible region, and not achieve the centralized target solution after iteration to convergence. But Exercise 6.30 discusses how to show that the equality constraints are, in fact, uncoupled. Also, the distributed target problem as expressed here may not have a unique solution when there are more manipulated variables than controlled variables. In such cases, a regularization term using the input setpoint can be added to the objective function. The controlled variable penalty can be converted to a linear penalty with a large penalty weight to ensure exact satisfaction of the controlled variable setpoint.

If the input inequality constraints are coupled, however, then the distributed target problem may indeed become stuck on the boundary of the feasible region and not eliminate offset in the controlled variables. If the input inequality constraints are coupled, we recommend using the centralized approach to computing the steady-state target. As discussed above, the centralized target problem eliminates offset in the controlled variables as long as it remains feasible given the disturbance estimates.

Zero offset. Finally we establish the zero offset property. As described in Chapter 1, the regulator is posed in deviation variables

$$\tilde{x}(k) = \hat{x}(k) - x_s(k) \qquad \tilde{u}(k) = u(k) - u_s(k) \qquad \tilde{\mathbf{u}} = \mathbf{u} - u_s(k)$$

in which the notation $\mathbf{u} - u_s(k)$ means to subtract $u_s(k)$ from each element of the $\mathbf{u}$ sequence. Player one then solves

$$\min_{\tilde{\mathbf{u}}_1} V(\tilde{x}_1(0), \tilde{x}_2(0), \tilde{\mathbf{u}}_1, \tilde{\mathbf{u}}_2)$$

$$\text{s.t.} \quad \begin{bmatrix} \tilde{x}_1 \\ \tilde{x}_2 \end{bmatrix}^+ = \begin{bmatrix} A_1 & 0 \\ 0 & A_2 \end{bmatrix} \begin{bmatrix} \tilde{x}_1 \\ \tilde{x}_2 \end{bmatrix} + \begin{bmatrix} \overline{B}_{11} \\ \overline{B}_{21} \end{bmatrix} \tilde{u}_1 + \begin{bmatrix} \overline{B}_{12} \\ \overline{B}_{22} \end{bmatrix} \tilde{u}_2$$

$$\tilde{\mathbf{u}}_1 \in \mathbb{U}_1 \ominus u_s(k)$$

$$S_{1u}' \tilde{x}_1(N) = 0$$

$$|\tilde{\mathbf{u}}_1| \leq d_1 \, |\tilde{x}_1(0)|$$

Notice that because the input constraint is shifted by the input target, we must retain feasibility of the regulation problem by restricting also the plant disturbance and its initial estimate error. If the two players' regulation problems remain feasible as the estimate error converges to zero, we have exponential stability of the zero solution from Lemma 6.13. Therefore we conclude

$(\tilde{x}(k), \tilde{u}(k)) \to (0, 0)$	Lemma 6.13
$\Longrightarrow (\hat{x}(k), u(k)) \to (x_s(k), u_s(k))$	definition of deviation variables
$\Longrightarrow (\hat{x}(k), u(k)) \to (x_s^*, u_s^*)$	target problem convergence
$\Longrightarrow x(k) \to x_s^*$	estimator stability
$\Longrightarrow r(k) \to r_{sp}$	target equality constraint

and we have *zero offset* in the plant controlled variable $r = Hy$. The rate of convergence of $r(k)$ to r_{sp} is also exponential. As we saw here, this convergence depends on maintaining feasibility in both the target problem and the regulation problem at all times.

6.4 Constrained M-Player Game

We have set up the constrained two-player game so that the approach generalizes naturally to the M-player game. We do not have a lot of work left to do to address this general case. Recall $\mathbb{I}_{1:M}$ denotes the set of integers $\{1, 2, \ldots, M\}$. We define the following systemwide variables

$$x(0) = \begin{bmatrix} x_1(0) \\ x_2(0) \\ \vdots \\ x_M(0) \end{bmatrix} \qquad \mathbf{u} = \begin{bmatrix} \mathbf{u}_1 \\ \mathbf{u}_2 \\ \vdots \\ \mathbf{u}_M \end{bmatrix} \qquad B_i = \begin{bmatrix} \overline{B}_{1i} \\ \overline{B}_{2i} \\ \vdots \\ \overline{B}_{Mi} \end{bmatrix} \qquad i \in \mathbb{I}_{1:M}$$

$$V(x(0), \mathbf{u}) = \sum_{j \in \mathbb{I}_{1:M}} \rho_j V_j(x_j(0), \mathbf{u})$$

Each player solves a similar optimization, so for $i \in \mathbb{I}_{1:M}$

$$\min_{\mathbf{u}_i} V(x(0), \mathbf{u})$$

$$\text{s.t. } x^+ = Ax + \sum_{j \in \mathbb{I}_{1:M}} B_j u_j$$

$$\mathbf{u}_i \in \mathbb{U}_i$$

$$S_{ji}^{u\,\prime} x_{ji}(N) = 0 \quad j \in \mathbb{I}_{1:M}$$

$$|\mathbf{u}_i| \leq d_i \sum_{j \in \mathbb{I}_{1:M}} \left| x_{ji}(0) \right| \qquad \text{if } x_{ji}(0) \in r\mathcal{B},\ j \in \mathbb{I}_{1:M}$$

This optimization can be expressed as a quadratic program, whose constraints and linear cost term depend affinely on parameter x. The warm start for each player at the next sample is generated from purely local information

$$\widetilde{\mathbf{u}}_i^+ = \{u_i(1), u_i(2), \ldots, u_i(N-1), 0\} \quad i \in \mathbb{I}_{1:M}$$

The controller iteration is given by

$$\mathbf{u}^{p+1} = \sum_{j \in \mathbb{I}_{1:M}} w_j \left(\mathbf{u}_1^p, \ldots, \mathbf{u}_j^0, \ldots, \mathbf{u}_M^p \right)$$

in which $\mathbf{u}_i^0 = \mathbf{u}_i^0\left(x(0), \mathbf{u}^p_{j\in\mathbb{I}_{1:M}, j\neq i}\right)$. The plantwide cost function then satisfies for any $p \geq 0$

$$V(x^+, \mathbf{u}^+) \leq V(x, \mathbf{u}) - \sum_{j\in\mathbb{I}_{1:M}} \rho_j \ell_j(x_j, u_j)$$

$$|\mathbf{u}| \leq d\,|x| \qquad x \in r\mathcal{B}$$

For the M-player game, we generalize Assumption 6.12 of the two-player game to the following.

Assumption 6.16 (Constrained M-player game).

(a) The systems (A_{ij}, B_{ij}), $i, j \in \mathbb{I}_{1:M}$ are stabilizable.

(b) The systems (A_{ij}, C_{ij}), $i, j \in \mathbb{I}_{1:M}$ are detectable.

(c) The input penalties $R_i, i \in \mathbb{I}_{1:M}$ are positive definite, and $Q_i, i \in \mathbb{I}_{1:M}$ are semidefinite.

(d) The systems $(A_i, Q_i), i \in \mathbb{I}_{1:M}$ are detectable.

(e) The horizon is chosen sufficiently long to zero the unstable modes; $N \geq \max_{j\in\mathbb{I}_{1:M}}(\sum_{i\in\mathbb{I}_{1:M}} n^u_{ij})$.

(f) Zero offset. For achieving zero offset, we augment the models with integrating disturbances such that

$$\text{rank}\begin{bmatrix} I - A_i & -B_{di} \\ C_i & C_{di} \end{bmatrix} = n_i + p_i \qquad i \in \mathbb{I}_{1:M}$$

Applying Theorem 6.4 then establishes exponential stability of the solution $x(k) = 0$ for all k. The region of attraction is the set of states for which the unstable modes of each subsystem can be brought to zero in N moves while satisfying the respective input constraints. These conclusions apply regardless of how many iterations of the players' optimizations are used in the control calculation. Although the closed-loop system is exponentially stable for both coupled and uncoupled constraints, the converged distributed controller is equal to the centralized controller only for the case of uncoupled constraints.

The exponential stability of the regulator implies that the states and inputs of the constrained M-player system converge to the steady-state target. The steady-state target can be calculated as a centralized or distributed problem. We assume the centralized target has a feasible, zero offset solution for the true plant disturbance. The initial state of the plant and the estimate error must be small enough that feasibility of the target is maintained under the nonzero estimate error.

6.5 Nonlinear Distributed MPC

In the nonlinear case, the usual model comes from physical principles and conservation laws of mass, energy and momentum. The state has a physical meaning and the measured outputs usually are a subset of the state. We assume the model is of the form

$$\begin{aligned}
\frac{dx_1}{dt} &= f_1(x_1, x_2, u_1, u_2) \\
y_1 &= C_1 x_1 \\
\frac{dx_2}{dt} &= f_2(x_1, x_2, u_1, u_2) \\
y_2 &= C_2 x_2
\end{aligned}$$

in which C_1, C_2 are matrices of zeros and ones selecting the part of the state that is measured in subsystems one and two. We generally cannot avoid state x_2 dependence in the differential equation for x_1. But often it is only a small subset of the entire state x_2 that appears in f_1, and vice versa. The reason in chemical process systems is that the two subsystems are generally coupled through a small set of process streams transferring mass and energy between the systems. These connecting streams isolate the coupling between the two systems and reduce the influence to a small part of the entire state required to describe each system.

Given these physical system models of the subsystems, the overall plant model is

$$\begin{aligned}
\frac{dx}{dt} &= f(x, u) \\
y &= Cx
\end{aligned}$$

in which

$$x = \begin{bmatrix} x_1 \\ x_2 \end{bmatrix} \quad u = \begin{bmatrix} u_1 \\ u_2 \end{bmatrix} \quad f = \begin{bmatrix} f_1 \\ f_2 \end{bmatrix} \quad y = \begin{bmatrix} y_1 \\ y_2 \end{bmatrix} \quad C = \begin{bmatrix} C_1 & \\ & C_2 \end{bmatrix}$$

Nonconvexity. The basic difficulty in both the theory and application of nonlinear MPC is the nonconvexity in the control objective function caused by the nonlinear dynamic model. This difficulty applies even to centralized nonlinear MPC as discussed in Section 2.8, and motivates the development of suboptimal MPC. In the distributed case, nonconvexity causes even greater difficulties. As an illustration, consider the

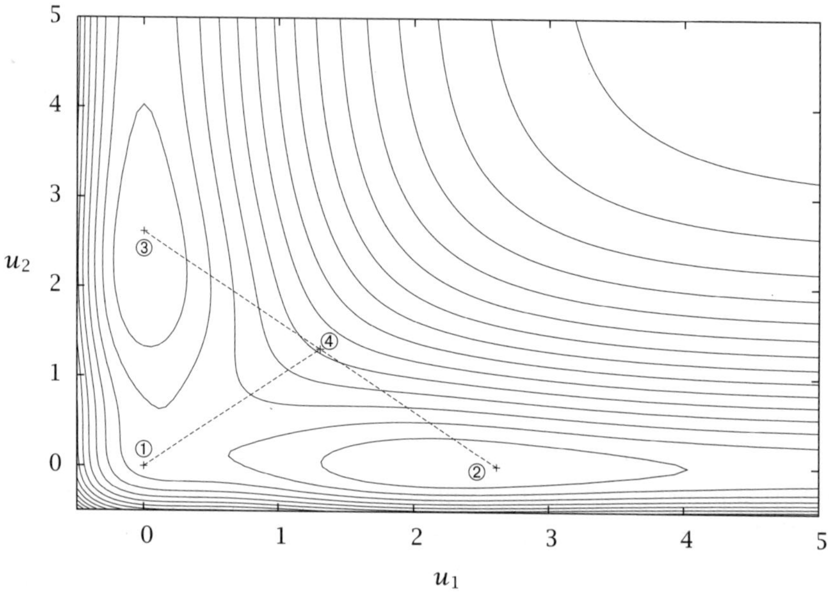

Figure 6.7: Cost contours for a two-player, nonconvex game; cost *increases* for the convex combination of the two players' optimal points.

simple two-player, nonconvex game depicted in Figure 6.7. The cost function is

$$V(u_1, u_2) = e^{-2u_1} - 2e^{-u_1} + e^{-2u_2} - 2e^{-u_2} \\ + a \exp(-\beta((u_1 + 0.2)^2 + (u_2 + 0.2)^2))$$

in which $a = 1.1$ and $\beta = 0.4$. Each player optimizes the cooperative objective starting at ① and produces the points, (u_1^0, u_2^p), denoted ② and (u_1^p, u_2^0), denoted ③. Consider taking a convex combination of the two players' optimal points for the next iterate

$$(u_1^{p+1}, u_2^{p+1}) = w_1(u_1^0, u_2^p) + w_2(u_1^p, u_2^0) \qquad w_1 + w_2 = 1, \quad w_1, w_2 \geq 0$$

We see in Figure 6.7 that this iterate causes the objective function to *increase* rather than decrease for most values of w_1, w_2. For $w_1 = w_2 = 1/2$, we see clearly from the contours that V at point ④ is greater than V at point ①. The values of the four points are given in the following table

Point	u_1	u_2	$V(u)$
①	0	0	−0.93
②	2.62	0	−1.10
③	0	2.62	−1.10
④	1.31	1.31	−0.76

The possibility of a cost increase leads to the possibility of closed-loop instability and precludes developing even a nominal control theory for this situation. In the centralized MPC problem, this nonconvexity issue can be addressed in the optimizer, which can move both inputs simultaneously and always avoid a cost increase. In the distributed case, the required information to avoid a cost increase is *by design* unavailable to the players.

One can of course consider adding another player to the game who has access to more systemwide information. This player takes the optimization results of the individual players and determines a search direction and step length that achieve a cost decrease for the overall system. This player is often known as a coordinator. The main challenge of this approach is that the design of the coordinator may not be significantly simpler than the design of the centralized controller. This issue remains a topic of current research.

6.6 Notes

At least three different fields have contributed substantially to the material presented in this chapter. We attempt here to point out briefly what each field has contributed, and indicate what literature the interested reader may wish to consult for further pursuing this and related subjects.

Game theory. Game theory emerged in the mid-1900s to analyze situations in which multiple players follow a common set of rules but have their own and different objectives that they try to optimize in competition with each other. Von Neumann and Morgenstern introduced the classic text on this subject, "Theory of Games and Economic Behavior," in 1944. A principle aim of game theory since its inception was to model and understand human *economic* behavior, especially as it arises in a capitalistic, free-market system. For that reason, much of the subsequent game theory literature was published in economics journals rather than systems theory journals. This field has contributed richly to the ideas and vocabulary used in this chapter to describe distributed control. For example, the game in which players have different objectives is termed *noncooperative*. The equilibrium of a noncooperative game is known as a *Nash equilibrium* (Nash, 1951). The Nash equilibrium is usually not Pareto optimal, which means that the outcomes for all players can be improved simultaneously from the Nash solution. A comprehensive overview of the game theory literature, especially the parts relevant to control theory, is provided by Başar and Olsder (1999, Chapter 1), which is a highly recommended reference. Analyzing the equilibria of a noncooperative game is usually more complex than the cooperative game (optimal control problem). The closed-loop properties of a receding horizon implementation of any of these game theory solutions is not addressed in game theory. That topic is addressed by control theory.

Distributed optimization. The optimization community has extensively studied the issue of solving large-scale optimization problems using distributed optimization methods. The primary motivation in this field is to exploit parallel computing hardware and distributed data communication networks to solve large optimization problems faster. Bertsekas and Tsitsiklis provide an excellent and comprehensive overview of this field focusing on numerical algorithms for implementing the distributed approaches. The important questions that are addressed in designing a distributed optimization are: task allocation, communication, and synchronization (Bertsekas and Tsitsiklis, 1997, Chapter 1).

These basic concepts arise in distributed problems of all types, and therefore also in the distributed MPC problem, which provides good synergy between these fields. But one should also be aware of the structural distinctions between distributed optimization and distributed MPC. The primary obstacle to implementing centralized MPC for large-scale plants is not *computational* but *organizational*. The agents considered in distributed MPC are usually existing MPC systems already built for units or subsystems within an existing large-scale process. The plant management often is seeking to improve the plant performance by better coordinating the behavior of the different agents already in operation. Ignoring these structural constraints and treating the distributed MPC problem purely as a form of distributed optimization, ignores aspects of the design that are critical for successful industrial application (Rawlings and Stewart, 2008).

Control theory. Researchers have long studied the issue of how to distribute control tasks in a complex large-scale plant (Mesarović, Macko, and Takahara, 1970; Sandell Jr., Varaiya, Athans, and Safonov, 1978). The centralized controller and decentralized controller define two limiting design extremes. Centralized control accounts for all possible interactions, large and small, whereas decentralized control ignores them completely. In decentralized control the local agents have no knowledge of each others' actions. It is well known that the nominal closed-loop system behavior under decentralized control can be arbitrarily poor (unstable) if the system interactions are not small. The following reviews provide general discussion of this and other performance issues involving decentralized control (Šiljak, 1991; Lunze, 1992; Larsson and Skogestad, 2000; Cui and Jacobsen, 2002).

The next level up in design complexity from decentralized control is noncooperative control. In this framework, the agents have interaction models and communicate at each iteration (Jia and Krogh, 2002; Motee and Sayyar-Rodsari, 2003; Dunbar and Murray, 2006). The advantage of noncooperative control over decentralized control is that the agents have accurate knowledge of the effects of all other agents on their local objectives. The basic issue to analyze and understand in this setup is the competition between the agents. Characterizing the noncooperative equilibrium is the subject of noncooperative game theory, and the impact of using that solution for feedback control is the subject of control theory. For example, Dunbar (2007) shows closed-loop stability for an extension of noncooperative MPC described in (Dunbar and Murray, 2006) that handles systems with interacting subsystem dynamics. The key assumptions are the existence of a stabilizing *decentralized* feedback law valid near the origin, and an inequality condition limiting the coupling between the agents.

Cooperative MPC was introduced by Venkat, Rawlings, and Wright (2007). They show that a receding horizon implementation of a cooperative game with any number of iterates of the local MPC controllers leads to closed-loop stability in the linear dynamics case. Venkat, Rawlings, and Wright (2006a,b) show that state estimation errors (output instead of state feedback) do not change the system closed-loop stability if the estimators are also asymptotically stable. Most of the theoretical results on cooperative MPC given in this chapter are presented in Venkat (2006) using an earlier, different notation. If implementable, this form of distributed MPC clearly has the best control properties. Although one can easily modify the agents' objective functions in a single large-scale process owned by a single company, this kind of modification may not be possible in other situations in which competing interests share critical infrastructure. The requirements of the many different classes of applications create exciting opportunities for continued research in this field.

6.7 Exercises

Exercise 6.1: Three looks at solving the LQ problem

In the following exercise, you will write three codes to solve the LQR using Octave or MATLAB. The objective function is the LQR with mixed term

$$V = \frac{1}{2}\sum_{k=0}^{N-1}\left(x(k)'Qx(k) + u(k)'Ru(k) + 2x(k)'Mu(k)\right) + (1/2)x(N)'P_f x(N)$$

First, implement the method described in Section 6.1.1 in which you eliminate the state and solve the problem for the decision variable

$$\mathbf{u} = \{u(0), u(1), \ldots, u(N-1)\}$$

Second, implement the method described in Section 6.1.1 in which you do *not* eliminate the state and solve the problem for

$$\mathbf{z} = \{u(0), x(1), u(1), x(2), \ldots, u(N-1), x(N)\}$$

Third, use backward dynamic programming (DP) and the Riccati iteration to compute the closed-form solution for $u(k)$ and $x(k)$.

(a) Let

$$A = \begin{bmatrix} 4/3 & -2/3 \\ 1 & 0 \end{bmatrix} \qquad B = \begin{bmatrix} 1 \\ 0 \end{bmatrix} \qquad C = \begin{bmatrix} -2/3 & 1 \end{bmatrix} \qquad x(0) = \begin{bmatrix} 1 \\ 1 \end{bmatrix}$$

$$Q = C'C + 0.001I \qquad P_f = \Pi \qquad R = 0.001 \qquad M = 0$$

in which the terminal penalty, P_f is set equal to Π, the steady-state cost to go. Compare the three solutions for $N = 5$. Plot $x(k)$, $u(k)$ versus time for the closed-loop system.

(b) Let $N = 50$ and repeat. Do any of the methods experience numerical problems generating an accurate solution? Plot the condition number of the matrix that is inverted in the first two methods versus N.

(c) Now consider the following unstable system

$$A = \begin{bmatrix} 27.8 & -82.6 & 34.6 \\ 25.6 & -76.8 & 32.4 \\ 40.6 & -122.0 & 51.9 \end{bmatrix} \qquad B = \begin{bmatrix} 0.527 & 0.548 \\ 0.613 & 0.530 \\ 1.06 & 0.828 \end{bmatrix} \qquad x(0) = \begin{bmatrix} 1 \\ 1 \\ 1 \end{bmatrix}$$

Consider regulator tuning parameters and constraints

$$Q = I \qquad P_f = \Pi \qquad R = I \qquad M = 0$$

Repeat parts 6.1a and 6.1b for this system. Do you lose accuracy in any of the solution methods? What happens to the condition number of $H(N)$ and $S(N)$ as N becomes large? Which methods are still accurate for this case? Can you explain what happened?

Exercise 6.2: LQ as least squares

Consider the standard LQ problem

$$\min_{\mathbf{u}} V = \frac{1}{2}\sum_{k=0}^{N-1}\left(x(k)'Qx(k) + u(k)'Ru(k)\right) + (1/2)x(N)'P_f x(N)$$

subject to

$$x^+ = Ax + Bu$$

(a) Set up the dense Hessian least squares problem for the LQ problem with a horizon of three, $N = 3$. Eliminate the state equations and write out the objective function in terms of only the decision variables $u(0), u(1), u(2)$.

(b) What are the conditions for an optimum, i.e., what linear algebra problem do you solve to compute $u(0), u(1), u(2)$?

Exercise 6.3: Lagrange multiplier method

Consider the general least squares problem

$$\min_{x} V(x) = \frac{1}{2}x'Hx + \text{const}$$

subject to

$$Dx = d$$

(a) What is the Lagrangian L for this problem? What is the dimension of the Lagrange multiplier vector, λ?

(b) What are necessary and sufficient conditions for a solution to the optimization problem?

(c) Apply this approach to the LQ problem of Exercise 6.2 using the equality constraints to represent the model equations. What are H, D, d for the LQ problem?

(d) Write out the linear algebra problem to be solved for the optimum.

(e) Contrast the two different linear algebra problems in these two approaches. Which do you want to use when N is large and why?

Exercise 6.4: Reparameterizing an unstable system

Consider again the LQR problem with cross term

$$\min_{\mathbf{u}} V = \frac{1}{2}\sum_{k=0}^{N-1}\left(x(k)'Qx(k) + u(k)'Ru(k) + 2x(k)'Mu(k)\right) + (1/2)x(N)'P_f x(N)$$

subject to

$$x^+ = Ax + Bu$$

and the three approaches of Exercise 6.1:

1. The method described in Section 6.1.1 in which you eliminate the state and solve the problem for the decision variable

$$\mathbf{u} = \{u(0), u(1), \ldots, u(N-1)\}$$

2. The method described in Section 6.1.1 in which you do *not* eliminate the state and solve the problem for

$$\mathbf{z} = \{u(0), x(1), u(1), x(2), \ldots, u(N-1), x(N)\}$$

3. The method of DP and the Riccati iteration to compute the closed-form solution for $u(k)$ and $x(k)$.

(a) You found that unstable A causes numerical problems in the first method using large horizons. So let's consider a fourth method. Reparameterize the input in terms of a state feedback gain via

$$u(k) = Kx(k) + v(k)$$

in which K is chosen so that $A + BK$ is a stable matrix. Consider the matrices in a transformed LQ problem

$$\min_{\mathbf{v}} V = \frac{1}{2} \sum_{k=0}^{N-1} \left(x(k)'\widetilde{Q}x(k) + v(k)'\widetilde{R}v(k) + 2x(k)'\widetilde{M}v(k) \right) + (1/2)x(N)'\widetilde{P}_f x(N)$$

subject to $x^+ = \widetilde{A}x + \widetilde{B}v$.

What are the matrices $\widetilde{A}, \widetilde{B}, \widetilde{Q}, \widetilde{P}_f, \widetilde{R}, \widetilde{M}$ such that the two problems give the same solution (state trajectory)?

(b) Solve the following problem using the first method and the fourth method and describe differences between the two solutions. Compare your results to the DP approach. Plot $x(k)$ and $u(k)$ versus k.

$$A = \begin{bmatrix} 27.8 & -82.6 & 34.6 \\ 25.6 & -76.8 & 32.4 \\ 40.6 & -122.0 & 51.9 \end{bmatrix} \qquad B = \begin{bmatrix} 0.527 & 0.548 \\ 0.613 & 0.530 \\ 1.06 & 0.828 \end{bmatrix} \qquad x(0) = \begin{bmatrix} 1 \\ 1 \\ 1 \end{bmatrix}$$

Consider regulator tuning parameters and constraints

$$Q = P_f = I \qquad R = I \qquad M = 0 \qquad N = 50$$

Exercise 6.5: Recursively summing quadratic functions

Consider generalizing Example 1.1 to an N-term sum. Let the N-term sum of quadratic functions be defined as

$$V(N, x) = \frac{1}{2} \sum_{i=1}^{N} (x - x(i))' X_i (x - x(i))$$

in which $x, x(i) \in \mathbb{R}^n$ are real n-vectors and $X_i \in \mathbb{R}^{n \times n}$ are positive definite matrices.

(a) Show that $V(N, x)$ can be found recursively

$$V(N, x) = (1/2)(x - v(N))' H(N)(x - v(N)) + \text{constant}$$

in which $v(i)$ and $H(i)$ satisfy the recursion

$$H(i+1) = H_i + X_{i+1} \qquad v(i+1) = H^{-1}(i+1)\left(H_i v_i + X_{i+1} x(i+1)\right)$$
$$H_1 = X_1 \qquad v_1 = x_1$$

Notice the recursively defined $v(m)$ and $H(m)$ provide the solutions and the Hessian matrices of the sequence of optimization problems

$$\min_x V(m, x) \qquad 1 \leq m \leq N$$

(b) Check your answer by solving the equivalent, but larger dimensional, constrained least squares problem (see Exercise 1.16)

$$\min_z (z - z_0)'\widetilde{H}(z - z_0)$$

subject to

$$Dz = 0$$

in which $z, z_0 \in \mathbb{R}^{nN}$, $\widetilde{H} \in \mathbb{R}^{nN \times nN}$ is a block diagonal matrix, $D \in \mathbb{R}^{n(N-1)\times nN}$

$$z_0 = \begin{bmatrix} x(1) \\ \vdots \\ x(N-1) \\ x(N) \end{bmatrix} \qquad \widetilde{H} = \begin{bmatrix} X_1 & & & \\ & \ddots & & \\ & & X_{N-1} & \\ & & & X_N \end{bmatrix} \qquad D = \begin{bmatrix} I & -I & & \\ & \ddots & \ddots & \\ & & I & -I \end{bmatrix}$$

(c) Compare the size and number of matrix inverses required for the two approaches.

Exercise 6.6: Why call the Lyapunov stability *nonuniform*?

Consider the following linear system

$$\begin{aligned} w^+ &= Aw \qquad w(0) = Hx(0) \\ x &= Cw \end{aligned}$$

with solution $w(k) = A^k w(0) = A^k Hx(0)$, $x(k) = CA^k Hx(0)$. Notice that $x(0)$ completely determines both $w(k)$ and $x(k)$, $k \geq 0$. Also note that zero is a solution, i.e., $x(k) = 0, k \geq 0$ satisfies the model.

(a) Consider the following case

$$A = \rho \begin{bmatrix} \cos\theta & -\sin\theta \\ \sin\theta & \cos\theta \end{bmatrix} \qquad H = \begin{bmatrix} 0 \\ -1 \end{bmatrix} \qquad C = \begin{bmatrix} 1 & -1 \end{bmatrix}$$

$$\rho = 0.925 \qquad \theta = \pi/4 \qquad x(0) = 1$$

Plot the solution $x(k)$. Does $x(k)$ converge to zero? Does $x(k)$ achieve zero exactly for finite $k > 0$?

(b) Is the zero solution $x(k) = 0$ Lyapunov stable? State your definition of Lyapunov stability, and prove your answer. Discuss how your answer is consistent with the special case considered above.

Exercise 6.7: Exponential stability of suboptimal MPC with unbounded feasible set

Consider again Lemma 6.4 when both $\mathbb{U}$ and $\mathcal{X}_N$ are unbounded. Show that the suboptimal MPC controller is exponentially stable on the following sets.

(a) Any sublevel set of $V(x, \mathbf{h}(x))$

(b) Any compact subset of $\mathcal{X}_N$

Exercise 6.8: A refinement to the warm start

Consider the following refinement to the warm start in the suboptimal MPC strategy. First add the requirement that the initialization strategy satisfies the following bound

$$\mathbf{h}(x) \leq \bar{d}\,|x| \qquad x \in \mathcal{X}_N$$

in which $\bar{d} > 0$. Notice that all initializations considered in the chapter satisfy this requirement.

Then, at time k and state x, in addition to the shifted input sequence from time $k-1$, $\tilde{\mathbf{u}}$, evaluate the initialization sequence applied to the current state, $\mathbf{u} = \mathbf{h}(x)$. Select whichever of these two input sequence has lower cost as the warm start for time k. Notice also that this refinement makes the constraint

$$|\mathbf{u}| \leq d\,|x| \qquad x \in r\mathcal{B}$$

redundant, and it can be removed from the MPC optimization.

Prove that this refined suboptimal strategy is exponentially stabilizing on the set $\mathcal{X}_N$. Notice that with this refinement, we do not have to assume that $\mathcal{X}_N$ is bounded or that $\mathbb{U}$ is bounded.

Exercise 6.9: Exponential stability with mixed powers of the norm

Prove Lemma 6.5.

Hints: exponential convergence can be established as in standard exponential stability theorems. To establish Lyapunov stability, consider sublevel sets of the function $|x|^\sigma + |e|^\gamma$

$$L_\rho = \{(x,e) \mid |x|^\sigma + |e|^\gamma \leq \rho\} \qquad \rho > 0$$

Choose the function $\overline{\rho}(R)$ to be the maximal ρ value such that $L_{\overline{\rho}(R)} \subset B_R$. Similarly, choose $\overline{r}(\rho)$ to be the maximal r value such that $B_{\overline{r}(\rho)} \subset L_\rho$. Use $\overline{\rho}$ and $\overline{r}$ to establish Lyapunov stability.

Exercise 6.10: Decentralized control of Examples 6.8–6.10

Apply decentralized control to the systems in Examples 6.8–6.10. Which of these systems are closed-loop unstable with decentralized control? Compare this result to the result for noncooperative MPC.

Exercise 6.11: Cooperative control of Examples 6.8–6.10

Apply cooperative MPC to the systems in Examples 6.8–6.10. Are any of these systems closed-loop unstable? Compare the closed-loop eigenvalues of converged cooperative control to centralized MPC, and discuss any differences.

Exercise 6.12: Adding norms

Establish the following result used in the proof of Lemma 6.13. Given that $w \in \mathbb{R}^m$, $e \in \mathbb{R}^n$

$$\frac{1}{\sqrt{2}}(|w| + |e|) \leq |(w,e)| \leq |w| + |e| \qquad \forall w, e$$

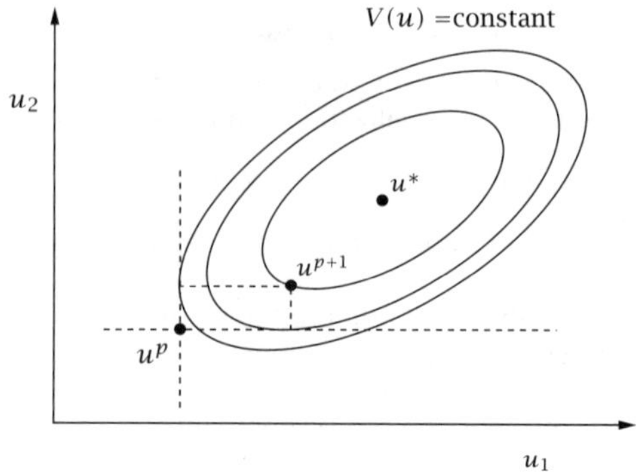

Figure 6.8: Optimizing a quadratic function in one set of variables at a time.

Exercise 6.13: Padding matrices

Given a vector $\mathbf{z}$ and subvector $\mathbf{u}$

$$\mathbf{z} = \begin{bmatrix} u(0) \\ x(1) \\ u(1) \\ x(2) \\ \vdots \\ u(N-1) \\ x(N) \end{bmatrix} \qquad \mathbf{u} = \begin{bmatrix} u(0) \\ u(1) \\ \vdots \\ u(N-1) \end{bmatrix} \qquad x \in \mathbb{R}^n \qquad u \in \mathbb{R}^m$$

and quadratic function of $\mathbf{u}$

$$(1/2)\mathbf{u}'H\mathbf{u} + h'\mathbf{u}$$

Find the corresponding quadratic function of $\mathbf{z}$ so that

$$(1/2)\mathbf{z}'H_z\mathbf{z} + h_z'\mathbf{z} = (1/2)\mathbf{u}'H\mathbf{u} + h'\mathbf{u} \qquad \forall \mathbf{z}, \mathbf{u}$$

Hint: first find the padding matrix E such that $\mathbf{u} = E\mathbf{z}$.

Exercise 6.14: A matrix inverse

Compute the four partitioned elements in the two-player feedback gain $(I - L)^{-1}\overline{K}$

$$\mathbf{u}^{\infty} = (I - L)^{-1}\overline{K}x(0) \qquad |\text{eig}(L)| < 1$$

in which

$$(I - L)^{-1}\overline{K} = \begin{bmatrix} I & -L_1 \\ -L_2 & I \end{bmatrix}^{-1} \begin{bmatrix} K_1 & 0 \\ 0 & K_2 \end{bmatrix}$$

Exercise 6.15: Optimizing one variable at a time

Consider the positive definite quadratic function partitioned into two sets of variables

$$V(u) = (1/2)u'Hu + c'u + d$$

$$V(u_1, u_2) = (1/2)\begin{bmatrix} u_1' & u_2' \end{bmatrix}\begin{bmatrix} H_{11} & H_{12} \\ H_{21} & H_{22} \end{bmatrix}\begin{bmatrix} u_1 \\ u_2 \end{bmatrix} + \begin{bmatrix} c_1' & c_2' \end{bmatrix}\begin{bmatrix} u_1 \\ u_2 \end{bmatrix} + d$$

in which $H > 0$. Imagine we wish to optimize this function by first optimizing over the u_1 variables holding u_2 fixed and then optimizing over the u_2 variables holding u_1 fixed as shown in Figure 6.8. Let's see if this procedure, while not necessarily efficient, is guaranteed to converge to the optimum.

(a) Given an initial point (u_1^p, u_2^p), show that the next iteration is

$$\begin{aligned} u_1^{p+1} &= -H_{11}^{-1}\left(H_{12}u_2^p + c_1\right) \\ u_2^{p+1} &= -H_{22}^{-1}\left(H_{21}u_1^p + c_2\right) \end{aligned} \tag{6.29}$$

The procedure can be summarized as

$$u^{p+1} = Au^p + b \tag{6.30}$$

in which the iteration matrix A and constant b are given by

$$A = \begin{bmatrix} 0 & -H_{11}^{-1}H_{12} \\ -H_{22}^{-1}H_{21} & 0 \end{bmatrix} \qquad b = \begin{bmatrix} -H_{11}^{-1}c_1 \\ -H_{22}^{-1}c_2 \end{bmatrix} \tag{6.31}$$

(b) Establish that the optimization procedure converges by showing the iteration matrix is stable

$$\left|\operatorname{eig}(A)\right| < 1$$

(c) Given that the iteration converges, show that it produces the same solution as

$$u^* = -H^{-1}c$$

Exercise 6.16: Monotonically decreasing cost

Consider again the iteration defined in Exercise 6.15.

(a) Prove that the cost function is monotonically decreasing when optimizing one variable at a time

$$V(u^{p+1}) < V(u^p) \qquad \forall u^p \neq -H^{-1}c$$

(b) Show that the following expression gives the size of the decrease

$$V(u^{p+1}) - V(u^p) = -(1/2)(u^p - u^*)'P(u^p - u^*)$$

in which

$$P = HD^{-1}\widetilde{H}D^{-1}H \qquad \widetilde{H} = D - N \qquad D = \begin{bmatrix} H_{11} & 0 \\ 0 & H_{22} \end{bmatrix} \qquad N = \begin{bmatrix} 0 & H_{12} \\ H_{21} & 0 \end{bmatrix}$$

and $u^* = -H^{-1}c$ is the optimum.

Hint: to simplify the algebra, first change coordinates and move the origin of the coordinate system to u^*.

Exercise 6.17: One variable at a time with convex step

Consider Exercise 6.15 but with the convex step for the iteration

$$\begin{bmatrix} u_1^{p+1} \\ u_2^{p+1} \end{bmatrix} = w_1 \begin{bmatrix} u_1^0(u_2^p) \\ u_2^p \end{bmatrix} + w_2 \begin{bmatrix} u_1^p \\ u_2^0(u_1^p) \end{bmatrix} \qquad 0 \leq w_1, w_2 \quad w_1 + w_2 = 1$$

(a) Show that the iteration for the convex step is also of the form

$$u^{p+1} = Au^p + b$$

and the A matrix and b vector for this case are

$$A = \begin{bmatrix} w_2 I & -w_1 H_{11}^{-1} H_{12} \\ -w_2 H_{22}^{-1} H_{21} & w_1 I \end{bmatrix} \qquad b = \begin{bmatrix} -w_1 H_{11}^{-1} & \\ & -w_2 H_{22}^{-1} \end{bmatrix} \begin{bmatrix} c_1 \\ c_2 \end{bmatrix}$$

(b) Show that A is stable.

(c) Show that this iteration also converges to $u^* = -H^{-1}c$.

Exercise 6.18: Monotonically decreasing cost with convex step

Consider again the problem of optimizing one variable at a time with the convex step given in Exercise 6.17.

(a) Prove that the cost function is monotonically decreasing

$$V(u^{p+1}) < V(u^p) \qquad \forall u^p \neq -H^{-1}c$$

(b) Show that the following expression gives the size of the decrease

$$V(u^{p+1}) - V(u^p) = -(1/2)(u^p - u^*)'P(u^p - u^*)$$

in which

$$P = HD^{-1}\widetilde{H}D^{-1}H \qquad \widetilde{H} = D - N$$

$$D = \begin{bmatrix} w_1^{-1} H_{11} & 0 \\ 0 & w_2^{-1} H_{22} \end{bmatrix} \qquad N = \begin{bmatrix} -w_1^{-1} w_2 H_{11} & H_{12} \\ H_{21} & -w_1 w_2^{-1} H_{22} \end{bmatrix}$$

and $u^* = -H^{-1}c$ is the optimum.

Hint: to simplify the algebra, first change coordinates and move the origin of the coordinate system to u^*.

Exercise 6.19: Splitting more than once

Consider the generalization of Exercise 6.15 in which we repeatedly decompose a problem into one-variable-at-a-time optimizations. For a three-variable problem we have the three optimizations

$$u_1^{p+1} = \arg\min_{u_1} V(u_1, u_2^p, u_3^p)$$

$$u_2^{p+1} = \arg\min_{u_2} V(u_1^p, u_2, u_3^p) \qquad u_3^{p+1} = \arg\min_{u_3} V(u_1^p, u_2^p, u_3)$$

Is it true that

$$V(u_1^{p+1}, u_2^{p+1}, u_3^{p+1}) \leq V(u_1^p, u_2^p, u_3^p)$$

Hint: you may wish to consider the following example, $V(u) = (1/2)u'Hu + c'u$, in which

$$H = \begin{bmatrix} 2 & 1 & 1 \\ 1 & 1 & 1 \\ 1 & 1 & 2 \end{bmatrix} \qquad c = \begin{bmatrix} 0 \\ 1 \\ 1 \end{bmatrix} \qquad u^p = \begin{bmatrix} 1 \\ 0 \\ 1 \end{bmatrix}$$

Exercise 6.20: Time-varying controller iterations

We let $p_k \geq 0$ be a time-varying integer-valued index representing the iterations applied in the controller at time k.

$$\begin{aligned} x_1(k+1) &= A_1 x_1(k) + \overline{B}_{11} u_1(0;k) + \overline{B}_{12} u_2(0;k) \\ x_2(k+1) &= A_2 x_2(k) + \overline{B}_{21} u_1(0;k) + \overline{B}_{22} u_2(0;k) \\ \mathbf{u}_1(k+1) &= g_1^{p_k}(x_1(k), x_2(k), \mathbf{u}_1(k), \mathbf{u}_2(k)) \\ \mathbf{u}_2(k+1) &= g_2^{p_k}(x_1(k), x_2(k), \mathbf{u}_1(k), \mathbf{u}_2(k)) \end{aligned}$$

Notice the system evolution is time-varying even though the models are time invariant because we allow a time-varying sequence of controller iterations.

Show that cooperative MPC is exponentially stabilizing for any $p_k \geq 0$ sequence.

Exercise 6.21: Stable interaction models

In some industrial applications it is preferable to partition the plant so that there are no unstable connections between subsystems. Any inputs u_j that have unstable connections to outputs y_i should be included in the ith subsystem inputs. Allowing an unstable connection between two subsystems may not be robust to faults and other kinds of system failures.[4] To implement this design idea in the two-player case, we replace Assumption 6.12 (b) with the following

Assumption 6.12 (Constrained two-player game).
(b') The interaction models $A_{ij}, i \neq j$ are stable.

Prove that Assumption 6.12 (b') implies Assumption 6.12 (b). It may be helpful to first prove the following lemma.

Lemma 6.17 (Local detectability). *Given partitioned system matrices*

$$\mathbf{A} = \begin{bmatrix} A & 0 \\ 0 & A_s \end{bmatrix} \qquad \mathbf{C} = \begin{bmatrix} C & C_s \end{bmatrix}$$

in which A_s is stable, the system $(\mathbf{A}, \mathbf{C})$ is detectable if and only if the system (A, C) is detectable.

Hint: use the Hautus lemma as the test for detectability.

Next show that this lemma and Assumption 6.12 (b') establishes the distributed detectability assumption 6.12 (b).

Exercise 6.22: Norm constraints as linear inequalities

Consider the quadratic program (QP) in decision variable u with parameter x

$$\begin{aligned} &\min_u (1/2)u'Hu + x'Du \\ &\text{s.t. } Eu \leq Fx \end{aligned}$$

---

[4]We are not considering the common instability of base-level inventory management in this discussion. It is assumed that level control in storage tanks (integrators) is maintained at all times with simple, local level controllers. The internal unit flowrates dedicated for inventory management are not considered available inputs in the MPC problem.

in which $u \in \mathbb{R}^m$, $x \in \mathbb{R}^n$, and $H > 0$. The parameter x appears linearly (affinely) in the cost function and constraints. Assume that we wish to add a norm constraint of the following form

$$|u|_\alpha \leq c\,|x|_\alpha \qquad \alpha = 2, \infty$$

(a) If we use the infinity norm, show that this problem can be posed as an equivalent QP with additional decision variables, and the cost function and constraints remain linear (affine) in parameter x. How many decision variables and constraints are added to the problem?

(b) If we use the two norm, show that this problem can be approximated by a QP whose solution does satisfy the constraints, but the solution may be suboptimal compared to the original problem.

Exercise 6.23: Steady-state noncooperative game

Consider again the steady-state target problem for the system given in Example 6.11.

(a) Resolve the problem for the choice of convex step parameters $w_1 = 0.2$, $w_2 = 0.8$. Does the iteration for noncooperative control converge? Plot the iterations for the noncooperative and cooperative cases.

(b) Repeat for the convex step $w_1 = 0.8, w_2 = 0.2$. Are the results identical to the previous part? If not, discuss any differences.

(c) For what choices of w_1, w_2 does the target iteration converge using noncooperative control for the target calculation?

Exercise 6.24: Optimality conditions for constrained optimization

Consider the convex quadratic optimization problem

$$\min_u V(u) \qquad \text{subject to} \quad u \in \mathbb{U}$$

in which V is a convex quadratic function and $\mathbb{U}$ is a convex set. Show that u^* is an optimal solution if and only if

$$\langle z - u^*, -\nabla V|_{u^*} \rangle \;\leq 0 \qquad \forall z \in \mathbb{U} \tag{6.32}$$

Figure 6.9(a) depicts this condition for $u \in \mathbb{R}^2$. This condition motivates defining the normal cone (Rockafellar, 1970) to $\mathbb{U}$ at u^* as follows

$$N(\mathbb{U}, u^*) = \{y \mid \langle z - u^*, y - u^* \rangle \;\leq 0 \quad \forall z \in \mathbb{U}\}$$

The optimality condition can be stated equivalently as u^* is an optimal point if and only if the negative gradient is in the normal cone to $\mathbb{U}$ at u^*

$$-\nabla V|_{u^*} \in N(\mathbb{U}, u^*)$$

This condition and the normal cone are depicted in Figure 6.9(b).

Exercise 6.25: Partitioned optimality conditions with constraints

Consider a partitioned version of the constrained optimization problem of Exercise 6.24 with uncoupled constraints

$$\min_{u_1, u_2} V(u_1, u_2) \qquad \text{subject to} \quad u_1 \in \mathbb{U}_1 \qquad u_2 \in \mathbb{U}_2$$

in which V is a quadratic function and $\mathbb{U}_1$ and $\mathbb{U}_2$ are convex and nonempty.

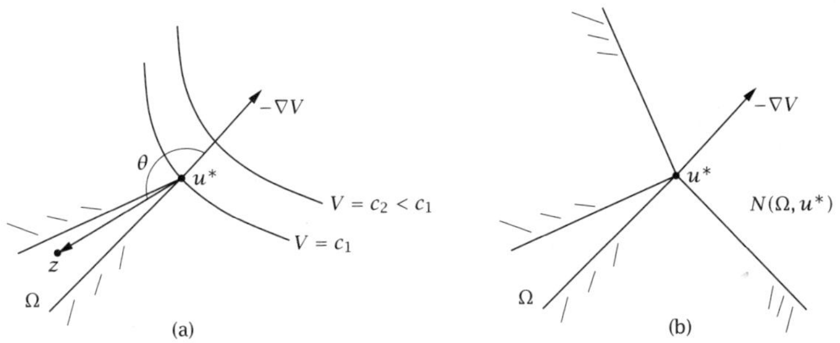

Figure 6.9: (a) Optimality of u^* means the angle between $-\nabla V$ and any point z in the feasible region must be greater than $90°$ and less than $270°$. (b) The same result restated: u^* is optimal if and only if the negative gradient is in the normal cone to the feasible region at u^*, $-\nabla V|_{u^*} \in N(\mathbb{U}, u^*)$.

(a) Show that (u_1^*, u_2^*) is an optimal solution if and only if

$$\begin{aligned} \langle z_1 - u_1^*, -\nabla_{u_1} V|_{(u_1^*, u_2^*)} \rangle &\leq 0 \qquad \forall z_1 \in \mathbb{U}_1 \\ \langle z_2 - u_2^*, -\nabla_{u_2} V|_{(u_1^*, u_2^*)} \rangle &\leq 0 \qquad \forall z_2 \in \mathbb{U}_2 \end{aligned} \tag{6.33}$$

(b) Extend the optimality conditions to cover the case

$$\min_{u_1, \ldots, u_M} V(u_1, \ldots, u_M) \qquad \text{subject to} \quad u_j \in \mathbb{U}_j \quad j = 1, \ldots, M$$

in which V is a quadratic function and the $\mathbb{U}_j$ are convex and nonempty.

Exercise 6.26: Constrained optimization of M variables

Consider an optimization problem with M variables and uncoupled constraints

$$\min_{u_1, u_2, \ldots, u_M} V(u_1, u_2, \ldots, u_M) \qquad \text{subject to} \quad u_l \in \mathbb{U}_j \quad j = 1, \ldots, M$$

in which V is a strictly convex function. Assume that the feasible region is convex and nonempty and denote the unique optimal solution as $(u_1^*, u_2^*, \ldots, u_M^*)$ having cost $V^* = V(u_1^*, \ldots, u_M^*)$. Denote the M one-variable-at-a-time optimization problems at iteration k

$$z_j^{p+1} = \arg\min_{u_j} V(u_1^p, \ldots, u_j, \ldots, u_M^p) \qquad \text{subject to } u_j \in \mathbb{U}_j$$

Then define the next iterate to be the following convex combination of the previous and new points

$$\begin{aligned} u_j^{p+1} &= \alpha_j^p z_j^{p+1} + (1 - \alpha_j^p) u_j^p \qquad j = 1, \ldots, M \\ \varepsilon &\leq \alpha_j^p < 1 \qquad 0 < \varepsilon \qquad j = 1, \ldots, M, \quad p \geq 1 \\ &\sum_{j=1}^{M} \alpha_j^p = 1, \qquad p \geq 1 \end{aligned}$$

Prove the following results.

(a) Starting with any feasible point, $(u_1^0, u_2^0, \ldots, u_M^0)$, the iterations $(u_1^p, u_2^p, \ldots, u_M^p)$ are feasible for $p \geq 1$.

(b) The objective function decreases monotonically from any feasible initial point

$$V(u_1^{p+1}, \ldots, u_M^{p+1}) \leq V(u_1^p, \ldots, u_M^p) \qquad \forall u_l^0 \in \mathbb{U}_j, j = 1, \ldots, M, \quad p \geq 1$$

(c) The cost sequence $V(u_1^p, u_2^p, \ldots, u_M^p)$ converges to the optimal cost V^* from any feasible initial point.

(d) The sequence $(u_1^p, u_2^p, \ldots, u_M^p)$ converges to the optimal solution $(u_1^*, u_2^*, \ldots, u_M^*)$ from any feasible initial point.

Exercise 6.27: The constrained two-variable special case

Consider the special case of Exercise 6.26 with $M = 2$

$$\min_{u_1, u_2} V(u_1, u_2) \qquad \text{subject to} \quad u_1 \in \mathbb{U}_1 \qquad u_2 \in \mathbb{U}_2$$

in which V is a strictly positive quadratic function. Assume that the feasible region is convex and nonempty and denote the unique optimal solution as (u_1^*, u_2^*) having cost $V^* = V(u_1^*, u_2^*)$. Consider the two one-variable-at-a-time optimization problems at iteration k

$$u_1^{p+1} = \arg\min_{u_1} V(u_1, u_2^p) \qquad\qquad u_2^{p+1} = \arg\min_{u_2} V(u_1^p, u_2)$$

$$\text{subject to } u_1 \in \mathbb{U}_1 \qquad\qquad \text{subject to } u_2 \in \mathbb{U}_2$$

We know from Exercise 6.15 that taking the full step in the unconstrained problem with $M = 2$ achieves a cost decrease. We know from Exercise 6.19 that taking the full step for an unconstrained problem with $M \geq 3$ does *not* provide a cost decrease in general. We know from Exercise 6.26 that taking a reduced step in the constrained problem for all M achieves a cost decrease. That leaves open the case of a full step for a constrained problem with $M = 2$.

Does the full step in the constrained case for $M = 2$ guarantee a cost decrease? If so, prove it. If not, provide a counterexample.

Exercise 6.28: Subsystem stability constraints

Show that the following uncoupled subsystem constraints imply an overall system constraint of the same type. The first is suitable for asymptotic stability and the second for exponential stability.

(a) Given $r_1, r_2 > 0$, and functions γ_1 and γ_2 of class $\mathcal{K}$, assume the following constraints are satisfied

$$|\mathbf{u}_1| \leq \gamma_1(|x_1|) \quad x_1 \in r_1\mathcal{B}$$
$$|\mathbf{u}_2| \leq \gamma_2(|x_2|) \quad x_2 \in r_2\mathcal{B}$$

Show that there exists $r > 0$ and function γ of class $\mathcal{K}$ such that

$$|(\mathbf{u}_1, \mathbf{u}_2)| \leq \gamma(|(x_1, x_2)|) \qquad (x_1, x_2) \in r\mathcal{B} \tag{6.34}$$

(b) Given $r_1, r_2 > 0$, and constants $c_1, c_2, \sigma_1, \sigma_2 > 0$, assume the following constraints are satisfied

$$\begin{aligned} |\mathbf{u}_1| &\leq c_1 \, |x_1|^{\sigma_1} & x_1 \in r_1\mathcal{B} \\ |\mathbf{u}_2| &\leq c_2 \, |x_2|^{\sigma_2} & x_2 \in r_2\mathcal{B} \end{aligned}$$

Show that there exists $r > 0$ and function $c, \sigma > 0$ such that

$$|(\mathbf{u}_1, \mathbf{u}_2)| \leq c \, |(x_1, x_2)|^{\sigma} \qquad (x_1, x_2) \in r\mathcal{B} \tag{6.35}$$

Exercise 6.29: Distributed disturbance detectability

Prove Lemma 6.15.

Hint: use the Hautus lemma as the test for detectability.

Exercise 6.30: Distributed target problem and uncoupled constraints

Player one's distributed target problem in the two-player game is given in (6.28)

$$\min_{x_{1s}, u_{1s}} (1/2) \begin{bmatrix} H_1 y_{1s} - z_{1sp} \\ H_2 y_{2s} - z_{2sp} \end{bmatrix}' \begin{bmatrix} T_{1s} & \\ & T_{2s} \end{bmatrix} \begin{bmatrix} H_1 y_{1s} - z_{1sp} \\ H_2 y_{2s} - z_{2sp} \end{bmatrix}$$

subject to:

$$\begin{bmatrix} I - A_1 & & -\overline{B}_{11} & -\overline{B}_{12} \\ & I - A_2 & -\overline{B}_{21} & -\overline{B}_{22} \end{bmatrix} \begin{bmatrix} x_{1s} \\ x_{2s} \\ u_{1s} \\ u_{2s} \end{bmatrix} = \begin{bmatrix} B_{1d}\hat{d}_1(k) \\ B_{2d}\hat{d}_2(k) \end{bmatrix}$$

$$E_1 u_{1s} \leq e_1$$

Show that the constraints can be expressed so that the target problem constraints are uncoupled.

Hint: in the modified form, player one solves for and exchanges with player two (x_{11s}, u_{1s}) instead of (x_{1s}, u_{1s}) as written above.

Bibliography

T. Başar and G. J. Olsder. *Dynamic Noncooperative Game Theory.* SIAM, Philadelphia, 1999.

D. P. Bertsekas and J. N. Tsitsiklis. *Parallel and Distributed Computation.* Athena Scientific, Belmont, Massachusetts, 1997.

A. E. Bryson and Y. Ho. *Applied Optimal Control.* Hemisphere Publishing, New York, 1975.

H. Cui and E. W. Jacobsen. Performance limitations in decentralized control. *J. Proc. Cont.*, 12:485–494, 2002.

W. B. Dunbar. Distributed receding horizon control of dynamically coupled nonlinear systems. *IEEE Trans. Auto. Cont.*, 52(7):1249–1263, 2007.

W. B. Dunbar and R. M. Murray. Distributed receding horizon control with application to multi-vehicle formation stabilization. *Automatica*, 42(4):549–558, 2006.

G. H. Golub and C. F. Van Loan. *Matrix Computations.* The Johns Hopkins University Press, Baltimore, Maryland, third edition, 1996.

D. Jia and B. H. Krogh. Min-max feedback model predictive control for distributed control with communication. In *Proceedings of the American Control Conference*, pages 4507–4512, Anchorage,Alaska, May 2002.

T. Larsson and S. Skogestad. Plantwide control- A review and a new design procedure. *Mod. Ident. Control*, 21(4):209–240, 2000.

J. Lunze. *Feedback Control of Large Scale Systems.* Prentice-Hall, London, U.K., 1992.

M. Mesarović, D. Macko, and Y. Takahara. *Theory of hierarchical, multilevel systems.* Academic Press, New York, 1970.

N. Motee and B. Sayyar-Rodsari. Optimal partitioning in distributed model predictive control. In *Proceedings of the American Control Conference*, pages 5300–5305, Denver,Colorado, June 2003.

J. Nash. Noncooperative games. *Ann. Math.*, 54:286–295, 1951.

J. B. Rawlings and B. T. Stewart. Coordinating multiple optimization-based controllers: new opportunities and challenges. *J. Proc. Cont.*, 18:839–845, 2008.

R. T. Rockafellar. *Convex Analysis.* Princeton University Press, Princeton, N.J., 1970.

N. R. Sandell Jr., P. Varaiya, M. Athans, and M. Safonov. Survey of decentralized control methods for larger scale systems. *IEEE Trans. Auto. Cont.*, 23(2):108–128, 1978.

P. O. M. Scokaert, D. Q. Mayne, and J. B. Rawlings. Suboptimal model predictive control (feasibility implies stability). *IEEE Trans. Auto. Cont.*, 44(3):648–654, March 1999.

D. D. Šiljak. *Decentralized Control of Complex Systems.* Academic Press, London, 1991.

A. N. Venkat. *Distributed Model Predictive Control: Theory and Applications.* PhD thesis, University of Wisconsin–Madison, October 2006.

A. N. Venkat, J. B. Rawlings, and S. J. Wright. Stability and optimality of distributed, linear MPC. Part 1: state feedback. Technical Report 2006–03, TWMCC, Department of Chemical and Biological Engineering, University of Wisconsin–Madison (Available at http://jbrwww.che.wisc.edu/tech-reports.html), October 2006a.

A. N. Venkat, J. B. Rawlings, and S. J. Wright. Stability and optimality of distributed, linear MPC. Part 2: output feedback. Technical Report 2006–04, TWMCC, Department of Chemical and Biological Engineering, University of Wisconsin–Madison (Available at http://jbrwww.che.wisc.edu/tech-reports.html), October 2006b.

A. N. Venkat, J. B. Rawlings, and S. J. Wright. Distributed model predictive control of large-scale systems. In *Assessment and Future Directions of Nonlinear Model Predictive Control*, pages 591–605. Springer, 2007.

M. Vidyasagar. *Nonlinear Systems Analysis.* Prentice-Hall, Inc., Englewood Cliffs, New Jersey, second edition, 1993.

J. von Neumann and O. Morgenstern. *Theory of Games and Economic Behavior.* Princeton University Press, Princeton and Oxford, 1944.

S. J. Wright. Applying new optimization algorithms to model predictive control. In J. C. Kantor, C. E. García, and B. Carnahan, editors, *Chemical Process Control–V*, pages 147–155. CACHE, AIChE, 1997.

7

Explicit Control Laws for Constrained Linear Systems

7.1 Introduction

In preceding chapters we show how model predictive control (MPC) can be derived for a variety of control problems with constraints. It is interesting to recall the major motivation for MPC; solution of a *feedback* optimal control problem yielding a stabilizing control *law* is often prohibitively difficult. MPC sidesteps the problem of determining a stabilizing control *law* $\kappa(\cdot)$ by determining, instead, at each state x encountered, a control *action* $u = \kappa(x)$ by solving a mathematical programming problem. This procedure, if repeated at *every* state x, yields an implicit control *law* $\kappa(\cdot)$ that solves the original feedback problem. In many cases, determining an explicit control law is impractical while solving a mathematical programming problem online for a given state is possible; this fact has led to the wide-scale adoption of MPC in the chemical process industry.

Some of the control problems for which MPC has been extensively used, however, have recently been shown to be amenable to analysis. One such problem is control of linear systems with polytopic constraints, for which determination of a stabilizing control law was thought in the past to be prohibitively difficult. It has been shown that it is possible, in principle, to determine a stabilizing control law for some of these control problems. Some authors have referred to these results as *explicit MPC* because they yield an explicit control law in contrast to MPC that yields a control action for each encountered state, thereby implicitly defining a control law. There are two objections to this terminology. First, determination of control laws for a wide variety of control problems has been the prime concern of control theory since its birth and certainly before the advent of MPC, an important tool in this endeavor being dynamic programming (DP). These new results merely show that classical control-theoretic tools, such as DP, can be successfully applied to a wider range of problems than was previously thought possible. MPC is a useful method for implementing a control law that can, in principle, be determined using control-theoretic tools.

Second, some authors using this terminology have, perhaps inadvertently, implied that these results can be employed in place of conventional MPC. This is far from the truth, since only relatively simple problems, far simpler than those routinely solved in MPC applications, can be solved. That said, the results may be useful in applications where models with low state dimension, say six or less, are sufficiently accurate and where it is important that the control be rapidly computed. A previously determined control law generally yields the control action more rapidly than solving an optimal control problem. Potential applications include vehicle control.

In the next section we give a few simple examples of parametric programming. In subsequent sections we show how the solutions to parametric linear and quadratic programs may be obtained, and also show how these solutions may be used to solve optimal control problems when the system is linear, the cost quadratic or affine, and the constraints polyhedral.

7.2 Parametric Programming

A conventional optimization problem has the form $V^0 = \min_u \{V(u) \mid u \in \mathcal{U}\}$ where u is the "decision" variable, $V(u)$ is the cost to be minimized, and $\mathcal{U}$ is the constraint set. The solution to a conventional optimization is a *point* or *set* in $\mathcal{U}$; the value V^0 of the problem satisfies $V^0 = V(u^0)$ where u^0 is the minimizer. A simple example of such a problem is $V^0 = \min_u \{a + bu + (1/2)cu^2 \mid u \in [-1, 1]\}$ where the solution is required for only *one* value of the parameters a, b and c. The solution to this problem $u^0 = -b/c$ if $|b/c| \leq 1$, $u^0 = -1$ if $b/c \geq 1$ and $u^0 = 1$ if $b/c \leq -1$. This may be written more compactly as $u^0 = -\text{sat}(b/c)$ where $\text{sat}(\cdot)$ is the saturation function. The corresponding value is $V^0 = a - b^2/2c$ if $|b/c| \leq 1$, $V^0 = a - b + c^2/2$ if $b/c \geq 1$ and $V^0 = a + b + c^2/2$ if $b/c \leq -1$.

A parametric programming problem $\mathbb{P}(x)$ on the other hand, takes the form $V^0(x) = \min_u \{V(x, u) \mid u \in \mathcal{U}(x)\}$ where x is a *parameter* so that problem, and its solution, depends of the value of the parameter. Hence, the solution to a parametric programming problem $\mathbb{P}(x)$ is not a point or set but a *function* $x \mapsto u^0(x)$ that may be set valued; similarly the value of the problem is a function $x \mapsto V^0(x)$. At each x, the minimizer $u^0(x)$ may be a point or a set. Optimal control problems often take this form, with x being the state, and u, in open-loop discrete time optimal control, being a control sequence; $u^0(x)$, the optimal control sequence, is a function of the initial state. In state feedback optimal control, necessary when uncertainty is present, DP is employed yielding a sequence of parametric optimization problems in each of which x is the state and u a control action; see Chapter 2. The programming problem in the first paragraph of this section may be regarded as a parametric programming problem with the parameter $x := (a, b, c)$, $V(x, u) := (x_1 + x_2 u + (1/2)x_3 u^2/2)$ and $\mathcal{U}(x) := [-1, 1]$; $\mathcal{U}(x)$, in this example, does not depend on x. The solution to this problem yields the functions $u^0(\cdot)$ and $V^0(\cdot)$ defined by $u^0(x) = -\text{sat}(x_2/x_3)$ and $V^0(x) = V(x, u^0(x)) = x_1 + x_2 u^0(x) + (x_3/2)(u^0(x))^2$.

Because the minimizer and value of a parametric programming problem are *functions* rather than points or sets, we would not, in general, expect to be able to compute a solution. Surprisingly, parametric programs are relatively easily solved when the cost function $V(\cdot)$ is affine ($V(x, u) = a + b'x + c'u$) or quadratic ($V(x, u) = (1/2)x'Qx + x'Su + (1/2)u'Ru$) and $\mathcal{U}(x)$ is defined by a set of linear inequalities: $\mathcal{U}(x) = \{u \mid Mu \leq Nx + p\}$. The parametric constraint $u \in \mathcal{U}(x)$ may be conveniently expressed as $(x, u) \in \mathbb{Z}$ where $\mathbb{Z}$ is a subset of (x, u)-space which we will take to be $\mathbb{R}^n \times \mathbb{R}^m$; for each x

$$\mathcal{U}(x) = \{u \mid (x, u) \in \mathbb{Z}\}$$

We assume that $x \in \mathbb{R}^n$ and $u \in \mathbb{R}^m$. Let $\mathcal{X} \subset \mathbb{R}^n$ be defined by

$$\mathcal{X} := \{x \mid \exists u \text{ such that } (x, u) \in \mathbb{Z}\} = \{x \mid \mathcal{U}(x) \neq \emptyset\}$$

The set $\mathcal{X}$ is the domain of $V^0(\cdot)$ and $u^0(\cdot)$ and is thus the set of points x for which a feasible solution of $\mathbb{P}(x)$ exists; it is the projection of $\mathbb{Z}$ (which is a set in (x, u)-space) onto x-space. See Figure 7.1, which illustrates $\mathbb{Z}$ and $\mathcal{U}(x)$ for the case when $\mathcal{U}(x) = \{u \mid Mu \leq Nx + p\}$; the set $\mathbb{Z}$ is thus defined by $\mathbb{Z} := \{(x, u) \mid Mu \leq Nx + p\}$. In this case, both $\mathbb{Z}$ and $\mathcal{U}(x)$ are polyhedral.

Before proceeding to consider parametric linear and quadratic programming, some simple examples may help the reader to appreciate the underlying ideas. Consider first a very simple parametric linear program $\min_u\{V(x, u) \mid (x, u) \in \mathbb{Z}\}$ where $V(x, u) := x + u$ and $\mathbb{Z} := \{(x, u) \mid u + x \geq 0,\ u - x \geq 0\}$ so that $\mathcal{U}(x) = \{u \geq -x,\ u \geq x\}$.

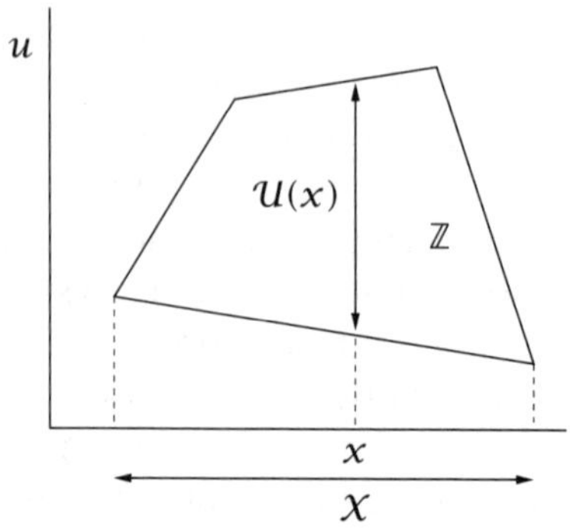

Figure 7.1: The sets $\mathbb{Z}$, $\mathcal{X}$ and $\mathcal{U}(x)$.

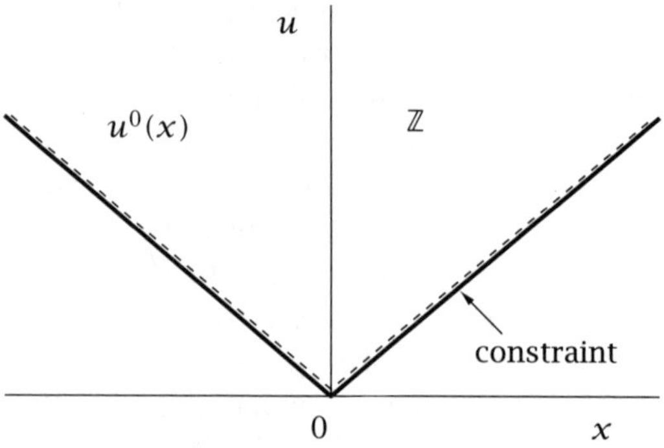

Figure 7.2: Parametric linear program.

The problem is illustrated in Figure 7.2. The set $\mathbb{Z}$ is the region lying above the two solid lines $u = -x$ and $u = x$, and is convex. The gradient $\nabla_u V(x, u) = 1$ everywhere, so the solution, at each x, to the parametric program is the smallest u in $\mathcal{U}(x)$, i.e., the smallest u lying above the two lines $u = -x$ and $u = x$. Hence $u^0(x) = -x$ if $x \leq 0$ and $u^0(x) = x$ if $x \geq 0$, i.e., $u^0(x) = |x|$; the graph of $u^0(\cdot)$ is the dashed line in Figure 7.2. Both $u^0(\cdot)$ and $V^0(\cdot)$, where $V^0(x) = x + u^0(x)$, are *piecewise affine*, being affine in each of the two regions $X_1 := \{x \mid x \leq 0\}$ and $X_2 := \{x \mid x \geq 0\}$.

Next we consider an unconstrained parametric quadratic program $\min_u V(x, u)$ where $V(x, u) := (1/2)(x - u)^2 + u^2/2$. The problem is illustrated in Figure 7.3. For each $x \in \mathbb{R}$, $\nabla_u V(x, u) = -x + 2u$ and $\nabla_{uu} V(x, u) = 2$ so that $u^0(x) = x/2$ and $V^0(x) = x^2/4$. Hence $u^0(\cdot)$

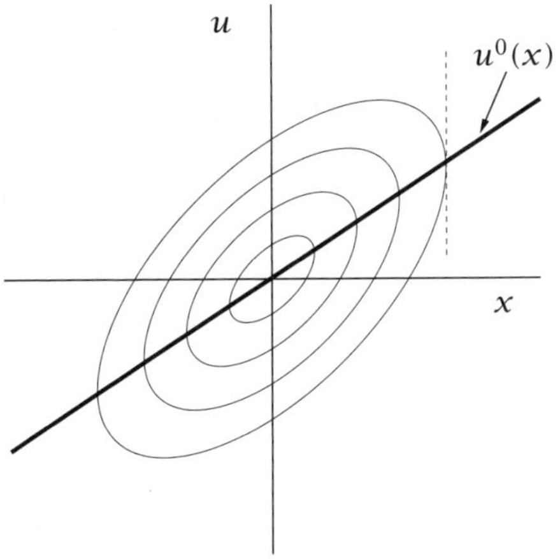

Figure 7.3: Unconstrained parametric quadratic program.

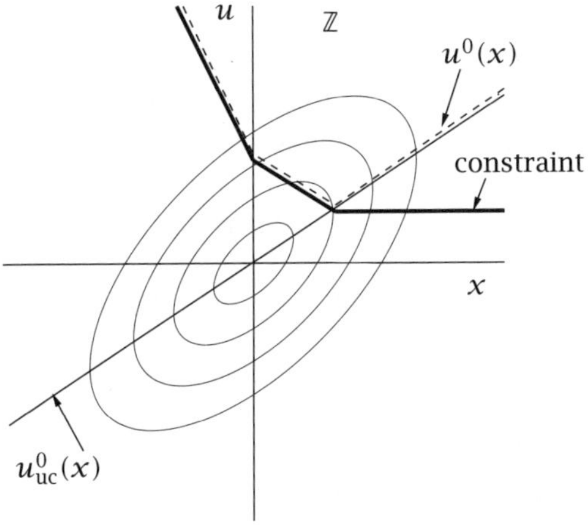

Figure 7.4: Parametric quadratic program.

is linear and $V^0(\cdot)$ is quadratic in $\mathbb{R}$.

We now add the constraint set $\mathbb{Z} := \{(x, u) \mid u \geq 1,\ u + x/2 \geq 2,\ u + x \geq 2\}$; see Figure 7.4. The solution is defined on three regions, $X_1 := (-\infty, 0]$, $X_2 := [0, 2]$, and $X_3 := [2, \infty)$. From the preceding example, the unconstrained minimum is achieved at $u^0_{\text{uc}}(x) = x/2$ shown by the solid straight line in Figure 7.4. Since $\nabla_u V(x, u) = -x + 2u$, $\nabla_u V(x, u) > 0$ for all $u > u^0_{\text{uc}}(x) = x/2$. Hence, in X_1, $u^0(x)$ lies on the boundary of $\mathbb{Z}$ and satisfies $u^0(x) = 2 - x$. Similarly, in X_2,

$u^0(x)$ lies on the boundary of $\mathbb{Z}$ and satisfies $u^0(x) = 2 - x/2$. Finally, in X_3, $u^0(x) = u^0_{\text{uc}}(x) = x/2$, the unconstrained minimizer, and lies in the interior of $\mathbb{Z}$ for $x > 1$. The third constraint $u \geq 2 - x$ is active in X_1, the second constraint $u \geq 2 - x/2$ is active in X_2, while no constraints are active in X_3. Hence the minimizer $u^0(\cdot)$ is piecewise affine, being affine in each of the regions X_1, X_2 and X_3. Since $V^0(x) = (1/2)|x - u^0(x)|^2 + u^0(x)^2/2$, the value function $V^0(\cdot)$ is piecewise quadratic, being quadratic in each of the regions X_1, X_2 and X_3.

We require, in the sequel, the following definitions:

Definition 7.1 (Polytopic (polyhedral) partition). A set $\mathcal{P} = \{\mathbb{Z}_i \mid i \in \mathcal{I}\}$, for some index set $\mathcal{I}$, is called a polytopic (polyhedral) partition of the polytopic (polyhedral) set $\mathbb{Z}$ if $\mathbb{Z} = \cup_{i\in\mathcal{I}}\mathbb{Z}_i$ and the sets $\mathbb{Z}_i$, $i \in \mathcal{I}$, are polytopes (polyhedrons) with nonempty interiors (relative to $\mathbb{Z}$)[1] that are nonintersecting: $\text{int}(\mathbb{Z}_i) \cap \text{int}(\mathbb{Z}_j) = \emptyset$ if $i \neq j$.

Definition 7.2 (Piecewise affine function). A function $f : \mathbb{Z} \to \mathbb{R}^m$ is said to be piecewise affine on a polytopic (polyhedral) partition $\mathcal{P} = \{\mathbb{Z}_i \mid i \in \mathcal{I}\}$ if it satisfies, for some K_i, k_i, $i \in \mathcal{I}$, $f(x) = K_i x + k_i$ for all $x \in \mathbb{Z}_i$, all $i \in \mathcal{I}$. Similarly, a function $f : \mathbb{Z} \to \mathbb{R}$ is said to be piecewise quadratic on a polytopic (polyhedral) partition $\mathcal{P} = \{\mathbb{Z}_i \mid i \in \mathcal{I}\}$ if it satisfies, for some Q_i, r_i, and s_i, $i \in \mathcal{I}$, $f(x) = (1/2)x'Q_i x + r_i'x + s_i$ for all $x \in \mathbb{Z}_i$, all $i \in \mathcal{I}$.

Note the piecewise affine and piecewise quadratic functions defined this way are not necessarily continuous and may, therefore, be set valued at the intersection of the defining polyhedrons. An example is the piecewise affine function $f(\cdot)$ defined by

$$\begin{aligned} f(x) &:= -x - 1 \quad & x \in (-\infty, 0] \\ &:= x + 1 \quad & x \in [0, \infty) \end{aligned}$$

This function is set valued at $x = 0$ where it has the value $f(0) = \{-1, 1\}$. We shall mainly be concerned with continuous piecewise affine and piecewise quadratic functions.

We now generalize the points illustrated by our example above and consider, in turn, parametric quadratic programming and parametric linear programming and their application to optimal control problems.

---

[1]The interior of a set $S \subseteq \mathbb{Z}$ relative to the set $\mathbb{Z}$ is the set $\{z \in S \mid \varepsilon(z)\mathcal{B} \cap \text{aff}(\mathbb{Z}) \subseteq \mathbb{Z} \text{ for some } \varepsilon > 0\}$ where $\text{aff}(\mathbb{Z})$ is the intersection of all affine sets containing $\mathbb{Z}$.

We deal with parametric quadratic programming first because it is more widely used and because, with reasonable assumptions, the minimizer is unique making the underlying ideas somewhat simpler to follow.

7.3 Parametric Quadratic Programming

7.3.1 Preliminaries

The parametric quadratic program $\mathbb{P}(x)$ is defined by

$$V^0(x) = \min_u \{V(x,u) | (x,u) \in \mathbb{Z}\}$$

where $x \in \mathbb{R}^n$ and $u \in \mathbb{R}^m$. The cost function $V(\cdot)$ is defined by

$$V(x,u) := (1/2)x'Qx + u'Sx + (1/2)u'Ru + q'x + r'u + c$$

and the polyhedral constraint set $\mathbb{Z}$ is defined by

$$\mathbb{Z} := \{(x,u) \mid Mx \leq Nu + p\}$$

where $M \in \mathbb{R}^{r \times n}$, $N \in \mathbb{R}^{r \times m}$ and $p \in \mathbb{R}^r$; thus $\mathbb{Z}$ is defined by r affine inequalities. Let $u^0(x)$ denote the solution of $\mathbb{P}(x)$ if it exists, i.e., if $x \in \mathcal{X}$, the domain of $V^0(\cdot)$; thus

$$u^0(x) := \arg\min_u \{V(x,u) \mid (x,u) \in \mathbb{Z}\}$$

The solution $u^0(x)$ is unique if $V(\cdot)$ is strictly convex in u; this is the case if R is positive definite. Let the matrix $\mathcal{Q}$ be defined by

$$\mathcal{Q} := \begin{bmatrix} Q & S' \\ S & R \end{bmatrix}$$

For simplicity we assume, in the sequel:

Assumption 7.3 (Strict convexity). The matrix $\mathcal{Q}$ is positive definite.

Assumption 7.3 implies that both R and Q are positive definite. The cost function $V(\cdot)$ may be written in the form

$$V(x,u) = (1/2)(x,u)'\mathcal{Q}(x,u) + q'x + r'u + c$$

where, as usual, the vector (x,u) is regarded as a column vector $(x',u')'$ in algebraic expressions. The parametric quadratic program may also be expressed as

$$V^0(x) := \min_u \{V(x,u) \mid u \in \mathcal{U}(x)\}$$

where the parametric constraint set $\mathcal{U}(x)$ is defined by

$$\mathcal{U}(x) := \{u \mid (x,u) \in \mathbb{Z}\} = \{u \in \mathbb{R}^m \mid Mu \leq Nx + p\}$$

For each x the set $\mathcal{U}(x)$ is polyhedral. The domain $\mathcal{X}$ of $V^0(\cdot)$ and $u^0(\cdot)$ is defined by

$$\mathcal{X} := \{x \mid \exists u \in \mathbb{R}^m \text{ such that } (\mathrm{x},\mathrm{u}) \in \mathbb{Z}\} = \{x \mid \mathcal{U}(x) \neq \emptyset\}$$

For all $(x,u) \in \mathbb{Z}$, let the index set $I(x,u)$ specify the constraints that are *active* at (x,u), i.e.,

$$I(x,u) := \{i \in \mathbb{I}_{1:r} \mid M_i u = N_i x + p_i\}$$

where M_i, N_i and p_i denote, respectively, the ith row of M, N and p. Similarly, for any matrix or vector A and any index set I, A_I denotes the matrix or vector with rows A_i, $i \in I$. For any $x \in \mathcal{X}$, the indices set $I^0(x)$ specifies the constraints that are active at $(x, u^0(x))$, i.e.,

$$I^0(x) := I(x, u^0(x)) = \{i \in \mathbb{I}_{1:r} \mid M_i u^0(x) = N_i x + p_i\}$$

Since $u^0(x)$ is unique, $I^0(x)$ is well defined. Thus $u^0(x)$ satisfies the equation

$$M_x^0 u = N_x^0 x + p_x^0$$

where

$$M_x^0 := M_{I^0(x)}, \; N_x^0 := N_{I^0(x)}, \; p_x^0 := p_{I^0(x)} \tag{7.1}$$

7.3.2 Preview

We show in the sequel that $V^0(\cdot)$ is piecewise quadratic and $u^0(\cdot)$ piecewise affine on a polyhedral partition of $\mathcal{X}$, the domain of both these functions. To do this, we take an arbitrary point x in $\mathcal{X}$, and show that $u^0(x)$ is the solution of an *equality* constrained quadratic program $\mathbb{P}(x) : \min_u \{V(x,u) \mid M_x^0 u = N_x^0 x + p_x^0\}$ in which the equality constraint is $M_x^0 u = N_x^0 x + p_x^0$. We then show that there is a polyhedral region $R_x^0 \subset \mathcal{X}$ in which x lies and such that, for all $w \in R_x^0$, $u^0(w)$ is the solution of the equality constrained quadratic program $\mathbb{P}(w) : \min_u \{V(w,u) \mid M_x^0 u = N_x^0 w + p_x^0\}$ in which the equality constraints are the same as those for $\mathbb{P}(x)$. It follows that $u^0(\cdot)$ is affine and $V^0(\cdot)$ is quadratic in R_x^0. We then show that there are only a finite number of such polyhedral regions so that $u^0(\cdot)$ is piecewise affine, and $V^0(\cdot)$ piecewise quadratic, on a polyhedral partition of $\mathcal{X}$. To carry out this program, we require a suitable characterization of optimality. We develop this in the next subsection. Some readers may prefer to jump to Proposition 7.8, which gives the optimality condition we employ in the sequel.

7.3.3 Optimality Condition for a Convex Program

Necessary and sufficient conditions for nonlinear optimization problems are developed in Section C.2 of Appendix C. Since we are concerned here with a relatively simple optimization problem where the cost is convex and the constraint set polyhedral, we give a self-contained exposition that uses the concept of a *polar cone*:

Definition 7.4 (Polar cone). The *polar cone* of a cone $C \subseteq \mathbb{R}^n$ is the cone C^* defined by

$$C^* := \{g \in \mathbb{R}^n \mid \langle g, h \rangle \leq 0 \ \forall h \in C\}$$

We recall that a set $C \subseteq \mathbb{R}^n$ is a cone if $0 \in C$ and that $h \in C$ implies $\lambda h \in C$ for all $\lambda > 0$. A cone C is said to be *generated* by $\{a_i \mid i \in I\}$ where I is an index set if $C = \sum_{i \in I}\{\mu_i a_i \mid \mu_i \geq 0, i \in I\}$ in which case we write $C = \text{cone}\{a_i \mid i \in I\}$. We need the following result:

Proposition 7.5 (Farkas's Lemma). *Suppose C is a polyhedral cone defined by*

$$C := \{h \mid Ah \leq 0\} = \{h \mid \langle a_i, h \rangle \leq 0 \mid i \in \mathbb{I}_{1:m}\}$$

where, for each i, a_i is the ith row of A. Then

$$C^* = \text{cone}\{a_i \mid i \in \mathbb{I}_{1:m}\}$$

A proof of this result is given in Section C.2 of Appendix C; that $g \in \text{cone}\{a_i \mid i \in \mathbb{I}_{1:m}\}$ implies $\langle g, h \rangle \leq 0$ for all $h \in C$ is easily shown. An illustration of Proposition 7.5 is given in Figure 7.5.

Next we make use of a standard necessary and sufficient condition of optimality for optimization problems in which the cost is convex and differentiable and the constraint set is convex:

Proposition 7.6 (Optimality conditions for convex set). *Suppose, for each $x \in \mathcal{X}$, $u \mapsto V(x,u)$ is convex and differentiable and $\mathcal{U}(x)$ is convex. Then u is optimal for* $\min_u\{V(x,u) \mid u \in \mathcal{U}(x)\}$ *if and only if*

$$u \in \mathcal{U}(x) \text{ and } \langle \nabla_u V(x,u), v - u \rangle \geq 0 \quad \forall v \in \mathcal{U}(x)$$

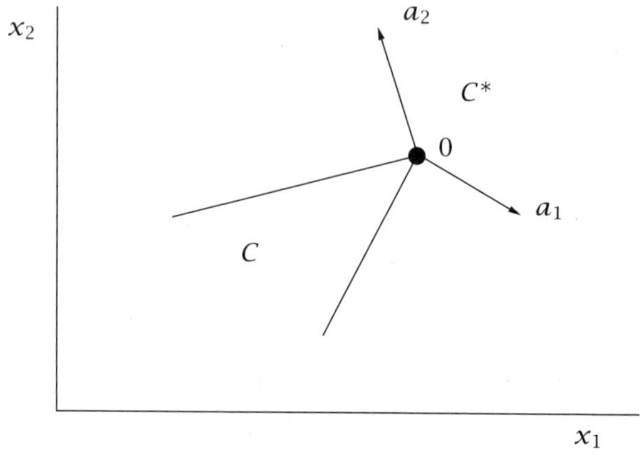

Figure 7.5: Polar cone.

Proof. This Proposition appears as Proposition C.9 in Appendix C where a proof is given. ∎

In our case $\mathcal{U}(x)$, $x \in \mathcal{X}$, is polyhedral and is defined by

$$\mathcal{U}(x) := \{v \in \mathbb{R}^m \mid Mv \leq Nx + p\} \tag{7.2}$$

so $v \in \mathcal{U}(x)$ if and only if, for all $u \in \mathcal{U}(x)$, $v - u \in \mathcal{U}(x) - \{u\} := \{v - u \mid v \in \mathcal{U}(x)\}$. With $h := v - u$

$$\mathcal{U}(x) - \{u\} = \left\{ h \in \mathbb{R}^m \;\middle|\; \begin{array}{ll} M_i h \leq 0, & i \in I(x,u) \\ M_j h < N_j x + p_j - M_j u, & j \in \mathbb{I}_{1:r} \setminus I(x,u) \end{array} \right\}$$

since $M_i u = N_i x + p_i$ for all $i \in I(x,u)$. For each $z = (x,u) \in \mathbb{Z}$, let $C(x,u)$ denote the cone of first-order feasible variations of u; $C(x,u)$ is defined by

$$C(x,u) := \{h \in \mathbb{R}^m \mid M_i h \leq 0,\ i \in I(x,u)\}$$

Clearly

$$\mathcal{U}(x) - \{u\} = C(x,u) \cap \{h \in \mathbb{R}^m \mid M_i h < N_i x + p_i - M_i u,\ i \in \mathbb{I}_{1:r} \setminus I(x,u)\}$$

so that $\mathcal{U}(x) - \{u\} \subseteq C(x,u)$; for any $(x,u) \in \mathbb{Z}$, any $h \in C(x,u)$, there exists an $\alpha > 0$ such that $u + \alpha h \in \mathcal{U}(x)$. Proposition 7.6 may be expressed as: u is optimal for $\min_u \{V(x,u) \mid u \in \mathcal{U}(x)\}$ if and only if

$$u \in \mathcal{U}(x) \text{ and } \langle \nabla_u V(x,u), h \rangle \geq 0 \quad \forall h \in \mathcal{U}(x) - \{u\}$$

We may now state a modified form of Proposition 7.6:

Proposition 7.7 (Optimality conditions in terms of polar cone). *Suppose for each $x \in \mathcal{X}$, $u \mapsto V(x,\cdot)$ is convex and differentiable, and $\mathcal{U}(x)$ is defined by* (7.2). *Then u is optimal for* $\min_u\{V(x,u) \mid u \in \mathcal{U}(x)\}$ *if and only if*

$$u \in \mathcal{U}(x) \text{ and } \langle \nabla_u V(x,u), h\rangle \geq 0 \quad \forall h \in C(x,u)$$

Proof. We show that the condition $\langle \nabla_u V(x,u), h\rangle \geq 0$ for all $h \in C(x,u)$ is equivalent to the condition $\langle \nabla_u V(x,u), h\rangle \geq 0$ for all $h \in \mathcal{U}(x) - \{u\}$ employed in Proposition 7.6. (i) Since $\mathcal{U}(x) - \{u\} \subseteq C(x,u)$, $\langle \nabla_u V(x,u), h\rangle \geq 0$ for all $h \in C(x,u)$ implies $\langle \nabla_u V(x,u), h\rangle \geq 0$ for all $h \in \mathcal{U}(x) - \{u\}$. (ii) $\langle \nabla_u V(x,u), h\rangle \geq 0$ for all $h \in \mathcal{U}(x) - \{u\}$ implies $\langle \nabla_u V(x,u), \alpha h\rangle \geq 0$ for all $h \in \mathcal{U}(x) - \{u\}$, all $\alpha > 0$. But, for any $h^* \in C(x,u)$, there exists an $\alpha \geq 1$ such that $h^* = \alpha h$ where $h := (1/\alpha)h^* \in \mathcal{U}(x) - \{u\}$. Hence $\langle \nabla_u V(x,u), h^*\rangle = \langle \nabla_u V(x,u), \alpha h\rangle \geq 0$ for all $h^* \in C(x,u)$. ■

We now make use of Proposition 7.7 to obtain the optimality condition in the form we use in the sequel. For all $(x,u) \in \mathbb{Z}$, let $C^*(x,u)$ denote the polar cone to $C(x,u)$.

Proposition 7.8 (Optimality conditions for linear inequalities). *Suppose, for each $x \in \mathcal{X}$, $u \mapsto V(x,u)$ is convex and differentiable, and $\mathcal{U}(x)$ is defined by* (7.2). *Then u is optimal for* $\min_u\{V(x,u) \mid u \in \mathcal{U}(x)\}$ *if and only if*

$$u \in \mathcal{U}(x) \text{ and } -\nabla_u V(x,u) \in C^*(x,u) = \text{cone}\{M_i' \mid i \in I(x,u)\}$$

Proof. The desired result follows from a direct application of Proposition 7.5 to Proposition 7.7. ■

Note that $C(x,u)$ and $C^*(x,u)$ are both cones so that each set contains the origin. In particular, $C^*(x,u)$ is generated by the gradients of the constraints active at $z = (x,u)$, and may be defined by a set of affine inequalities: for each $z \in \mathbb{Z}$, there exists a matrix L_z such that

$$C^*(x,u) = C^*(z) = \{g \in \mathbb{R}^m \mid L_z g \leq 0\}$$

The importance of this result for us lies in the fact that the necessary and sufficient condition for optimality is satisfaction of two polyhedral constraints, $u \in \mathcal{U}(x)$ and $-\nabla_u V(x,u) \in C^*(x,u)$. Proposition 7.8 may also be obtained by direct application of Proposition C.12 of Appendix C; $C^*(x,u)$ may be recognized as $\mathcal{N}_{\mathcal{U}(x)}(u)$, the regular normal cone to the set $\mathcal{U}(x)$ at u.

7.3.4 Solution of the Parametric Quadratic Program

For the parametric programming problem $\mathbb{P}(x)$, the parametric cost function is

$$V(x,u) := (1/2)x'Qx + u'Sx + (1/2)u'Ru + q'x + r'u + c$$

and the parametric constraint set is

$$\mathcal{U}(x) := \{u \mid Mu \leq Nx + p\}$$

Hence, the cost gradient is

$$\nabla_u V(x,u) = Ru + Sx + r$$

where, because of Assumption 7.3, R is positive definite. Hence, the necessary and sufficient condition for the optimality of u for the parametric quadratic program $\mathbb{P}(x)$ is

$$\begin{aligned} &Mu \leq Nx + p \\ &-(Ru + Sx + r) \in C^*(x,u) \end{aligned}$$

where $C^*(x,u) = \text{cone}\{M_i' \mid i \in I(x,u)\}$, the cone generated by the gradients of the active constraints, is polyhedral. We cannot use this characterization of optimality directly to solve the parametric programming problem since $I(x,u)$ and, hence, $C^*(x,u)$, varies with (x,u). Given any $x \in \mathcal{X}$, however, there exists the possibility of a region containing x such that $I^0(x) \subseteq I^0(w)$ for all w in this region. We make use of this observation as follows. It follows from the definition of $I^0(x)$ that the unique solution $u^0(x)$ of $\mathbb{P}(x)$ satisfies the equation

$$\begin{aligned} M_i u &= N_i x + p_i, \quad i \in I^0(x), \text{ i.e.,} \\ M_x^0 u &= N_x^0 x + p_x^0 \end{aligned}$$

where M_x^0, N_x^0 and p_x^0 are defined in (7.1). Hence $u^0(x)$ is the solution of the equality constrained problem

$$V^0(x) = \min_u \{V(x,u) \mid M_x^0 u = N_x^0 x + p_x^0\}$$

If the active constraint set remains constant near the point x or, more precisely, if $I^0(x) \subseteq I^0(w)$ for all w in some region in $\mathbb{R}^n$ containing x, then, for all w in this region, $u^0(w)$ satisfies the equality constraint

$M_x^0 u = N_x^0 w + p_x^0$. This motivates us to consider the simple equality constrained problem $\mathbb{P}_x(w)$ defined by

$$V_x^0(w) = \min_u \{V(w,u) \mid M_x^0 u = N_x^0 w + p_x^0\}$$
$$u_x^0(w) = \arg\min_u \{V(w,u) \mid M_x^0 u = N_x^0 w + p_x^0\}$$

The subscript x indicates that the equality constraints in $\mathbb{P}_x(w)$ depend on x. Problem $\mathbb{P}_x(w)$ is an optimization problem with a quadratic cost function and linear equality constraints and is, therefore, easily solved; see the exercises at the end of this chapter. Its solution is

$$V_x^0(w) = (1/2)w'Q_x w + r_x' w + s_x \tag{7.3}$$
$$u_x^0(w) = K_x w + k_x \tag{7.4}$$

for all w such that $I^0(w) = I^0(x)$ where $Q_x \in \mathbb{R}^{n\times n}$, $r_x \in \mathbb{R}^n$, $s_x \in \mathbb{R}$, $K_x \in \mathbb{R}^{m\times n}$ and $k_x \in \mathbb{R}^m$ are easily determined. Clearly, $u_x^0(x) = u^0(x)$; but, is $u_x^0(w)$, the optimal solution to $\mathbb{P}_x(w)$, the optimal solution $u^0(w)$ to $\mathbb{P}(w)$ in some region containing x and, if it is, what is the region? Our optimality condition answers this question. For all $x \in \mathcal{X}$, let the region R_x^0 be defined by

$$R_x^0 := \left\{ w \;\middle|\; \begin{matrix} u_x^0(w) \in \mathcal{U}(w) \\ -\nabla_u V(w, u_x^0(w)) \in C^*(x, u^0(x)) \end{matrix} \right\} \tag{7.5}$$

Because of the equality constraint $M_x^0 u = N_x^0 w + P_x^0$ in problem $\mathbb{P}_x(w)$, it follows that $I(w, u_x^0(w)) \supseteq I(x, u^0(x))$, and that $C(w, u_x^0(w)) \subseteq C(x, u^0(x))$ and $C^*(w, u_x^0(w)) \supseteq C^*(x, u^0(x))$ for all $w \in R_x^0$. Hence $w \in R_x^0$ implies $u_x^0(w) \in \mathcal{U}(w)$ and $-\nabla_u V(w, u_x^0(w)) \in C^*(w, u_x^0(w))$ for all $w \in R_x^0$ which, by Proposition 7.8, is a necessary and sufficient condition for $u_x^0(w)$ to be optimal for $\mathbb{P}(w)$. In fact, $I(w, u_x^0(w)) = I(x, u^0(x))$ so that $C^*(w, u_x^0(w)) = C^*(x, u^0(x))$ for all w in the interior of R_x^0. The obvious conclusion of this discussion is

Proposition 7.9 (Solution of $\mathbb{P}(w)$, $w \in R_x^0$). *For any $x \in \mathcal{X}$, $u_x^0(w)$ is optimal for $\mathbb{P}(w)$ for all $w \in R_x^0$.*

The constraint $u_x^0(w) \in \mathcal{U}(w)$ may be expressed as

$$M(K_x w + k_x) \le Nw + p$$

which is a linear inequality in w. Similarly, since $\nabla_u V(w,u) = Ru + Sw + r$ and since $C^*(x, u^0(x)) = \{g \mid L_x^0 g \le 0\}$ where $L_x^0 = L_{(x,u^0(x))}$, the constraint $-\nabla_u V(x, u_x^0(w)) \in C(x, u^0(x))$ may be expressed as

$$-L_x^0(R(K_x w + k_x) + Sw + r) \le 0$$

which is also an affine inequality in the variable w. Thus, for each x, there exists a matrix F_x and vector f_x such that

$$R_x^0 = \{w \mid F_x w \leq f_x\}$$

so that R_x^0 is polyhedral. Since $u_x^0(x) = u^0(x)$, it follows that $u_x^0(x) \in \mathcal{U}(x)$ and $-\nabla_u V(x, u_x^0(x)) \in C^*(x, u^0(x))$ so that $x \in R_x^0$.

Our next task is to bound the number of distinct regions R_x^0 that exist as we permit x to range over $\mathcal{X}$. We note, from its definition, that R_x^0 is determined, through the constraint $M_x^0 u = N_x^0 w + p_x^0$ in $\mathbb{P}_x(w)$, through $u_x^0(\cdot)$ and through $C^*(x, u^0(x))$, by $I^0(x)$ so that $R_{x_1}^0 \neq R_{x_2}^0$ implies that $I^0(x_1) \neq I^0(x_2)$. Since the number of subsets of $\{1, 2, \ldots, p\}$ is finite, the number of distinct regions R_x^0 as x ranges over $\mathcal{X}$ is finite. Because each $x \in \mathcal{X}$ lies in the set R_x^0, there exists a discrete set of points $X \subset \mathcal{X}$ such that $\mathcal{X} = \cup\{R_x^0 \mid x \in X\}$. We have proved:

Proposition 7.10 (Piecewise quadratic (affine) cost (solution)).

(a) There exists a set X of a finite number of points in $\mathcal{X}$ such that $\mathcal{X} = \cup\{R_x^0 \mid x \in X\}$ and $\{R_x^0 \mid x \in X\}$ is a polyhedral partition of $\mathcal{X}$.

(b) The value function $V^0(\cdot)$ of the parametric piecewise quadratic program $\mathbb{P}$ is piecewise quadratic in $\mathcal{X}$, being quadratic and equal to $V_x^0(\cdot)$, defined in (7.3) in each polyhedron R_x, $x \in X$. Similarly, the minimizer $u^0(\cdot)$ is piecewise affine in $\mathcal{X}$, being affine and equal to $u_x^0(\cdot)$ defined in (7.4) in each polyhedron R_x^0, $x \in X$.

Example 7.11: Parametric QP

Consider the example in Section 7.2. This may be expressed as

$$V^0(x) = \min_u V(x,u), \quad V(x,u) := \{(1/2)x^2 - ux + u^2 \mid Mu \leq Nx + p\}$$

where

$$M = \begin{bmatrix} -1 \\ -1 \\ -1 \end{bmatrix}, \quad N = \begin{bmatrix} 0 \\ 1/2 \\ 1 \end{bmatrix}, \quad p = \begin{bmatrix} -1 \\ -2 \\ -2 \end{bmatrix}$$

At $x = 1$, $u^0(x) = 3/2$ and $I^0(x) = \{2\}$. The equality constrained optimization problem $\mathbb{P}_x(w)$ is

$$V_x^0(w) = \min_u \{(1/2)w^2 - uw + u^2 \mid -u = (1/2)w - 2\}$$

so that $u^0(w) = 2 - w/2$. Hence

$$R_x^0 := \left\{ w \;\middle|\; \begin{array}{c} Mu_x^0(w) \leq Nw + p(w) \\ -\nabla_u V(w, u_x^0(w)) \in C^*(x, u^0(x)) \end{array} \right\}$$

Since $M_2 = -1$, $C^*(x) = \text{cone}\{M_i' \mid i \in I^0(x)\} = \text{cone}\{M_2'\} = \{h \in \mathbb{R} \mid h \leq 0\}$; also

$$\nabla_u V(w, u_x^0(w)) = -w + 2u^0(w) = -w + 2(2 - w/2) = -2w + 4$$

so that R_x^0 is defined by the following inequalities:

$$\begin{aligned} (1/2)w - 2 &\leq -1 && \text{or } w \leq 2 \\ (1/2)w - 2 &\leq (1/2)w - 2 && \text{or } w \in \mathbb{R} \\ (1/2)w - 2 &\leq w - 2 && \text{or } w \geq 0 \\ 2w - 4 &\leq 0 && \text{or } w \leq 2 \end{aligned}$$

which reduces to $w \in [0, 2]$ so $R_x^0 = [0, 2]$ when $x = 1$; $[0, 2]$ is the set X_2 determined in Section 7.2. □

Example 7.12: Explicit optimal control

We return to the MPC problem presented in Example 2.5 of Chapter 2

$$\begin{aligned} V^0(x, \mathbf{u}) &= \min_{\mathbf{u}} \{V(x, \mathbf{u}) \mid \mathbf{u} \in \mathcal{U}\} \\ V(x, \mathbf{u}) &:= (3/2)x^2 + [2x, x]\mathbf{u} + (1/2)\mathbf{u}' H \mathbf{u} \\ H &:= \begin{bmatrix} 3 & 1 \\ 1 & 2 \end{bmatrix} \\ \mathcal{U} &:= \{\mathbf{u} \mid M\mathbf{u} \leq p\} \end{aligned}$$

where

$$M := \begin{bmatrix} 1 & 0 \\ -1 & 0 \\ 0 & 1 \\ 0 & -1 \end{bmatrix}, \qquad p := \begin{bmatrix} 1 \\ 1 \\ 1 \\ 1 \end{bmatrix}$$

It follows from the solution to Example 2.5 that

$$u^0(2) = \begin{bmatrix} -1 \\ -(1/2) \end{bmatrix}$$

and $I^0(x) = \{2\}$. The equality constrained optimization problem at $x = 2$ is

$$V_x^0(w) = \min_{\mathbf{u}} \{(3/2)w^2 + 2wu_1 + wu_2 + (1/2)\mathbf{u}' H \mathbf{u} \mid u_1 = -1\}$$

so that

$$u_x^0(w) = \begin{bmatrix} -1 \\ (1/2) - (1/2)w \end{bmatrix}$$

Hence $u_x^0(2) = [-1, -1/2]' = u^0(2)$ as expected. Since $M_x^0 = M_2 = [-1, 0]$, $C^*(x, u^0(x)) = \{g \in \mathbb{R}^2 \mid g_1 \leq 0\}$. Also

$$\nabla_{\mathbf{u}} V(w, \mathbf{u}) = \begin{bmatrix} 2w + 3u_1 + u_2 \\ w + u_1 + 2u_2 \end{bmatrix}$$

so that

$$\nabla_{\mathbf{u}} V(w, \mathbf{u}_x^0(w)) = \begin{bmatrix} (3/2)w - (5/2) \\ 0 \end{bmatrix}$$

Hence R_x^0, $x = 2$ is the set of w satisfying the following inequalities

$$\begin{aligned} (1/2) - (1/2)w &\leq 1 \quad \text{or } w \geq -1 \\ (1/2) - (1/2)w &\geq -1 \text{ or } w \leq 3 \\ -(3/2)w + (5/2) &\leq 0 \quad \text{or } w \geq (5/3) \end{aligned}$$

which reduces to $w \in [5/3, 3]$; hence $R_x^0 = [5/3, 3]$ when $x = 2$ as shown in Example 2.5. □

7.3.5 Continuity of $V^0(\cdot)$ and $u^0(\cdot)$

Continuity of $V^0(\cdot)$ and $u^0(\cdot)$ follows from Theorem C.34 in Appendix C. We present here a simpler proof, however, based on the above analysis. We use the fact that the parametric quadratic problem is strictly convex, i.e., for each $x \in \mathcal{X}$, $u \mapsto V(x, u)$ is strictly convex and $\mathcal{U}(x)$ is convex, so that the minimizer $u^0(x)$ is unique as shown in Proposition C.8 of Appendix C.

Let $X = \{x_i \mid i \in \mathbb{I}_{1:I}\}$ denote the set defined in Proposition 7.10(a). For each $i \in \mathbb{I}_{i:I}$, let $R_i := R_{x_i}^0$, $V_i(\cdot) := V_{x_i}^0(\cdot)$ and $u_i(\cdot) := u_{x_i}^0(\cdot)$. From Proposition 7.10, $u^0(x) = u_i(x)$ for each $x \in R_i$, each $i \in \mathbb{I}_{1:I}$ so that $u^0(\cdot)$ is affine and hence continuous in the interior of each R_i, and also continuous at any point x on the boundary of $\mathcal{X}$ such that x lies in a single region R_i. Consider now a point x lying in the intersection of several regions, $x \in \cap_{i \in J} R_i$, where J is a subset of $\mathbb{I}_{1:I}$. Then, by Proposition 7.10, $u_i(x) = u^0(x)$ for all $x \in \cap_{i \in J} R_i$, all $i \in J$. Each $u_i(\cdot)$ is affine and, therefore, continuous, so that $u^0(\cdot)$ is continuous in $\cap_{i \in J} R_i$. Hence $u^0(\cdot)$ is continuous in $\mathcal{X}$. Because $V(\cdot)$ is continuous and $u^0(\cdot)$ is continuous in $\mathcal{X}$, the value function

$V^0(\cdot)$ defined by $V^0(x) = V(x, u^0(x))$ is also continuous in $\mathcal{X}$. Let S denote any bounded subset of $\mathcal{X}$. Then, since $V^0(x) = V_i(x) = (1/2)x'Q_i x + r_i' x + s_i$ for all $x \in R_i$, all $i \in \mathbb{I}_{1:I}$ where $Q_i := Q_{x_i}$, $r_i := r_{x_i}$ and $s_i := s_{x_i}$, it follows that $V^0(\cdot)$ is Lipschitz continuous in each set $R_i \cap S$ and, hence, Lipschitz continuous in $\mathcal{X} \cap S$. We have proved the following.

Proposition 7.13 (Continuity of cost and solution). *The value function $V^0(\cdot)$ and the minimizer $u^0(\cdot)$ are continuous in $\mathcal{X}$. Moreover, the value function is Lipschitz continuous on bounded sets.*

7.4 Constrained Linear Quadratic Control

We now show how parametric quadratic programming may be used to solve the optimal receding horizon control problem when the system is linear, the constraints polyhedral, and the cost is quadratic. The system is described, as before, by

$$x^+ = Ax + Bu \tag{7.6}$$

and the constraints are, as before,

$$x \in \mathbb{X}, \quad u \in \mathbb{U} \tag{7.7}$$

where $\mathbb{X}$ is a polyhedron containing the origin in its interior and $\mathbb{U}$ is a polytope also containing the origin in its interior. There may be a terminal constraint of the form

$$x(N) \in \mathbb{X}_f \tag{7.8}$$

where $\mathbb{X}_f$ is a polyhedron containing the origin in its interior. The cost is

$$V_N(x, \mathbf{u}) = \left[\sum_{i=0}^{N-1} \ell(x(i), u(i)) \right] + V_f(x(N)) \tag{7.9}$$

in which, for all i, $x(i) = \phi(i; x, \mathbf{u})$, the solution of (7.6) at time i if the initial state at time 0 is x and the control sequence is $\mathbf{u} := \{u(0), u(1), \ldots, u(N-1)\}$. The functions $\ell(\cdot)$ and $V_f(\cdot)$ are quadratic

$$\ell(x, u) := (1/2)x'Qx + (1/2)u'Ru, \quad V_f(x) := (1/2)x'Q_f x \tag{7.10}$$

The state and control constraints (7.7) induce, via the difference equation (7.6), an implicit constraint $(x, \mathbf{u}) \in \mathcal{Z}$ where

$$\mathbb{Z} := \{(x, \mathbf{u}) \mid x(i) \in \mathbb{X}, \ u(i) \in \mathbb{U}, i \in \mathbb{I}_{0:N-1}, x(N) \in \mathbb{X}_f\} \tag{7.11}$$

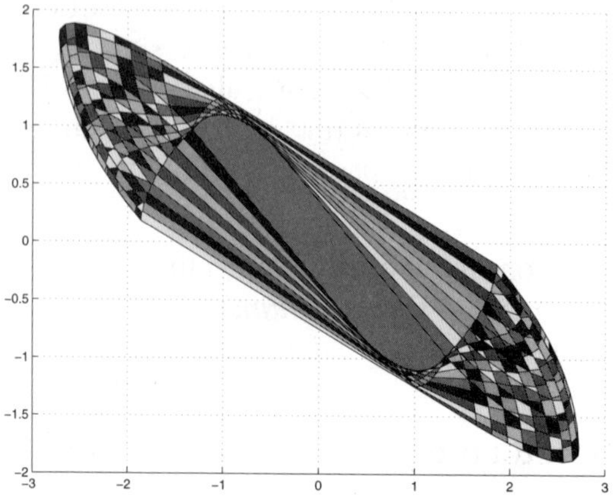

Figure 7.6: Regions R_x, $x \in X$ for a second-order example.

where, for all i, $x(i) = \phi(i; x, \mathbf{u})$. It is easily seen that $\mathbb{Z}$ is polyhedral since, for each i, $x(i) = A^i x + M_i \mathbf{u}$ for some matrix M_i in $\mathbb{R}^{n \times Nm}$; here $\mathbf{u}$ is regarded as the column vector $\begin{bmatrix} u(0)' & u(1)' & \cdots & u(N-1)' \end{bmatrix}'$. Clearly $x(i) = \phi(i; x, \mathbf{u})$ is linear in $(x, \mathbf{u})$. The constrained linear optimal control problem may now be defined by

$$V_N^0(x) = \min_{\mathbf{u}} \{V_N(x, \mathbf{u}) \mid (x, \mathbf{u}) \in \mathbb{Z}\}$$

Using the fact that for each i, $x(i) = A^i x + M_i \mathbf{u}$, it is possible to determine matrices $\mathbf{Q} \in \mathbb{R}^{n \times n}$, $\mathbf{R} \in \mathbb{R}^{Nm \times Nm}$, and $\mathbf{S} \in \mathbb{R}^{Nm \times n}$ such that

$$V_N(x, \mathbf{u}) = (1/2)x'\mathbf{Q}x' + (1/2)\mathbf{u}'\mathbf{R}\mathbf{u} + \mathbf{u}'\mathbf{S}x \tag{7.12}$$

Similarly, as shown above, there exist matrices $\mathbf{M}$, $\mathbf{N}$ and a vector $\mathbf{p}$ such that

$$\mathbb{Z} = \{(x, \mathbf{u}) \mid \mathbf{M}\mathbf{u} \leq \mathbf{N}x + \mathbf{p}\} \tag{7.13}$$

This is precisely the parametric problem studied in Section 7.3, so that the solution $\mathbf{u}^0(x)$ to $\mathbb{P}(x)$ is piecewise affine on a polytopic partition $\mathcal{P} = \{R_x \mid x \in X\}$ of X the projection of $\mathbb{Z} \subset \mathbb{R}^n \times \mathbb{R}^{Nm}$ onto $\mathbb{R}^n$, being affine in each of the constituent polytopes of $\mathcal{P}$. The receding horizon control law is $x \mapsto u^0(0; x)$, the first element of $\mathbf{u}^0(x)$. Two examples are shown in Figures 7.6 and 7.7.

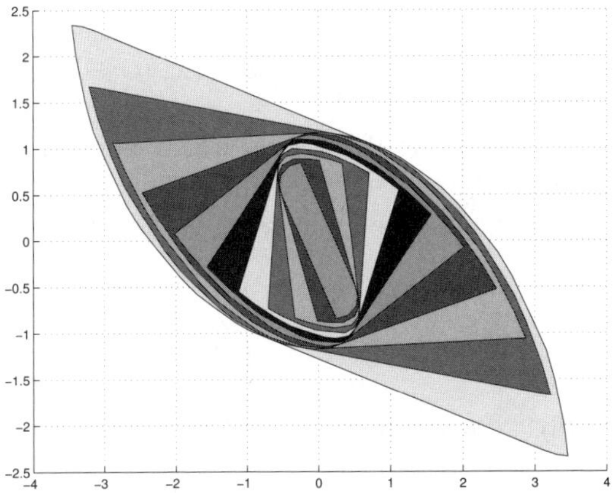

Figure 7.7: Regions R_x, $x \in X$ for a second-order example.

7.5 Parametric Piecewise Quadratic Programming

The parametric quadratic program $\mathbb{P}(x)$ is defined, as before, by

$$V^0(x) = \min_u \{V(x,u) \mid (x,u) \in \mathbb{Z}\} \tag{7.14}$$

where $x \in \mathcal{X} \subset \mathbb{R}^n$ and $u \in \mathbb{R}^m$, but now the cost function $V(\cdot)$ is assumed to be continuous, strictly convex, and piecewise quadratic on a polytopic partition $\mathcal{P} = \{\mathbb{Z}_i \mid i \in \mathcal{I}\}$ of the set $\mathbb{Z}$ so that

$$V(z) = V_i(z) = (1/2)z'\mathcal{Q}_i z + s_i'z + c_i$$

for all $z \in \mathbb{Z}_i$, all $i \in \mathcal{I}$ where $\mathcal{I}$ is an index set. In (7.14), the matrix $\mathcal{Q}_i$ and the vector s_i have the structure

$$\mathcal{Q}_i = \begin{bmatrix} Q_i & S_i' \\ S_i & R_i \end{bmatrix} \qquad s_i = \begin{bmatrix} q_i \\ r_i \end{bmatrix}$$

so that for all $i \in \mathcal{I}$,

$$V_i(x,u) = (1/2)x'Q_i x + u'S_i x + (1/2)u'R_i u + q_i'x + r_i'u + c$$

For each x, the function $u \mapsto V_i(x,u)$ is quadratic and depends on x. The constraint set $\mathbb{Z}$ is defined, as above, by

$$\mathbb{Z} := \{(x,u) \mid Mu \leq Nx + p\}$$

Let $u^0(x)$ denote the solution of $\mathbb{P}(x)$, i.e.,

$$u^0(x) = \arg\min_u \{V(x,u) \mid (x,u) \in \mathbb{Z}\}$$

The solution $u^0(x)$ is unique if $V(\cdot)$ is strictly convex in u at each x; this is the case if each R_i is positive definite. The parametric piecewise quadratic program may also be expressed, as before, as

$$V^0(x) = \min_u \{V(x,u) \mid u \in \mathcal{U}(x)\}$$
$$u^0(x) = \arg\min_u \{V(x,u) \mid u \in \mathcal{U}(x)\}$$

where the parametric constraint set $\mathcal{U}(x)$ is defined by

$$\mathcal{U}(x) := \{u \mid (x,u) \in \mathbb{Z}\} = \{u \mid Mu \leq Nx + p\}$$

Let $\mathcal{X} \subset \mathbb{R}^n$ be defined by

$$\mathcal{X} := \{x \mid \exists u \text{ such that } (x,u) \in \mathbb{Z}\} = \{x \mid \mathcal{U}(x) \neq \emptyset\}$$

The set $\mathcal{X}$ is the domain of $V^0(\cdot)$ and of $u^0(\cdot)$ and is thus the set of points x for which a feasible solution of $\mathbb{P}(x)$ exists; it is the projection of $\mathbb{Z}$, which is a set in (x,u)-space, onto x-space as shown in Figure 7.1. We make the following assumption in the sequel.

Assumption 7.14 (Continuous, piecewise quadratic function). The function $V(\cdot)$ is continuous, strictly convex, and piecewise quadratic on the polytopic partition $\mathcal{P} = \{\mathbb{Z}_i \mid i \in \mathcal{I} := \mathbb{I}_{1:q}\}$ of the polytope $\mathbb{Z}$ in $\mathbb{R}^n \times \mathbb{R}^m$; $V(x,u) = V_i(x,u)$ where $V_i(\cdot)$ is a positive definite quadratic function of (x,u) for all $(x,u) \in \mathbb{Z}_i$, all $i \in \mathcal{I}$, and q is the number of constituent polytopes in $\mathcal{P}$.

The assumption of continuity places restrictions on the quadratic functions $V_i(\cdot)$, $i \in \mathcal{I}$. For example, we must have $V_i(z) = V_j(z)$ for all $z \in \mathbb{Z}_i \cap \mathbb{Z}_j$. Assumption 7.14 implies that the piecewise quadratic programming problem $\mathbb{P}(x)$ satisfies the hypotheses of Theorem C.34 so that the value function $V^0(\cdot)$ is continuous. It follows from Assumption 7.14 and Theorem C.34 that $V^0(\cdot)$ is strictly convex and continuous and that the minimizer $u^0(\cdot)$ is continuous. Assumption 7.14 implies that Q_i is positive definite for all $i \in \mathcal{I}$. For each x, let the set $\mathcal{U}(x)$ be defined by

$$\mathcal{U}(x) := \{u \mid (x,u) \in \mathbb{Z}\}$$

Thus $\mathcal{U}(x)$ is the set of admissible u at x, and $\mathbb{P}(x)$ may be expressed in the form $V^0(x) = \min_u \{V(x,u) \mid u \in \mathcal{U}(x)\}$.

For each $i \in \mathcal{I}$, we define an "artificial" problem $\mathbb{P}_i(x)$ as follows

$$V_i^0(x) := \min_u \{V_i(x,u) \mid (x,u) \in \mathbb{Z}_i\}$$
$$u_i^0(x) := \arg\min_u \{V_i(x,u) \mid (x,u) \in \mathbb{Z}_i\}$$

The cost $V_i(x,u)$ in the above equations may be replaced by $V(x,u)$ since $V(x,u) = V_i(x,u)$ in $\mathbb{Z}_i$. The problem is artificial because it includes constraints (the boundaries of $\mathbb{Z}_i$) that are not necessarily constraints of the original problem. We introduce this problem because it helps us to understand the solution of the original problem. For each $i \in \mathbb{I}_{1:p}$, let the set $\mathcal{U}_i(x)$ be defined as follows

$$\mathcal{U}_i(x) := \{u \mid (x,u) \in \mathbb{Z}_i\}$$

Thus the set $\mathcal{U}_i(x)$ is the set of admissible u at x, and problem $\mathbb{P}_i(x)$ may be expressed as $V_i^0(x) := \min_u\{V_i(x,u) \mid u \in \mathcal{U}_i(x)\}$; the set $\mathcal{U}_i(x)$ is polytopic. For each i, problem $\mathbb{P}_i(x)$ may be recognized as a standard parametric quadratic program discussed in Section 7.4. Because of the piecewise nature of $V(\cdot)$, we require another definition.

Definition 7.15 (Active polytope (polyhedron)). A polytope (polyhedron) $\mathbb{Z}_i$ in a polytopic (polyhedral) partition $\mathcal{P} = \{\mathbb{Z}_i \mid i \in \mathcal{I}\}$ of a polytope (polyhedron) $\mathbb{Z}$ is said to be *active* at $z \in \mathbb{Z}$ if $z = (x,u) \in \mathbb{Z}_i$. The index set specifying the polytopes active at $z \in \mathbb{Z}$ is

$$S(z) := \{i \in \mathcal{I} \mid z \in \mathbb{Z}_i\}$$

A polytope $\mathbb{Z}_i$ in a polytopic partition $\mathcal{P} = \{\mathbb{Z}_i \mid i \in \mathcal{I}\}$ of a polytope $\mathbb{Z}$ is said to be *active* for problem $\mathbb{P}(x)$) if $(x, u^0(x)) \in \mathbb{Z}_i$. The index set specifying polytopes active at $(x, u^0(x))$ is $S^0(x)$ defined by

$$S^0(x) := S(x, u^0(x)) = \{i \in \mathcal{I} \mid (x, u^0(x)) \in \mathbb{Z}_i\}$$

Because we know how to solve the "artificial" problems $\mathbb{P}_i(x)$, $i \in \mathcal{I}$ that are parametric quadratic programs, it is natural to ask if we can recover the solution of the original problem $\mathbb{P}(x)$ from the solutions to these simpler problems. This question is answered by the following proposition.

Proposition 7.16 (Solving $\mathbb{P}$ using $\mathbb{P}_i$). *For any $x \in X$, u is optimal for $\mathbb{P}(x)$ if and only if u is optimal for $\mathbb{P}_i(x)$ for all $i \in S(x,u)$.*

Proof. (i) Suppose u is optimal for $\mathbb{P}(x)$ but, contrary to what we wish to prove, there exists an $i \in S(x,u) = S^0(x)$ such that u is not optimal for $\mathbb{P}_i(x)$. Hence there exists a $v \in \mathbb{R}^m$ such that $(x,v) \in \mathbb{Z}_i$ and $V(x,v) = V_i(x,v) < V_i(x,u) = V(x,u) = V^0(x)$, a contradiction of the optimality of u for $\mathbb{P}(x)$. (ii) Suppose u is optimal for $\mathbb{P}_i(x)$ for all $i \in S(x,u)$ but, contrary to what we wish to prove, u is not optimal for $\mathbb{P}(x)$. Hence $V^0(x) = V(x,u^0(x)) < V(x,u)$. If $u^0(x) \in \mathbb{Z}^{(x,u)} := \cup_{i \in S(x,u)} \mathbb{Z}_i$, we have a contradiction of the optimality of u in $\mathbb{Z}^{(x,u)}$. Assume then that $u^0(x) \in \mathbb{Z}_j$, $j \notin S(x,u)$; for simplicity, assume further that $\mathbb{Z}_j$ is adjacent to $\mathbb{Z}^{(x,u)}$. Then, there exists a $\lambda \in (0,1]$ such that $u^\lambda := u + \lambda(u^0(x) - u) \in \mathbb{Z}^{(x,u)}$; if not, $j \in S(x,u)$, a contradiction. Since $V(\cdot)$ is strictly convex, $V(x,u^\lambda) < V(x,u)$, which contradicts the optimality of u in $\mathbb{Z}^{(x,u)}$. The case when $\mathbb{Z}_j$ is not adjacent to $\mathbb{Z}^{(x,u)}$ may be treated similarly. ∎

To obtain a parametric solution, we proceed as before. We select a point $x \in \mathcal{X}$ and obtain the solution $u^0(x)$ to $\mathbb{P}(x)$ using a standard algorithm for convex programs. The solution $u^0(x)$ satisfies an equality constraint $E_x u = F_x x + g_x$, which we employ to define, for any $w \in \mathcal{X}$ near x an easily solved equality constrained optimization problem $\mathbb{P}_x(w)$ that is derived from the problems $\mathbb{P}_i(x)$, $i \in S^0(x)$. Finally, we show that the solution to this simple problem is also a solution to the original problem $\mathbb{P}(w)$ at all w in a polytope $R_x \subset \mathcal{X}$ in which x lies.

For each $i \in \mathcal{I}$, $\mathbb{Z}_i$ is defined by

$$\mathbb{Z}_i := \{(x,u) \mid M^i u \leq N^i x + p^i\}$$

Let M_j^i, N_j^i and q_j^i denote, respectively, the jth row of M^i, N^i and q^i, and let $I_i(x,u)$ and $I_i^0(x)$, defined by

$$I_i(x,u) := \{j \mid M_j^i u = N_j^i x + p_j^i\}, \quad I_i^0(x) := I_i(x, u_i^0(x))$$

denote, respectively, the active constraint set at $(x,u) \in \mathbb{Z}_i$ and the active constraint set for $\mathbb{P}_i(x)$. Because we now use subscript i to specify $\mathbb{Z}_i$, we change our notation slightly and now let $C_i(x,u)$ denote the cone of first-order feasible variations for $\mathbb{P}_i(x)$ at $u \in \mathcal{U}_i(x)$, i.e.,

$$C_i(x,u) := \{h \in \mathbb{R}^m \mid M_j^i h \leq 0 \ \forall j \in I_i(x,u)\}$$

Similarly, we define the polar cone $C_i^*(x,u)$ of the cone $C_i(x,u)$ at

$h = 0$ by

$$C_i^*(x,u) := \{v \in \mathbb{R}^m \mid v'h \leq 0 \ \forall h \in C_i(x,u)\}$$
$$= \left\{ \sum_{j \in I_i(x,u)} (M_j^i)'\lambda_j \;\middle|\; \lambda_j \geq 0, j \in I_i(x,u) \right\}$$

As shown in Proposition 7.7, a necessary and sufficient condition for the optimality of u for problem $\mathbb{P}_i(x)$ is

$$-\nabla_u V_i(x,u) \in C_i^*(x,u), \quad u \in \mathcal{U}_i(x) \tag{7.15}$$

If u lies in the interior of $\mathcal{U}_i(x)$ so that $I_i^0(x) = \emptyset$, condition (7.15) reduces to $\nabla_u V_i(x,u) = 0$. For any $x \in \mathcal{X}$, the solution $u^0(x)$ of the piecewise parametric program $\mathbb{P}(x)$ satisfies

$$M_j^i u = N_j^i x + p_j^i, \ \forall j \in I_i^0(x), \ \forall i \in S^0(x) \tag{7.16}$$

To simplify our notation, we rewrite the equality constraint (7.16) as

$$E_x u = F_x x + g_x$$

where the subscript x denotes the fact that the constraints are precisely those constraints that are active for the problems $\mathbb{P}_i(x)$, $i \in S^0(x)$. The fact that $u^0(x)$ satisfies these constraints and is, therefore, the unique solution of the optimization problem

$$V^0(x) = \min_u \{V(x,u) \mid E_x u = F_x x + g_x\}$$

motivates us to define the equality constrained problem $\mathbb{P}_x(w)$ for $w \in \mathcal{X}$ near x by

$$V_x^0(w) = \min_u \{V_x(w,u) \mid E_x u = F_x w + g_x\}$$

where $V_x(w,u) := V_i(w,u)$ for all $i \in S^0(x)$ and is, therefore, a positive definite quadratic function of (x,u). The notation $V_x^0(w)$ denotes the fact that the parameter in the parametric problem $\mathbb{P}_x(w)$ is now w but the data for the problem, namely (E_x, F_x, g_x), is derived from the solution $u^0(x)$ of $\mathbb{P}(x)$ and is, therefore, x-dependent. Problem $\mathbb{P}_x(w)$ is a simple equality constrained problem in which the cost $V_x(\cdot)$ is quadratic and the constraints $E_x u = F_x w + g_x$ are linear. Let $V_x^0(w)$ denote the value of $\mathbb{P}_x(w)$ and $u_x^0(w)$ its solution. Then

$$V_x^0(w) = (1/2)w'Q_x w + r_x' w + s_x$$
$$u_x^0(w) = K_x w + k_x \tag{7.17}$$

where Q_x, r_x, s_x, K_x and k_x are easily determined. It is easily seen that $u_x^0(x) = u^0(x)$ so that $u_x^0(x)$ is optimal for $\mathbb{P}(x)$. Our hope is that $u_x^0(w)$ is optimal for $\mathbb{P}(w)$ for all w in some neighborhood R_x of x. We now show this is the case.

Proposition 7.17 (Optimality of $u_x^0(w)$ in R_x). *Let x be an arbitrary point in $\mathcal{X}$. Then,*

(a) $u^0(w) = u_x^0(w)$ and $V^0(w) = V_x^0(w)$ for all w in the set R_x defined by

$$R_x := \left\{ w \in \mathbb{R}^n \,\middle|\, \begin{array}{c} u_x^0(w) \in \mathcal{U}_i(w)\ \forall i \in S^0(x) \\ -\nabla_u V_i(w, u_x^0(w)) \in C_i^*(x, u^0(x))\ \forall i \in S^0(x) \end{array} \right\}$$

(b) R_x is a polytope

(c) $x \in R_x$

Proof.

(a) Because of the equality constraint 7.16 it follows that $I_i(w, u_x(w)) \supseteq I_i(x, u^0(x))$ and that $S(w, u_x^0(w)) \supseteq S(x, u^0(x))$ for all $i \in S(x, u^0(x)) = S^0(x)$, all $w \in R_x$. Hence $C_i(w, u_x^0(w)) \subseteq C_i(x, u^0(x))$, which implies $C_i^*(w, u_x^0(w)) \supseteq C_i^*(x, u^0(x))$ for all $i \in S(x, u^0(x)) \subseteq S(w, u_x^0(w))$. It follows from the definition of R_x that $u_x^0(w) \in \mathcal{U}_i(w)$ and that $-\nabla_u V_i(w, u_x^0(w)) \in C_i^*(w, u_x^0(w))$ for all $i \in S(w, u_x^0(w))$. Hence $u = u_x^0(w)$ satisfies necessary and sufficient for optimality for $\mathbb{P}_i(w)$ for all $i \in S(w, u)$, all $w \in R_x$ and, by Proposition 7.16, necessary and sufficient conditions of optimality for $\mathbb{P}(w)$ for all $w \in R_x$. Hence $u_x^0(w) = u^0(w)$ and $V_x^0(w) = V^0(w)$ for all $w \in R_x$.

(b) That R_x is a polytope follows from the facts that the functions $w \mapsto u_x^0(w)$ and $w \mapsto \nabla_u V_i(w, u_x^0(w))$ are affine, the sets $\mathbb{Z}_i$ are polytopic and the sets $C_i^0(x, u^0(x))$ are polyhedral; hence $(w, u_x^0(w)) \in \mathbb{Z}_i$ is a polytopic constraint and $-\nabla_u V_i(w, u_x^0(w)) \in C_i^*(x, u^0(x))$ a polyhedral constraint on w.

(c) That $x \in R_x$ follows from Proposition 7.16 and the fact that $u_x^0(x) = u^0(x)$. ■

Reasoning as in the proof of Proposition 7.10, we obtain:

Proposition 7.18 (Piecewise quadratic (affine) solution). *There exists a finite set of points X in $\mathcal{X}$ such that $\{R_x \mid x \in X\}$ is a polytopic partition of $\mathcal{X}$. The value function $V^0(\cdot)$ for $\mathbb{P}(x)$ is strictly convex and* *piecewise quadratic and the minimizer $u^0(\cdot)$ is piecewise affine in X being equal, respectively, to the quadratic function $V_x^0(\cdot)$ and the affine function $u_x^0(\cdot)$ in each region R_x, $x \in \mathcal{X}$.*

7.6 DP Solution of the Constrained LQ Control Problem

A disadvantage in the procedure described in Section 7.4 for determining the piecewise affine receding horizon control law is the dimension Nm of the decision variable $\mathbf{u}$. It seems natural to inquire whether or not DP, which replaces a multistage decision problem by a sequence of relatively simple single-stage problems, provides a simpler solution. We answer this question by showing how DP may be used to solve the constrained linear quadratic (LQ) problem discussed in Section 7.4. For all $j \in \mathbb{I}_{1:N}$, let $V_j^0(\cdot)$, the optimal value function at time-to-go j, be defined by

$$V_j^0(x) := \min_u \{V_j(x,u) \mid (x,\mathbf{u}) \in \mathbb{Z}_j\}$$

$$V_j(x,\mathbf{u}) := \sum_{i=0}^{j-1} \ell(x(i),u(i)) + V_f(x(j))$$

$$\mathbb{Z}_j := \{(x,\mathbf{u}) \mid x(i) \in \mathbb{X}, u(i) \in \mathbb{U}, i \in \mathbb{I}_{0:j-1},\ x(j) \in \mathbb{X}_f\}$$

with $x(i) := \phi(i;x,\mathbf{u})$; $V_j^0(\cdot)$ is the value function for $\mathbb{P}_j(x)$. As shown in Chapter 2, the constrained DP recursion is

$$V_{j+1}^0(x) = \min_u \{\ell(x,u) + V_j^0(f(x,u)) \mid u \in \mathbb{U}, f(x,u) \in \mathcal{X}_j\} \tag{7.18}$$

$$\mathcal{X}_{j+1} = \{x \in \mathbb{X} \mid \exists\, u \in \mathbb{U} \text{ such that } f(x,u) \in \mathcal{X}_j\} \tag{7.19}$$

where $f(x,u) := Ax + Bu$ with boundary condition

$$V_0^0(\cdot) = V_f(\cdot), \quad \mathcal{X}_0 = \mathbb{X}_f$$

The minimizer of (7.18) is $\kappa_{j+1}(x)$. In the equations, the subscript j denotes time to go, so that current time $i = N - j$. For each j, $\mathcal{X}_j$ is the domain of the value function $V_j^0(\cdot)$ and of the control law $\kappa_j(\cdot)$, and is the set of states that can be steered to the terminal set $\mathbb{X}_f$ in j steps or less by an admissible control that satisfies the state and control constraints. The time-invariant receding horizon control law for horizon j is $\kappa_j(\cdot)$ whereas the optimal policy for problem $\mathbb{P}_j(x)$ is $\{\kappa_j(\cdot), \kappa_{j-1}(\cdot), \ldots, \kappa_1(\cdot)\}$. The data of the problem are identical to the data in Section 7.4.

We know from Section 7.4 that $V_j^0(\cdot)$ is continuous, strictly convex and piecewise quadratic, and that $\kappa_j(\cdot)$ is continuous and piecewise affine on a polytopic partition $\mathcal{P}_{\mathcal{X}_j}$ of $\mathcal{X}_j$. Hence the function $(x,u) \mapsto V(x,u) := \ell(x,u) + V_j^0(Ax+Bu)$ is continuous, strictly convex and piecewise quadratic on a polytopic partition $\mathcal{P}_{\mathbb{Z}_{j+1}}$ of the polytope $\mathbb{Z}_{j+1}$ defined by

$$\mathbb{Z}_{j+1} := \{(x,u) \mid x \in \mathbb{X}, u \in \mathbb{U}, Ax + Bu \in \mathcal{X}_j\}$$

The polytopic partition $\mathcal{P}_{\mathbb{Z}_{j+1}}$ of $\mathbb{Z}_{j+1}$ may be computed as follows: if X is a constituent polytope of $\mathcal{X}_j$, then, from (7.19), the corresponding constituent polytope of $\mathcal{P}_{\mathbb{Z}_{j+1}}$ is the polytope Z defined by

$$Z := \{z = (x,u) \mid x \in \mathbb{X}, u \in \mathbb{U}, Ax + Bu \in X\}$$

Thus Z is defined by a set of linear inequalities; also $\ell(x,u)+V_j^0(f(x,u))$ is quadratic on Z. Thus the techniques of Section 7.5 can be employed for its solution, yielding the piecewise quadratic value function $V_{j+1}^0(\cdot)$, the piecewise affine control law $\kappa_{j+1}(\cdot)$, and the polytopic partition $\mathcal{P}_{\mathcal{X}_{j+1}}$ on which $V_{j+1}^0(\cdot)$ and $\kappa_{j+1}(\cdot)$ are defined. Each problem (7.18) is much simpler than the problem considered in Section 7.4 since m, the dimension of u, is much less than Nm, the dimension of $\mathbf{u}$. Thus, the DP solution is preferable to the direct method described in Section 7.4.

7.7 Parametric Linear Programming

7.7.1 Preliminaries

The parametric linear program $\mathbb{P}(x)$ is

$$V^0(x) = \min_u \{V(x,u) \mid (x,u) \in \mathbb{Z}\}$$

where $x \in \mathcal{X} \subset \mathbb{R}^n$ and $u \in \mathbb{R}^m$, the cost function $V(\cdot)$ is defined by

$$V(x,u) = q'x + r'u$$

and the constraint set $\mathbb{Z}$ is defined by

$$\mathbb{Z} := \{(x,u) \mid Mu \leq Nx + p\}$$

Let $u^0(x)$ denote the solution of $\mathbb{P}(x)$, i.e.,

$$u^0(x) = \arg\min_u \{V(x,u) \mid (x,u) \in \mathbb{Z}\}$$

The solution $u^0(x)$ may be set valued. The parametric linear program may also be expressed as

$$V^0(x) = \min_u \{V(x,u) \mid u \in \mathcal{U}(x)\}$$

where, as before, the parametric constraint set $\mathcal{U}(x)$ is defined by

$$\mathcal{U}(x) := \{u \mid (x,u) \in \mathbb{Z}\} = \{u \mid Mu \leq Nx + p\}$$

Also, as before, the domain of $V^0(\cdot)$ and $u^0(\cdot)$, i.e., the set of points x for which a feasible solution of $\mathbb{P}(x)$ exists, is the set $\mathcal{X}$ defined by

$$\mathcal{X} := \{x \mid \exists u \text{ such that } (x,u) \in \mathbb{Z}\} = \{x \mid \mathcal{U}(x) \neq \emptyset\}$$

The set $\mathcal{X}$ is the projection of $\mathbb{Z}$ (which is a set in (x,u)-space) onto x-space; see Figure 7.1. We assume in the sequel that the problem is well posed, i.e., for each $x \in \mathcal{X}$, $V^0(x) > -\infty$. This excludes problems like $V^0(x) = \inf_u\{x + u \mid -x \leq 1, x \leq 1\}$ for which $V^0(x) = -\infty$ for all $x \in \mathcal{X} = [-1,1]$.

Let $\mathbb{I}_{1:p}$ denote, as usual, the index set $\{1,2,\ldots,p\}$. For all $(x,u) \in \mathbb{Z}$, let $I(x,u)$ denote the set of active constraints at (x,u), i.e.,

$$I(x,u) := \{i \in \mathbb{I}_{1:p} \mid M_i u = N_i x + p_i\}$$

where A_i denotes the ith row of any matrix (or vector) A. Similarly, for any matrix A and any index set I, A_I denotes the matrix with rows A_i, $i \in I$. If, for any $x \in \mathcal{X}$, $u^0(x)$ is unique, the set $I^0(x)$ of constraints active at $(x, u^0(x))$ is defined by

$$I^0(x) := I(x, u^0(x))$$

When $u^0(x)$ is unique, it is a vertex (a face of dimension zero) of the polyhedron $\mathcal{U}(x)$ and is the *unique* solution of

$$M_x^0 u = N_x^0 x + p_x^0$$

where

$$M_x^0 := M_{I^0(x)}, \; N_x^0 := N_{I^0(x)}, \; p_x^0 := p_{I^0(x)}$$

In this case, the matrix M_x^0 has rank m.

Any face F of $\mathcal{U}(x)$ with dimension $d \in \{1,2,\ldots,m\}$ satisfies $M_i u = N_i x + p_i$ for all $i \in I_F$, all $u \in F$ for some index set $I_F \subseteq \mathbb{I}_{1:p}$. The matrix M_{I_F} with rows M_i, $i \in I_F$, has rank $m - d$, and the face F is defined by

$$F := \{u \mid M_i u = N_i x + p_i, i \in I_F\} \cap \mathcal{U}(x)$$

When $u^0(x)$ is not unique, it is a face of dimension $d \geq 1$ and the set $I^0(x)$ of active constraints is defined by

$$I^0(x) := \{i \mid M_i u = N_i x + p_i \ \forall u \in u^0(x)\} = \{i \mid i \in I(x,u) \ \forall u \in u^0(x)\}$$

The set $\{u \mid M_i u = N_i x + p_i,\ i \in I^0(x)\}$ is an affine hyperplane in which $u^0(x)$ lies. See Figure 7.8 where $u^0(x_1)$ is unique, a vertex of $\mathcal{U}(x_1)$, and $I^0(x_1) = \{2,3\}$. If, in Figure 7.8, $r = -e_1$, then $u^0(x_1) = F_2(x_1)$, a face of dimension 1; $u^0(x_1)$ is, therefore, set valued. Since $u \in \mathbb{R}^m$ where $m = 2$, $u^0(x_1)$ is a facet, i.e., a face of dimension $m-1 = 1$. Thus $u^0(x_1)$ is a set defined by $u^0(x_1) = \{u \mid M_1 u \leq N_1 x_1 + p_1,\ M_2 u = N_2 x_1 + p_2,\ M_3 u \leq N_3 x_1 + p_3\}$.

At each $z = (x,u) \in \mathbb{Z}$, i.e., for each (x,u) such that $x \in \mathcal{X}$ and $u \in \mathcal{U}(x)$, the cone $C(z) = C(x,u)$ of first-order feasible variations is defined, as before, by

$$C(z) := \{h \in \mathbb{R}^m \mid M_i h \leq 0,\ i \in I(z)\} = \{h \in \mathbb{R}^m \mid M_{I(z)} h \leq 0\}$$

If $I(z) = I(x,u) = \emptyset$ (no constraints are active), $C(z) = \mathbb{R}^m$ (all variations are feasible).

Since $u \mapsto V(x,\cdot)$ is convex and differentiable, and $\mathcal{U}(x)$ is polyhedral for all x, the parametric linear program $\mathbb{P}(x)$ satisfies the assumptions of Proposition 7.8. Hence, repeating Proposition 7.8 for convenience, we have

Proposition 7.19 (Optimality conditions for parametric linear program). *A necessary and sufficient condition for u to be a minimizer for the parametric linear program $\mathbb{P}(x)$ is*

$$u \in \mathcal{U}(x) \text{ and } -\nabla_u V(x,u) \in C^*(x,u)$$

where $\nabla_u V(x,u) = r$ and $C^*(x,u)$ is the polar cone of $C(x,u)$.

An important difference between this result and that for the parametric quadratic program is that $\nabla_u V(x,u) = r$ and, therefore, does not vary with x or u. We now use this result to show that both $V^0(\cdot)$ and $u^0(\cdot)$ are piecewise affine. We consider the simple case when $u^0(x)$ is unique for all $x \in \mathcal{X}$.

7.7.2 Minimizer $u^0(x)$ Is Unique for all $x \in \mathcal{X}$

Before proceeding to obtain the solution to a parametric linear program when the minimizer $u^0(x)$ is unique for each $x \in \mathcal{X}$, we look first at

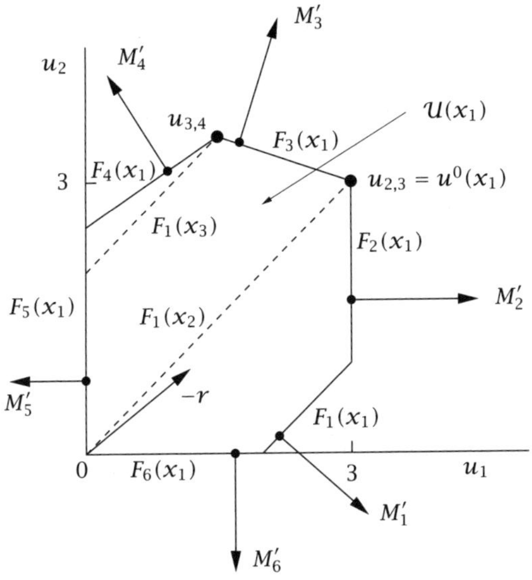

Figure 7.8: Solution to a parametric linear program.

the simple example illustrated in Figure 7.8, which shows the constraint set $\mathcal{U}(x)$ for various values of the parameter x in the interval $[x_1, x_3]$. The set $\mathcal{U}(x_1)$ has six faces: $F_1(x_1)$, $F_2(x_1)$, $F_3(x_1)$, $F_4(x_1)$, $F_5(x_1)$ and $F_6(x_1)$. Face $F_1(x)$ lies in the hyperplane $\mathcal{H}_1(x)$ that varies linearly with x; each face $F_i(x)$, $i = 2, \ldots, 6$, lies in the hyperplane $\mathcal{H}_i$ that does *not* vary with x. All the faces vary with x as shown so that $\mathcal{U}(x_2)$ has four faces: $F_1(x_2)$, $F_3(x_2)$, $F_4(x_2)$ and $F_5(x_2)$; and $\mathcal{U}(x_3)$ has three faces: $F_1(x_3)$, $F_4(x_3)$ and $F_5(x_3)$. The face $F_1(x)$ is shown for three values of x: $x = x_1$ (the bold line), and $x = x_2$ and $x = x_3$ (dotted lines). It is apparent that for $x \in [x_1, x_2]$, $u^0(x) = u_{2,3}$ in which $u_{2,3}$ is the intersection of $\mathcal{H}_2$ and $\mathcal{H}_3$, and $u^0(x_3) = u_{3,4}$, in which $u_{3,4}$ is the intersection of $\mathcal{H}_3$ and $\mathcal{H}_4$. It can also be seen that $u^0(x)$ is unique for all $x \in \mathcal{X}$.

We now return to the general case. Suppose, for some $\in \mathcal{X}$, $u^0(x)$ is the unique solution of $\mathbb{P}(x)$; $u^0(x)$ is the unique solution of

$$M_x^0 u = N_x^0 x + p_x^0$$

It follows that $u^0(x)$ is the trivial solution of the simple equality constrained problem defined by

$$V^0(x) = \min_u \{V(x,u) \mid M_x^0 u = N_x^0 x + p_x^0\} \tag{7.20}$$

The solution $u^0(x)$ of this equality constrained problem is trivial because it is determined entirely by the equality constraints; the cost plays no part.

The optimization problem (7.20) motivates us, as in parametric quadratic programming, to consider, for any parameter w "close" to x, the simpler equality constrained problem $\mathbb{P}_x(w)$ defined by

$$V_x^0(w) = \min_u \{V(w,u) \mid M_x^0 u = N_x^0 w + p_x^0\}$$
$$u_x^0(w) = \arg\min_u \{V(w,u) \mid M_x^0 u = N_x^0 w + p_x^0\}$$

Let $u_x^0(w)$ denote the solution of $\mathbb{P}_x(w)$. Because, for each $x \in \mathcal{X}$, the matrix M_x^0 has full rank m, there exists an index set I_x such that $M_{I_x} \in \mathbb{R}^{m \times m}$ is invertible. Hence, for each w, $u_x^0(w)$ is the unique solution of

$$M_{I_x} u = N_{I_x} w + p_{I_x}$$

so that for all $x \in \mathcal{X}$, all $w \in \mathbb{R}^m$

$$u_x^0(w) = K_x w + k_x \tag{7.21}$$

where $K_x := (M_{I_x})^{-1} N_{I_x}$ and $k_x := (M_{I_x})^{-1} p_{I_x}$. In particular, $u^0(x) = u_x^0(x) = K_x x + k_x$. Since $V_x^0(x) = V_x(x, u_x^0(w)) = q'x + r'u_x^0(w)$, it follows that

$$V_x^0(x) = (q' + r'K_x)x + r'k_x$$

for all $x \in \mathcal{X}$, all $w \in \mathbb{R}^m$. Both $V_x^0(\cdot)$ and $u_x^0(\cdot)$ are affine in x.

It follows from Proposition 7.19 that $-r \in C^*(x, u^0(x)) = \text{cone}\{M_i' \mid i \in I^0(x) = I(x, u^0(x))\} = \text{cone}\{M_i' \mid i \in I_x\}$. Since $\mathbb{P}_x(w)$ satisfies the conditions of Proposition 7.8, we may proceed as in Section 7.3.4 and define, for each $x \in \mathcal{X}$, the set R_x^0 as in (7.5)

$$R_x^0 := \left\{ w \in \mathbb{R}^n \;\middle|\; \begin{array}{c} u_x^0(w) \in \mathcal{U}(w) \\ -\nabla_u V(w, u_x^0(w)) \in C^*(x, u^0(x)) \end{array} \right\}$$

It then follows, as shown in Proposition 7.9, that for any $x \in \mathcal{X}$, $u_x^0(w)$ is optimal for $\mathbb{P}(w)$ for all $w \in R_x^0$. Because $\mathbb{P}(w)$ is a parametric linear program, however, rather than a parametric quadratic program, it is possible to simplify the definition of R_x^0. We note that $\nabla_u V(w, u_x^0(w)) = r$ for all $x \in \mathcal{X}$, all $w \in \mathbb{R}^m$. Also, it follows from Proposition 7.8, since $u^0(x)$ is optimal for $\mathbb{P}(x)$, that $-\nabla_u V(x, u^0(x)) = -r \in C^*(x)$ so that the second condition in the definition above for

R_x^0 is automatically satisfied. Hence we may simplify our definition for R_x^0; for the parametric linear program, R_x^0 may be defined by

$$R_x^0 := \{w \in \mathbb{R}^n \mid u_x^0(w) \in \mathcal{U}(w)\} \tag{7.22}$$

Because $u_x^0(\cdot)$ is affine, it follows from the definition of $\mathcal{U}(w)$ that R_x^0 is polyhedral. The next result follows from the discussion in Section 7.3.4:

Proposition 7.20 (Solution of $\mathbb{P}$). *For any $x \in \mathcal{X}$, $u_x^0(w)$ is optimal for $\mathbb{P}(w)$ for all w in the set R_x^0 defined in* (7.22).

Finally, the next result characterizes the solution of the parametric linear program $\mathbb{P}(x)$ when the minimizer is unique.

Proposition 7.21 (Piecewise affine cost and solution).

(a) There exists a finite set of points X in $\mathcal{X}$ such that $\{R_x^0 \mid x \in X\}$ is a polyhedral partition of $\mathcal{X}$.

(b) The value function $V^0(\cdot)$ for $\mathbb{P}(x)$ and the minimizer $u^0(\cdot)$ are piecewise affine in $\mathcal{X}$ being equal, respectively, to the affine functions $V_x^0(\cdot)$ and $u_x^0(\cdot)$ in each region R_x, $x \in \mathcal{X}$.

(c) The value function $V^0(\cdot)$ and the minimizer $u^0(\cdot)$ are continuous in $\mathcal{X}$.

Proof. The proof of parts (a) and (b) follows, apart from minor changes, the proof of Proposition 7.10. The proof of part (c) uses the fact that $u^0(x)$ is unique, by assumption, for all $x \in \mathcal{X}$ and is similar to the proof of Proposition 7.13. ■

7.8 Constrained Linear Control

The previous results on parametric linear programming may be applied to obtain the optimal receding horizon control law when the system is linear, the constraints polyhedral, and the cost linear as is done in a similar fashion in Section 7.4 where the cost is quadratic. The optimal control problem is therefore defined as in Section 7.4, except that the stage cost $\ell(\cdot)$ and the terminal cost $V_f(\cdot)$ are now defined by

$$\ell(x,u) := q'x + r'u, \quad V_f(x) := q_f'x$$

As in Section 7.4, the optimal control problem $\mathbb{P}_N(x)$ may be expressed as

$$V_N^0(x) = \min_{\mathbf{u}}\{V_N(x,\mathbf{u}) \mid \mathbf{M}\mathbf{u} \leq \mathbf{N}x + \mathbf{p}\}$$

where, now

$$V_N(x, \mathbf{u}) = \mathbf{q}'x + \mathbf{r}'\mathbf{u}$$

Hence the problem has the same form as that discussed in Section 7.7 and may be solved as shown there.

It is possible, using a simple transcription, to use the solution of $\mathbb{P}_N(x)$ to solve the optimal control problem when the stage cost and terminal cost are defined by

$$\ell(x,u) := |Qx|_p + |Ru|_p, \quad V_f(x) := |Q_f x|_p$$

where $|\cdot|_p$ denotes the p-norm and p is either 1 or ∞.

7.9 Computation

Our main purpose above was to establish the structure of the solution of parametric linear or quadratic programs and, hence, of the solutions of constrained linear optimal control problems when the cost is quadratic or linear and we have not presented algorithms for solving these problems. A naive approach to computation would be to generate points x in $\mathcal{X}$ randomly and to compute the corresponding polyhedral sets R_x using the formula given previously. But, due to numerical errors, these regions would either overlap or leave gaps or both, rendering the solution useless. Methods for low-dimensional problems are described in the survey paper by Alessio and Bemporad (2008). The preferred methods generate new regions adjacent to a set of regions already determined. In this class is the lexicographic perturbation algorithm described by Jones, Kerrigan, and Maciejowski (2007) for parametric linear programs and by Jones and Morari (2006) for parametric linear complementarity problems and for quadratic programs. The toolboxes (Bemporad, 2004) and (Kvasnica, Grieder, and Baotić, 2006) provide tools for the determination of feedback control laws for relatively simple linear systems with polyhedral constraints.

7.10 Notes

Early work on parametric programming, e.g. (Dantzig, Folkman, and Shapiro, 1967) and (Bank, Guddat, Klatte, Kummer, and Tanner, 1983), was concerned with the sensitivity of optimal solutions to parameter variations. Solutions to the parametric linear programming problem were obtained relatively early (Gass and Saaty, 1955) and (Gal and Nedoma, 1972). Solutions to parametric quadratic programs were obtained in (Seron, De Doná, and Goodwin, 2000) and (Bemporad, Morari, Dua, and Pistikopoulos, 2002) and applied to the determination of optimal control laws for linear systems with polyhedral constraints. Since then a large number of papers on this topic have appeared, many of which are reviewed in (Alessio and Bemporad, 2008). Most papers employ the Kuhn-Tucker conditions of optimality in deriving the regions R_x, $x \in X$. Use of the polar cone condition was advocated in (Mayne and Raković, 2002) in order to focus on the geometric properties of the parametric optimization problem and avoid degeneracy problems. Section 7.5, on parametric piecewise quadratic programming, is based on (Mayne, Raković, and Kerrigan, 2007). The examples in Section 7.5 were computed by Raković. That uniqueness of the minimizer can be employed, instead of maximum theorems, to establish, as in Section 7.3.5, continuity of $u^0(\cdot)$ and, hence, of $V^0(\cdot)$, was pointed out by Bemporad et al. (2002) and Borrelli (2003, p. 37).

7.11 Exercises

Exercise 7.1: Quadratic program with equality constraints

Obtain the solution u^0 and the value V^0 of the equality constrained optimization problem $V^0 = \min_u\{V(u) \mid h(u) = 0\}$ where $V(u) = (1/2)u'Ru + r'u + c$ and $h(u) := Mu - p$.

Exercise 7.2: Parametric quadratic program with equality constraints

Show that the solution $u^0(x)$ and the value $V^0(x)$ of the parametric optimization problem $V^0(x) = \min_u\{V(x,u) \mid h(x,u) = 0\}$ where $V(x,u) := (1/2)x'Qx + u'Sx + (1/2)u'Ru + q'x + r'u + c$ and $h(x,u) := Mu - Nx - p$ have the form $u^0(x) = Kx + k$ and $V^0(x) = (1/2)x'\bar{Q}x + \bar{q}'x + s$. Determine $\bar{Q}$, $\bar{q}$, s, K and k.

Exercise 7.3: State and input trajectories in constrained LQ problem

For the constrained linear quadratic problem defined in Section 7.4, show that $\mathbf{u} := \{u(0), u(1), \ldots, u(N-1)\}$ and $\mathbf{x} := \{x(0), x(1), \ldots, x(N)\}$, where $x(0) = x$ and $x(i) = \phi(i; x, \mathbf{u})$, $i = 0, 1, \ldots, N$, satisfy:

$$\mathbf{x} = \mathbf{F}x + \mathbf{G}\mathbf{u}$$

and determine the matrices $\mathbf{F}$ and $\mathbf{G}$; in this equation $\mathbf{u}$ and $\mathbf{x}$ are column vectors. Hence show that $V_N(x, \mathbf{u})$ and $\mathbb{Z}$, defined respectively in (7.9) and (7.11), satisfy (7.12) and (7.13), and determine $\mathbf{Q}$, $\mathbf{R}$, $\mathbf{M}$, $\mathbf{N}$ and $\mathbf{p}$.

Exercise 7.4: The parametric linear program with unique minimizer

For the example of Figure 7.8, determine $u^0(x)$, $V^0(x)$, $I^0(x)$ and $C^*(x)$ for all x in the interval $[x_1, x_3]$. Show that $-r$ lies in $C^*(x)$ for all x in $[x_1, x_3]$.

Exercise 7.5: Cost function and constraints in constrained LQ control problem

For the constrained linear control problem considered in Section 7.8, determine the matrices $\mathbf{M}$, $\mathbf{N}$ and $\mathbf{p}$ that define the constraint set $\mathbb{Z}$, and the vectors $\mathbf{q}$ and $\mathbf{r}$ that define the cost $V_N(\cdot)$.

Exercise 7.6: Cost function in constrained linear control problem

Show that $|x|_p$, $p = 1$ and $p = \infty$, may be expressed as $\max_j\{s_j'x \mid j \in J\}$ and determine s_i, $i \in I$ for the two cases $p = 1$ and $p = \infty$. Hence show that the optimal control problem in Section 7.8 may be expressed as

$$V_N^0(x) = \min_{\mathbf{v}}\{V_N(x, \mathbf{v}) \mid \mathbf{M}\mathbf{v} \leq \mathbf{N}x + \mathbf{p}\}$$

where, now, $\mathbf{v}$ is a column vector whose components are $u(0), u(1), \ldots, u(N-1)$, $\ell_x(0), \ell_x(1), \ldots, \ell_x(N)$, $\ell_u(0), \ell_u(1), \ldots, \ell_u(N-1)$ and f; the cost $V_N(x, \mathbf{v})$ is now defined by

$$V_N(x, \mathbf{v}) = \sum_{i=0}^{N-1} (\ell_x(i) + \ell_u(i)) + f$$

Finally, $\mathbf{Mv} \leq \mathbf{N}x + \mathbf{p}$ now specifies the constraints $u(i) \in \mathbb{U}$ and $x(i) \in \mathbb{X}$, $|Ru(i)|_p \leq \ell_u(i)$, $|Qx(i)|_p \leq \ell_x(i)$, $i = 0, 1, \ldots, N-1$, $x(N) \in \mathbb{X}_f$, and $|Q_f x(N)| \leq f$. As before, $\mathbf{x}^+ = \mathbf{Fx} + \mathbf{Gu}$.

Exercise 7.7: Is QP constraint qualification relevant to MPC?

Continuity properties of the MPC control law are often used to establish robustness properties of MPC such as robust asymptotic stability. In early work on continuity properties of linear model MPC, Scokaert, Rawlings, and Meadows (1997) used results on continuity of QPs with respect to parameters to establish MPC stability under perturbations. For example, Hager (1979) considered the following quadratic program

$$\min_u (1/2)u'Hu + h'u + c$$

subject to

$$Du \leq d$$

and established that the QP solution u^0 and cost V^0 are Lipschitz continuous in the data of the QP, namely the parameters H, h, D, d. To establish this result Hager (1979) made the following assumptions.

- The solution is unique for all H, h, D, d in a chosen set of interest.
- The rows of D corresponding to the constraints active at the solution are linearly independent. The assumption of linear independence of active constraints is a form of *constraint qualification*.

(a) First we show that some form of constraint qualification is required to establish continuity of the QP solution with respect to matrix D. Consider the following QP example that does not satisfy Hager's constraint qualification assumption.

$$H = \begin{bmatrix} 1 & 0 \\ 0 & 1 \end{bmatrix} \qquad D = \begin{bmatrix} 1 & 1 \\ -1 & -1 \end{bmatrix} \qquad d = \begin{bmatrix} 1 \\ -1 \end{bmatrix} \qquad h = \begin{bmatrix} -1 \\ -1 \end{bmatrix} \qquad c = 1$$

Find the solution u^0 for this problem.

Next perturb the D matrix to

$$D = \begin{bmatrix} 1 & 1 \\ -(1+\epsilon) & -1 \end{bmatrix}$$

in which $\epsilon > 0$ is a small perturbation. Find the solution to the perturbed problem. Are V^0 and u^0 continuous in parameter D for this QP? Draw a sketch of the feasible region and cost contours for the original and perturbed problems. What happens to the feasible set when D is perturbed?

(b) Next consider MPC control of the following system with state inequality constraint and no input constraints

$$A = \begin{bmatrix} -1/4 & 1 \\ -1 & 1/2 \end{bmatrix} \qquad B = \begin{bmatrix} 1 & 1 \\ -1 & -1 \end{bmatrix} \qquad x(k) \leq \begin{bmatrix} 1 \\ 1 \end{bmatrix} \qquad k \in \mathbb{I}_{0:N}$$

Using a horizon $N = 1$, eliminate the state $x(1)$ and write out the MPC QP for the input $u(0)$ in the form given above for $Q = R = I$ and zero terminal penalty. Find an initial condition x_0 such that the MPC constraint matrix D and vector d are identical to those given in the previous part. Is this $x_0 \in \mathcal{X}_N$?

Are the rows of the matrix of active constraints linearly independent in this MPC QP on the set $\mathcal{X}_N$? Are the MPC control law $\kappa_N(x)$ and optimal value function $V_N^0(x)$ Lipschitz continuous on the set $\mathcal{X}_N$ for this system? Explain the reason if these two answers differ.

Bibliography

A. Alessio and A. Bemporad. A survey on explicit model predictive control. In L. Magni, editor, *Proceedings of International Workshop on Assessment and Future Directions of Model Predictive Control*, 2008.

B. Bank, J. Guddat, D. Klatte, B. Kummer, and K. Tanner. *Non-linear parametric optimization.* Birkhäuser Verlag, Basel, Boston, Stuttgart, 1983.

A. Bemporad. *Hybrid Toolbox - User's Guide*, 2004. `http://www.dii.unisi.it/hybrid/toolbox`.

A. Bemporad, M. Morari, V. Dua, and E. N. Pistikopoulos. The explicit linear quadratic regulator for constrained systems. *Automatica*, 38(1):3–20, 2002.

F. Borrelli. *Constrained Optimal Control of Linear and Hybrid Systems.* Springer, 2003.

G. B. Dantzig, J. Folkman, and N. Z. Shapiro. On the continuity of the minimum set of a continuous function. *Journal of Mathematical Analysis and Applications*, 17(3):519–548, 1967.

T. Gal and J. Nedoma. Multiparametric linear programming. *Management Science*, 18(7):406–422, 1972.

S. I. Gass and T. L. Saaty. The computational algorithm for the parametric objective function. *Naval Research Logistics Quarterly*, 2:39–45, 1955.

W. W. Hager. Lipschitz continuity for constrained processes. *SIAM J. Cont. Opt.*, 17(3):321–338, 1979.

C. N. Jones and M. Morari. Multiparametric linear complementarity problems. In *Proceedings 45th IEEE Conference on Decision and Control*, pages 5687–5692, San Diego, California, USA, 2006.

C. N. Jones, E. C. Kerrigan, and J. M. Maciejowski. Lexicographic perturbation for multiparametric linear programming with applications to control. *Automatica*, 43:1808–1816, 2007.

M. Kvasnica, P. Grieder, and M. Baotić. *Multi-Parametric Toolbox (MPT)*, 2006.

D. Q. Mayne and S. V. Raković. Optimal control of constrained piecewise affine discrete-time systems using reverse transformation. In *Proceedings of the IEEE 2002 Conference on Decision and Control*, volume 2, pages 1546 - 1551 vol.2, Las Vegas, USA, 2002.

D. Q. Mayne, S. V. Raković, and E. C. Kerrigan. Optimal control and piecewise parametric programming. In *Proceedings of the European Control Conference 2007*, pages 2762–2767, Kos, Greece, July 2–5 2007.

P. O. M. Scokaert, J. B. Rawlings, and E. S. Meadows. Discrete-time stability with perturbations: Application to model predictive control. *Automatica*, 33(3): 463–470, 1997.

M. M. Seron, J. A. De Doná, and G. C. Goodwin. Global analytical model predictive control with input constraints. In *Proceedings of the 39th IEEE Conference on Decision and Control*, pages 154–159, Sydney, Australia, December 2000.

Author Index

Citation Index

Subject Index

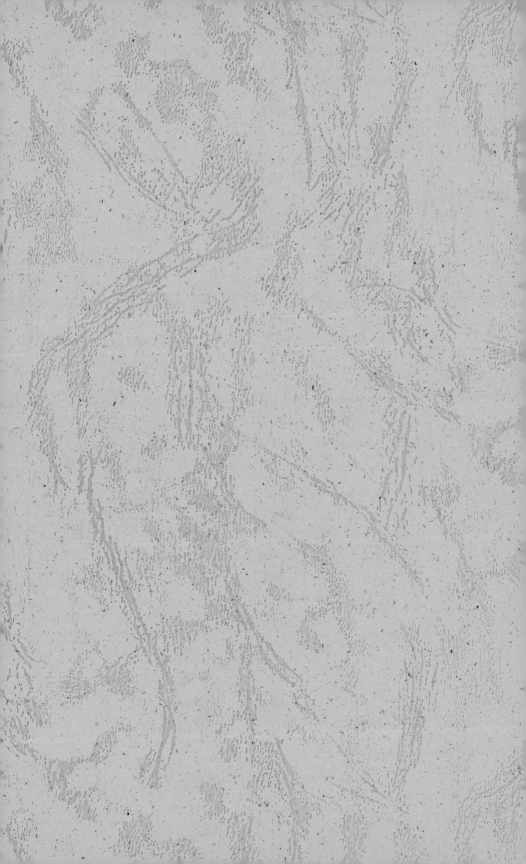